Graduate Texts in Mathematics **221**

T0202415

Springer
New York
Berlin
Heidelberg
Hong Kong
London
Milan
Paris
Tokyo

Graduate Texts in Mathematics

(continued after index)

Branko Grünbaum

Convex Polytopes

Second Edition

First Edition Prepared
with the Cooperation of
Victor Klee, Micha Perles, and Geoffrey C. Shephard

Second Edition Prepared by
Volker Kaibel, Victor Klee, and Günter M. Ziegler

With 162 Illustrations

 Springer

Branko Grünbaum
Department of Mathematics
University of Washington, Seattle
Seattle, WA 98195-4350
USA
grunbaum@math.washington.edu

Second edition prepared by:

Volker Kaibel
MA 6-2, Institute of
  Mathematics
TU Berlin
D-10623 Berlin
Germany
kaibel@math.tu-berlin.de

Victor Klee
Department of Mathematics
University of Washington
Seattle, Washington, 98195
USA
klee@math.washington.edu

Günter M. Ziegler
MA 6-2, Institute of
  Mathematics
TU Berlin
D-10623 Berlin
Germany
ziegler@math.tu-berlin.de

Mathematics Subject Classification (2000): 52-xx

Library of Congress Cataloging-in-Publication Data
Grünbaum, Branko.
    Convex polytopes / Branko Grünbaum. — 2nd ed.
      p. cm. — (Graduate texts in mathematics ; 221)
    Includes bibliographical references and index.
    ISBN 0-387-00424-6 (alk. paper)—ISBN 0-387-40409-0 (pbk.: alk. paper)
      1. Convex polytopes. I. Title. II. Series.
    QA482.G7 2003
    516.3′5—dc21   2003042435

ISBN 0-387-00424-6 (hardcover)   Printed on acid-free paper.
ISBN 0-387-40409-0 (softcover)

9 8 7 6 5 4 3 2 1   SPIN 10908334 (hardcover)   SPIN 10938768 (softcover)

www.springer-ny.com

Springer-Verlag   New York Berlin Heidelberg
A member of BertelsmannSpringer Science+Business Media GmbH

In humble homage
dedicated to the memory of the outstanding geometer

Ernst Steinitz
(1871–1928)

PREFACE

Convex polytopes—as exemplified by convex polygons and some three-dimensional solids—have been with us since antiquity. However, hardly any results worth mentioning and dealing specifically with the *combinatorial properties* of convex polytopes were discovered prior to Euler's famous theorem concerning the number of vertices, edges, and faces of three-dimensional polytopes. Euler's relation, hailed by Klee as "the first landmark" in the theory of convex polytopes, served as the starting point of a multitude of investigations which led to the determination of its limits of validity, and helped focus attention on the notion of convexity. Additional ideas and results came from such mathematicians as Cauchy, Steiner, Sylvester, Cayley, Möbius, Kirkman, Schläfli, and Tait. Since the middle of the last century, polytopes of four or more dimensions attracted interest; crystallography, generalizations of Euler's theorem, the search for polytopes exhibiting regularity features, and many other fields provided added impetus to the investigation of convex polytopes.

About the turn of the century, however, a steep decline in the interest in convex polytopes was produced by two causes working in the same direction. Efforts at enumerating the different combinatorial types of polytopes, started by Euler and pursued with much patience and ingenuity during the second half of the XIXth century, failed to produce any significant results even in the three-dimensional case; this lead to a widespread feeling that the interesting problems concerning polytopes are hopelessly hard. Simultaneously, the ascendance of Klein's "Erlanger Program" and the spread of its normative influence tended to cast the preoccupation with the combinatorial theory of convex polytopes into a rather disreputable rôle—and that at a time when such "legitimate" fields as algebraic geometry and in particular topology started their spectacular development.

Due to this combination of circumstances and pressures it is probably not too surprising that only few specialized directions of research in polytopes remained active during the first half of the present century. Stretching slightly the time limits, the most prominent examples of those efforts were: Minkowski's fundamental contributions, related to his

vii

work on convexity in general, and applications to number theory in particular; Coxeter's work on regular polytopes; A. D. Aleksandrov's investigations in the metric theory of polytopes.

Nevertheless, as far as "main-line mathematics" is concerned, the combinatorial theory of convex polytopes was "out". Despite the appreciable number of published papers dealing with isolated (mostly extremal) problems, the whole area was relegated to the borderline between serious research and amateurish curiosity. The one notable exception in this respect among first-rank mathematicians was Ernst Steinitz, who devoted a sizable part of his life and efforts to the combinatorial theory of polytopes. Unfortunately, his beautiful results did not become as well known as they deserve, and till very recently did not stimulate additional research.

It was mainly under the influence of computational techniques (in particular, linear programming) that a renewed interest in the combinatorial theory of convex polytopes became evident slightly more than ten years ago. The phenomenon of "neighborly polytopes" was rediscovered by Gale in 1955 (the rather involved history of this concept is related in Section 7.4). Neighborly polytopes, and Motzkin's "upper-bound conjecture" (1957) served as focal points for many investigations (see Chapters 9 and 10). Despite many scattered results on the upper-bound conjecture and other combinatorial problems about convex polytopes, obtained by different authors in the first few years of the 1960's, the emergence of a theory proper began only with Klee's work, starting in 1962. Klee's results on the Dehn–Sommerville equations (the interesting history of this topic is given in Section 9.8) and his almost complete solution of the upper-bound conjecture were the source and basis for many of the subsequent developments.

During the last three years, research into the combinatorial structure of convex polytopes has grown at an astonishing rate. It would be premature to attempt to give here even the briefest historic outline of this period. Instead, detailed bibliographic references are supplied with each topic discussed in the book.

The present book grew out of lecture notes prepared by the author for a course on the combinatorial theory of convex polytopes given at the Hebrew University of Jerusalem in 1964/65. The main part of the final version was written while the author was lecturing on the same topic at the Michigan State University in East Lansing during 1965/66. The various parts of the book may be described briefly as follows:

The first four chapters are introductory and are meant to acquaint the reader with some basic facts on convex sets in general, and polytopes in particular; as well as to provide "experimental material" in the form of examples.

Some basic tools for the investigation of polytopes are described in detail in Chapter 5; most of them are used in different subsequent sections. In Chapters 6 and 7 some of those techniques are applied to polytopes with "few" vertices, and to neighborly polytopes.

Chapters 8, 9, and 10 have as common topic the relations between the numbers of faces of different dimensions. Starting with Euler's equation, the Dehn–Sommerville equations for simplicial polytopes (and for certain other families) are discussed and used in the (partial) solution of the upper-bound conjecture. Chapter 14 is related to Chapter 9 by the similarity of the equations involved.

Chapters 11 and 12 deal with problems of a more topological flavor, while Chapter 13 discusses the much more detailed results known about 3-dimensional convex polytopes.

Chapter 15 contains a survey of the known results concerning the representation of polytopes as sums of other polytopes.

A summary of the available results about graphs of polytopes and paths in those graphs, as well as their relation to various problems that arose in applications, forms the topic of Chapters 16 and 17.

Chapter 18 deals with a topic related to convex polytopes more by the spirit of the problems considered than by actual interdependence: partitions of the (projective) space by hyperplanes.

In the last chapter a number of unrelated areas is surveyed. Their inclusion—at the expense of other topics which could have been included —is due to the author's interest in them.

It is hoped that parts of the book will prove suitable as texts for a number of different courses. On the other hand, the book is meant to serve as a ready reference for research workers; hence an attempt at completeness was made both in the coverage of the topics discussed, and in the bibliography. While the author is confident that the current surge of interest and research in the combinatorial properties of convex polytopes will continue and will render the book obsolete within a few years, he may only hope that the book itself will contribute to the revitalization of the field and act as a stimulant to further research. (Some of the results that came to the author's attention after completion of the manuscript in August 1966 are mentioned in the Addendum on pp. 426–428.)

It was the author's good fortune to obtain the cooperation of his friends and colleagues Victor Klee, M. A. Perles, and G. C. Shephard. Professor Klee wrote Chapters 16 and 17, while Professor Shephard contributed Chapter 15, Section 14.3, and part of Section 14.4. Professor Perles permitted the inclusion of many of his unpublished results; they are reproduced in Sections 5.1, 5.4, 5.5, 6.3, 11.1, 12.3, and in many other places throughout the book. In addition, Perles corrected many of the errors contained in the various preliminary versions, and contributed a large number of exercises. The author's indebtedness to Klee, Perles, and Shephard, hardly needs elaboration.

Thanks are also due to many other colleagues who contributed to the effort through discussions, suggestions, corrections etc. It would not be feasible to mention them all here. Particular thanks are due to W. E. Bonnice, L. M. Kelly, J. R. Reay, V. P. Sreedharan, and B. M. Stewart, all colleagues at Michigan State University during 1965/66, whose patience, encouragement and help during the most exasperating stages are gladly acknowledged and deeply appreciated.

The author gratefully acknowledges the financial support obtained at various times from the National Science Foundation and from the Air Force Office of Scientific Research, U.S. Air Force. Much of the research that is being published for the first time in the present book was conducted under the sponsorship of those agencies. Professor Klee acknowledges some helpful suggestions from David Barnette, and financial support from the University of Washington, the National Science Foundation, the RAND Corporation, and especially from the Boeing Scientific Research Laboratories; Chapters 16 and 17 appeared in a slightly different form as a BSRL Report.

The author's most particular thanks go to his wife Zdenka; without her encouragement and patience the book would have never been completed.

University of Washington, Seattle
December 31, 1966   BRANKO GRÜNBAUM

Preface to the 2002 edition

There is no such thing as an "updated classic"—so this is not what you have in hand.

In his 1966 preface, Branko Grünbaum expressed confidence "that the current surge of interest and research in the combinatorial properties of convex polytopes will continue and will render the book obsolete in a few years." He also stated his "hope that the book itself will contribute to the revitalization of the field and act as a stimulant to further research."

This hope has been realized. The combinatorial study of convex polytopes is today an extremely active and healthy area of mathematical research, and the number and depth of its relationships to other parts of mathematics have grown astonishingly. To some extent, Branko's confidence in the obsolescence of his book was also justified, for some of the most important open problems mentioned in it have by now been solved. However, the book is still an outstanding compendium of interesting and useful information about convex polytopes, containing many facts not found elsewhere.

Major topics, from Gale diagrams to cubical polytopes, have their beginnings in this book. The book is comprehensive in a sense that was never achieved (or even attempted) again. So it is still a major reference for polytope theory (without needing any changes).

Unfortunately, the book went out of print as early as 1970, and some of our colleagues have been looking for "their own copy" since then. Thus, responding to "popular demand", there have been continued efforts to make the book accessible again. Now we are happy to say: *Here it is!*

The present new edition contains the full text of the original, in the original typesetting, and with the original page numbering—except for the table of contents and the index, which have been expanded. You will see yourself all that has been added: The notes that we provide are meant to help to bridge the thirty-five years of intensive research on polytopes that were to a large extent initiated, guided, motivated, and fuelled by this book. However, to make this edition feasible, we had to restrict these notes severely, and there is no claim or even attempt for any complete coverage. The notes that we provide for the individual chapters try to summarize a few important developments with respect to the topics treated by Grünbaum, quite a remarkable number of them triggered by his exposition. Nevertheless, the selection of topics for these notes is clearly biased by our own interests.

xi

The material that we have added provides a direct guide to more than 400 papers and books that have appeared since 1967; thus references like "Grünbaum [a]" refer to the additional bibliography which starts on page 448a. Many of those publications are themselves surveys, so there is also much work to which the reader is guided indirectly. However, there remain many gaps that we would have liked to fill if space permitted, and we apologize to fellow researchers whose favorite polytopal papers are not mentioned here.

Principal references to "polytope theory since Grünbaum's book" that we have relied on include the books by McMullen–Shephard [b], Brøndsted [a], Yemelichev–Kovalev–Kravtsov [a], Ziegler [a], and Ewald [a], as well as the surveys by Grünbaum–Shephard [a], Grünbaum [d], Bayer–Lee [a], and Klee–Kleinschmidt [b]. Furthermore, we want to direct the readers' attention to Croft–Falconer–Guy [a] for (more) unsolved problems about polytopes.

We have taken advantage of some tools available in 2002 (but not in 1967), in order to compute and to visualize examples. In particular, the figures that appear in the additional notes were computed in the `polymake` framework by Gawrilow–Joswig [a, b], and were visualized using `javaview` by Polthier et al. [a].

Moreover and most of all, we are indebted to a great number of very helpful and supportive colleagues—among them Marge Bayer, Lou Billera, Anders Björner, David Bremner, Christoph Eyrich (LATEX with style!), Branko Grünbaum, Torsten Heldmann, Martin Henk, Michael Joswig, Gil Kalai, Peter Kleinschmidt, Horst Martini, Jirka Matoušek, Peter McMullen, Micha Perles, Julian Pfeifle, Elke Pose, Thilo Schröder, Egon Schulte, and Richard Stanley—who have provided information and assistance on the way to completion of this long-planned "Grünbaum reissue" project.

Berlin/Seattle, September 2002,

Volker Kaibel · Victor Klee · Günter M. Ziegler

CONTENTS

CHAPTER 1

Notation and Prerequisites

1.1 Algebra

With few exceptions, we shall be concerned with convexity in R^d, the d-dimensional real Euclidean space. Lower case characters such as a, b, x, y, z shall denote points of R^d, as well as the corresponding vectors; 0 is the origin as well as the number zero. Capitals like A, B, C, K shall denote sets; occasionally single points, if considered as one-pointed sets, shall be denoted by capitals. Greek characters $\alpha, \beta, \lambda, \mu$, etc., shall denote reals, while n, k, i, j shall be used for integers. The coordinate representation of a point $a \in R^d$ shall be $a = (\alpha_i) = (\alpha_1, \alpha_2, \cdots, \alpha_d)$.

Sets defined explicitly by specifying their elements will be written in the forms $\{a_1, \cdots, a_k\}$, $\{a_1, \cdots, a_n, \cdots\}$, or $\{a \in A \mid a$ has property $\mathscr{S}\}$, the last expression indicating all those elements of a set A which have a certain property $\mathscr{S}$. Finite or infinite *sequences* (of not necessarily different elements) will be denoted by $(a_1, \cdots, a_k)$ or $(a_1, \cdots, a_n, \cdots)$; the first expression will also be called a *k-tuple*. For the set-theoretic notions of *union, intersection, difference, subset* we shall use the symbols $\cup, \cap, \sim$, and $\subset$. The empty set will be denoted by $\varnothing$, while card A will denote the cardinality of the set A.

The algebraic signs are reserved for algebraic operations; thus

$$a \pm b = (\alpha_i) \pm (\beta_i) = (\alpha_i \pm \beta_1)$$

$$\lambda a = \lambda(\alpha_i) = (\lambda \alpha_i)$$

$$A \pm B = \{a \pm b \mid a \in A, b \in B\}$$

$$\lambda A = \{\lambda a \mid a \in A\}.$$

If a set A consists of a single point a we shall use the simplified notation $a + B$ instead of $\{a\} + B = A + B$. The set $(-1)A$ will be denoted $-A$. The set $x + \lambda B$, for $\lambda \neq 0$, is said to be *homothetic* to B, and *positively homothetic* if $\lambda > 0$.

The *scalar product* $\langle a, b \rangle$ of vectors $a, b \in R^d$ is the real number defined by

$$\langle a, b \rangle = \sum_{i=1}^{d} \alpha_i \beta_i.$$

The most important properties of the scalar product are

$$\langle a, b \rangle = \langle b, a \rangle$$

$$\langle \lambda a, b \rangle = \lambda \langle a, b \rangle$$

$$\langle a + b, c \rangle = \langle a, c \rangle + \langle b, c \rangle$$

$$\langle a, a \rangle \geq 0 \text{ with equality if and only if } a = 0.$$

If $\langle a, b \rangle = 0$ then a and b are said to be *orthogonal* to each other. If $\langle a, a \rangle = 1$ then a is called a *unit vector*. In the sequel, the letter u (with or without indices) shall be used *only* for unit vectors.

A *hyperplane H* is a set which may be defined as $H = \{x \in R^d \mid \langle x, y \rangle = \alpha\}$ for suitable $y \in R^d$, $y \neq 0$, and α. An *open halfspace* [*closed halfspace*] is defined as $\{x \in R^d \mid \langle x, y \rangle > \alpha\}$ [respectively $\{x \in R^d \mid \langle x, y \rangle \geq \alpha\}$] for suitable $y \in R^d$, $y \neq 0$, and α. Clearly, $\{x \in R^d \mid \langle x, y \rangle < \alpha\}$ is also an open halfspace for $y \neq 0$; similarly for closed halfspaces.

Each hyperplane has a translate which is (isomorphic to) a $(d-1)$-dimensional Euclidean space R^{d-1}. For each hyperplane H which does not contain the origin 0 there exists a unique representation $H = \{x \in R^d \mid \langle x, u \rangle = \alpha\}$ in which u is a unit vector and $\alpha > 0$.

If $x, x_i \in R^d$, we shall say that x is a *linear combination* of the x_i's provided

$$x = \sum_{i=1}^{k} \lambda_i x_i$$

for suitable real numbers λ_i.

If $x = \sum_{i=1}^{k} \lambda_i x_i$ for reals λ_i satisfying $\sum_{i=1}^{k} \lambda_i = 1$ we shall say that x is an *affine combination* of the x_i's.

A set $X = \{x_1, \cdots, x_k\}$ is *linearly* [respectively *affinely*] *dependent* provided 0 is representable as a linear combination $0 = \sum_{i=1}^{k} \lambda_i x_i$ in which some $\lambda_i \neq 0$ [and $\sum_{i=1}^{k} \lambda_i = 0$]. If a set X fails to be linearly [affinely] dependent we call it *linearly* [*affinely*] *independent*. In any linearly

·[affinely] dependent set some point is a linear [affine] combination of the remaining points. The d-dimensional space contains d-membered sets which are linearly independent, but every $(d + 1)$-membered set in R^d is linearly dependent. A set $X = \{x_0, x_1, \cdots, x_k\}$ is affinely dependent [independent] if and only if the set $(X \sim \{x_0\}) - x_0 = \{x_1 - x_0, x_2 - x_0, \cdots, x_k - x_0\}$ is linearly dependent [independent]. For any $x \in R^d$ the sets X and $x + X$ are simultaneously affinely dependent or independent.

The set of all affine combinations of two different points $x, y \in R^d$ is the *line* $L(x, y) = \{(1 - \lambda)x + \lambda y | \lambda \text{ real}\}$. If $x', y' \in L(x, y)$ and $x' \neq y'$ then $L(x', y') = L(x, y)$.

If a set H has the property that $L(x, y) \subset H$ whenever $x, y \in H, x \neq y$, we call H a *flat* (or an *affine variety*). Clearly, the set of all affine [linear] combinations formed from all finite subsets of a given set A is a flat [subspace]; it is denoted by aff A [lin A] and is called the *affine hull* [*linear hull*] of A. The family of all flats in R^d contains R^d, $\emptyset$, all one-pointed sets, all lines, all hyperplanes; also, it is *intersectional*: if all H_α's are flats, so is $\underset{\alpha}{\cap} H_\alpha$. The affine hull aff A of a set A may equivalently be defined as the intersection of all flats which contain A. Similar statements hold for linear hulls. The formation of the affine hull is *translation invariant*, i.e. aff $(x + A) = x + $ aff A.

Every nonempty flat H is a translate $H = x + V$ of some subspace V of R^d, and is therefore isomorphic to a Euclidean space of a certain dimension $r \leq d$; the *dimension* of H (and of V) is then $r = \dim H = \dim V$. A flat of dimension r will be called an r-flat. We agree to put dim $\emptyset = -1$. In general, instead of saying 'an object of dimension r' we shall use the shorter term 'r-object'; for example, d-space, r-subspace, etc. If A is any subset of R^d, its *dimension* dim A is defined by dim $A = \dim$ aff A.

Each r-flat contains $r + 1$ affinely independent points, but each $(r + 2)$-membered set of its points is affinely dependent.

If $A = \{a_1, a_2, \cdots, a_k\}$, where $a_i = \{\alpha_{i1}, \alpha_{i2}, \cdots, \alpha_{id}\}$, then the maximal number of linearly independent members of A equals the rank of the matrix

$$\begin{pmatrix} \alpha_{11} & \alpha_{12} & \cdots & \alpha_{1d} \\ \alpha_{21} & \alpha_{22} & \cdots & \alpha_{2d} \\ & \cdots\cdots\cdots\cdots \\ \alpha_{k1} & \alpha_{k2} & \cdots & \alpha_{kd} \end{pmatrix}$$

while the maximal number of affinely independent members of A equals the rank of the matrix

$$\begin{pmatrix} 1 & \alpha_{11} & \alpha_{12} & \cdots & \alpha_{1d} \\ 1 & \alpha_{21} & \alpha_{22} & \cdots & \alpha_{2d} \\ & & \cdots\cdots\cdots \\ 1 & \alpha_{k1} & \alpha_{k2} & \cdots & \alpha_{kd} \end{pmatrix}.$$

A finite set $X \subset R^d$ is said to be in *general position* provided each subset of X containing at most $d + 1$ members is affinely independent.

The following remark is sometimes useful: Given positive integers d and k, there exists an integer $n(d, k)$ with the property that whenever $A \subset R^d$ satisfies card $A \geq n(d, k)$, there exists a subset B of A such that card $B = k$ and the points of B are in general position in aff B.

Let a transformation T from R^d to R^d be defined by

$$Tx = \frac{Ax + b}{\langle c, x \rangle + \delta},$$

where A is a linear transformation of R^d into itself, b and c are d-dimensional vectors, and δ is a real number, at least one of c and δ being different from 0. Any transformation of this type is called a *projective transformation** from R^d into R^d. Note that T is not defined for x in $N(T) = \{y \mid \langle c, y \rangle + \delta = 0\}$. The set $N(T)$ may be empty (in which case T is an affine transformation); if A is regular and $c \neq 0$, $N(T)$ is a hyperplane (which, in the *projective space*, is mapped by T into the 'hyperplane at infinity'). The reader is invited to verify that collinear points are mapped by projective transformations onto collinear points. A projective transformation T is *nonsingular* provided the matrix $\begin{pmatrix} A' & b' \\ c & \delta \end{pmatrix}$ is regular (here A' is the matrix of A, and b' the transposed vector b); in this case T has an inverse which is again a projective transformation. If $(x_0, \cdots, x_{d+1})$ and $(y_0, \cdots, y_{d+1})$ are two $(d + 2)$-tuples of points in general position in R^d, there exists a unique projective transformation T such that $Tx_i = y_i$ for $i = 0, \cdots, d + 1$; moreover, this T is nonsingular. If K is a subset of R^d, T is said to be *permissible* for K provided $K \cap N(T) = \emptyset$. If $K_i \subset R^d$

---

* In case of need, the reader should consult a suitable textbook on projective geometry. However, he should bear in mind that we are dealing with Euclidean (or affine) spaces, and nonhomogeneous coordinates, while the most natural setting for projective transformations are projective spaces and homogeneous coordinates.

and T_i is a projective transformation permissible for K_i, $i = 1, 2$, and if $T_1 K_1 \subset K_2$, then $T_2 T_1$ is a projective transformation permissible for K_1.

Subsets A, B of R^d will be called affinely [projectively] equivalent provided there exists a nonsingular affine [permissible for A projective] transformation T such that $TA = B$.

1.2 Topology

A set X is a *metric space* provided a real valued *metric function* (or *distance*) ρ is defined for all pairs of points of X satisfying the conditions:

(i) $\rho(x, y) \geq 0$, with equality if and only if $x = y$;

(ii) $\rho(x, y) = \rho(y, x)$;

(iii) $\rho(x, y) \leq \rho(x, z) + \rho(z, y)$.

In the remaining part of this section X shall denote a metric space with a given distance ρ.

For any $x \in X$ and $\delta > 0$ the *open ball* $\overset{\circ}{B}(x; \delta)$, the *closed ball* $B(x; \delta)$, and the *sphere* $S(x; \delta)$ with *center* x and *radius* δ are defined by

$$\overset{\circ}{B}(x; \delta) = \{y \in X \mid \rho(y, x) < \delta\}$$

$$B(x; \delta) = \{y \in X \mid \rho(y, x) \leq \delta\}$$

$$S(x; \delta) = \{y \in X \mid \rho(y, x) = \delta\}.$$

A set $A \subset X$ is *open* provided every $a \in A$ is the center of some open ball $\overset{\circ}{B}(a; \delta)$ which is contained in A. It is easily shown that open balls, the whole space X, the empty set $\varnothing$, are open sets. The union of any family of open sets is an open set; the intersection of any finite family of open sets is open.

A set $A \subset X$ is *closed* provided its complement $\sim A = X \sim A$ is open. All closed balls, all spheres, all finite sets of points, $\varnothing$, and X, are closed. The family of closed sets is *intersectional*, i.e. the intersection of any family of closed sets is itself closed; the union of any finite family of closed sets is closed. A set $A \subset X$ is *bounded* provided there exists $\delta > 0$ and $x \in X$ such that $\rho(a, x) < \delta$ for all $a \in A$. The *diameter* diam A is defined by diam $A = \sup\{\rho(x, y) \mid x, y \in A\}$.

A sequence of $(x_1, x_2, \cdots, x_n, \cdots)$ of points of X is said to *converge* to $x \in X$ (or to have x as *limit*) provided $\lim_{n \to \infty} \rho(x_n, x) = 0$. A sequence $(x_1, x_2, \cdots, x_n, \cdots) \subset X$ is a *Cauchy sequence* provided for every $\varepsilon > 0$ there exists $k = k(\varepsilon)$ such that $\rho(x_i, x_j) < \varepsilon$ whenever $i, j > k$. The metric space X is *complete* provided every Cauchy sequence in X converges to some point of X.

An alternative definition of closed sets is: A set $A \subset X$ is closed provided the limit of every convergent sequence of points of A belongs to A.

A set $A \subset X$ is *compact* provided every infinite sequence of points of A contains a subsequence which has a point of A as limit.

The union of all open sets contained in a set $A \subset X$ is an open set, the *interior* of A; it is denoted by int A. The intersection of all closed sets containing A is a closed set, the *closure* of A; it is denoted by cl A. The *boundary* of A, denoted bd A, is defined by bd $A = $ cl $A \cap$ cl$(\sim A)$. Clearly bd A is closed (possibly empty) for every A.

The metric space which will be most important in the sequel is the d-dimensional real Euclidean space R^d. For $a, b \in R^d$ we define

$$\rho(a, b) = \left(\sum_{i=1}^{d} (\alpha_i - \beta_i)^2 \right)^{\frac{1}{2}} = (\langle a - b, a - b \rangle)^{\frac{1}{2}}.$$

It is easily shown that all flats and all closed halfspaces are closed sets, and that open halfspaces are open sets.

The metric function of R^d has also the following properties:

$$\rho(\lambda a, \lambda b) = |\lambda| \, \rho(a, b)$$

$$\rho(a + c, b + c) = \rho(a, b).$$

Using the notation $\|x\| = \rho(x, 0)$, this becomes $\rho(a, b) = \|a - b\|$.

A set $A \subset R^d$ is compact if and only if A is closed and bounded.

If $A, B \subset R^d$ are closed sets and at least one of them is compact, then $A + B$ is closed; if both are compact, so is $A + B$.

If $A \subset R^d$ is open, then $A + B$ is open for every B.

The *distance* $\delta(A, B)$ between sets $A, B \subset R^d$ is defined by

$$\delta(A, B) = \inf\{\rho(a, b) \mid a \in A, b \in B\}.$$

The family $\mathscr{S}$ of all compact subsets of R^d is a metric space with the *Hausdorff distance* $\rho(A_1, A_2)$ defined by

$$\rho(A_1, A_2) = \inf\{\alpha > 0 \mid A_1 \subset A_2 + \alpha B, A_2 \subset A_1 + \alpha B\},$$

where $B = B(0; 1)$ is the closed ball of unit radius centered at the origin 0. An equivalent definition is

$$\rho(A_1, A_2) = \max\left\{ \sup_{x_1 \in A_1} \inf_{x_2 \in A_2} \|x_1 - x_2\|, \sup_{x_2 \in A_2} \inf_{x_1 \in A_1} \|x_2 - x_1\| \right\}.$$

$\mathscr{S}$ is a complete metric space, with the following local compactness property:

Every subfamily of $\mathscr{S}$ which is bounded and closed in the Hausdorff metric, is compact in this metric.

Convergence of closed subsets of R^d may be defined by stipulating that a sequence $(A_1, A_2, \cdots, A_n, \cdots)$ of closed sets in R^d converges to a closed set A provided

(i) for every $a \in A$ there exists a sequence $a_n \in A_n$ such that $a = \lim a_n$; and

(ii) for every convergent sequence (a_n), where $a_n \in A_n$, we have $\lim a_n \in A$.

1.3   Additional notes and comments

The first sentence on page 1 is
"With a few exceptions, we shall be concerned with convexity in R^d, the d-dimensional Euclidean space."

For the study of convex polytopes, this is justified by the observation that in Euclidean d-space one encounters the same (combinatorial types of) polytopes as those met in elliptic/spherical space or in hyperbolic space. Indeed, with any polytope $P \subset R^d$ one may associate the pointed cone $C_P \subset R^{d+1}$ that is spanned by all points $(1, x)$ with $x \in P$. The intersection of this cone with the unit sphere $S^d = \{x \in R^{d+1} : x_0^2 + \cdots + x_d^2 = 1\}$ yields a spherical polytope; furthermore, any spherical polytope (distinct from S^d) may be transformed to lie in the open hemisphere $\{x \in S^d : x_0 > 0\}$, and then determines a cone C_P and a convex polytope $P = \{x \in R^d : (1, x) \in C_P\}$. (See pages 10 and 30 for discussions of spherical convexity.) Similarly, if we scale $P \subset R^d$ to lie in the interior of the unit ball $B^d \subset R^d$, then the intersection of the cone C_P with the hyperboloid $H^d = \{x \in R^{d+1} : x_0^2 = 1 + x_1^2 + \cdots + x_d^2\}$ yields a hyperbolic polytope, and conversely any hyperbolic polytope (in the sheet of H^d given by $x_0 > 0$) determines a Euclidean polytope contained in the unit ball.

One may also note that orthogonal transformations that keep a polytope in the positive hemisphere of S^d correspond to admissible projective transformations (as discussed on page 4). The use of homogeneous coordinates gives correspondences between affine, spherical, and hyperbolic polytopes.

Nevertheless, it has turned out to be very useful at times to view polytopes in spherical resp. hyperbolic space, for arguments or computations that would exploit aspects that are specific to the geometry (metric, angles, volumes) of spherical or hyperbolic space.

A remark on page 4.
Grünbaum's "useful observation" may be proved by induction on k and d: One obtains recursions of the type

$$n(d, k) \leq \binom{k-1}{d} n(d-1, k),$$

based at $n(1, k) = k$. Here one may assume that the given set A has dimension d, otherwise the claim is true by induction. Then we consider a d-dimensional general position subset $B \subset A$ of maximal cardinality card $B \leq k - 1$; if the $(d-1)$-flats it spans all contain fewer than $n(d-1, k)$ points from A, then there are points from A that extend B.

The observation has been considerably strengthened: The subset $B \subset A$ of cardinality k can be found to lie on a curve of order d' in d'-dimensional affine space; thus it will give a cyclic oriented matroid. For $d' > 1$ the set B will be in convex position, forming the vertex set of a cyclic polytope $C(k, d')$—this is obtained by combination of Grünbaum's remark with results of Duchet–Roudneff [a, Cor. 3.8] and Sturmfels [a] (see also Björner et al. [a, pp. 398–399]).

The observation is in essence a Ramsey-theoretic result, see Graham–Rothschild–Spencer [a]. Correspondingly, the bounds that follow from recursions of the type given above grow extremely fast. More precise versions for small corank are discussed in exercises 2.4.12 and 6.5.6.

The footnote on page 4 ...

... asks the reader to consult, if necessary, "a suitable textbook on projective geometry". Classical accounts of projective geometry include Veblen–Young [a] and Hodge–Pedoe [a]. A treatment of projective transformations using homogenization for the manipulation of convex polytopes, as suggested by Grünbaum, is Ziegler [a, Sect. 2.6].

CHAPTER 2

Convex Sets

The present chapter deals with some fundamental notions and facts on convex sets. It serves a double purpose: we establish certain properties of convex sets which shall be used later in the special case of convex polytopes; certain other properties are investigated which do not hold for all convex sets but are valid (and important) for more restricted families such as compact convex sets, polyhedral sets, or polytopes. We discuss these properties in order to enable the reader to place the convex polytopes in a better perspective among all convex sets.

Though readers familiar with the theory of convex sets may omit chapter 2, it is the author's hope that some of the facts and approaches presented will be of interest even for the specialist.

2.1 Definition and Elementary Properties

A set $K \subset R^d$ is *convex* if and only if for each pair of distinct points $a, b \in K$ the closed segment with endpoints a and b is contained in K.

Equivalently, K is convex if its intersection with every straight line is either empty, or a connected set.

Examples of convex subsets of R^d: the empty set $\varnothing$; any single point; any linear subspace (including R^d) of R^d; any flat (affine variety) and any (closed or open) halfspace of R^d; the interior of any triangle, or simplex in general; the interior of a circle (or k-dimensional sphere), together with any subset of the circle resp. sphere.

Using the vectorial notation, the definition of convexity may be reformulated as follows:

K is convex if and only if $a, b \in K$ and $0 \leq \lambda \leq 1$ imply $\lambda a + (1 - \lambda)b \in K$.

The following statements are almost obvious (and should become completely obvious after the reader proves them):

1. *If* $\{K_v\}$ *is any (finite or infinite, denumerable or not) family of convex sets in* R^d, *then their intersection* $\bigcap_v K_v$ *is also convex.*

2. *If A and B are convex then A + B and A − B are convex, and for any real λ, the set λA is convex.*

3. *If A is convex, $a_i \in A$ and $\lambda_i \geq 0$ for i = 1, 2, ..., k, and*

$$\sum_{i=1}^{k} \lambda_i = 1, \quad then \quad \sum_{i=1}^{k} \lambda_i a_i \in A.$$

4. *If $A \subset R^d$ is convex, the sets* cl *A and* int *A are also convex.* (Hint for cl *A*: Use exercise 2 and cl $A = \bigcap_{\mu > 0} (A + \mu B)$, where *B* is the unit ball of R^d.)

5. *If $A \subset R^d$ is convex, $x \in A$, and $y \in$ int A, then all points of the line segment between x and y belong to* int *A*.

6. *If T is an affine transformation of R^d into itself, and if $A \subset R^d$ is convex, then T(A) is convex.*

7. *For a convex set $A \subset R^d$ let H = aff A be the affine hull of A. The relative interior* relint *A of A as a subset of H is never empty, and* relint cl $A \subset A \subset$ cl relint *A. The relative boundary* relbd *A of A with respect to H is empty if and only if A is an affine variety (i.e. A = H).*

Exercises

1. For subsets *A* and *B* of R^d let $A \simeq B$ mean: there exist $x \in R^d$ and $\alpha > 0$ such that $B = x + \alpha A$. (This symbol will be used only in the present exercise.) This is an equivalence relation. Prove: The convex subsets of R^1 (including $\varnothing$ and R^1) form eleven distinct classes with respect to the relation $\simeq$. Describe these classes. How many classes are there if in the definition of $\simeq$ the only restriction on α is $\alpha \neq 0$? What is the number of classes in R^2?

2. Determine all subsets *A* of R^1 such that both *A* and its complement $\sim A$ are convex; the same for R^2 and R^3. Try to generalize to R^d.

3. Let $\{K_v\}$ be a family of convex sets in R^d. If every denumerable subfamily of $\{K_v\}$ has a non-empty intersection, then $\bigcap_v K_v \neq \varnothing$. (For generalizations and a survey of related results see Klee [7].)

4. For any pair of distinct points $a, b \in R^d$ let $[a, b[$ denote the (closed) halfline with endpoint *a*, passing through *b*. For any set $K \subset R^d$, and any $a \in R^d$ we define the *cone* $\text{cone}_a K$ generated by *K*, with *apex* a, by $\text{cone}_a K = \bigcup_{\substack{b \in K \\ b \neq a}} [a, b[$. We also define $\text{cone}_a \varnothing = \{a\}$. For convex $K \subset R^d$ prove the following statements:

(i) $\text{cone}_a K$ is convex;

(ii) if K is open, $(\text{cone}_a K) \sim \{a\}$ is open;

(iii) the assertion 'if K is closed then $\text{cone}_a K$ is closed' is false;

(iv) if K is compact and if $a \notin K$, then $\text{cone}_a K$ is closed.

5. Determine which of the statements in exercises 3 and 4 remain valid if K is not assumed to be convex.

6. Let S^{d-1} be the unit sphere in R^d, centered at the origin 0. A set $A \subset S^{d-1}$ is *spherically convex* provided $\text{cone}_0 A$ is a convex set. Prove that this definition is equivalent with the following: $A \subset S^{d-1}$ is spherically convex if and only if for every $x, y \in A$, $y \neq \pm x$, the set A contains the small arc of the great circle determined on S^{d-1} by x and y.

7. Prove Blaschke's 'selection theorem' (Blaschke [1], Eggleston [3]; compare p. 7): Every infinite sequence of compact convex sets which is bounded in the Hausdorff metric, contains a subsequence which converges (in the Hausdorff metric) to a compact, convex set.

8. Let $(A_1, \cdots, A_n, \cdots)$ be a sequence of closed convex sets in R^d. Show that the sequence (A_n) converges to the closed set A if and only if for every sufficiently large λ the sequence $(A_i \cap \lambda B, A_2 \cap \lambda B, \cdots, A_n \cap \lambda B, \cdots)$ converges to $A \cap \lambda B$ (where B denotes the solid unit ball in R^d).

2.2   Support and Separation

Let A be a subset of R^d. We shall say that a hyperplane

$$H = \{x \in R^d \,|\, \langle x, u \rangle = \alpha\}$$

cuts A provided both open halfspaces determined by H contain points of A. In other words, H cuts A provided there exist $x_1, x_2 \in A$ such that $\langle x_1, u \rangle < \alpha$ and $\langle x_2, u \rangle > \alpha$.

We shall say that a hyperplane H *supports* A provided H does not cut A but the distance between A and H is 0, $\delta(A, H) = 0$. In other words, H supports A if either $\sup\{\langle x, u \rangle | x \in A\} = \alpha$ or else $\inf\{\langle x, u \rangle \,|\, x \in A\} = \alpha$.

Since bounded and closed subsets of R^d are compact, this implies:

A bounded set $A \subset R^d$ is supported by a hyperplane H if and only if H does not cut A and $H \cap \text{cl } A \neq \emptyset$.

Two subsets A and A' of R^d are said to be *separated* by a hyperplane H provided A is contained in one of the closed halfspaces determined by H while A' is contained in the other. The sets A and A' are *strictly separated* by H if they are separated and $A \cap H = A' \cap H = \emptyset$. In other words, A and A' are strictly separated by H provided they are contained in different open halfspaces determined by H.

The following results are of fundamental importance:

1. *If A and A' are convex subsets of R^d such that A' is bounded and* $\operatorname{cl} A \cap \operatorname{cl} A' = \varnothing$, *then A and A' may be strictly separated by a hyperplane.*

2. *If A and A' are convex subsets of R^d such that* $\operatorname{aff}(A \cup A') = R^d$ *then A and A' may be separated by a hyperplane if and only if.*

$$\operatorname{relint} A \cap \operatorname{relint} A' = \varnothing.$$

PROOF OF THEOREM 1 Since the distance between two sets is the same as the distance between their closures, it is obviously enough to consider the case in which A and A' are closed sets. Let $\delta = \delta(A, A')$ be the distance between A and A'; by the hypothesis $\delta > 0$. Clearly, the function $\delta(x, A') = \inf\{\rho(x, y) \mid y \in A'\}$ depends continuously on x. If $B(\varepsilon)$ denotes the closed d-dimensional ball centered at x, with radius $\varepsilon + \delta(x, A')$, then $A' \cap B(\varepsilon)$ is, for $\varepsilon > 0$, a nonempty compact set, and $A' \cap B(\varepsilon) \subset A' \cap B(\varepsilon')$ whenever $0 < \varepsilon < \varepsilon'$. Therefore $\bigcap\limits_{\varepsilon > 0} (A' \cap B(\varepsilon)) = A' \cap B(0)$ $\neq \varnothing$. In other words, there exists a point $y = y(x) \in A'$ such that $\delta(x, A') = \rho(x, y(x))$. Moreover, the point $y(x)$ is unique, since if there would exist distinct $y_1, y_2 \in A'$ with $\rho(x, y_1) = \rho(x, y_2) = \delta(x, A')$ then $\frac{1}{2}(y_1 + y_2) \in A'$ would satisfy $\rho(x, \frac{1}{2}(y_1 + y_2)) < \delta(x, A')$, contradicting the definition of $\delta(x, A')$.

Since $\delta(x, A')$ is a continuous function of x, it assumes a minimum on the compact set A; thus there exists an $x_0 \in A$ such that $\delta = \delta(x_0, A') = \rho(x_0, y(x_0))$. The hyperplanes H_1 and H_2, orthogonal to $[x_0, y(x_0)]$ and passing through x_0 respectively $y(x_0)$, have the following property: The open slab Q of width $\delta > 0$, determined by H_1 and H_2, contains no point of $A \cup A'$. Indeed, let $z \in Q$ and consider the open intervals $]z, x_0[$ and $]z, y(x_0)[$. Each point of the first interval, sufficiently near to x_0, is at a distance less than δ from $y(x_0)$ and thus can not belong to A; similarly, each point of the second interval, sufficiently near $y(x_0)$ is at a distance less than δ from x_0 and therefore does not belong to A'. Since both A and A' are convex, it follows that z belongs to neither of them. In other words, we have established that H_1 (as well as H_2) separates A and A', and each hyperplane parallel to H_1 and intersecting the open interval $]x_0, y(x_0)[$ strictly separates A and A'. This completes the proof of theorem 1.

PROOF OF THEOREM 2 The 'only if' part is obvious. We shall establish the other part of the theorem assuming, without loss of generality, that A and A' are closed. Let $x_0 \in \operatorname{relint} A$, $y_0 \in \operatorname{relint} A'$, and let $0 < \varepsilon < 1$.

Denoting by B the solid d-dimensional unit ball centered at the origin, let

$$A_\varepsilon = x_0 + \frac{1}{\varepsilon} B \cap ((1 - \varepsilon)(-x_0 + A))$$

and

$$A'_\varepsilon = y_0 + (1 - \varepsilon)(-y_0 + A').$$

Then A'_ε is homothetic to A', and A_ε is homothetic to a compact subset of A. Note that for $\varepsilon' > \varepsilon'' > 0$, $A_{\varepsilon'} \subset A_{\varepsilon''}$ and $A'_{\varepsilon'} \subset A'_{\varepsilon''}$, and that relint $A = \bigcup_{0 < \varepsilon < 1} A_\varepsilon$ and relint $A' = \bigcup_{0 < \varepsilon < 1} A'_\varepsilon$. Since $A_\varepsilon \subset$ relint A and $A'_\varepsilon \subset$ relint A', it follows that $A_\varepsilon \cap A'_\varepsilon = \varnothing$. By theorem 1, there exists a hyperplane $H_\varepsilon = \{x \in R^d \,|\, \langle x, u_\varepsilon \rangle = \alpha_\varepsilon\}$ strictly separating A_ε and A'_ε. Since each H_ε meets the segment $[x_0, y_0]$, the set $\{\alpha_\varepsilon \,|\, \varepsilon > 0\}$ is bounded. The set $\{u_\varepsilon \,|\, \varepsilon > 0\}$ being contained in the compact unit sphere S^{d-1}, this implies the existence of a sequence $(\varepsilon_n \,|\, n = 1, 2, \cdots)$ with $\varepsilon_n > 0$ and $\lim_{n \to \infty} \varepsilon_n = 0$ such that the sequences (u_{ε_n}) and (α_{ε_n}) converge to u and α. Let H be the hyperplane $H = \{x \,|\, \langle x, u \rangle = \alpha\}$; then H clearly separates A and A', and the proof of theorem 2 is completed.

The reader is invited to derive the following propositions from theorems 1 and 2.

3. *Each closed, convex subset of R^d is the intersection of all the closed (or of all the open) halfspaces of R^d which contain the set. Each open convex subset of R^d is the intersection of all the open halfspaces containing it.*

4. *If K is a convex set in R^d and if C is a convex subset of $\mathrm{bd}\, K$ (in particular, if C is a single point of $\mathrm{bd}\, K$) then there is a hyperplane separating K and C. In other words, there exists a supporting hyperplane of K which contains C.*

Exercises

1. In the proof of theorem 1 the uniqueness of the point x_0 was not asserted; could it have been asserted? Is the boundedness of A' essential for the validity of theorem 1?

2. If A is a bounded set in R^d and if H is a given hyperplane, there exists a supporting hyperplane of A parallel to H. If A is, moreover, convex and int $A \neq \varnothing$, there exist exactly two such hyperplanes. If A is convex and bounded, and $x \notin \mathrm{int}\, A$, there exists a supporting hyperplane of A which contains x. Are all the conditions mentioned in the above statements necessary?

3. If A and A' are disjoint compact convex sets in R^d, then the set $\mathscr{H} = \{H(u, \alpha)\}$ of all hyperplanes $H(u, \alpha) = \{x \in R^d \mid \langle x, u \rangle = \alpha\}$ which strictly separate A and A' is *open* in the sense that $\{(u, \alpha) \mid H(u, \alpha) \in \mathscr{H}\}$ is an open subset of the product $S^{d-1} \times R$.

4. If A is a closed subset of R^d with int $A \neq \varnothing$, such that each boundary point of A is on a supporting hyperplane of A, then A is convex.

5. Determine all the *semispaces* of R^d, that is, the maximal (with respect to inclusion) convex sets which do not contain a given point. (Motzkin [2], Hammer [1, 2], Klee [1]). Prove:

(i) The complement (in R^d) of a semispace is a convex set.

(ii) The family of all semispaces of R^d is an *intersectional basis* for all convex sets in R^d (that is, every convex subset of R^d is the intersection of all the semispaces containing it).

(iii) The family of all semispaces in R^d is a *minimal* intersectional basis for the convex subsets of R^d (that is, none of its proper subfamilies is an intersectional basis). It is a minimal intersectional basis for all bounded convex sets.

6. Characterize those subsets of R^d which are obtainable as intersections of d-dimensional solid balls.

7. Let $K \subset R^d$ be a closed convex set and $L \subset R^d$ a flat such that dim $L < d$ and $L \cap K = \varnothing$. Show that there exists a hyperplane H such that $K \cap H = \varnothing$ and $L \subset H$.

8. Let $K \subset R^d$ be a nonempty set. The *supporting function* $H(K, x)$ of K is defined for all $x \in R^d$ by

$$H(K, x) = \sup\{\langle y, x \rangle \mid y \in K\}.$$

If for some nonzero $x \in R^d$ we have $H(K, x) < \infty$, the hyperplane

$$L(K, x) = \{y \in R^d \mid \langle y, x \rangle = H(K, x)\}$$

is obviously a supporting hyperplane of K; $L(K, x)$ is called the *supporting hyperplane* of K with *outward normal* x. The following facts will be used mainly in Chapter 15; the reader is urged to provide their proofs, or to look them up in the literature (see, for example, Bonnesen–Fenchel [1]).

(i) The supporting function $H(K, x)$ of a convex set $K \neq \varnothing$ is *positively homogeneous* and *convex*, that is, it satisfies

$$H(K, \lambda x) = \lambda H(K, x) \qquad \text{for all } \lambda \geq 0, \quad x \in R^d;$$

$$H(K, x + y) \leq H(K, x) + H(K, y) \qquad \text{for all } x, y \in R^d.$$

On the other hand, if $H(x)$ is any function defined on R^d such that $H(0) = 0$, $H(\lambda x) = \lambda H(x)$ and $H(x + y) \le H(x) + H(y)$ for all $\lambda \ge 0$ and $x, y \in R^d$, then there exists a nonempty closed convex set K such that $H(x) = H(K, x)$ for all $x \in R^d$.

(ii) If K is a nonempty set, if $\lambda \ge 0$, and if $y \in R^d$, then $H(y + K, x) = \langle y, x \rangle + H(K, x)$, $H(\lambda K, x) = \lambda H(K, x)$, and $H(\text{cl } K, x) = H(K, x)$ for all $x \in R^d$.

(iii) If K_1, K_2 are nonempty and $0 \ne x \in R^d$, then $H(K_1 + K_2, x) = H(K_1, x) + H(K_2, x)$ and $(K_1 + K_2) \cap L(K_1 + K_2, x) = (K_1 \cap L(K_1, x)) + (K_2 \cap L(K_2, x))$.

(iv) If K_1, K_2 are nonempty, closed convex sets in R^d such that $H(K_1, x) = H(K_2, x)$ for every $x \in R^d$, then $K_1 = K_2$.

2.3   Convex Hulls

The space R^d is convex [and closed], and the intersection of any family of convex [and closed] sets is again convex [and closed]. Therefore the following definitions make sense:

The *convex hull* conv A of a subset A of R^d is the intersection of all the convex sets in R^d which contain A. The *closed convex hull* clconv A of $A \subset R^d$ is the intersection of all the closed convex subsets of R^d which contain A.

Clearly, if A is bounded so are conv A and clconv A.

An immediate consequence of the definitions is

1. *For every $A \subset R^d$ we have* clconv $A = \text{cl}(\text{conv } A)$.

Proposition 3 from the preceding section implies

2. clconv A *is the intersection of all the closed halfspaces which contain A.*

A useful representation of conv A is given by

3. *The convex hull* conv A *of a nonempty set $A \subset R^d$ is the set of all points which may be represented as convex combinations of points of A; that is, points which can be written in the form $\sum_{i=1}^{n} \alpha_i x_i$, where $x_i \in A$, $\alpha_i \ge 0, \sum_{i=1}^{n} \alpha_i = 1, n = 1, 2, \cdots$.*

In many applications the following result of Carathéodory [1] is very important.

4. *If A is a compact subset of R^d then* conv A *is closed; in other words, for compact A we have* clconv $A = $ conv A.

Using the results of the preceding section it is not hard to give a direct proof of theorem 4, by induction on the dimension d. Since a much simpler proof results from Carathéodory's theorem, we defer the proof of theorem 4 till we establish theorem 5.

The following theorem, known as *Carathéodory's theorem*, is one of the basic results in convexity, and has important application in other fields.

5. (Carathéodory [2]) *If A is a subset of R^d then every $x \in$ conv A is expressible in the form*

$$x = \sum_{i=0}^{d} \alpha_i x_i \quad where \quad x_i \in A, \alpha_i \geq 0, \quad and \quad \sum_{i=0}^{d} \alpha_i = 1.$$

PROOF   Let $x \in$ conv A be given; let $x = \sum_{i=0}^{p} \alpha_i x_i$ (with $x_i \in A$, $\alpha_i \geq 0$, $\sum_{i=0}^{p} \alpha_i = 1$) be a representation of x as a convex combination of points of A, involving the smallest possible number $p + 1$ of points of A. We shall prove Carathéodory's theorem by showing $p \leq d$. Indeed, assuming $p \geq d + 1$ it follows that the set $\{x_0, \cdots, x_p\}$ is affinely dependent. Thus there exists $\beta_i, 0 \leq i \leq p$, not all equal to 0, such that $\sum_{i=0}^{p} \beta_i x_i = 0$ and $\sum_{i=0}^{p} \beta_i = 0$. Without loss of generality we assume the notation such that $\beta_p > 0$ and $\alpha_p/\beta_p \leq \alpha_i/\beta_i$ for all those i ($0 \leq i \leq p - 1$) for which $\beta_i > 0$. For $0 \leq i \leq p - 1$, let $\gamma_i = \alpha_i - (\alpha_p/\beta_p)\beta_i$. Then

$$\sum_{i=0}^{p-1} \gamma_i = \sum_{i=0}^{p} \alpha_i - \frac{\alpha_p}{\beta_p} \sum_{i=0}^{p} \beta_i = 1.$$

Moreover, $\gamma_i \geq 0$; indeed, if $\beta_i \leq 0$ then $\gamma_i \geq \alpha_i \geq 0$; if $\beta_i > 0$ then

$$\gamma_i = \beta_i \left(\frac{\alpha_i}{\beta_i} - \frac{\alpha_p}{\beta_p} \right) \geq 0.$$

Thus

$$\sum_{i=0}^{p-1} \gamma_i x_i = \sum_{i=0}^{p-1} \left(\alpha_i - \frac{\alpha_p}{\beta_p} \beta_i \right) x_i = \sum_{i=0}^{p} \alpha_i x_i = x$$

is a representation of x as a convex combination of less than $p + 1$ points of A, contradicting the assumed minimality of p. This completes the proof of Carathéodory's theorem.

The proof of theorem 4 is now immediate. Indeed, if $x \in$ clconv A there exists a sequence $x_n \in$ conv A such that $x = \lim_{n \to \infty} x_n$.

By Carathéodory's theorem $x_n = \sum_{i=0}^{d} \lambda_{n,i} x_{n,i}$, where $x_{n,i} \in A$,

$0 \leq \lambda_{n,i} \leq 1, \sum_{i=0}^{d} \lambda_{n,i} = 1$ for each n. The compactness of $[0,1]$ and of A guarantees the existence of converging subsequences $(\lambda_{n_k,i})$ and $(x_{n_k,i})$ such that $\lim_{k \to \infty} \lambda_{n_k,i} = \lambda^{(i)}$ and $\lim_{k \to \infty} x_{n_k,i} = x^{(i)}$. Then obviously $0 \leq \lambda^{(i)} \leq 1$, $\sum_{i=0}^{d} \lambda^{(i)} = 1$, $x^{(i)} \in A$ and $x = \sum_{i=0}^{d} \lambda^{(i)} x^{(i)}$, as claimed.

A result closely related to Carathéodory's theorem in the sense that either is easily derived from the other, is *Radon's theorem*:

6. (Radon [1]) *If A is a $(d + 2)$-pointed subset of R^d, it is possible to find disjoint subsets A', A'' of A such that* conv $A' \cap$ conv $A'' \neq \emptyset$.

A direct proof of Radon's theorem is very easy. Let $A = \{x_0, \cdots, x_{d+1}\}$; since $d + 2$ points in d-space are affinely dependent there exist α_i, not all equal 0, such that $\sum_{i=0}^{d+1} \alpha_i = 0$ and $\sum_{i=0}^{d+1} \alpha_i x_i = 0$. Without loss of generality we assume the notation such that $\alpha_0, \cdots, \alpha_p$ are positive, $\alpha_{p+1}, \cdots, \alpha_{d+1}$ non positive. Then $0 \leq p \leq d$. Let $\alpha = \sum_{i=0}^{p} \alpha_i > 0$, and define $\beta_i = \alpha_i/\alpha$ for $0 \leq i \leq p$, and $\gamma_i = -\alpha_i/\alpha$ for $p + 1 \leq i \leq d + 1$. The affine dependence of A can be rewritten in the form

$$\sum_{i=0}^{p} \beta_i x_i = \sum_{i=p+1}^{d+1} \gamma_i x_i.$$

Since $\beta_i \geq 0, \gamma_i \geq 0$, and $\sum_{i=0}^{p} \beta_i = \sum_{i=p+1}^{d+1} \gamma_i = 1$, this relation expresses conv $\{x_0, \cdots, x_p\} \cap$ conv $\{x_{p+1}, \cdots, x_{d+1}\} \neq \emptyset$, as claimed by Radon's theorem.

For far-reaching generalizations of Radon's theorem see Tverberg [1] and Reay [3].

Exercises

1. Show that a hyperplane $H \subset R^d$ supports [cuts] a set $A \subset R^d$ if and only if H supports [cuts] conv A.

2. Proposition 4 states that the convex hull of a compact set is compact; show that the convex hull of an open set is open. The convex hull of a closed set is not necessarily closed; find a closed set $A \neq \emptyset$ such that conv A is an open proper subset of the whole space.

3. For $A \subset R^d$ let $\tau(A) = \{\frac{1}{2}(x_1 + x_2) \,|\, x_1, x_2 \in A\}$; let $\tau^1(A) = \tau(A)$, and for $n \geq 1$ let $\tau^{n+1}(A) = \tau(\tau^n(A))$. Denote $\tau^*(A) = \bigcup_{n \geq 1} \tau^n(A)$. Show that cl $\tau^*(A) = $ clconv A, although in general $\tau^*(A) \neq$ conv A. If $A = $ bd K where K is a bounded convex set in $R^d, d \geq 2$, show that cl $K = \tau(A)$.

4. For $A \subset R^d$ let $\vartheta(A) = \{\lambda x_1 + (1 - \lambda)x_2 \,|\, x_1, x_2 \in A, 0 \leq \lambda \leq 1\}$. Define $\vartheta^1(A) = \vartheta(A)$ and $\vartheta^{n+1}(A) = \vartheta(\vartheta^n(A))$ for $n \geq 1$. Show that

conv $A = \bigcup_{n \geq 1} \mathcal{S}^n(A)$. Characterize those convex sets $K \subset R^d$ for which
$K = \mathcal{S}(\text{bd } K)$.

5. Prove Steinitz's theorem (Steinitz [5]; Rademacher–Schoenberg [1]; generalizations in Bonnice–Klee [1], Reay [1, 2]): If $x \in \text{int conv } A \subset R^d$, there exists a subset A' of A, containing at most $2d$ points such that $x \in \text{int conv } A'$. Show that the number $2d$ may not be decreased in general, and characterize those A and x for which $2d$ points are needed in A'.

6. Let $A \subset R^d$ be a finite set. Then $x \in \text{relint conv } A$ if and only if x is representable as a convex combination of *all* points of A, with all coefficients positive.

7. Show that in Radon's theorem 2.3.6 the sets A' and A'' are unique if and only if every $d + 1$ points of A are affinely independent. Show also that in this case two points of A belong to the same set if and only if they are separated by the hyperplane determined by the remaining d points. (Proskuryakov [1], Kosmák [1]).

8. Let A be a nonempty subset of R^d and let $\mathcal{S}(A)$ denote the family of all subsets S of A with the property card $S = 1 + \dim S$. Show that
conv $A = \bigcup_{S \in \mathcal{S}(A)} \text{relint conv } S$.

9. Using the notation of section 2.2, show that for every nonempty set $A \subset R^d$ and every $x \in R^d$,

$$H(A, x) = H(\text{conv } A, x)$$

and

$$\text{conv}(A \cap L(A, x)) = (\text{conv } A) \cap L(A, x).$$

2.4 Extreme and Exposed Points; Faces and Poonems

Let K be a convex subset of R^d. A point $x \in K$ is an *extreme point* of K provided $y, z \in K$, $0 < \lambda < 1$, and $x = \lambda y + (1 - \lambda)z$ imply $x = y = z$. In other words, x is an extreme point of K if it does not belong to the relative interior of any segment contained in K. The set of all extreme points of K is denoted by ext K. Clearly, if $x \in \text{ext } K$ then $x \notin \text{conv}(K \sim \{x\})$.

Let K be a convex subset of R^d. A set $F \subset K$ is a *face* of K if either $F = \varnothing$ or $F = K$, or if there exists a supporting hyperplane H of K such that $F = K \cap H$. $\varnothing$ and K are called the *improper* faces of K. The set of all faces of K is denoted by $\mathcal{F}(K)$. A point $x \in K$ is an *exposed point* of K if the set $\{x\}$ consisting of the single point x is a face of K. The set of all exposed points of K is denoted by exp K. If K is a closed convex set, it is obvious that each $F \in \mathscr{F}(K)$ is closed. Throughout the sequel, the notations ext K, exp K, and $\mathscr{F}(K)$ will be used *only* for closed convex sets K.

The following statements result at once from the definitions:

1. *If $F \in \mathscr{F}(K)$ and if $K' \subset K$ is a closed convex set, then $F \cap K' \in \mathscr{F}(K')$.*

2. *If $F \in \mathscr{F}(K)$ and if $x \in F$, then $x \in$ ext K if and only if $x \in$ ext F; thus, if $F \in \mathscr{F}(K)$ then ext $F = F \cap$ ext K.*

3. *For every convex $K \subset R^d$ we have exp $K \subset$ ext K.*

4. *Let K be a closed convex set in R^d, let $x \in K$, and let B be a solid ball centered at x. Then $x \in$ ext K if and only if $x \in$ ext$(K \cap B)$, while $x \in$ exp K if and only if $x \in$ exp$(K \cap B)$.*

The next two results explain the role of the extreme points.

5. *Let K be a compact convex subset of R^d. Then $K = $ conv ext K. Moreover, if $K = $ conv A then $A \supset$ ext K.*

PROOF Clearly $K \supset$ conv ext K. In order to establish $K \subset$ conv ext K, we use induction on the dimension of the convex set K, the assertion being obvious in case dim K is -1, or 0, or 1. Without loss of generality we assume $R^d = $ aff K. Let $x \in K$. If $x \notin$ ext K, let L be a line such that $x \in$ relint $(L \cap K)$. Then $L \cap K$ is a segment $[y, z]$, where obviously $y, z \in$ bd K. Since through each boundary point of the convex set K there passes a supporting hyperplane, there exist faces F_y and F_z of K containing y respectively z. Now, the dimensions of F_y and F_z are smaller than dim K; by the inductive assumption, $F_y = $ conv ext F_y and $F_z = $ conv ext F_z. Using statement 2 (above) we have

$$x \in \text{conv}\{y, z\} \subset \text{conv}(F_y \cup F_z) \subset \text{conv}(\text{conv ext } F_y \cup \text{conv ext } F_z)$$
$$\subset \text{conv}(\text{ext } F_y \cup \text{ext } F_z) \subset \text{conv ext } K,$$

as claimed. The last assertion of the theorem being obvious, this completes the proof of theorem 5.

An analogous inductive proof yields also

6. *Let K be a closed convex subset of R^d, which contains no line. Then* ext $K \neq \emptyset$.

Regarding exposed points, we have

7. *Let $K \subset R^d$ be a compact set and let $H = \{x \in R^d \mid (x, u) > \alpha\}$ (where u is a unit vector) be an open halfspace such that $H \cap K \neq \emptyset$. Then $H \cap$ exp $K \neq \emptyset$.*

Proof Let $K' = H \cap K$, let $y \in K'$, and denote by ε the distance from y to bd H and by δ the number $\delta = \max\{\rho(x, y - \varepsilon u) \mid x \in K \cap \text{bd } H\}$. Let $z = y - \beta u$, where β is some fixed number satisfying $\beta > (\delta^2 + \varepsilon^2)/2\varepsilon$. Denoting by B the solid unit ball centered at the origin, let $\mu = \inf\{\lambda > 0 \mid z + \lambda B \supset K'\}$. Clearly $\mu \geq \beta$. Then, by the compactness of cl K', we have $z + \mu B \supset K'$ and $C = (\text{cl } K') \cap \text{bd}(z + \mu B) \neq \varnothing$. We claim that $C \cap \text{bd } H = \varnothing$. Indeed, assuming the existence of a point $v \in C \cap \text{bd } H$, we would have $\delta^2 \geq (\rho(v, y - \varepsilon u))^2 = \mu^2 - (\beta - \varepsilon)^2 \geq 2\beta\varepsilon - \varepsilon^2$ which implies $2\beta\varepsilon \leq \delta^2 + \varepsilon^2$, in contradiction to the choice of β. Therefore $C \subset K'$; but clearly each point of C is an exposed point of $z + \mu B$ and therefore also of cl K' and of K, as claimed.

Lemma 7, together with theorems 4, 5, and 6 above, 3 from section 2.2, and 4 from section 2.3, imply *Straszewicz'* [1] *theorem*:

8. *If* $K \subset R^d$ *is a compact convex set then* cl conv exp $K = K$.

Indeed, let $K' = $ cl conv exp K; obviously $K' \subset K$. If $K' \neq K$, then there exists an $x \in K$ such that $x \notin K'$. Since the compact convex sets x and K' may be strictly separated, there exists an open halfspace H such that $H \cap K \neq \varnothing$ and $H \cap K' = \varnothing$. But then $H \cap \exp K \neq \varnothing$ by theorem 7, contradicting the definition of K'.

The reader is invited to prove

9. *If* $K \subset R^d$ *is a closed convex set then* ext $K \subset $ cl exp K; *therefore if K is line-free then* exp $K \neq \varnothing$.

Regarding the family $\mathscr{F}(K)$ of all faces of a closed convex set K we have:

10. *The intersection* $F = \overset{r}{\underset{i=1}{\cap}} F_i$ *of any family* $\{F_i\}$ *of faces of a closed convex set K is itself a face of K.*

Proof If $F = \varnothing$ the assertion is true according to our definitions; thus we shall consider only the case $F \neq \varnothing$. Without loss of generality we may assume that the origin 0 belongs to F and that each F_i is a proper face of K. Then the face F_i is given by $F_i = K \cap \{x \mid \langle x, u_i \rangle = 0\}$ where u_i is some unit vector such that $K \subset \{x \mid \langle x, u_i \rangle \geq 0\}$. Let

$$H = \{x \mid \langle x, v \rangle = 0\} \qquad \text{where} \quad v = \sum_{i=1}^{r} u_i;$$

then clearly $K \subset \{x \mid \langle x, v \rangle \geq 0\}$. Since $0 \in K \cap H$ this implies that H is a supporting hyperplane of K. Now, if $x \in F$ then $\langle x, u_i \rangle = 0$ for all i and therefore $\langle x, v \rangle = 0$; hence $x \in H \cap K$ and thus $F \subset H \cap K$.

On the other hand, if $x \in K \sim F$ then $\langle x, u_j \rangle > 0$ for at least one j and $\langle x, v \rangle \geq \langle x, u_j \rangle > 0$; thus $x \notin H \cap K$. Therefore $F = H \cap K$ and F is a face of K, as claimed.

The family $\{F_i\}$ in theorem 10 may be infinite; in this case the face of smallest dimension obtainable as intersection of finite subfamilies of $\{F_i\}$ equals the intersection of all members of $\{F_i\}$.

It is easy to find examples (in each $R^d, d \geq 2$) which show that the situation is possible: K is a compact convex set, $C \in \mathscr{F}(K)$ and $F \in \mathscr{F}(C)$, but $F \notin \mathscr{F}(K)$.

This observation leads to the following definition:

A set F is called a *poonem** of the closed convex set K provided there exist sets $F_0, \cdots, F_k$ such that $F_0 = F$, $F_k = K$, and $F_{i-1} \in \mathscr{F}(F_i)$ for $i = 1, \cdots, k$.

By this definition, each face of K is also a poonem of K, but the converse is not true in general. Clearly, each poonem F is a closed convex set, and ext $F = F \cap$ ext K. The set of all poonems of a closed convex set K shall be denoted by $\mathscr{P}(K)$.

The reader is invited to deduce from theorem 10 the analogous result

11. *The intersection $F = \cap_i F_i$ of any family $\{F_i\}$ of poonems of a closed convex set K is in $\mathscr{P}(K)$.*

12. *If $F \in \mathscr{P}(K)$ then $\mathscr{P}(F) = \{P \in \mathscr{P}(K) \,|\, P \subset F\}$.*

13. *If $F \in \mathscr{F}(K)$ and $P \in \mathscr{P}(K)$ then $P \cap F \in \mathscr{P}(F)$ and $P \cap F \in \mathscr{F}(P)$.*

Exercises

1. A convex cone has at most one exposed point.

2. Let K denote a compact convex set. Show that if dim $K \leq 2$ then ext K is closed, but exp K is not necessarily closed. Find a $K \subset R^3$ such that exp $K \neq$ ext $K \neq$ cl exp K.

3. If (A_n) is a sequence of sets in R^d let the set lim sup A_n consist of all $x \in R^d$ such that for every open set V containing x, the intersection $V \cap A_n$ is nonempty for infinitely many n. Prove the following result (Jerison [1]): Let (K_n) be a sequence of compact convex sets in R^d, and let K be a compact convex set such that $K =$ lim sup K_n. If $E_n =$ ext K_n then $K =$ conv(lim sup E_n).

4. Extending the definition given above, a point x of a compact convex set K is called *k-exposed* [*k-extreme*] provided for some $j \leq k$, x belongs

---

* 'Poonem' is derived from the Hebrew word for 'face'. Klee [2] uses 'face' for this notion; however, it seems worthwhile to reserve 'face' for the different notion considered at the beginning of the present section.

to a j-face [j-poonem] of K. Clearly, the case $k = 0$ corresponds to the previously considered notions of exposed and extreme points. Denoting the set of all k-exposed points of K by $\exp_k K$ and that of all k-extreme points by $\text{ext}_k K$, the following generalization of theorem 2.4.9 holds (Asplund [1]; See Karlin–Shapley [1] for some related notions): If K is a closed convex set and if $k \geq 0$, then $\exp_k K \subset \text{ext}_k K \subset \text{cl} \exp_k K$.

5. Let K be a closed convex set. Show that $\exp_k K = \bigcup_{\substack{F \in \mathscr{F}(K) \\ \dim F \leq k}} F$ and $\text{ext}_k K = \bigcup_{\substack{F \in \mathscr{P}(K) \\ \dim F \leq k}} F$.

6. If the family $\mathscr{F}(K)$ of all faces of a closed convex set K is partially ordered by inclusion, then $\mathscr{F}(K)$ is a complete lattice. (For lattice-theoretic notions see, for example, Birkhoff [1], Szász [1].) The same is true for the family $\mathscr{P}(K)$ of all poonems of K. (In both cases the greatest lower bound of a family of elements is their intersection.)

7. If K is a closed convex set and if C is a subset of K, show that $C \in \mathscr{P}(K)$ is equivalent to each of the following conditions:

(i) C is convex and for every pair x, y of points of K either the closed segment $[x, y]$ is contained in C, or else the open interval $]x, y[$ does not meet C.

(ii) $C = K \cap \text{aff } C$ and $K \sim \text{aff } C$ is convex.

(iii) $C = K \cap L$, where L is a flat, and $K \sim L$ is convex.

(iv) There exists an $x \in K$ such that C is the maximal convex subset of K satisfying $x \in \text{relint } C$.

8. If $F_i \in \mathscr{F}(K)$ for $0 \leq i \leq n$ and if $F_0 \subset \bigcup_{i=1}^{n} F_i$, then there exists $i_0, 1 \leq i_0 \leq n$, such that $F_0 \subset F_{i_0}$. The same is true if all F_i belong to $\mathscr{P}(K)$.

9. Let K_1 and K_2 be closed convex sets. Prove:

(i) If $F_i \in \mathscr{F}(K_i)$ for $i = 1, 2$, then $F_1 \cap F_2 \in \mathscr{F}(K_1 \cap K_2)$.

(ii) If $F_i \in \mathscr{P}(K_i)$ for $i = 1, 2$, then $F_1 \cap F_2 \in \mathscr{P}(K_1 \cap K_2)$.

(iii) If $F \in \mathscr{P}(K_1 \cap K_2)$ there exist $F_1 \in \mathscr{P}(K_1)$ and $F_2 \in \mathscr{P}(K_2)$ such that $F = F_1 \cap F_2$.

(iv) If $\text{relint } K_1 \cap \text{relint } K_2 \neq \varnothing$ and if $F \in \mathscr{F}(K_1 \cap K_2)$, there exist $F_1 \in \mathscr{F}(K_1)$ and $F_2 \in \mathscr{F}(K_2)$ such that $F = F_1 \cap F_2$.

(v) Find examples showing that (iv) is not true if $\text{relint } K_1 \cap \text{relint } K_2 = \varnothing$.

10. Let T be a nonsingular projective transformation,
$$Tx = \frac{Ax + b}{\langle c, x \rangle + \delta},$$

and let H^+ be the open halfspace $H^+ = \{x \in R^d \mid \langle c, x \rangle + \delta > 0\}$. Prove:

(i) If A is any subset of H^+ then $T(\operatorname{conv} A) = \operatorname{conv} TA$.

(ii) For every closed convex set K for which T is permissible, $\mathscr{F}(TK) = \{TF \mid F \in \mathscr{F}(K)\}$ and $\mathscr{P}(TK) = \{TF \mid F \in \mathscr{P}(K)\}$.

11. A *Helly-type theorem* (see Danzer–Grünbaum–Klee [1], p. 127) is a statement of the following general type: A family of elements has a certain property whenever each of its subfamilies, containing not more than a fixed number of elements, has this property. Prove the following Helly-type theorems (see also exercise 7.3.5):

(i) A compact set $A \subset R^d$ has the property $A = \operatorname{ext} \operatorname{conv} A$ if and only if for every $B \subset A$ such that card $B \leq d + 2$ we have $B = \operatorname{ext} \operatorname{conv} B$.

(ii) A set $A \subset R^d$ satisfies $A \subset \operatorname{bd} \operatorname{conv} A$ if and only if for every $B \subset A$ with card $B \leq 2d + 1$ we have $B \subset \operatorname{bd} \operatorname{conv} B$.

(iii) Find examples showing that the 'Helly-numbers' $d + 2$ and $2d + 1$ of (i) and (ii) are best possible.

12. Show that if $A \subset R^d$ is any set of $d + 3$ points in general position, there exists a $B \subset A$ with card $B = d + 2$ such that $B = \operatorname{ext} \operatorname{conv} B$. (See Danzer–Grünbaum–Klee [1], p. 119; for $d = 2$ see Erdös–Szekeres [1].)

13. The following statement is a particular case of the result known as *Ramsey's theorem* (see Ramsey [1], Skolem [1], Erdös–Rado [1], Ryser [1]): Given positive integers p_1, p_2, q there exists an integer $r(p_1, p_2; q)$ with the property: If A is a set of elements such that card $A \geq r(p_1, p_2; q)$ and if all the q-tuples of elements of A are partitioned into two families $\mathscr{A}_1$ and $\mathscr{A}_2$, then either there exists in A a set A_1 containing p_1 elements such that all q-tuples of elements of A_1 are in $\mathscr{A}_1$, or there exists a subset A_2 of A containing p_2 elements such that all the q-tuples of elements of A_2 are in $\mathscr{A}_2$.

Use Ramsey's theorem and exercises 11 and 12 to prove the following results, which generalize a theorem of Erdös–Szekeres [1]:

(i) Given integers d and v, with $2 \leq d < v$, there exists an integer $e(d, v)$ with the following property: Whenever $A \subset R^d$ consists of $e(d, v)$ or more points in general position, there exists $B \subset A$ such that card $B = v$ and $B = \operatorname{ext} \operatorname{conv} B$. (Hint: Apply Ramsey's theorem, with $q = d + 2$, $p_1 = v$, $p_2 = d + 3$, taking as $\mathscr{A}_1$ the set of all $(d + 2)$-tuples C with $C = \operatorname{ext} \operatorname{conv} C$.)

Exercise 12 shows that we may take $e(d, d + 2) = d + 3$; the least possible values for $e(d, d + 3)$ are not known except for $d = 2$, in which case $e(2, 5) = 9$ (Erdös–Szekeres [1]); for additional results in this direction see Erdös–Szekeres [2].

(ii) Given integers d and v, with $2 \leq d < v$, there exists an integer $e'(d, v)$ with the following property: Whenever $A \subset R^d$ satisfies card $A \geq e'(d, v)$ and dim aff $A = d$, there exists $B \subset A$ such that card $B = v$, dim conv $B = d$, and $B \subset$ bd conv B.

Note that (ii) is a weaker version of exercise 7.3.5(ii); it would be interesting to find a direct proof of (ii) paralleling that of (i); the only direct proof known to the author uses (i) and the remark on p. 4.

2.5   Unbounded Convex Sets

The present section deals with some important properties of unbounded convex sets.

1. *A closed convex set $K \subset R^d$ is unbounded if and only if K contains a ray.*

PROOF   We shall consider only the nontrivial part of the assertion. Let $x_0 \in K$, and let $S = \text{bd } B$ denote the unit sphere of R^d centered at the origin. For each $\lambda > 0$ we consider the radial projection $P_\lambda = \pi(K \cap (x_0 + \lambda S))$ of the compact set $K \cap (x_0 + \lambda S)$ onto $x_0 + S$, the point x_0 serving as center of projection.† Since radial projection is obviously a homeomorphism between $x_0 + \lambda S$ and $x_0 + S$, the set P_λ is compact. If K is unbounded then $P_\lambda \neq \varnothing$ for every $\lambda > 0$. Since K is convex and $x_0 \in K$, we have $P_\lambda \subset P_{\lambda*}$ for $\lambda* \leq \lambda$. Therefore $\underset{\lambda > 0}{\cap} P_\lambda \neq \varnothing$. If y_0 is any point of this intersection, the ray $\{\lambda y_0 + (1 - \lambda)x_0 \mid \lambda \geq 0\}$ is clearly contained in K. This completes the proof of lemma 1.

2. *Let $K \subset R^d$ be closed and convex, let $L = \{\lambda z \mid \lambda \geq 0\}$ be a ray emanating from the origin, and let $x, y \in K$. Then $x + L \subset K$ if and only if $y + L \subset K$.*

PROOF   Let $x + L \subset K$, and let $y + \lambda z \in y + L$ be given, $\lambda \geq 0$. For $0 < \mu < 1$, consider the point $v_\mu = (1 - \mu)y + \mu(x + (\lambda/\mu)z) \in K$. Since $\rho(v_\mu, y + \lambda z) = \rho(0, \mu(x - y))$, the distance between $y + \lambda z$ and v_μ is arbitrarily small provided $\mu > 0$ is sufficiently small. But K is closed, and therefore $v_\mu \in K$ implies $y + \lambda z \in K$. Since x and y play equivalent roles, the proof of lemma 2 is completed.

A convex set $C \subset R^d$ is a *cone with apex* 0 provided $\lambda x \in C$ whenever $x \in C$ and $\lambda \geq 0$. A set C is a *cone with apex* x_0 provided $-x_0 + C$ is a

† If $x_0 \in R^d$, the *radial projection* π, with *center of projection* x_0, of $R^d \sim \{x_0\}$ onto the unit sphere $x_0 + S$ is defined by $\pi(x + x_0) = x_0 + x/\|x\|$.

cone with apex 0. A cone C with apex x_0 is *pointed* provided $x_0 \in \text{ext } C$. Let C be a closed cone with apex 0. The following assertions are easily verified:

(i) The apices of C form a subspace $C \cap -C$ of R^d; therefore either C is pointed, or there exists a line all points of which are apices of C.

(ii) $C = C + C = \lambda C$ for every $\lambda > 0$.

Conversely, if a nonempty closed set $C \subset R^d$ has property (ii) then C is a cone with apex 0.

The intersection of any family of cones with common apex x_0 is a cone with apex x_0. Therefore it is possible to define the cone with apex x_0 *spanned* by a set $A \subset R^d$ as the intersection of all cones in R^d which have apex x_0 and contain A. Though this notion is rather important in different investigations, we shall be more interested in another construction of cones from convex sets.

Let K be a convex set and let $x \in K$. We define $\text{cc}_x K = \{y \mid x + \lambda y \in K$ for all $\lambda \geq 0\}$. Clearly $\text{cc}_x K$ is a convex cone which has the origin as an apex. Lemma 2 implies that for closed K we have $\text{cc}_x K = \text{cc}_y K$ for all $x, y \in K$. Thus the index x is unnecessary and may be omitted. The convex cone $\text{cc } K$ is called the *characteristic cone* of K. Using lemmas 1 and 2 we obtain the following result:

3. *If $K \subset R^d$ is a closed convex set then $\text{cc } K$ is a closed convex cone; moreover, $\text{cc } K \neq \{0\}$ if and only if K is unbounded.*

A closed convex set K shall be called *line-free* provided no (straight) line is contained in K. Using this terminology, theorems 2.4.6 and 2.4.9 may be formulated as: If K is line-free then $\text{ext } K \neq \varnothing \neq \text{exp } K$. It is also clear that every line-free cone is pointed.

Returning to lemma 2 we note that it immediately implies: If L is a linear subspace of R^d such that $x + L \subset K$ for some x, then $y + L \subset K$ for every $y \in K$. Therefore the following decomposition theorem results:

4. *If $K \subset R^d$ is a closed convex set there exists a unique linear subspace $L \subset R^d$ of maximal dimension such that a translate of L is contained in K. Moreover, denoting by L^* any linear subspace of R^d complementary to L, we have $K = L + (K \cap L^*)$, where $K \cap L^*$ is a line-free set.*

Some information on the structure of line-free sets is given by the following theorem.

5. *Let $K \subset R^d$ be an unbounded, line-free, closed convex set. Then $K = P + \text{cc } K$, where P is the union of all bounded poonems of K.*

PROOF   We use induction on the dimension of K, the assertion being obvious if $\dim K = 1$. If $\dim K > 1$ and if $x \in K$, let $y \in$ relbd K and $z \in$ cc K be such that $x = y + z$. (Since K is line-free such a choice is possible; indeed, for any $t \in$ cc K, $t \neq 0$, there exists a $\lambda \geq 0$ such that $x - \lambda t \in$ relbd K.) Let F be any proper face of K such that $y \in F$. If F is bounded then $F \subset P$ and $x \in P + $ cc K. If F is not bounded, the inductive assumption and $\dim F < \dim K$ imply that $y = v + w$, where $w \in$ cc F and v belongs to P', the union of the bounded poonems of F. Since $P' \subset P$, cc $F \subset$ cc K, and cc K is convex, it follows that $x = y + z = v + w + z \in P' + $ cc $K + $ cc $K \subset P + $ cc K. Since obviously $K \supset P + $ cc K, this completes the proof of theorem 5.

Since for each bounded poonem F of K we have ext $F = F \cap$ ext K and $F = $ conv ext F, theorem 5 implies

6. *Let* $K \subset R^d$ *be a line-free, closed convex set. Then* $K = $ cc K + conv ext K.

Exercises

1. Show that lemma 1 is valid even without the assumption that K is closed.

2. If K is any convex set in R^d, show that $x, y \in$ relint K implies $cc_x K = cc_y K$. Moreover, for $x \in$ relint K the characteristic cone $cc_x K$ is closed.

3. Show that the decomposition theorem 4 holds also if K is a relatively open convex set.

4. Let $K \subset R^d$ be a closed convex set; then cc K is *the maximal* (with respect to inclusion) subset $T \subset R^d$ with the property: For every $x \subset K$, $x + T \subset K$.

5. Let $K \subset R^d$ be a closed convex set; then cc $K = \{x \in R^d \mid \langle x, u \rangle \geq 0$ for all u such that there exists an α with $K \subset \{z \mid \langle x, u \rangle \geq \alpha\}\}$.

6. Let $K \subset R^d$ be a closed convex set such that $0 \in$ relint K. Prove that

$$\text{cc } K = \bigcap_{n=1}^{\infty} \left(\frac{1}{n} K \right).$$

7. Using the notation of the decomposition theorem 4, let L^{**} denote another linear subspace of R^d complementary to L. Show that $L^{**} \cap K$ is an affine image of $L^* \cap K$.

8. If $K \subset R^d$ is a closed, convex, line-free set, then there exists a hyperplane H such that $H \cap K$ is compact and $\dim K = 1 + \dim (H \cap K)$.

9. If $K \subset R^d$ is a closed pointed cone with apex x_0, there exists a hyperplane H such that $H \cap K$ is compact and K is the cone with apex x_0 spanned by $H \cap K$.

10. Prove the following results converse to theorems 5 and 6.

(i) If K is an unbounded, line-free, closed convex set and if $K = C + P$, where C is a cone with apex 0, then P contains all bounded poonems of K.

(ii) If K is a line-free, closed, convex set and if $K = C + P$, where C is a cone with apex 0 and P is a closed, bounded convex set, then $C = \operatorname{cc} K$ and $P \supset \operatorname{conv} \operatorname{ext} K$.

11. A convex set K is called *reducible* (Klee [2]; this notion of reducibility will be used *only* in the present exercise) provided $K = \operatorname{conv} \operatorname{relbd} K$. Prove the following results:

(i) If K is a closed convex set then K is the convex hull of the union of all irreducible members of $\mathscr{P}(K)$.

(ii) Each irreducible closed convex set is either a flat, or a closed halfflat.

12. Show that each d-dimensional closed convex set is homeomorphic with one of the following $d + 2$ sets: (i) a closed halfspace of R^d; (ii) the product $R^{d-k} \times B^k$ for some k with $0 \leq k \leq d$, where B^k denotes the k-dimensional (solid) unit ball.

2.6   Polyhedral Sets

A set $K \subset R^d$ is called a *polyhedral set* provided K is the intersection of a finite family of closed halfspaces of R^d.

Polyhedral sets have many properties which are not shared by all closed convex sets. One of the most important such properties is

1. *Each poonem of a polyhedral set K is a face of K.*

Before proving theorem 1 we note a few facts about polyhedral sets.

Let $H_i^+ = \{x \in R^d \mid \langle x, u_i \rangle \geq \alpha_i\}$, $1 \leq i \leq n$, be halfspaces, and let $K = \bigcap_{i=1}^{n} H_i^+$. Without loss of generality we shall in the present section assume that $\dim K = d$; we shall also say that a maximal proper face of K is a *facet* of K. (Note that if K is a flat, then K has no facets.) The family $\{H_i^+ \mid 1 \leq i \leq n\}$ is called *irredundant* provided $K_i = \bigcap_{\substack{1 \leq j \leq n \\ j \neq i}} H_j^+ \neq K$

for each $i = 1, 2, \cdots, n$.

Denoting $H_i = \operatorname{bd} H_i^+ = \{x \in R^d \mid \langle x, u_i \rangle = \alpha_i\}$, we have

2. *If* $K = \bigcap\limits_{j=1}^{n} H_j^+$, *where* $\{H_j^+ \mid 1 \leq j \leq n\}$ *is irredundant, then* $F_i = H_i \cap K$ *is a facet of* K.

This follows at once from the observation that $H_i \cap \operatorname{int} K_i \neq \varnothing$ which, in turn, is a reformulation of the irredundancy assumption. The same assumption also implies

3. $\operatorname{bd} K = \bigcup\limits_{i=1}^{n} F_i$, *where* $F_i = H_i \cap K$, $i = 1, \cdots, n$, *are all the facets of* K.

In particular, for each proper face F of K there exists a facet F_i of K such that $F \subset F_i$ (see exercise 2.4.8).

Let $F_i = H_i \cap K$ be a facet of K. Then

$$F_i = H_i \cap \left(\bigcap\limits_{\substack{1 \leq j \leq n \\ j \neq i}} H_j^+ \right) = \bigcap\limits_{\substack{1 \leq j \leq n \\ j \neq i}} (H_i \cap H_j^+).$$

Thus F_i is a polyhedral set, namely the intersection of the sets $H_i \cap H_j^+$, $1 \leq j \leq n$, each of which is either H_i, or a halfspace of the $(d-1)$-dimensional space H_i. Therefore, by theorem 3, each facet F of F_i is of the form

$$F = F_i \cap \operatorname{relbd}(H_i \cap H_j^+) = F_i \cap H_i \cap H_j = K \cap H_i \cap H_j = F_i \cap F_j,$$

for a suitable j. Thus

4. *Each facet of a facet of a polyhedral set* K *is the intersection of two facets of* K.

Now we are ready for the proof of the following theorem which, in view of theorem 2.4.10, clearly implies theorem 1.

5. *Every nonempty proper poonem* F *of a polyhedral set* K *is an intersection of facets of* K.

PROOF We shall use induction on $\dim K$, the assertion being obvious if $\dim K = 1$. If $\dim K > 1$, let $x \in \operatorname{relint} F$. By theorem 3, there exists a facet F_i of K such that $x \in F_i$, i.e. $F \in F_i$. Theorem 2.4.12 then implies that F is a poonem of F_i. Using the inductive assumption we see that F is an intersection of facets of F_i. Since each facet of F_i is the intersection of 2 facets of K, this completes the proof of theorem 5.

We mention also the following immediate consequence of theorem 5:

6. *If* K *is a polyhedral set then the family* $\mathscr{F}(K)$ *is finite.*

Exercises

1. (See theorem 2.5.4 for the notation.) Show that if K is a nonempty polyhedral set, $K = \bigcap_{i=1}^{n} \{x \in R^d \mid \langle x, u_i \rangle \geq \alpha_i\}$, then

$$L = \bigcap_{i=1}^{n} \{x \in R^d \mid \langle x, u_i \rangle = 0\}.$$

2. Show that if K is as above, then

$$\mathrm{cc}\, K = \bigcap_{i=1}^{n} \{x \in R^d \mid \langle x, u_i \rangle \geq 0\}.$$

3. Show that if K is as above, and if $p \in K$ satisfies

$$\langle p, u_i \rangle = \alpha_i \qquad \text{for } 1 \leq i \leq m$$

and

$$\langle p, u_i \rangle > \alpha_i \qquad \text{for } m < i \leq n,$$

then

$$\mathrm{cone}_p\, K = \bigcap_{i=1}^{m} \{x \in R^d \mid \langle x, u_i \rangle \geq \alpha_i\}.$$

4. Show that every affine map of a polyhedral set is a polyhedral set. Find a polyhedral set K and a projective transformation T permissible for K, such that TK is not a polyhedral set.

5. As converses of exercise 4, prove the following results:

(i) (Klee [3]) If K is a convex subset of R^d, $d \geq 3$, and if all projections of K into 3-dimensional subspaces of R^d are polyhedral sets, then K is a polyhedral set.

(ii) (Mirkil [1], Klee [3]) If K is a convex cone in R^d, $d \geq 3$, and if all projections of K into 2-dimensional subspaces of R^d are closed, then K is a polyhedral cone.

6. Let $K_1, \cdots, K_n$ be polyhedral sets in R^d, and let C be a convex set such that $C \subset \bigcup_{i=1}^{n} K_i$. Prove that there exists a polyhedral set K such that $C \subset K \subset \bigcup_{i=1}^{n} K_i$.

2.7 Remarks

An adequate account of the history of the main results on convex sets would require much more room than we have at our disposal; therefore we shall limit ourselves to just a few remarks.

Though quite a few notions and facts related to convexity have been considered appreciably earlier, it was mainly through the pioneering work of Minkowski (see Minkowski [2]) that convexity became a well-known subject of research, applicable to many other disciplines. The scope of research greatly expanded during the first quarter of the present century; most influential on the other workers were probably the papers of Carathéodory [1, 2] and Steinitz [5], and the book of Blaschke [1]. An extremely useful review of results on convexity up to 1933 is the book Bonnesen–Fenchel [1].

A complete bibliography of papers dealing with various aspects of convexity would contain several thousand entries. We shall mention here as general references only some of the books published recently (though some of them do not have much bearing on polytopes): Aleksandrov [2, 3], Busemann [2], Eggleston [2, 3], Fejes-Tóth [1, 3], Fenchel [4], Hadwiger [3, 5], Hadwiger–Debrunner [1], Klee [8], Kuhn–Tucker [1], Lyusternik [1], Rogers [1], Valentine [1], Yaglom–Boltyanskiĭ [1].

Most results of the present chapter are well known, though the formulations used by different authors often vary, and various settings and degrees of generality are considered. A survey of known results and an extensive bibliography* on the material of sections 2.1, 2.2, and 2.3, may be found in Danzer–Grünbaum–Klee [1]. For the facts dealt with in sections 2.4, 2.5, and 2.6, and for related material and additional references the reader may consult, for example, Weyl [1], Motzkin [1, 2], Fenchel [4], Klee [4], Gale–Klee [1], and, in particular, Klee [2].

With suitable changes, many results of the present chapter have valid analogues for convex sets in vector spaces over any ordered field, or for convex sets which are not necessarily closed (in Euclidean spaces). In many cases, the proofs of such generalizations are much more elaborate (see, e.g., Weyl [1], Motzkin [1], Klee [10].

Convexity has been studied—and is a natural and interesting notion—in many settings different from the Euclidean (or affine) spaces. We shall briefly explain two such variants, since they will be mentioned in the sequel.

If P^d is the d-dimensional (real) projective space, we shall say that a set $K \subset P^d$ is *convex* provided

---

* The reader should be aware of the fact that in some of the papers the presentations of definitions or theorems are rather careless, to the extent of being ambiguous (e.g. the definition of spherically convex polygons in Aleksandrov [2, p. 13]) or false (e.g. the separation theorem in Karlin [1, p. 356]).

(i) For each line L in P^d, the intersection $L \cap K$ is either empty, or else a connected subset of L;

(ii) There exists a $(d-1)$-dimensional subspace H of P^d such that $H \cap K = \varnothing$.

It is obvious that if K is a convex subset of P^d and if H is as in (ii), K may be considered as a subset of the affine d-space obtained from P^d by assigning to H the role of the 'hyperplane at infinity'. In this interpretation, K becomes a convex subset of the affine space. Hence most of the notions and results of the present chapter may be reformulated for convex subsets of projective spaces. One important exception derives from the possibility that the intersection of two or more convex sets may fail to be convex. (However, if $\{K_\alpha\}$ is a family of convex sets such that the intersection of each two sets is convex, then $\bigcap_\alpha K_\alpha$ is convex.) This implies that the convex hull of a set $A \subset P^d$ (which exists only if some hyperplane misses A) is in general not unique. For more detailed accounts of convexity in projective spaces see, for example, Steinitz [5], Veblen-Young [1], p. 386, Motzkin [2], Fenchel [4], Sinden [1]; additional references are given in Danzer–Grünbaum–Klee [1]. (In certain investigations it seems to be more convenient to define convex sets in projective spaces by the single condition (i) (see, for example, Kneser [1], Marchaud [1]; we shall not use this terminology.)

For subsets of the d-sphere S^d a number of different definitions of convexity are frequently used; they coincide for sets contained in an open hemisphere, but differ in the treatment of larger sets. For our purposes, the most suitable definition results by taking S^d as the unit sphere of R^{d+1} with center at the origin 0 and calling a set $K \subset S^d$ *convex* if and only if $\text{cone}_0 K$ is a convex subset of R^{d+1}. For a discussion of other definitions, and for references to the rather voluminous literature, see Danzer–Grünbaum–Klee [1].

2.8 Additional notes and comments

An example.
In the figure below, the point x is an extreme point that is not exposed; also, $\{x\}$ is a poonem (a face of a face) that is not a face itself. However, such simple examples cannot display the full complexity of the facial/extremal structure of general convex bodies. For example, Grünbaum (see Lindenstrauss–Phelps [a]) produced a 3-dimensional body with uncountably many extreme points but only countably many exposed points.

Face functions on general convex bodies.
Let K be a general d-dimensional convex body, let B be its boundary, and for each $x \in B$ let $F(x)$ denote the union of all segments in B that have x as an inner point. The set-valued function F is called the *face function* of K, and when K is a polytope it behaves very simply: It is lower semicontinuous at each point of B, and is upper semicontinuous precisely on the relative interiors of K's facets. The analogue of this for a general K is as follows: F is lower semicontinuous almost everywhere on B in the sense of Baire category (i. e., at the points of a dense G_δ subset of B), and is upper semicontinuous on B almost everywhere in the sense of measure (i. e., at the points of a subset of B whose complement is of zero $(d-1)$-dimensional measure). However, for $d \geq 3$ there exists a d-dimensional K whose face-function is lower semicontinuous *almost nowhere* in the sense of measure and is upper semicontinuous *almost nowhere* in the sense of category. For these results, see Klee–Martin [a] and Larman [b], and also Corson's paper [a] on which the example is based.

Convex bodies—geometric and algorithmic aspects.
In this chapter, general convex sets and their faces and poonems are presented as *foundational* material, whose specific "pathologies" disappear in the much more special, discrete setting of convex polytopes.

Nevertheless, the geometry of general convex sets is important, in particular in view of the manifold connections and applications to fields such as functional analysis (Banach space theory), the geometry of numbers, etc. Key references to access this theory are the "Handbook of Convex Geometry" edited by Gruber and Wills [a] and the book by Schneider [b]. We refer to Thompson [a] for the geometry of finite dimensional normed spaces, to Gruber–Lekkerkerker [a] for the geometry of numbers, to Leichtweiß [a] for the theory of affine convex geometry, and to Ball [a], Matoušek [b, Chap. 14], and Giannopoulos–Milman [a] as guides to some recent developments such as the "concentration of measure" phenomenon, which still waits for more impact on the combinatorial theory of polytopes.

Algorithmic aspects have emerged and gradually become more influential in the theory of convex bodies (i. e., full-dimensional, closed convex sets). Thus, on the one hand, the geometry of convex bodies rules the field of convex optimization—see Rockafellar [a] and Stoer–Witzgall [a]; on the other hand, non-linear optimization concepts such as the ellipsoid method have had tremendous impact on the "algorithmic model" of a convex set, starting with the fundamental problem of how we can be "given" a convex set. We refer to Grötschel–Lovász–Schrijver [a] and to the introduction by Lovász [a].

Tverberg, Helly, Ramsey, and Erdős–Szekeres.
Tverberg's theorem, pointed to on page 16, has turned out to be a driving force for discrete geometry and combinatorial convexity. This led to new proofs (by Tverberg and by others—see, e. g., Sarkaria [a]), to far-reaching extensions such as the so-called "colored Tverberg theorem", and to the development of new tools and methods, in particular from equivariant topology. Živaljević [a] is a guide to the current discussion.

For Helly type theorems (as in exercise 2.4.11), surveys are Eckhoff [a] and Wenger [a]. In exercise 2.4.13 one meets Ramsey theory and the Erdős–Szekeres theorem as an application. We refer to Matoušek [b, Chap. 3].

Generalizations of convexity.
In addition to projective and spherical convexity (see pages 29–30), the case of convexity in *hyperbolic* space has turned out to be particularly interesting again and again. Highlights include the work by Sleator–Tarjan–Thurston [a] on rotation distance of trees and triangulations of n-gons, and Smith's [a] lower bounds for the number of simplices needed to triangulate the d-cube.

We refer to Boltyanski–Martini–Soltan [a] for a survey and geometric study of various generalized convexity models. See also Coppel [a] [b], Edelman–Jamison [a], and Prenowitz–Jantosciak [a].

A combinatorial model for the convexity structure of finite sets of points (such as the vertices of a polytope) was provided by the theory of oriented matroids (see Björner et al. [a], Ziegler [a, Lect. 6]), which emerged in the late seventies and has produced substantial tools and results for polytope theory; see in particular the notes in section 5.6 (on Gale-diagrams).

CHAPTER 3

Polytopes

The present chapter contains the fundamental concepts and facts on which we rely in the sequel. Polytopes, their faces and combinatorial types, complexes, Schlegel diagrams, combinatorial equivalence, duality, and polarity are the main topics discussed.

3.1 Definition and Fundamental Properties of Polytopes

A compact convex set $K \subset R^d$ is a *polytope* provided ext K is a finite set. From the results of section 2.4 and theorem 2.3.4 it follows that polytopes may equivalently be defined as convex hulls of finite sets. Also, if K is a polytope then, by theorem 2.4.9, exp K = ext K; in other words, each point of ext K is a face of K. For a polytope (or polyhedral set) K, it is customary to call the points of ext K *vertices*, and to denote their totality by vert K; 1-faces of K are called *edges*, while maximal proper faces are *facets* of K.

Clearly each face F of a polytope K is itself a polytope, and vert F = vert $K \cap$ aff F. We shall use *d-polytope* and *k-face* as abbreviations for 'polytope of dimension d' and 'face of dimension k'. Since each k-face of a d-polytope K contains $k + 1$ affinely independent vertices of K, and since different faces of K have different affine hulls, it follows that the number of different k-faces of a polytope is finite for each k. Moreover, denoting by $f_k(K)$ the number of different k-faces of a d-polytope (or polyhedral set) K, we have $f_k(K) \leq \binom{f_0(K)}{k + 1}$; with the plausible convention $f_k(K) = 0$ for $k > d$ or $k < -1$, this relation holds for all k.

The following theorem is of fundamental importance in the theory of polytopes. It may be considered as a sharpening of theorem 2.2.3 for the special case of polytopes, showing that polytopes are polyhedral sets.

1. *Each d-polytope $K \subset R^d$ is the intersection of a finite family of closed halfspaces; the smallest such family consists of those closed halfspaces containing K whose boundaries are the affine hulls of the facets of K.*

31

PROOF   Let $\mathscr{H} = \{H_j \mid 1 \le j \le f_{d-1}(K)\}$ be the set of hyperplanes determined by the facets of K, and let a point $y \notin K$ be given. We shall show that there exists an H_j such that y does not belong to the closed halfspace determined by H_j and containing K. We denote by L the set of all affine combinations of at most $d - 1$ points of vert K. By Carathéodory's theorem 2.3.5 L contains all the faces of K which have dimension at most $d - 2$. Let M denote the cone spanned by L with vertex y; then M is contained in the union of finitely many hyperplanes through y. Since finitely many hyperplanes do not cover any nonempty open set, int K is not contained in M. Let x be any point of (int K) $\sim M$; we consider the ray $N = \{\lambda x + (1 - \lambda)y \mid \lambda \ge 0\}$ with endpoint y determined by x; clearly $N \cap \text{int } K \ne \varnothing$. Let $\lambda_0 = \inf\{\lambda > 0 \mid \lambda x + (1 - \lambda)y \in K\}$. Since K is compact and $y \notin K$, the greatest lower bound is attained, $0 < \lambda_0 < 1$, and $x_0 = \lambda_0 x + (1 - \lambda_0)y \in \text{bd } K$. It follows that x_0 belongs to some proper face F of K; but $x \notin M$ implies $x_0 \notin L$ and therefore F is not of dimension less than or equal to $d - 2$. Thus F is a facet and the hyperplane aff F has all the desired properties. The assertion about the minimality of $\mathscr{H}$ being obvious, this completes the proof of theorem 1.

A partial converse of theorem 1 is given by

2. *Every bounded polyhedral set K is a polytope.*

The proof follows at once from the previous results. Indeed, the assumptions imply that K is compact and therefore (by theorem 2.4.5) $K = \text{conv ext } K$. By theorem 2.6.1, ext $K = \exp K$, and by theorem 2.6.6 exp K is a finite set; hence K is a polytope.

The last two results may be combined to yield the following theorem.

3. *A set $P \subset R^d$ is a polytope if and only if P is a bounded polyhedral set.*

The reader is now invited to establish the following assertions which provide a number of methods for generating new polytopes from given ones; some of the proofs use theorem 3, others follow directly from the definitions.

4. *The convex hull, the vector sum, and the intersection of finitely many polytopes is a polytope. The intersection of a polytope with an affine variety, or with any polyhedral set, is a polytope. Any affine image, and any permissible projective image of a polytope is a polytope.*

We shall next consider in some detail the family $\mathscr{F}(P)$ of all (proper and improper) faces of a polytope P.

Theorem 2.4.1 implies that the intersection of any family of faces of a polytope K is itself a face of K. Trivially, it is also true that if F_1 and F_2 are faces of K and $F_2 \subset F_1$, then F_2 is a face of F_1. Theorems 3.1.3 and 2.6.1 imply for polytopes the transitivity of the property 'is a face of':

5. *If F_1 is a face of the polytope P and if F_2 is a face of the polytope F_1, then F_2 is a face of P.*

We find it interesting to give a direct proof of theorem 5, independent of the results of section 2.6. For such a proof, it is clearly enough to consider proper faces; without loss of generality we may assume that the origin 0 belongs to F_2 and that P is a d-polytope in R^d. Let u_1 and u_2 be unit vectors such that, denoting $H_1 = \{x \in R^d \mid \langle x, u_1 \rangle = 0\}$ we have: H_1 is a supporting hyperplane of P with $F_1 = H_1 \cap P$ and $P \subset \{x \mid \langle x, u_1 \rangle \geq 0\}$; $u_2 \in H_1$, $F_1 \subset \{x \in H_1 \mid \langle x, u_2 \rangle \geq 0\}$ and $F_2 = F_1 \cap H_2$, where H_2 is the $(d-2)$ — flat $\{x \in H_1 \mid \langle x, u_2 \rangle = 0\}$. Let $H(\varepsilon) = \{x \in R^d \mid \langle x, u_1 + \varepsilon u_2 \rangle = 0\}$; then $H(\varepsilon) \supset H_2 \supset F_2$ for every ε. Let $\alpha = \max\{|\langle v, u_2 \rangle| \mid v \in \text{vert } P \sim \text{vert } F_1\}$ and $\beta = \min\{\langle v, u_1 \rangle \mid v \in \text{vert } P \sim \text{vert } F_1\} > 0$. We claim that if ε satisfies $0 < \varepsilon < \beta/2\alpha$ (or just $\varepsilon > 0$ if $\alpha = 0$) then $H(\varepsilon)$ is a supporting hyperplane of P and $F_2 = P \cap H(\varepsilon)$. Indeed, if $v \in \text{vert } P \sim \text{vert } F_1$ then $\langle v; u_1 + \varepsilon u_2 \rangle \geq \beta - \varepsilon\alpha > \beta/2 > 0$; if $v \in \text{vert } F_1 \sim \text{vert } F_2$ then $\langle v; u_1 + \varepsilon u_2 \rangle = \varepsilon \langle v; u_2 \rangle > 0$ by the definition of u_2; finally, for $v \in \text{vert } F_2$ we have $\langle v; u_1 + \varepsilon u_2 \rangle = 0$, i.e. $v \in H(\varepsilon)$. This completes the direct proof of theorem 5.

Some remarks of a methodological nature seem indicated in view of the proofs given in the present Section. It is hoped that readers who worked their way through the proofs are by now ready to accept the validity of the results proved. The author doubts, however, that the above formal proofs give a good idea of *why* the proofs work. In a subject as elementary and intuitively as comprehensible as the theory of polytopes, it seems a pity to obscure the simple idea of a proof by the—almost necessarily—involved and complicated notation and symbolism. As an example, consider the following formulation of the idea of the direct proof of theorem 5. If H_1 is a hyperplane determining F_1, and if H_2 is a $(d-2)$-subflat of H_1 determining F_2, any sufficiently small rotation of H_1 about H_2 in the proper direction ('away' from vert $F_1 \sim$ vert F_2) will yield a hyperplane $H(\varepsilon)$ whose intersection with P is F_2.

In this context, as in many other cases, the idea of the proof becomes clearly comprehensible with the help of a graphic representation of the two- or three-dimensional case (see figures 3.1.1 and 3.1.2). The formal proof is necessary as a guarantee that no unwarranted simplifications have been made in the intuitive examination of the problem, and that all the choices, positions, and other aspects, are as imagined. But the formal proof should be carried out after the idea of the proof has been found and understood. The reader is most insistently advised to reread the proof of theorem 1 and to formulate for himself the intuitive ideas involved.

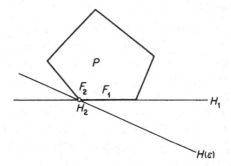

Figure 3.1.1

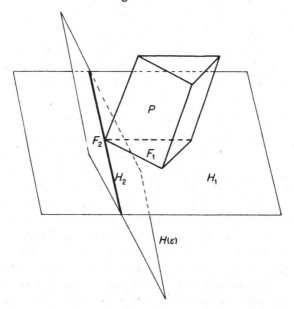

Figure 3.1.2

In the opposite direction, the reader is invited to expand the ideas of the direct proofs of the statements given below to formal proofs.

6. *If P is a d-polytope, each $(d-2)$-face F of P is contained in precisely two facets F_1 and F_2 of P, and $F = F_1 \cap F_2$.*

Indeed, if H is a hyperplane such that $H \cap P = F$, a rotation of H about the $(d - 2)$-flat aff F yields the two (extremal) positions H_1 and H_2 for which $H_i \cap P = F_i$.

Using theorems 5 and 6 it easily follows by induction that

7. *If $-1 \leq h < k \leq d - 1$ and if P is a d-polytope, each h-face of P is the intersection of the family (containing at least $k - h + 1$ members) of k-faces of P containing it.*

The following statement is rather obvious, but nevertheless occasionally useful:

8. *If P is a d-polytope and if F is a k-face of P, there exists a $(d - k - 1)$-face F^* of P such that $\dim \operatorname{conv}(F \cup F^*) = d$. (Then necessarily $F \cap F^* = \varnothing$.)*

Indeed, if $k = 0$ (that is, F is a vertex), let x be a point of bd P such that the segment $[F, x]$ meets int P. Then any facet of P which contains x can serve as F^*. The proof is easily completed by induction.

Exercises

1. Let P be a polytope and let $A \subset$ vert P. Then conv A is a face of P if and only if aff $A \cap \operatorname{conv}((\operatorname{vert} P) \sim A) = \varnothing$.

2. Let P be a d-polytope, F a proper face of P, and F_0 a proper k-face of F. Prove that there exists a $(k + 1)$-face F_1 of P such that $F_0 = F_1 \cap F$. Prove also the sharper result: If P is a d-polytope, F an h-face of P, and F_0 a k-face of F, where $-1 \leq k < h \leq d$, then there exists a $(d - h + k)$-face F_1 of P such that $F_0 = F \cap F_1$ and $P = \operatorname{conv}(F \cup F_1)$.

3. Let F_{k-1} be a $(k - 1)$-face and let F_{k+1} be a $(k + 1)$-face of the d-polytope P, $0 \leq k < d$. There exist precisely two distinct k-faces of P each of which is contained in F_{k+1} and contains F_{k-1}. Does this result remain valid if P is assumed to be a polyhedral set?

4. Let V be a vertex of a polytope $P \subset R^d$ and let H^* be a closed half-space with bounding hyperplane H, such that $V \in H$ and all the edges of P which contain V are contained in H^*. Prove that $P \subset H^*$, and therefore H is a supporting hyperplane of P.

5. Prove directly, or derive from theorem 3.1.8, the following fact: If P is a d-polytope and if k vertices $V_1, \cdots, V_k$ of P are given, $1 \leq k \leq d$, there exists a $(d - k)$-face of P which contains none of the vertices $V_1, \cdots, V_k$.

6. Let P be a polytope and T a projective transformation (not necessarily regular) permissible for P. Let $P' = T(P)$ and let F' be a face of P'. We have seen in exercise 2.4.9 that there exists a face F of P such that

$F' = TF$. Find examples which show that it is possible that every F such that $F' = T(F)$ satisfies dim $F >$ dim F'.

7. Let $P \subset R^d$ be a d-polytope and let L be an m-flat such that $P \cap L \neq \varnothing$. Prove that $F \cap L \neq \varnothing$ for some $(d - m)$-face F of P.

8. If P is a d-polytope then $f_k(P) \geq \begin{pmatrix} d + 1 \\ k + 1 \end{pmatrix}$ for all k with $-1 \leq k \leq d$; if $f_0(P) > d + 1$ then $f_k(P) > \begin{pmatrix} d + 1 \\ k + 1 \end{pmatrix}$ whenever $0 \leq k \leq d - 1$.

9. Let $0 \leq i, j \leq d - 1$; prove the existence of numbers $\varphi_{ij}(k, d)$ such that every d-polytope P with $f_i(P) \leq k$ satisfies $f_j(P) \leq \varphi_{ij}(k, d)$.

10. Let $(P_i | i = 1, 2, \cdots)$ be a sequence of polytopes which is convergent in the Hausdorff metric to the compact set K. Prove that if $(f_0(P_i) | i = 1, 2, \cdots)$ is a bounded sequence then K is a polytope. (Hint: Use exercise 2.4.4).

11. Show that the results of theorems 6 and 7 and exercises 2 and 4 generalize to polyhedral sets (with the restrictions: $d > 1$ for theorem 6, $h \geq 0$ for theorem 7, and $k \geq 0$ for exercise 2). Does theorem 8 generalize?

12. A set $K \subset R^d$ is called a *quasi-polyhedral set* provided $K \cap P$ is a polytope whenever P is a polytope. Show that each quasi-polyhedral set is closed and convex, and that the results mentioned in exercise 11 are valid for quasi-polyhedral sets. Prove that if K is quasi-polyhedral and $0 < \text{card ext } K < \infty$ then K is a polyhedral set.

13. Let K be a convex set and let $x \in K$. We shall say that K is *polyhedral at x* provided there exists a polytope P such that $x \in \text{int } P$ and $K \cap P$ is a polytope.

(i) Show that K is polyhedral at x if and only if $\text{cone}_x K$ is a polyhedral cone.

(ii) Show that if K is compact [closed] and polyhedral at each point, then K is a polytope [a quasi-polyhedral set].

14. Let $K \subset R^d$ be a polytope and let $\text{vert } K = \{v_1, \cdots, v_r\}$. Then each $x \in K$ is expressible—in general in many ways—in the form

$$x = \sum_{i=1}^{r} \lambda_i(x) v_i$$

with $\lambda_i(x) \geq 0$ and $\sum_{i=1}^{r} \lambda_i(x) = 1$. Show that it is always possible to choose the numbers $\lambda_i(x)$ in such a way that all the functions $\lambda_i(x), 1 \leq i \leq r$, depend continuously on $x \in K$. (This is a result of Kalman [1]. Hint: On 0- and 1-dimensional faces of P the functions $\lambda_i(x)$ are uniquely determined. If the $\lambda_i(x)$ are already defined for x belonging to p-dimensional faces for some $p \leq d - 1$, we extend the definition to $(p + 1)$-dimensional faces in the following manner. Let F be a $(p + 1)$-dimensional face of K, and $w = (\text{card vert } F)^{-1} \cdot \sum_{v \in \text{vert } F} v$. Then $w \in \text{relint } F$ and each $x \in F$ has a unique decomposition $x = (1 - \mu)w + \mu y$, where $y \in \text{relbd } F$. By assumption $y = \sum_{v_i \in \text{vert } F} \lambda_i(y)v_i$ and therefore

$$x = \sum \lambda_i(x)v_i = \sum_{v_i \in \text{vert } F} \left(\frac{1 - \mu}{\text{card vert } F} + \mu\lambda_i(y) \right)v_i.$$

Show that $\lambda_i(x)$ defined in this fashion satisfies all the requirements.) Show also that the $\lambda_i(x)$ may be chosen so that they are all continuous and that, for one preassigned i, $\lambda_i(x)$ is a *convex function* of $x \in K$. (A function $\varphi(x)$ is *convex* on the convex set K provided $\varphi(\lambda x_1 + (1 - \lambda)x_2) \leq \lambda\varphi(x_1) + (1 - \lambda)\varphi(x_2)$ whenever $x_1, x_2 \in K$ and $0 \leq \lambda \leq 1$.) Also, by considering the case in which K is a square in R^2, show that it is not always possible to have all the $\lambda_i(x)$ convex. (This provides a negative answer to a problem of Kalman [1]; compare Wiesler [1].)

15. The lattice $\mathscr{F}(P)$ of all faces of a d-polytope P has various interesting properties (see Perles [1,2].) Let a *tower* in P be a family $\mathscr{M} = \{M(i) \mid 0 \leq i \leq d - 1\}$ of faces of P such that dim $M(i) = i$ for all i, and $M(i) \subset M(j)$ for $0 \leq i \leq j \leq d - 1$. Denote also $M(i) = \varnothing$ for $i < 0$ and $M(i) = P$ for $i \geq d$. For a tower $\mathscr{M}$ define the tower $T\mathscr{M} = \mathscr{N}$ by putting $N(-1) = \varnothing$ and by taking as $N(i)$, for $0 \leq i \leq d - 1$, the unique i-face of P different from $M(i)$ which contains $N(i - 1)$ and is contained in $M(i + 1)$.

(i) Prove that T is a one-to-one mapping of the set of all towers in P onto itself; define the inverse mapping T^{-1}.

(ii) Let $-\infty < r \leq s < \infty$ and $0 \leq k \leq d - 1$. Prove that

$$T^r M(k + s - r) = \bigvee_{i=r}^{s} T^i M(k)$$

and

$$T^s M(k - s + r) = \bigcap_{i=r}^{s} T^i M(k).$$

If, moreover, $s \neq r$ and $s - r \leq d$, then $T^r M(k) \neq T^s M(k)$.

16. Let $K \subset R^d$ be a line-free polyhedral set. Show that there exists a nonsingular projective transformation P permissible for K such that $\text{cl}(PK)$ is a polytope, $(\text{cl}(PK)) \sim PK$ being one of its faces.

17. (Klee [2]) A compact [closed] set $K \subset R^d$ is a polytope [a quasi-polyhedral set] if and only if $\text{cone}_p K$ is closed for every $p \in K$.

18. Let P be a polytope and let T be a (not necessarily regular) projective transformation permissible for P. Prove that $f_k(TP) \leq f_k(P)$ for all k.

19. If P is a polytope in R^d, prove that the supporting function $H(P, x)$ (see exercise 2.2.8) is a piecewise linear function of x (that is, R^d is the union of a finite number of convex cones $C_1, \cdots, C_r$, such that for a suitable a_i and all $x \in C_i$ we have $H(P, x) = \langle x, a_i \rangle$, for $i = 1, \cdots, r$). Conversely, show that every piecewise linear function satisfying the conditions given in exercise 2.2.8(i) is the supporting function of some polytope.

20. If K_1, K_2 are polytopes show that

$$K_1 + K_2 = \text{conv}((\text{vert } K_1) + (\text{vert } K_2)).$$

21. (Motzkin [7]) Let P be a d-polytope, let $d_0, \cdots, d_k$ be nonnegative integers such that $k + \sum_{i=0}^k d_i = d$, and let $x \in P$. Prove that there exist faces F_i of P such that $\dim F_i = d_i$ for $i = 0, \cdots, k$, and $x \in \text{conv } \bigcup_{i=0}^k F_i$.

3.2 Combinatorial Types of Polytopes; Complexes

Two polytopes P and P' are said to be *combinatorially equivalent* (or *isomorphic*, or of the same *combinatorial type*) provided there exists a one-to-one correspondence φ between the set $\{F\}$ of all faces of P and the set $\{F'\}$ of all faces of P', such that φ is inclusion-preserving (i.e. such that $F_1 \subset F_2$ if and only if $\varphi(F_1) \subset \varphi(F_2)$.) Equivalently, one could say that the lattices $\mathcal{F}(P)$ and $\mathcal{F}(P')$ are isomorphic. Clearly, combinatorial equivalence is an equivalence relation; if P and P' are combinatorially equivalent we shall write $P \approx P'$.

The following assertions are easily established.

1. *If $P \approx P'$ then $\dim F = \dim \varphi(F)$ and $F \approx \varphi(F)$; also, $f_k(P) = f_k(P')$ for all k.*

2. *If $P \approx P'$ and if $\{F_1, \cdots, F_n\}$ is any family of faces of P, then*

$$\varphi \left(\bigcap_{i=1}^n F_i \right) = \bigcap_{i=1}^n \varphi(F_i) \quad \text{and} \quad \varphi \left(\bigvee_{i=1}^n F_i \right) = \bigvee_{i=1}^n \varphi(F_i).$$

3. *If T is a nonsingular affine map of R^d onto itself and if $P \subset R^d$ is a polytope, then $P \approx TP$. If T is a nonsingular projective mapping permissible for P, then $P \approx TP$.*

In particular, all d-simplices are of the same combinatorial type.

The concept of combinatorial equivalence of polytopes is of fundamental importance in many questions of the theory of polytopes, since many properties of a polytope depend only on its combinatorial type. The intrinsic difficulty of many problems on polytopes is intimately related to the fact that the combinatorial equivalence of polytopes is not endowed with properties usually encountered when dealing with equivalence relations in other mathematical disciplines. For example, combinatorial equivalence is neither a *closed* relation in the topological sense, nor an *open* one (that is, a limit of polytopes, all of which are of the same combinatorial type, is not necessarily of the same type; in every neighborhood of a polytope there are polytopes of a different combinatorial type.) Also it is impossible to define the 'sum-type', 'intersection-type', etc., of given combinatorial types. Following the procedure useful in many other disciplines, it would be desirable to find characteristics, easily computable for every polytope, such that the equality of the characteristics of the polytopes would indicate their combinatorial equivalence.* Unfortunately no such invariants of combinatorial types are known. Accordingly, very little is known about the combinatorial types of d-polytopes for $d \geq 3$. For $d = 1$ the problem is trivial, since all 1-polytopes are segments. For $d = 2$ the combinatorial types may be characterized by the number of vertices, since two polygons are of the same combinatorial type if and only if they have the same number of vertices. The number of different combinatorial types of d-polytopes with v vertices shall be denoted by $c(v, d)$; the number of simplicial d-polytopes (see section 4.5) with v vertices by $c_s(v, d)$. For the known results on $c(v, d)$ and $c_s(v, d)$ see chapters 6 and 13, and tables 1 and 2.

We shall return to some problems of classification of polytopes according to combinatorial type later on; presently we turn to certain notions belonging to combinatorial topology.

A finite family $\mathscr{C}$ of polytopes in R^d will be called a *complex*† provided

(i) Every face of a member of $\mathscr{C}$ is itself a member of $\mathscr{C}$;

(ii) The intersection of any two members of $\mathscr{C}$ is a face of each of them.

If a polytope P is a member of a complex $\mathscr{C}$ we shall call P a face of $\mathscr{C}$ and write $P \in \mathscr{C}$. The number of k-faces of $\mathscr{C}$ will be denoted by $f_k(\mathscr{C})$.

Among the simplest complexes we mention the following two which are associated with a k-polytope P:

---

* For two given polytopes it is, in principle, easy to determine whether they are combinatorially equivalent or not. It is enough to find all the faces of each of the polytopes and to check whether there exists any inclusion preserving one-to-one correspondence between the two sets of faces. However, this procedure is practically feasible only if the number of faces is rather small.

† Note that we depart here from the usual topological terminology. Our 'complexes' are commonly referred to as 'polyhedral complexes', 'convex complexes', or 'geometric cell complexes' (see, for example, Alexandroff–Hopf [1, p. 126], Lefschetz [1, p. 60]). When considering the more general topological objects we shall specify that the reference is to 'topological complexes'.

(1) *The boundary complex* $\mathcal{B}(P)$ of P, which is the complex consisting of all the faces of P which have dimension at most $k - 1$.

(2) The complex $\mathcal{C}(P) = \mathcal{B}(P) \cup \{P\}$ consisting of all the faces of P.

Note that the complex $\mathcal{C}(P)$ contains the same elements as the lattice $\mathcal{F}(P)$.

A complex $\mathcal{C}$ is said to be *k-dimensional*, or a *k-complex*, provided some member of $\mathcal{C}$ is a k-polytope but no member of $\mathcal{C}$ has dimension exceeding k.

Obviously, if P is a k-polytope then $\mathcal{B}(P)$ is a $(k - 1)$-complex, and $\mathcal{C}(P)$ is a k-complex.

Let $\mathcal{C}$ be a complex and let $C \in \mathcal{C}$. We define:

The *star* st$(C ; \mathcal{C})$ of C in $\mathcal{C}$ is the smallest subcomplex of $\mathcal{C}$ containing all the members of $\mathcal{C}$ which contain C.

The *antistar* ast$(C ; \mathcal{C})$ of C in $\mathcal{C}$ is the subcomplex of $\mathcal{C}$ consisting of all the members of $\mathcal{C}$ which do not intersect C.

The *linked complex* link$(C ; \mathcal{C})$ of C in $\mathcal{C}$ is the complex consisting of all polytopes of st$(C ; \mathcal{C})$ disjoint from C. Thus

$$\text{link}(C ; \mathcal{C}) = \text{st}(C ; \mathcal{C}) \cap \text{ast}(C ; \mathcal{C}).$$

Obviously, $\mathcal{C} = \text{st}(C ; \mathcal{C}) \cup \text{ast}(C ; \mathcal{C})$ whenever $C \in \mathcal{C}$ is 0-dimensional.

In order to illustrate the above notions, let T^d be a d-simplex, and V a vertex of T^d (see section 4.1). Then st$(V ; \mathcal{C}(T^d)) = \mathcal{C}(T^d)$, st$(V ; \mathcal{B}(T^d))$ is obtained from $\mathcal{B}(T^d)$ by omitting the $(d - 1)$-face T^{d-1} of T^d opposite to V; ast$(V, \mathcal{B}(T^d)) = \text{ast}(V, \mathcal{C}(T^d)) = \mathcal{C}(T^{d-1})$; link$(V ; \mathcal{B}(T^d)) = \mathcal{B}(T^{d-1})$, while link$(V ; \mathcal{C}(T^d)) = \mathcal{C}(T^{d-1})$.

To a complex $\mathcal{C}$ in R^d there is associated the subset of R^d consisting of all the points of members of $\mathcal{C}$; we shall denote it by set $\mathcal{C}$. Thus set $\mathcal{C} = \bigcup_{P \in \mathcal{C}} P$.

For example, if $P \subset R^d$ is a d-polytope then set $\mathcal{C}(P) = P$ and set $\mathcal{B}(P) = \text{bd } P$.

It is easy to establish the following assertions, the second of which is a refinement of the first and fits more naturally in the elementary-geometric theory of complexes.

4. *Let P be a d-polytope and $V \in \text{vert } P$. Then* set st$(V ; \mathcal{B}(P))$ *and* set ast$(V ; \mathcal{B}(P))$ *are each homeomorphic with the $(d - 1)$-dimensional solid ball B^{d-1} (and therefore with any $(d - 1)$-dimensional compact convex set.) Also,* set link$(V ; \mathcal{B}(P))$ *is homeomorphic with $S^{d-2} = \text{relbd } B^{d-1}$.*

5. *Let P be a d-polytope and $V \in$ vert P. Then* set st$(V; \mathcal{B}(P))$ *and* set ast$(V; \mathcal{B}(P))$ *are piecewise affinely† homeomorphic with the $(d-1)$-simplex T^{d-1}; also,* set link$(V; \mathcal{B}(P))$ *is piecewise affinely homeomorphic with* relbd T^{d-1}.

Substituting 'member of $\mathscr{C}$ or $\mathscr{C}'$' for 'face of P or P'' the definition of combinatorially equivalent polytopes may be generalized to that of combinatorially equivalent complexes. Clearly, properties 1, 2, and 3 hold for combinatorially equivalent complexes.

Entities more general than complexes are obtained by substituting 'polyhedral sets' for 'polytopes' in the definition of complexes. We shall not deal with those entities, though some of the results mentioned in the sequel are valid for them (for example, exercise 3.2.6), while others have to be only slightly modified (compare exercise 8.5.2).

Exercises

1. If P and P' are line-free polyhedral sets, show that the definition of $P \approx P'$ used for polytopes is suitable in the sense that it is intuitively acceptable and satisfies properties 1, 2, and 3. Show that this is no longer the case if P and P' are allowed to be any polyhedral sets, but that even in this case an acceptable definition is obtained if the additional requirement dim $F = $ dim $\varphi(F)$ is imposed. Show also that if K_1 and K_2 are polyhedral sets in R^d such that the lattices $\mathscr{F}(K_1)$ and $\mathscr{F}(K_2)$ are isomorphic, then $K_1 \cap L_1^* \approx K_2 \cap L_2^*$, where L_1^* is a linear subspace of R^d complementary to the maximal flat contained in K_1 (see theorem 2.5.4).

2. Let P and P' be two polytopes and let Ψ be a biunique correspondence between vert P and vert P' with the following property: For a set $A \subset$ vert P, there exists a face F of P such that $A = $ vert F if and only if there exists a face F' of P' such that $\Psi(A) = $ vert F'. Show that Ψ can be extended to a biunique correspondence between all the faces of P and P', under which $P \approx P'$.

3. Let P and P' be two polytopes and let there exist a one-to-one correspondence Ψ which maps vertices of P onto vertices of P' and facets of P onto facets of P', in such a way that incidence relations between vertices and facets are preserved (i.e. if V is a vertex of P and F a facet of P, then $V \in F$ if and only if $\Psi(V) \in \Psi(F)$.) Then $P \approx P'$.

† A mapping T defined on R^d is *piecewise* affine [projective] if it is possible to represent R^d in the form $R^d = \bigcup_{i=1}^{n} K_i$, where the K_i are closed convex sets, so that the restriction of T to each K_i is an affine map [is a permissible projective map of K_i]. For interesting results and problems concerning piecewise affine, convex functions see Davis [4].

4. Let $(P_i \mid i = 1, 2, \cdots)$ be a sequence of d-polytopes converging (in the Hausdorff metric) to the k-polytope P, where $k < d$. Prove that the sequence (bd $P_i \mid i = 1, 2, \cdots$) converges to P.

5. Let $(P_i \mid i = 1, 2, \cdots)$ be a sequence of d-polytopes of the same combinatorial type, such that the corresponding vertices form convergent sequences. Let P be the limit, in the Hausdorff metric, of the sequence (P_i), and let K_i^m denote the union of all the m-faces of P_i. Using exercise 4, show that for every m, dim $P \le m \le d$, the polytope P is the limit of the sequence $(K_i^m \mid i = 1, 2, \cdots)$.

6. Generalizing the above, prove the following: Let $(K_i \mid i = 1, 2, \cdots)$ be a sequence of compact convex sets in R^d, converging in the Hausdorff metric to the compact convex set K. Let $\mathrm{ext}_k\, K_j$ be the set of k-extreme points of K_j (see section 2.4). For every k such that dim $K \le k \le d$, the sequence $(\mathrm{cl}\, \mathrm{ext}_k\, K_i \mid i = 1, 2, \cdots)$ converges to K.

7. If $\mathscr{C}, \mathscr{D}$ are complexes in R^d then $\{C \cap D \mid C \in \mathscr{C}, D \in \mathscr{D}\}$ is a complex. If P is a polytope and L a flat, then $\mathscr{C}(P \cap L) = \{F \cap L \mid F \in \mathscr{C}(P)\}$.

8. Let P and P' be combinatorially equivalent polytopes in R^d the face of P' corresponding to the face F of P being $\varphi(F)$. Show that there exists a piecewise affine mapping T of R^d onto itself such that $TF = \varphi(F)$ for every face F of P. Find examples which show that this assertion may fail if P, P' are polyhedral sets, even if T is allowed to be piecewise projective.

9. Let K, K' be unbounded, line-free polyhedral sets which are combinatorially equivalent under the mapping φ of $\mathscr{F}(K)$ onto $\mathscr{F}(K')$. Show the equivalence of the following assertions:

(i) There exists a piecewise projective homeomorphism T of K onto K' such that $TF = \varphi(F)$ for each $F \in \mathscr{F}(K)$.

(ii) There exists a piecewise affine homeomorphism T of K onto K' such that $TF = \varphi(F)$ for each $F \in \mathscr{F}(K)$.

(iii) If T, T' are projective transformations permissible for K, K', such that TK and $T'K'$ are bounded, there exists an isomorphism ψ between $\mathscr{F}(\mathrm{cl}\, TK)$ and $\mathscr{F}(\mathrm{cl}\, T'K')$ such that $\psi(\mathrm{cl}\, TF) = \mathrm{cl}\, T'(\varphi(F))$ for each $F \in \mathscr{F}(K)$.

3.3 Diagrams and Schlegel Diagrams

Let $P \subset R^d$ be a d-polytope, F_0 a facet of P, and let $x_0 \notin P$ be a point* of R^d such that among all the affine hulls of the facets of P only that of F_0 separates x_0 and P. Let $\mathscr{P}$ denote the complex $\mathscr{B}(P) \sim \{F_0\}$. The projection of P onto aff F_0 by rays issuing from x_0 yields, when restricted to

* Prove the existence of such x_0 by showing that for each $y_0 \in \mathrm{relint}\, F_0$ all $x_0 \notin P$ sufficiently close to y_0 satisfy those assumptions.

set $\mathscr{P}$, a mapping of the $(d-1)$-complex $\mathscr{B}$ onto a $(d-1)$-complex $\mathscr{S}_0$ contained in aff F_0. The projection is a homeomorphism between set $\mathscr{P}$ and set $\mathscr{S}_0$ and shows that $\mathscr{S}_0$ is combinatorially equivalent to $\mathscr{P}$. Moreover, obviously $F_0 = \text{set } \mathscr{S}_0$. The family $\mathscr{S} = \{F_0\} \cup \mathscr{S}_0$ is called a *Schlegel diagram* of *P, based* on F_0. (Schlegel [1]; Sommerville [2, p. 100]).

As illustrations of the formation of Schlegel diagrams, figure 3.3.1 represents a pentagon and its Schlegel diagram, while figures 3.3.2 and 3.3.3 represent a square pyramid and two of its Schlegel diagrams. Figures 3.3.4 and 3.3.5 represent Schlegel diagrams of the 4-simplex and the 4-cube, respectively.

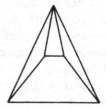

Figure 3.3.1

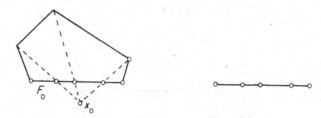

Figure 3.3.2

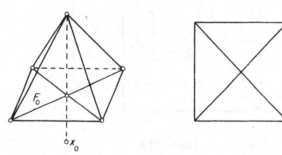

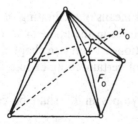

Figure 3.3.3

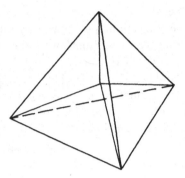

Figure 3.3.4

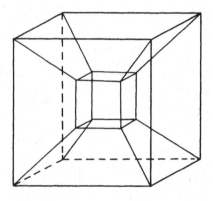

Figure 3.3.5

Schlegel diagrams are mainly used as a means of facilitating the representation of 3- and 4-polytopes by complexes in the plane or in R^3; but they are useful in some 'theoretical' questions as well (see chapter 11.)

Schlegel diagrams are a special case of certain complex-like families which we call *diagrams*.

A finite family $\mathscr{D} = \{D_0\} \cup \mathscr{C}$ of polytopes in R^d shall be called a *d-diagram* provided

    (i) $\mathscr{C}$ is a complex;

(ii) D_0 is a d-polytope such that D_0 = set $\mathscr{C}$ and each proper face of D_0 is a member of $\mathscr{C}$;

(iii) $C \cap \text{bd } D_0$ is a member of $\mathscr{C}$ whenever $C \in \mathscr{C}$.

It is remarkable that even at the beginning of the present century, when Schlegel diagrams were extensively used in the study of polytopes, no distinction was made between Schlegel diagrams and d-diagrams (that is, complexes which 'look like' Schlegel diagrams). It seems that the reason for this attitude is to be found in the (usually tacit) assumption that every d-diagram is (combinatorially equivalent to) a Schlegel diagram (see, for example, Brückner (2, 3].) This is indeed trivially the case for 1-diagrams. In what is probably the deepest result to date in the theory of polytopes, Steinitz proved that every 2-diagram is combinatorially equivalent to a Schlegel diagram of a 3-polytope. (We shall present a proof of Steinitz's theorem in chapter 13). However, already in the case of 3-diagrams the situation is different. As we shall see in chapter 11, there exist 3-diagrams (even simplicial ones—see section 4.5) which are not combinatorially equivalent to any Schlegel diagram of a 4-polytope.

Thus we are presented with the problem how to define d-diagrams 'correctly' for $d \geq 3$. In other words, what conditions must a complex satisfy in order to be identifiable with the Schlegel diagram of some polytope.

Conceivably, some clues to the solution of the problem may be derived from the observation that each Schlegel diagram has the following two properties:

(i) it may be 'inverted' in the sense that any maximal face of the polytope may be taken as the 'basic' face, into which the polytope is projected;

(ii) there exists a 'dual' Schlegel diagram (a Schlegel diagram of any dual polytope; see section 3.4).

At present it is not known whether any or all of these properties may be used in order to characterize Schlegel diagrams among diagrams.

Exercises

1. By considering 2-diagrams and by checking whether they are Schlegel diagrams, show that there exist two combinatorial types of 3-polytopes with 5 vertices, and 7 types with 6 vertices. There exist 5 types of 3-polytopes with 7 vertices having as faces only triangles.

2. With reference to the Schlegel diagrams in figure 3.3.6 determine which of them represent the same combinatorial type of 3-polytopes. Find all the other Schlegel diagrams of these polytopes.

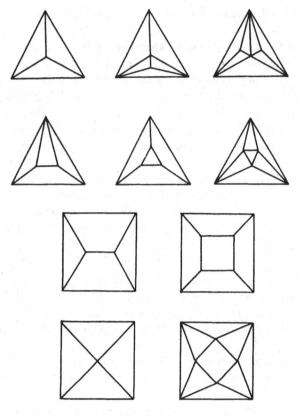

Figure 3.3.6

3.4 Duality of Polytopes

Two d-polytopes P and P^* are said to be *dual* to each other provided there exists a one-to-one mapping Ψ between the set of all faces of P and the set of all faces of P^* such that Ψ is inclusion-reversing (i.e. faces F_1 and F_2 of P satisfy $F_1 \subset F_2$ if and only if the faces $\Psi(F_1)$ and $\Psi(F_2)$ of P^* satisfy $\Psi(F_1) \supset \Psi(F_2)$.) Clearly this implies $\Psi(\varnothing) = P^*$, $\Psi(P) = \varnothing$, and in general $\dim F + \dim \Psi(F) = d - 1$ for every face F of P.

It is also obvious that if $P \approx P_1$, $P^* \approx P_1^*$ and P is dual to P^*, then P_1 is dual to P_1^*. Thus it is meaningful to define two combinatorial types to be dual to each other provided there exist polytopes, one of each of the types, which are dual one to the other.

If P_1 is dual to P^*, and if P_2 is also dual to P^*, then it is immediate that P_1 and P_2 are combinatorially equivalent.

As examples of dual 3-polytopes we mention: the cube and the octahedron, and in general the n-sided prism and the n-sided bipyramid; the n-sided pyramid is dual to itself; the dodecahedron and the icosahedron. The d-dimensional simplex is dual to itself.

Later on we shall encounter different applications of duality. Now we shall only settle one question arising naturally in connection with the notion of duality: Has each d-polytope a dual d-polytope? The answer is affirmative, and it may be deduced from various well-known geometric constructions. We shall describe only one of them, which has well-known analogues, variants and generalizations in many other fields.

Let A be a subset of R^d; the *polar set* A^* of A is defined by

$$A^* = \{y \in R^d \mid \langle x, y \rangle \le 1 \quad \text{for all} \quad x \in A\}.$$

It is easily verified that

1. *If $K \subset R^d$ is a compact convex set such that $0 \in \text{int } K$, then the polar set K^* is also a compact convex set and $0 \in \text{int } K^*$.*

2. *If $K_1, K_2 \subset R^d$ are compact convex sets such that $0 \in \text{int } K_1 \subset K_2$, then $K_1^* \supset K_2^*$.*

Using the notation $K^{**} = (K^*)^*$, we have

3. *If $K \subset R^d$ is a compact convex set with $0 \in \text{int } K$, then $K^{**} = K$.*

Indeed, it is immediate from the definitions that $K \subset K^{**}$. On the other hand, if $x_0 \notin K$ there exists a hyperplane $H = \{z \in R^d \mid \langle z, y_0 \rangle = 1\}$ strictly separating x_0 and K, i.e. such that $\langle x, y_0 \rangle < 1$ for all $x \in K$ and $\langle x_0, y_0 \rangle > 1$. Therefore $y_0 \in K^*$ and then $\langle x_0, y_0 \rangle > 1$ shows that $x_0 \notin K^{**}$.

Let F be a face of K; define the set $\hat{F}$ by

$$\hat{F} = \{y \in K^* \mid \langle x, y \rangle = 1 \quad \text{for all} \quad x \in F\}.$$

4. *If K is a compact convex set with $0 \in \text{int } K$, the mapping Ψ defined by $\Psi(F) = \hat{F}$ is a one-to-one inclusion-reversing correspondence between $\mathscr{F}(K)$ and $\mathscr{F}(K^*)$; moreover, $\Psi(\Psi(F)) = F$ for every face F of K.*

PROOF Clearly $\hat{\varnothing} = K^*$ and $\hat{K} = \varnothing$; thus we may restrict our attention to proper faces F. We shall first prove that $\hat{F}$ is a face of K^*. Let $x_0 \in \text{relint } F$ and consider the face F^* of K^* defined by

$$F^* = \{y \in K^* \mid \langle y, x_0 \rangle = 1\}.$$

Obviously $\hat{F} \subset F^*$; we shall show that $\hat{F} = F^*$. Indeed, assume $y_0 \in K^*$, $y_0 \notin \hat{F}$. Then there exists $x_1 \in F$ such that $\langle x_1, y_0 \rangle < 1$. Since $x_0 \in$ relint F, there exists $x_2 \in F$ such that $x_0 = \lambda x_1 + (1 - \lambda)x_2$, where $0 < \lambda < 1$. Since $y_0 \in K^*$ we have $\langle x_2, y_0 \rangle \leq 1$ and therefore $\langle x_0, y_0 \rangle < 1$. Thus $y_0 \notin F^*$ and $\hat{F} = F^*$ as claimed. This shows that Ψ maps faces of K onto faces of K^*; since the mapping is obviously inclusion-reversing the theorem will be completely proved if we show that $\Psi(\Psi(F)) = F$. Now, by the definition we have $\Psi(\Psi(F)) = \{z \in K^{**} \mid \langle y, z \rangle = 1 \text{ for all } y \in \hat{F}\}$; since $K^{**} = K$, it is obvious that $F \subset \Psi(\Psi(F))$. Let $z_0 \in K$, $z_0 \notin F$; since F is a face of K there exists a hyperplane $H = \{x \in R^d \mid \langle x, y_0 \rangle = 1\}$ such that $F = K \cap H$. This clearly implies that $y_0 \in \hat{F}$ and $\langle z_0, y_0 \rangle < 1$; therefore $z \notin \Psi(\Psi(F))$ and the proof of theorem 4 is completed.

In the special case that K is a d-polytope with $0 \in$ int K, theorem 4 shows that the polar set K^* is a polytope, and Ψ is a dual correspondence between the faces of K and the faces of K^*.

Exercises

1. Show that each combinatorial type of 2-polytopes is dual to itself.

2. Show that the 3-polytopes with at most 9 edges form 4 combinatorial types. Determine those types and their duals.

3. Determine all self-dual types of 3-polytopes with at most 7 vertices. (Hermes [1] has determined the number $t(v)$ of different combinatorial types of self-dual 3-polytopes with v vertices, for $v \leq 9$; by a different method, these numbers were determined by Jucovič [1], who discovered an error in Hermes' work; Bouwkamp–Duijvestyn–Medema [1] determined $t(v)$ for $v \leq 10$. The known values of $t(v)$ are: $t(4) = t(5) = 1$, $t(6) = 2$, $t(7) = 6$, $t(8) = 16$, $t(9) = 50$, $t(10) = 165$.)

4. Let $K \subset R^d$ be a compact convex set with $0 \in$ int K, and let $p \in$ int K. Show that $(-p + K)^*$ is a projective image of K^*, with $z \in (-p + K)^*$ if and only if $z = y/(1 - \langle p, y \rangle)$ for some $y \in K^*$.

5. Generalizing the properties of the polarity mapping $A \to A^*$ mentioned in the text, prove that if A, B are any subsets of R^d and if $\lambda \neq 0$, then:

    (i) $A^{**} = \text{cl conv}(A \cup \{0\})$

    (ii) $A^{***} = A^*$

    (iii) if $A \subset B$ then $A^* \supset B^*$

    (iv) $(\lambda A)^* = \dfrac{1}{\lambda} A^*$

    (v) $(A \cup B)^* = A^* \cap B^*$

(vi) A is bounded if and only if $0 \in \text{int } A^*$

(vii) A^* is bounded if and only if $0 \in \text{int conv } A$

(viii) if A is a polyhedral set then A^* is a polyhedral set

(ix) if A is a cone with apex 0 then A^* is a cone with apex 0

(x) if A and B are closed convex sets containing 0, then $(A \cap B)^*$
$= \text{cl conv}(A^* \cup B^*)$.

6. Let $C \subset R^d$ be a closed convex cone with apex at the origin; the *polar cone* C^* is defined as $C^* = \{x \in R^d \mid \langle x, y \rangle \le 0 \text{ for all } y \in C\}$. Establish the following assertions:

(i) This definition of C^* coincides with the definition on page 47.

(ii) If C is pointed and d-dimensional then C^* has the same properties.

(iii) More generally,

$$\dim C^* + \dim(C \cap -C) = d \quad \text{and} \quad \dim C + \dim(C^* \cap -C^*) = d.$$

(iv) If C is a pointed polyhedral cone of dimension d, and if H, H' are hyperplanes which do not contain 0 such that the sets $P = H \cap C$ and $P^* = H' \cap C^*$ are bounded, then the $(d-1)$-polytopes P and P^* are dual to each other.

7. If the convex polytope $P \subset R^d$ is the intersection of the closed halfspaces $\{x \in R^d \mid \langle x, y_i \rangle \le 1\}$, where $1 \le 1 \le n$, show that

$$P^* = \text{conv}\{y_1, \cdots, y_n\}.$$

More generally, if $K = \{x \in R^d \mid \langle x, y_1 \rangle \le 1, \ 1 \le i \le n, \ \langle x, z_j \rangle \le 0, \ 1 \le j \le m\}$, show that

$$K^* = \text{cl conv}\left(\{0, y_1, \cdots, y_n\} \cup \bigcup_{j=1}^{m} R^* z_j\right)$$

$$= \text{conv}\{0, y_1, \cdots, y_n\} + \text{conv}\left(\bigcup_{j=1}^{m} R^* z_j\right),$$

where $R^* z = \{\lambda z \mid 0 \le \lambda\}$.

8. Let $K \subset R^d$ be a d-polytope and $V \in \text{vert } K$. A *vertex figure of K at V* (compare Coxeter [1]) is the intersection of K by a hyperplane which strictly separates V from vert $K \sim \{V\}$.

(i) Show that any two vertex figures of K at V are projectively equivalent.

(ii) Let $0 \in \text{int } K$. Show that any vertex figure of K at V is dual (in the sense of duality of $(d-1)$-polytopes) to the facet $\hat{V}$ of K^*.

9. Let P be a polytope. We shall denote by $\mathcal{R}(P)$ the *set of cones associated with P*,

$$\mathcal{R}(P) = \{-V + \text{cone}_V P \mid V \in \text{vert } P\}.$$

If P_1 and P_2 are polytopes we shall say that P_2 is *related to* P_1 provided each member of $\mathscr{R}(P_2)$ is representable as an intersection of members of $\mathscr{R}(P_1)$; we shall say that P_1 and P_2 are *related* provided each of them is related to the other.

(i) Show that for each $d \geq 2$ there exists d-polytopes P_1, P_2, P_3 such that P_2, P_3 are related to P_1, but none of P_2 and P_3 is related to the other.

(ii) Show that polytopes P_1 and P_2 are related if and only if $\mathscr{R}(P_1) = \mathscr{R}(P_2)$.

(iii) Show that if P_1 and P_2 are related, then they are combinatorially equivalent.

(iv) Let $P \subset R^d$ be a d-polytope, $P = \{x \in R^d \mid \langle x, u_i \rangle \leq \alpha_i, 1 \leq i \leq n\}$. Show that there exists an $\varepsilon = \varepsilon(P) > 0$ such that P is related to every P' of the type $P' = \{x \in R^d \mid \langle x, u_i \rangle \leq \beta_i, 1 \leq i \leq n\}$, where $|\alpha_i - \beta_i| < \varepsilon$ for $i = 1, 2, \cdots, n$.

10. As in section 3.2, let $\mathscr{F}(K)$ denote the lattice of all faces of the compact convex set $K \subset R^d$.

(i) If $0 \in \operatorname{int} K$ show that $\mathscr{F}(K)$ is antiisomorphic to the lattice $\mathscr{F}(K^*)$ under the mapping $\psi(F) = \hat{F}$.

(ii) If K is, moreover, a d-polytope show that $\dim F + \dim \psi(F) = d - 1$.

(iii) Let K be a d-polytope, $F_1, F_2 \in \mathscr{F}(K)$, and $F_1 \subset F_2$. Show that $\{F \in \mathscr{F}(K) \mid F_1 \subset F \subset F_2\}$ is a sublattice of $\mathscr{F}(K)$, isomorphic to the lattice $\mathscr{F}(P)$ of some polytope P of dimension $d' = \dim F_2 - \dim F_1 - 1$, and antiisomorphic to the lattice $\mathscr{F}(P')$ of another d'-polytope P'.

(iv) Generalize (iii) to the case where K is a quasi-polyhedral set.

11. The notion of duality of two polytopes in R^d may easily be generalized to the notion of duality of two complexes in R^d, by requiring the existence of an inclusion-reversing correspondence between the *nonempty* faces of the complexes.

(i) Find examples of $(d - 1)$-complexes in R^d which have no dual complexes.

(ii) Let a k-complex $\mathscr{C}$ be called *boundary-free* if every member of $\mathscr{C}$ of dimension less than k belongs to at least two k-dimensional members of $\mathscr{C}$. Prove that if a $(d - 1)$-complex $\mathscr{C}$ in R^d, $d \geq 2$, has a dual then $\mathscr{C}$ is boundary-free and each $(d - 2)$-dimensional element of $\mathscr{C}$ belongs to precisely two $(d - 1)$-dimensional elements of $\mathscr{C}$. Find examples which show that those properties are not sufficient for the existence of a dual complex if $d \geq 3$.

12. Let A be a linear transformation from R^d into R^e. The *adjoint transformation* A^* from R^e onto R^d is defined by the condition: $y^* = A^* y$

is that point of R^d for which $\langle Ax, y \rangle = \langle x, y^* \rangle$ for all $x \in R^d$. Prove that A^* is a linear transformation.

13. Let $K_1 \subset R^d$ and $K_2 \subset R^e$ be compact convex sets such that $0 \in \text{int } K_1$ and $0 \in \text{int } K_2$, and let A be a linear transformation from R^d to R^e such that $AK_1 = K_2$. Show that the adjoint transformation A^* is one-to-one and maps K_2^* onto $K_1^* \cap A^*R^e$.

14. Using theorems 3.4.3 and 3.1.2, give a short proof of theorem 3.1.1. (Compare Glass [1].)

3.5 Remarks

The use of the term 'polytope' in this book—as well as the use of some other terms such as 'polyhedral set', 'complex', etc.—differs from the generally accepted one (though Klee [18] uses 'polytope' in the sense adopted here). Our 'polytopes' are in the literature mostly referred to as 'convex polytopes' (or 'convex polyhedra'; compare the preface to the second edition of Coxeter [1]); however, since the only 'polytopes' considered here are the convex ones, we felt that the omission of a few thousands of repetitions of the word 'convex' is justified. The weight of this decision was considerably lightened by the observation that in most instances in which the term 'polytope' is used in the literature for not necessarily convex objects of dimension exceeding 3, a precise definition is lacking (see, for example, chapter 7 of Sommerville [2]); and if given, varies according to the author's aims (compare, for example, Coxeter [1], pp. 126, 288, and N. W. Johnson [3]).

The main aim of the present chapter was to obtain a number of fundamental notions and results on polytopes. All the results in the main text and in most of the exercises are well known, though in some cases it is rather hard to find definite references. The reader wishing to pursue the historical aspect in more detail is referred, in a general way, to Schläfli [1], Minkowski [2], Weyl [1], and Coxeter [1].

The justification for our use of the term 'complex' lies only in its brevity. Though many books on topology define the objects we call complexes (using various terms, such as 'polyhedral complexes', 'geometric cell complexes', etc.), this appearance is mostly of marginal interest and consequence. Indeed, a topologist's aims are mostly invariant under subdivision, hence the study of 'polyhedral complexes' may be reduced to the study of 'simplicial polyhedral complexes'; but every such entity is a complex in our sense (see exercise 25 in section 4.8). In other words, the question whether a given topological complex is, or is not representable by a complex with convex cells is rather uninteresting from a topological point of view, and has consequently received little attention (see chapter 11 for a more detailed discussion). For us, however, the complexes as defined above are a natural generalization of polytopes. Moreover, even though our main interest are polytopes, valuable insights are reached by studying the more general case of complexes (see, for example, chapters 9, 11, 12, 13).

Polarity—for polytopes, or for more general sets—has been studied by many authors. As references for various approaches and for additional facts and references we mention Weyl [1], Motzkin [1,2], Fenchel [3,4].

Various notions of infinite-dimensional polytopes have been considered. Though some of the results obtained have important analytic content, there seems to be rather little that may be said about combinatorial properties of such polytopes and they will not be considered in the sequel. The interested reader should consult, for example, Choquet [1], Bastiani [1], Eggleston–Grünbaum–Klee [1], Maserik [1], Alfsen [1].

The notion of a polytope 'related' to another polytope (exercise 3.4.9) is rather recent (Grünbaum [11]); a similar concept was introduced somewhat earlier by Shephard [2] (see section 15.1; a polytope A is related to a polytope B if and only if, in the notation of section 15.1, $\lambda A \leq B$ for some $\lambda > 0$). This notion seems to be quite natural in various combinatorial problems (see, for example, Hadwiger–Debrunner [1], Asplund–Grünbaum [1], Grünbaum [11], Rado [1]); probably many results known at present only for some special families (such as parallelotopes having parallel edges) may be meaningfully extended to families of polytopes related to a given polytope.

3.6   Additional notes and comments

The "main theorem".
Theorems 3.1.1 and 3.1.2 together make up the "main theorem about poly-topes": Any polytope may be defined as the convex hull of a finite set of points (i.e., by a $\mathcal{V}$-*description*), or as a bounded intersection of finitely many closed half-spaces (an $\mathcal{H}$-*description*). In the case of a full-dimensional polytope, the minimal such descriptions are in fact unique: The minimal $\mathcal{V}$-representation of a polytope is given by the vertices, while the minimal $\mathcal{H}$-representation consists of the facet-defining halfspaces.

There are several different methods of proof available, some of which also lead to algorithms for the conversion between $\mathcal{V}$- and $\mathcal{H}$-descriptions (see be-low). Grünbaum's argument of theorem 3.1.1 reminds one of the "ray shoot-ing" techniques in computational geometry (de Berg et al. [a, Chap. 8]). Still an alternative proof is from metric aspects, see Ewald [a, Sect. II.1].

Rational Polytopes.
A polytope is *rational* if it has rational vertex coordinates or, equivalently, if it has rational facet-defining inequalities. For every rational polytope the facet complexity is bounded by a cubic polynomial of the vertex complexity, and conversely; here *facet complexity* refers to the maximal coding size of a facet-defining inequality, while *vertex complexity* is the maximal coding size of a vertex. See Schrijver [a, Sect. 10.2].

Under suitable magnification of coordinates, each rational polytope is equiv-alent to one whose vertices all have integral coordinates. The theory of such *lattice polytopes* is extensive, and makes many contacts with algebraic geome-try and with the geometry of numbers. See Barvinok [b] for a short survey.

Algorithmic aspects.
From the algorithmic point of view of computational convexity (see Gritz-mann–Klee [e]), it makes a great difference which type of representation of a polytope is given: First, one representation may be very large even though the other one is small—e. g. for the d-cube, which may be given by $2d$ half-spaces or by 2^d vertices. Moreover, even if this is not the case, the problem of *computing* one representation from the other—known as the convex hull problem—is non-trivial, both in theory and in practice.

In theory, an asymptotically optimal algorithm for polytopes in any fixed dimension was provided by Chazelle [a]. Similarly, in the case of input "in general position", the reverse search techniques of Avis–Fukuda [a] solve the problem in polynomial time. However, without any of these restrictions, the convex hull problem is not solved at all. Indeed, Avis–Bremner–Seidel [a] have demonstrated that the known methods for the convex hull problem have no polynomial bounds for their running times in terms of the size of "input plus output" (see also Bremner [a]). Check Fukuda [b] and Kaibel–Pfetsch [a] for current discussions about polyhedral computation.

Non-trivial isomorphism problems are raised, e. g., by the footnote on p. 39, where Grünbaum talks about a "given" convex polytope, and says that checking combinatorial equivalence "is, in principle, easy"; see Kaibel–Schwartz [a].

In practice, there are several reasonable algorithmic methods available to attack convex hull problems of moderate size:

o Fourier–Motzkin elimination/the double-description method are described in Ziegler [a, Lect. 1]; an implementation is cdd by Fukuda [a].

o Lexicographic reverse search is implemented in lrs by Avis [a].

These convex hull codes are integrated in the polymake system by Gawrilow–Joswig [a] [b]. In most cases, a solution of the convex hull problem for a given example is the first step for all further analysis of any "given" example. It is also often the computational bottleneck: If the convex hull problem part can be solved, then many other questions may be answered "easily".

Further representations.
$\mathcal{V}$- and $\mathcal{H}$-descriptions are the standard ways to represent polytopes. However, alternative representations have been studied; we describe two of them in the following.

A result of Bröcker and Scheiderer on semi-algebraic sets (see Bochnak–Coste–Roy [a] for references) implies that each d-polytope $P \subset R^d$ can be presented as the solution set of a system of $d(d+1)/2$ polynomial inequalities; for P's interior, d polynomial inequalities suffice. These striking results are nonconstructive, and it is at present unknown whether one can algorithmically convert an $\mathcal{H}$-description of a d-polytope into a *polynomial representation* in which the number $\mu(d)$ of polynomials depends only on d. Grötschel–Henk [a] show that for simple d-polytopes this can be done with $\mu(2) = 3$, $\mu(3) = 6$, and in general, $\mu(d) \leq d^d$.

In the *oracle* approach pioneered by Grötschel–Lovász–Schrijver [a], a suppliant attempts to determine the structure of a polytope P by means of the answers to a sequence of questions posed to some sort of oracle. For a d-polytope P whose interior is known to contain the origin, each query to the *ray-oracle* consists of a ray issuing from the origin, and the oracle responds by telling where the ray intersects P's boundary. For this oracle, Gritzmann–Klee–Westwater [a] show that the entire face lattice of a d-polytope P can be reconstructed with the aid of at most $f_0(P) + (d-1)f_{d-1}^2(P) + (5d-4)f_{d-1}(P)$

queries. For similar results involving other oracles, see Dobkin–Edelsbrunner–Yap [a]. Oracle representations are particularly important with respect to volume computation (see the notes in section 15.5).

Diagrams vs. Schlegel diagrams.
A review of diagrams and Schlegel diagrams appears in Ziegler [a, Lect. 5]. Schlegel diagrams of dimension d are most useful in the case of $d = 3$, where they provide a tool to visualize 4-dimensional polytopes. Such 3-dimensional diagrams may be visualized via cardboard or wire models, but also electronically using polymake by Gawrilow–Joswig [a] [b].

The decision whether a given d-diagram *is* a Schlegel diagram is easily reduced to linear programming. However, the question whether a given diagram is *combinatorially equivalent to* a Schlegel diagram seems to be very hard in general; see Richter-Gebert [b, Chap. 10]. A remarkable theorem, however, states that all *simple d-diagrams* are Schlegel diagrams, for $d \geq 3$; for this we refer to Rybnikov [a], who derives it from rather powerful, general criteria for liftability of polyhedral cell complexes.

In view of the question posed on page 45, we now know many examples of non-polytopal 3-diagrams. In particular, examples constructed and analyzed by Schulz [a] [b] show that neither invertible nor dualizable 3-diagrams are necessarily Schlegel diagrams. (Apparently it has not been proved explicitly that the combination of both properties is not sufficient.)

"Related" polytopes and their fans.
For exercise 3.4.9 one can, equivalently, consider the normal fan of the polytope, a concept that first arose in the theory of toric varieties (see Ewald [a], Fulton [a]): A *(complete) fan* is a complex of pointed polyhedral cones in R^d whose union is all of R^d. The *normal fan* $\mathcal{N}(P)$ of a d-polytope $P \subset R^d$ contains, for each non-empty face $F \subset P$, the collection C_F of all vectors $a \in R^d$ such that the linear function $x \mapsto \langle a,x \rangle$ on P is maximized by all points in F. (Thus for each vertex V, the cone C_V is dual to the cone $-V + \text{cone}_V P$ considered by Grünbaum.) By exercise 3.4.9(ii), two d-polytopes are "related" if their normal fans coincide. See also Ziegler [a, Lect. 7].

Related polytopes are also called *strongly isomorphic*, *normally equivalent*, *analogous*, and *locally similar* (see Schneider [b, Notes for Section 2.4]).

Two exercises.
In exercise 3.1.14, continuous functions are most easily derived from a triangulation (without new vertices) of the polytope.

Exercise 3.4.3 is to enumerate self-dual 3-polytopes—here the study of Dillencourt [a] has produced the following table:

n	4	5	6	7	8	9	10	11	12	13	14
$f(n)$	1	1	2	6	16	50	165	554	1908	6667	23556

It seems to be a recent insight that for self-dual polytopes, a self-duality of order 2 need not exist; we refer to a thorough discussion and survey by Ashley et al. [a]. See also Jendrol' [a].

CHAPTER 4

Examples

The aim of the present chapter is to describe in some detail certain polytopes and families of polytopes. This should serve the double purpose of familiarizing the reader with geometric relationships in higher-dimensional spaces, as well as providing factual material which will be used later on.

4.1 The d-Simplex

The simplest type of d-polytopes is the *d-simplex* T^d. A d-simplex is defined as the convex hull of some $d + 1$ affinely independent points. Since any affinely independent $(d + 1)$-tuple of points is affinely equivalent to every other $(d + 1)$-tuple of affinely independent points, and since affine transformations commute with the operation of forming convex hulls, it follows that each two d-simplices are nonsingular affine images of each other. Therefore, in particular:

1. *All d-simplices are of the same combinatorial type.*

Let T^d be a d-simplex and $V = \text{vert } T^d$. Each face of T^d is obviously the convex hull of some subset of V, and—being the convex hull of an affinely independent set—is itself a simplex of appropriate dimension. Since any d-pointed V' subset of V spans a supporting hyperplane of T^d, conv V' is a $(d - 1)$-simplex which is a face of T^d. Using theorem 3.1.5, or a direct argument, there follows:

2. *All the k-faces, $0 \leq k \leq d - 1$, of the d-simplex T^d are k-simplices, and any $k + 1$ vertices of T^d determine a k-face of T^d. The number of k-faces of T^d is therefore given by $f_k(T^d) = \binom{d + 1}{k + 1}$ for all k.*

The d-simplex $T^d \subset R^d$ is clearly the intersection of the $d + 1$ closed halfspaces determined by the $d + 1$ $(d - 1)$-faces of T^d and containing T^d. Thus the polytope dual to T^d is again a d-simplex, and the combinatorial type of the d-simplex is self-dual. Evidently this implies that each $(d - k)$-face of T^d is the intersection of the k $(d - 1)$-faces of T^d containing it—a fact which can easily be proved also by a direct argument.

A particular d-simplex, often very convenient from a computational point of view, is the convex hull of the $d + 1$ 'unit points' $(1, 0, \cdots, 0)$, $(0, 1, 0, \cdots, 0), \cdots, (0, \cdots, 0, 1)$ in R^{d+1}.

4.2 Pyramids

The d-simplex T^d may obviously be considered as the convex hull of the union of a $(d - 1)$-simplex T^{d-1} and a point $A \notin$ aff T^{d-1}. In analogy to the well-known solids in 3-space, this construction may be generalized as follows.

A *d-pyramid* P^d is the convex hull of the union of a $(d - 1)$-polytope K^{d-1} (*basis* of P^d) and a point A (*apex* of P^d), where A does not belong to aff K^{d-1}.

Let F^k be a k-face of P^d determined by the hyperplane H, $F^k = P^d \cap H$. Then there are two possibilities: either (i) $A \notin$ vert F^k, or (ii) $A \in$ vert F^k. In case (i), theorem 2.4.1 implies that F^k is a k-face of K^{d-1}. In case (ii), the vertices of F^k different from A are in K^{d-1}, and are exactly the vertices of the $(k - 1)$-face $H \cap P^d \cap$ aff K^{d-1} of K^{d-1}. Hence F^k is a k-pyramid with apex A and basis $H \cap K^{d-1}$.

On the other hand, theorem 3.1.5 implies that, for $0 \le k \le d - 1$, each k-face of K^{d-1} (including the improper face K^{d-1} of K^{d-1}) is a face of P^d; also, the convex hull of the union of any proper face of K^{d-1} and A is a proper face of P^d. Therefore we have

1. *If P^d is a d-pyramid with $(d - 1)$-dimensional basis of K^{d-1} then*

$$f_0(P^d) = f_0(K^{d-1}) + 1$$

$$f_k(P^d) = f_k(K^{d-1}) + f_{k-1}(K^{d-1}) \qquad \text{for} \quad 1 \le k \le d - 2$$

$$f_{d-1}(P^d) = 1 + f_{d-2}(K^{d-1}).$$

Using the extended notation $f_{-1}(P^d) = f_d(P^d) = 1$, $f_k(P^d) = 0$ for $k < -1$ and $k > d$, the above relations can be formulated as

$$f_k(P^d) = f_k(K^{d-1}) + f_{k-1}(K^{d-1}) \qquad \text{for all } k.$$

The above reasoning proves also that, as far as the combinatorial type is concerned, we may speak about *the* pyramid with a given basis. A similar remark applies to most classes of polytopes mentioned in the present chapter.

If P^d is a d-pyramid with basis P^{d-1}, where P^{d-1} is a $(d-1)$-pyramid with $(d-2)$-dimensional basis K^{d-2}, we shall say that P^d is a *two-fold d-pyramid* with basis K^{d-2}. In general, for a positive integer r we shall say that P^d is an *r-fold d-pyramid* with $((d-r)$-dimensional) basis K^{d-r} provided P^d is a d-pyramid with basis P^{d-1}, where P^{d-1} is an $(r-1)$-fold $(d-1)$-pyramid with basis K^{d-r}. A d-pyramid as defined earlier is a 1-fold d-pyramid. Any d-polytope is a 0-fold d-pyramid.

It is easily seen by induction that

2. *If P^d is an r-fold d-pyramid with basis K^{d-r} then*

$$f_k(P^d) = \sum_i \binom{r}{i} f_{k-i}(K^{d-r}) \qquad \text{for all } k.$$

Note that a $(d-1)$-fold d-pyramid has as basis a segment—which is itself a 1-fold 1-pyramid; thus every $(d-1)$-fold d-pyramid is also a d-fold d-pyramid, or in other words, it is a d-simplex.

4.3 Bipyramids

Let K^{d-1} be a $(d-1)$-polytope and let I be a segment such that $I \cap K^{d-1}$ is a single point belonging to relint $I \cap$ relint K^{d-1}. Then $B^d =$ conv$(K^{d-1} \cup I)$ is called a *d-bipyramid* with basis K^{d-1}. By a reasoning analogous to that used in section 4.2, the numbers $f_k(B^d)$ are easily determined. We have

1. *If B^d is a d-bipyramid with basis K^{d-1} then*

$$f_k(B^d) = 2f_{k-1}(K^{d-1}) + f_k(K^{d-1}) \qquad \text{for all } k \le d-2$$

and

$$f_{d-1}(B^d) = 2f_{d-2}(K^{d-1}).$$

For a positive integer r we may define *r-fold d-bipyramids* in analogy to r-fold d-pyramids. We shall not dwell here on the details of the general case, but a few words seem to be called for in the extreme case of $(d-1)$-fold d-bipyramids which, again in analogy to pyramids, are necessarily also d-fold d-bipyramids. The easiest way to study them is by considering the simplest representative of the type—the d-dimensional *crosspolytope* or *d-octahedron* Q^d. The d-crosspolytope Q^d may be defined as the convex hull of d segments $[v_i; w_i]$, $1 \leq i \leq d$, mutually orthogonal and having coinciding midpoints. As is easily checked, if $1 \leq k \leq d$, for each k different indices $i_1, \cdots, i_k$, and points $z_j \in \{v_{i_j}, w_{i_j}\}$, the points $\{z_j \mid 1 \leq j \leq k\}$ are the vertices of a $(k-1)$-face conv$\{z_j\}$ of Q^d, and each proper face of Q^d may be obtained in this way. From this, or from the observation that Q^d is a d-bipyramid with basis Q^{d-1}, we find

2. *For $d \geq 1$ and $-1 \leq k < d$,*

$$f_k(Q^d) = 2^{k+1} \binom{d}{k+1}.$$

4.4. Prisms

Let K^{d-1} be a $(d-1)$-polytope and let $I = [0, x]$ be a segment not parallel to aff K^{d-1}. Then the vector-sum $P^d = K^{d-1} + I$ is a d-polytope, the *d-prism* with basis K^{d-1}. Clearly P^d is also definable as the convex hull of K^{d-1} and its translate $x + K^{d-1}$. A k-face of P^d is either a k-face of K^{d-1} or of $x + K^{d-1}$, or it is the vector-sum of I with some $(k-1)$-face of K^{d-1}. Also, each face of K^{d-1} and of $x + K^{d-1}$ (including the improper ones) is a face of P^d, and the vector-sum of I with any face of K^{d-1} is a face of P^d. Therefore we have

1. *If P^d is a d-prism with basis K^{d-1} then*

$$f_0(P^d) = 2f_0(K^{d-1})$$

and

$$f_k(P^d) = 2f_k(K^{d-1}) + f_{k-1}(K^{d-1}) \qquad \text{for} \quad k > 0.$$

Agreeing that 1-fold d-prism means the same as d-prism, we shall say that a d-polytope P^d is an *r-fold d-prism* with basis K^{d-r} provided P^d is a prism with basis P^{d-1}, where P^{d-1} is an $(r-1)$-fold $(d-1)$-prism with basis K^{d-r}.

The $(d-1)$-fold d-prisms coincide with the d-fold d-prisms; they are the parallelotopes, a *d-parallelotope* being the vector-sum of d segments with a common point, such that none is parallel to (i.e. contained in) the affine hull of all the others. The simplest d-parallelotope is the *d-cube* C^d (also called the *measure polytope* (Coxeter [1])), which is the vector-sum of d mutually orthogonal segments of equal length. In a suitable Cartesian system of coordinates C^d is the set of all points $x = (x_1, \cdots, x_d)$ for which $0 \leq x_i \leq 1$ for $1 \leq i \leq d$.

Using theorem 1 it is easily seen by induction (or directly from the definition) that

2. $f_k(C^d) = 2^{d-k}\binom{d}{k}$ for $0 \le k \le d$.

Prismoids (Sommerville [2]) are a family of polytopes the definition of which generalizes that of the prisms. If P_1 and P_2 are polytopes contained in parallel, distinct $(d-1)$-hyperplanes, and such that $P^d = \text{conv}(P_1 \cup P_2)$ is d-dimensional, then P^d is called a *d-prismoid* with bases P_1 and P_2. In certain contexts (see, e.g., chapter 8) the d-prismoids are convenient building-blocks for the construction of all d-polytopes. Note that the number of faces of a d-prismoid is not determined by the numbers $f_i(P_1)$ and $f_i(P_2)$, and not even by the combinatorial type of P_1 and P_2, but depends on the polytopes P_1 and P_2 themselves, and on their mutual position.

4.5 Simplicial and Simple Polytopes

In general, $d + 1$ points in R^d are affinely independent; similarly, a finite subset of R^d will 'in general' be in 'general position', that is, no $d + 1$ of its points will belong to the same hyperplane.* In particular, considering a d-polytope P as the convex hull of its vertices, 'in general' no $d + 1$ vertices of P will belong to the same facet of the polytope. Thus all the facets of P will 'in general' be $(d-1)$-simplices. Any polytope not satisfying this condition is singular in the sense that it exhibits the 'unusual' incidence of more than d of its vertices in the same supporting hyperplane of P.

Thus we are naturally led to the 'general' family $\mathscr{P}_s^d$ of *simplicial d-polytopes*. A d-polytope P is called simplicial provided all its facets are $(d-1)$-simplices.† As examples of simplicial polytopes we mention: the d-simplex, d-bipyramids having as basis any simplicial $(d-1)$-polytope,

---

* The aim of the present section being only to make the family of simplicial polytopes appear a natural object of study, we do not wish to discuss the precise meaning of 'general position'. Such a meaning may easily be derived by considering the $(d + 1)$-tuples of points of R^d as elements of a $d(d + 1)$-dimensional space, and using an appropriate category classification, or a measure on this space, in order to proclaim nowhere dense sets, or sets of measure 0, as 'special' and their complements as general.

† It should be noted that in this class of polytopes the vertices do not necessarily form a 'general' set of points. More than d vertices of such a polytope P may belong to the same hyperplane provided the hyperplane does not support P. Thus, the regular 3-octahedron Q^3 is a 'general' polytope though it has quadruples of coplanar vertices.

the d-octahedron. Additional examples shall be provided by the cyclic d-polytopes (section 4.7).

In every d-polytope each $(d-2)$-face is incident with two facets (theorem 3.1.6); in a simplicial d-polytope each facet, being a $(d-1)$-simplex, is incident with $d(d-2)$-faces. Thus, for each $P \in \mathscr{P}_s^d$, we have $df_{d-1} = 2f_{d-2}$. In chapter 9 we shall see that the numbers $f_k(P)$, for $P \in \mathscr{P}_s^d$, satisfy additional linear relations.

The family $\mathscr{P}_s^d$ of simplicial polytopes is not only 'natural' but it turns out to be rather important. Not only is the family of simplicial polytopes from certain points of view more tractable than the family $\mathscr{P}^d$ of all d-polytopes, but a number of results bearing on all polytopes are at present obtainable only via simplicial polytopes (see, for example, chapter 10).

It is to be noted, however, that the identification of 'general' polytopes with the simplicial polytopes is quite arbitrary in at least one sense: According to section 3.1 a polytope—that is, the convex hull of a finite set of points—may as well be defined as a bounded intersection of finitely many closed halfspaces. The 'general' position of $d+1$ hyperplanes is *not to be incident* with one point. Therefore, from this point of view, the 'general' polytope has at most, and thus exactly, d facets incident with each of its vertices. In other words, 'general' from this point of view are polytopes which are usually called 'simple.'

The two points of view, and the classes of simplicial and simple polytopes, are obviously dual to each other and there is no intrinsic advantage of one of them over the other. For different reasons (none very important) we shall in chapters 8, 9, and 10 prefer to deal with simplicial polytopes. Naturally, each of the results may be dualized to the corresponding statement about simple polytopes. By reason of easier imagination some problems about 3-polytopes are usually treated in the setting of simple polytopes.

The notion of simple or simplicial polytopes may be generalized as follows. Let k and h be integers such that $1 \le k, h \le d-1$. A d-polytope P shall be called k-*simplicial* provided each k-face of P is a simplex; P shall be called h-*simple* provided each $(d-1-h)$-face of P is contained in $h+1$ facets of P. We shall denote by $\mathscr{P}^d(k, h)$ the family of all d-polytopes which are k-simplicial and h-simple; such polytopes are also said to be of type (k, h). Clearly $\mathscr{P}^d(1, 1) = \mathscr{P}^d$, and the dual of a d-polytope of type (k, h) is of type (h, k). The simplicial d-polytopes are obviously of the type $(d-1, 1)$, the simple ones of type $(1, d-1)$. The d-polytopes of type

$(d - 2, 1)$ are called *quasi-simplicial*, and their totality is denoted by $\mathcal{P}^d_q = \mathcal{P}^d(d - 2, 1)$. It is immediate that every 3-polytope is quasi-simplicial, and so is every d-pyramid having as basis a simplicial $(d - 1)$-polytope. We shall consider quasi-simplicial polytopes in section 9.3.

A complex $\mathscr{C}$ is called simplicial provided all its members are simplices. Simplicial complexes are in many respects easier to manage than complexes in general, in particular from the point of view of algebraic topology.

The definitions of simplicial complexes and of simplicial polytopes are concordant in the sense that a polytope P is simplicial if and only if its boundary complex $\mathscr{B}(P)$ is simplicial.

4.6 Cubical Polytopes

In section 9.3 we shall discuss the interesting family $\mathcal{P}^d_c$ of *cubical d-polytopes*. A d-polytope P is called cubical provided each of its $(d - 1)$-faces is combinatorially equivalent to the $(d - 1)$-cube C^{d-1}.

Here we shall describe some special cubical d-polytopes, the *cuboids* C^d_k, where $0 \leq k \leq d$. The cuboid C^d_k may be imagined as obtained by 'pasting together' 2^k d-cubes in the following fashion:

C^d_0 is the d-cube C^d;

C^d_1 is the union of two d-cubes which have a common $(d - 1)$-face;

C^d_2 is the union of two C^d_1's pasted together along a C^{d-1}_1 common to both (this, naturally, requires that the C^d_1's be deformed beforehand); and so forth.

In figure 4.6.1 the four 3-cuboids are represented.

Alternatively, C^d_k may be described in terms of its boundary complex as follows:

$\mathscr{B}(C^d_k)$ is isomorphic to the complex which is obtained from the boundary complex of the cube $C^d = \{(x_1, \cdots, x_d) \in R^d \mid |x_i| \leq 1, 1 \leq i \leq d\}$ by subdividing all its cells along the k coordinate hyperplanes $\{(x_1, \cdots, x_d) \in R^d \mid x_j = 0\}$ for $j = 1, \cdots, k$. Naturally, it is possible to give explicit formulae for the coordinates of the vertices of C^d_k (see exercise 4.8.20), but this approach is not very helpful for obtaining an intuitive picture of the polytope.

Since for $0 \leq i \leq d - 1$ an i-face of a d-cube is incident to $d - i$ facets of the cube, each i-face of the d-cubes forming C^d_k according to the above description will be an i-face of C^d_k, provided $i + k \leq d - 1$. Therefore, for $i + k \leq d - 1$ we have the recursion relation

$$f_i(C^d_k) = 2f_i(C^d_{k-1}) - f_i(C^{d-1}_{k-1}).$$

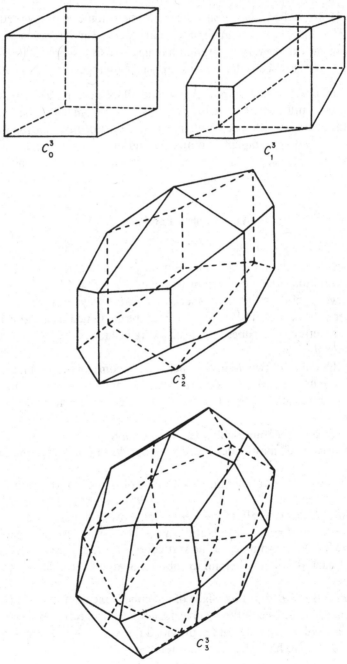

C_0^3

C_1^3

C_2^3

C_3^3

Figure 4.6.1

Using this relation and the formulae $f_i(C^d) = f_i(C_0^d) = 2^{d-1}\binom{d}{i}$ from section 4.4, it is easily verified by induction on k that:

1. *For $i, k \geq 0$ and $i + k \leq d - 1$*

$$f_i(C_k^d) = \sum_{j=0}^{k} (-1)^j \binom{k}{j} \binom{d-j}{i} 2^{d+k-i-2j}.$$

This expression is, for $i = 0$, easily evaluated in closed form and yields $f_0(C_k^d) = 2^{d-k}3^k$ for $0 \leq k \leq d - 1$, a result which may as well be obtained directly from the definition of C_k^d. For $k = d$ there results $f_0(C_d^d) = 3^d - 1$.

4.7 Cyclic Polytopes

Though only a relatively recent discovery (see the historical comments in section 7.4), the *cyclic polytopes* $C(v, d)$ play a very important role in many questions of the combinatorial theory of polytopes. Following Gale [4] and Klee [9] they are defined by the following simple procedure:

In R^d consider the *moment curve* M_d defined parametrically by $x(t) = (t, t^2, \cdots, t^d)$. A cyclic d-polytope $C(v, d)$ is the convex hull of $v \geq d + 1$ points $x(t_i)$ on M_d, with $t_1 < t_2 < \cdots < t_v$.

Let $V = \{x(t_i) \mid 1 \leq i \leq v\}$, where $t_1 < t_2 < \cdots < t_v$, be the v-pointed subset of M_d used in the definition of $C(v, d)$, and let k be a positive integer such that $2k \leq d$. We shall now show that any k-pointed subset $V_k = \{x(t_i^*) \mid i = 1, \cdots, k\}$ of V determines a face of $C(v, d)$. In order to find the equation of a supporting hyperplane H of $C(v, d)$ such that $H \cap C(v, d) = \operatorname{conv} V_k$ we consider the polynomial

$$p(t) = \prod_{i=1}^{k} (t - t_i^*)^2 = \beta_0 + \beta_1 t + \cdots + \beta_{2k} t^{2k},$$

where the coefficients β_j depend only on the t_i^*'s. Let

$$b = (\beta_1, \beta_2, \cdots, \beta_{2k}, 0, \cdots, 0)$$

and

$$H = \{x \in R^d \mid \langle x, b \rangle = -\beta_0\}.$$

Then clearly $x(t_i^*) \in H$ for $1 \leq i \leq k$, while for any $x(t) \in M_d \sim V_k$ we have

$$\langle x(t), b \rangle = -\beta_0 + \prod_{i=1}^{k} (t - t_i^*)^2 > -\beta_0.$$

Thus H is a supporting hyperplane of $C(v, d)$ and $H \cap V = V_k$, as claimed. It follows easily that

1. *If k is a positive integer such that $2k \leq d$, every k vertices of $C(v, d)$ determine a $(k - 1)$-face of $C(v, d)$; therefore*

$$f_i(C(v, d)) = \binom{v}{i+1} \qquad \text{for} \quad 0 \leq i \leq [\tfrac{1}{2}d].$$

We shall show next that $C(v, d)$ is a simplicial d-polytope. An easy way to do this is by showing that every $(d + 1)$-tuple of points in M_d is affinely independent. It follows that each proper face of $C(v, d)$ has at most d vertices, and therefore $C(v, d) \in \mathscr{P}_s^d$.

The affine independence of $d + 1$ points $x(t_i) \in M_d, t_0 < t_1 < \cdots < t_d$, is equivalent with the non-vanishing of the determinant

$$\Delta = \begin{vmatrix} 1 & t_0 & t_0^2 & \cdots & t_0^d \\ 1 & t_1 & t_1^2 & \cdots & t_1^d \\ & & \cdots\cdots\cdots & & \\ 1 & t_d & t_d^2 & \cdots & t_d^d \end{vmatrix}.$$

But, as is well known, $\Delta = \prod_{0 \leq i < j \leq d} (t_j - t_i) > 0$. This completes the proof of our assertion.

Let $V_d \{x(t_i^*) \mid 1 \leq i \leq d\}$, where $t_1^* < t_2^* < \cdots t_d^*$, be a subset of V. We consider the polynomial

$$p^*(t) = \prod_{i=1}^{d} (t - t_i^*) = \sum_{j=0}^{d} \gamma_j t^j$$

and define $c = (\gamma_1, \cdots, \gamma_d)$. Let H^* be the hyperplane

$$H^* = \{x \in R^d \mid \langle x, c \rangle = -\gamma_0\}; \qquad \text{then} \quad V_d \subset H^*.$$

The function $\langle x(t), c \rangle + \gamma_0$, defined for $x(t) \in M_d$, is clearly different from 0 for each $x(t) \notin V_d$ and it changes sign whenever the variable t increases and passes through one of the values t_i^*. Therefore, since every d-pointed subset of M_d is affinely independent, we have

2. *A d-tuple V_d of points of $V \subset M_d$ determines a facet* conv V_d *of* conv $V = C(v, d)$ *if and only if every two points of $V \sim V_d$ are separated on M_d by an even number of points of V_d.*

We shall call the criterion of theorem 2 'Gale's evenness condition'. It obviously makes the determination of $f_{d-1}(C(v, d))$ quite easy. We

state here only the final result, delaying the proof to Chapter 9 where
the expression for $f_k(C(v,d))$ will be found for all k.

3.
$$f_{d-1}(C(v,d)) = \binom{v - \left[\frac{d+1}{2}\right]}{v-d} + \binom{v - \left[\frac{d+2}{2}\right]}{v-d}.$$

This may be reformulated as

$$f_{d-1}(C(v,d)) = \begin{cases} \dfrac{v}{v-n}\binom{v-n}{n} & \text{for even } d = 2n, \\[2ex] 2\binom{v-n-1}{n} & \text{for odd } d = 2n+1. \end{cases}$$

Another consequence of Gale's evenness condition is that each two
cyclic polytopes $C(v,d)$ are combinatorially equivalent, the correspondence between vertices being given by their order on M_d. Thus we may
speak about the combinatorial type $C(v,d)$.

An analysis of the proofs of the present section shows that the construction of cyclic polytopes use only very few of the properties of the
moment curve M_d. It is therefore not surprising that it is possible to
develop the theory of cyclic polytopes using other curves in their
definition; the curves have many other interesting properties. For
additional results and references see, for example, Derry [1], Motzkin [4],
Fabricius–Bjerre [1], and Cairns [4] (the curves used being distinguished
by their geometric properties), by Carathéodory [1, 2] and Gale [4] (using
the curve $(\cos t, \sin t, \cos 2t, \sin 2t, \cdots, \cos nt, \sin nt)$), and by Šaškin [1]
(using analytic conditions). For some related results see Karlin–Shapley
[1]. Some of the above authors have a number of other papers on moment
curves and their relatives; we do not list them since their connection with
polytopes is rather tenuous.

4.8 Exercises

1. The definition of pyramids may be generalized in the following
fashion: Let P^s and P^t be two polytopes (of dimensions s and t, respectively) in R^d, $d = s + t + 1$, such that aff $P^s \cap$ aff $P^t = \emptyset$. Let $P^d =$
conv$(P^s \cup P^t)$; if dim $P^d = d$ then P^d is a d-pyramidoid (Sommerville [2],
p. 115) with bases P^s and P^t. Show that in this case

$$f_k(P^d) = \sum_i f_i(P^s)f_{k-i-1}(P^t).$$

Show that each r-fold d-pyramid is a d-pyramidoid and that the above formula for $f_k(P^d)$ reduces to theorem 4.2.2 if P^d is an r-fold d-pyramid.

2. Determine $f_k(B^d)$ if B^d is a 2-fold d-bipyramid with $(d-2)$-dimensional basis K^{d-2}; generalize.

3. Show that for $1 \leq r \leq d-2$ there exist d-polytopes combinatorially equivalent to an r-fold d-bipyramid, which are not r-fold d-bipyramids themselves.

4. The construction of bipyramids may be generalized as follows: Let P^s and P^t be polytopes of dimensions s respectively $t, d = s + t$, with $P^s \cap P^t$ a single point belonging to relint $P^s \cap$ relint P^t. Let $P^d = \text{conv}(P^s \cup P^t)$; clearly, if P^t is a segment then P^d is a bipyramid with basis P^s. Determine the value of $f_k(P^d)$ in terms of $f_i(P^s)$ and $f_j(P^t)$.

5. As a special case of the construction mentioned in exercise 4, let T_r^d denote the convex hull of the union of an r-simplex T^r and a $(d-r)$-simplex T^{d-r}, where $T^r \cap T^{d-r}$ is a single point belonging to relint $T^r \cap$ relint T^{d-r}, and $0 \leq 2r \leq d$. Show that

(i) $$f_k(T_r^d) = \sum_{0 \leq i \leq k+1} \binom{r+1}{i}\binom{d+1-r}{k+1-i}$$

$$= \binom{d+2}{k+1} \quad \text{for} \quad 0 \leq k < r.$$

(ii) $$f_r(T_r^d) = \binom{d+2}{r+1} - 1 \quad \text{for} \quad 0 \leq r < d-r.$$

(iii) $$f_r(T_r^{2r}) = \binom{2r+2}{r+1} - 2.$$

Determine $f_k(T_r^d)$ for $r < k < d$.

6. Let P be the 4-polytope defined as the convex hull of the points $(-1, -1, -2, 0), (-1, -1, 2, 0), (-1, 0, -1, 1), (-1, 0, 1, 1), (-1, 1, -2, 0),$ $(-1, 1, 2, 0), (1, -2, -2, 0), (1, -2, 2, 0), (1, 0, 0, 2), (1, 2, -2, 0), (1, 2, 2, 0).$ Show that the vertices of P may be assigned symbols A, B, C, D, E, F, G, H, I, J, K, in such a way that its 3-faces have the following sets of vertices: $\{A, B, C, D, E, F, G, H\}$, $\{A, B, C, D, I, K\}$, $\{A, B, E, F, I, J, K\}$, $\{A, D, E, H, I, J\}$, $\{B, C, F, G, K\}$, $\{C, D, G, H, I, J, K\}$, $\{E, F, G, H, J, K\}$. Construct a Schlegel diagram of P.

7. Theorems 4.3.1 and 4.4.1 indicate some connection between bipyramids and prisms. Elaborate this relationship.

8. Show that the d-octahedron Q^d is a d-prismoid.

9. Show that the family of all d-prismoids is projectively equivalent to the family of all 'generalized d-prismoids'. A d-polytope $P^d \subset R^d$ is a generalized d-prismoid provided $P = \text{conv}(P_1 \cup P_2)$, where P_1 and P_2 are polytopes contained respectively in hyperplanes H_1 and H_2 such that $H_1 \cap H_2 \cap (P_1 \cup P_2) = \emptyset$.

10. Explain the relationship between d-pyramidoids and d-prismoids.

11. Show that if P is a d-polytope of type (k, h) with $k + h \geq d + 1$, then P is the d-simplex.

12. Show that an i-face of a simple d-polytope P is contained in $\begin{pmatrix} d - i \\ d - j \end{pmatrix}$ j-faces of P whenever $-1 \leq i \leq j \leq d - 1$.

13. Consider the 3-polytopes Schlegel diagrams of which are given in Figure 3.4.6. Determine which of them represent pyramids, k-fold pyramids, bipyramids, k-fold bipyramids, prisms, k-fold prisms, prismoids, pyramidoids; consider the possibility of a polytope belonging to more than one of the classes, and of belonging to the same class in more than one way. Determine which types are dual to each other or to themselves. Find the duals of those for which a dual is not given.

14. Draw, or better still, construct cardboard models of, the different Schlegel diagrams of the 4-pyramid with basis a 3-prism based on a triangle, on a square, or on a pentagon; how many different Schlegel diagrams are there in each of the cases. Perform the same tasks for the cyclic polytopes $C(v, 4)$ where $v = 5, 6, 7$.

15. Determine the faces of the 4-polytope $P \subset R^4$, defined as the convex hull of the ten points $(0, 0, 0, 0)$, $(0, 1, 0, 1)$, $(0, 1, 1, 0)$, $(0, 0, 1, 1)$, $(\pm 1, 1, 0, 0)$, $(\pm 1, 0, 1, 0)$, $(\pm 1, 0, 0, 1)$. Show that all facets of the polytope dual to P are combinatorially equivalent; determine the type (k, h) of P.

16. Let K_k^d, where $1 \leq k \leq d$ and $d \geq 3$, denote the polytope

$$K_k^d = \{(x_1, \cdots, x_{d+1}) \in R^{d+1} \mid 0 \leq x_i \leq 1, i = 1, \cdots, d + 1, \sum_{i=1}^{d+1} x_i = k\}.$$

Prove the following assertions (compare Coxeter [1]):

(i) $K_k^d \in \mathscr{P}^d(2, d - 2)$.

(ii) K_k^d is a translate of K_{d+1-k}^d.

(iii) K_k^d is the convex hull of the centroids of $(k - 1)$-faces of a d-simplex T^d.

(iv) If the centroid of T^d is at the origin 0, then K_k^d is combinatorially equivalent to $kT^d \cap (k - d - 1)T^d$.

(v) K_k^d has $2(d + 1)$ facets, $d + 1$ combinatorially equivalent to K_k^{d-1}, the others to K_{k-1}^{d-1}.

17. Let

$$M^d = \{(x_1, \cdots, x_d) \in R^d \mid \sum_{i=1}^{d} |x_1| \leq d - 2, |x_1| \leq 1, i = 1, \cdots, d\}.$$

Prove:

(i) $M^d \in \mathscr{P}^d(2, d - 2)$.

(ii) M^d is the convex hull of the centroids of the 2-faces of a d-cube.

18. Let $N^d = \{(x_1, \cdots, x_d) \in R^d \mid \sum_{i=1}^{d} \varepsilon_1 x_1 \leq d - 2$ for all $\varepsilon_1, \cdots, \varepsilon_d$ such that $\varepsilon_i = \pm 1 (i = 1, \cdots, d)$, and the number of ε_i equal to $+1$ is odd$\}$. Prove that $N^d \in \mathscr{P}^d(3, d - 3)$.

19. A d-antiprism P is a d-prismoid in which the two bases P_1 and P_2 are $(d - 1)$-polytopes dual to each other, having such shapes and position that the facets of P are precisely those obtained as $\text{conv}(F_1 \cup F_2)$, where F_i is a face of P_i and F_1 is dual to F_2. Construct a 3-antiprism with basis P_1 a triangle, a square, or a pentagon. Construct the Schlegel diagram of 4-antiprisms with basis a three- or four-sided 3-pyramid. Determine in each case the number of faces of different dimensions. It is remarkable that it is not known whether each combinatorial type of $(d - 1)$-polytopes contains members which can serve as bases of d-antiprisms; the problem is open even for $d = 4$.

20. Let T_k^d denote the piecewise projective transformation mapping $x = (x_1, \cdots, x_d)$ into

$$T_k^d(x) = \frac{x}{1 + \sum_{i=1}^{k} |x_i|}.$$

Show that for each k satisfying $0 \leq k \leq d$, the image $T_k^d(C^d)$ of the 'unit cube' $C^d = \{x \in R^d \mid |x_i| \leq 1, 1 \leq i \leq d\}$ is the d-cuboid C_k^d.

21. Let $0 \leq k \leq d$ and $0 \leq i \leq d - 1$. Prove

$$f_i(C_k^d) = \sum_{j=0}^{d-i} \binom{d-k}{j} \binom{k}{d-i-j} 3^{d-i-j} 2^{k-d+i+2j} - \binom{k}{d-i} 2^{k-d+i}.$$

22. Using Gale's evenness condition show that each edge of the cyclic polytope $C(v, 4)$, $v \geq 6$, is incident with either three, or four, or $v - 2$ 3-faces of $C(v, 4)$, and that edges of all these types occur. Similarly, show that each edge of $C(v, 4)$ is incident with either 3, or 4, or $v - 2$ 2-faces of $C(v, 4)$, and that edges of all these types occur.

23. Let $p(t) = (\cos t, \sin t, \cos 2t, \sin 2t, \cdots, \cos nt, \sin nt) \in R^{2n}$, and let $K = \text{conv}\{p(t_i) \mid 1 \le i \le v\}$, where $v \ge 2n + 1$ and

$$0 \le t_1 < t_2 < \cdots < t_v < 2\pi.$$

Prove that K is combinatorially equivalent to $\text{conv}\{x(t_i) \mid 1 \le i \le v\}$, where (as in section 4.7) $x(t) = (t, t^2, \cdots, t^{2n})$. (Hint: Use the identity

$$\begin{vmatrix} 1 & \cos t_0 & \sin t_0 & \cdots & \cos nt_0 & \sin nt_0 \\ \cdot & \cdot & & & \cdot & \cdot \\ \cdot & \cdot & & \cdots & \cdot & \cdot \\ 1 & \cos t_{2n} & \sin t_{2n} & \cdots & \cos nt_{2n} & \sin nt_{2n} \end{vmatrix}$$

$$= 4^{n^2} \prod_{0 \le i < j \le 2n} \sin \tfrac{1}{2}(t_j - t_i),$$

due to Scott [1].)

24. Let $C = C(2n + 2, 2n)$. Prove that the vertices of C may be divided into two groups, each containing $n + 1$ of the vertices, in such a way that $2n$ vertices determine a facet of C if and only if n of them belong to one group, and n to the other. Show that this criterion is equivalent to Gale's evenness condition.

25. Using cyclic polytopes and Schlegel diagrams give an elementary proof of the following well-known theorem:

Every n-dimensional simplicial complex is combinatorially equivalent to a complex in R^{2n+1}.

It will be shown in chapter 11 that this theorem remains valid even if the complex is not assumed to be simplicial.

26. Let $T = \text{conv}\{x_i \mid 0 \le i \le 2n\}$ be a $2n$-simplex in R^{2n} such that $0 \in \text{int } T$. Let $\mathscr{C}$ denote the family consisting of all sets $C(I, J) = \text{conv } V(I, J)$, where I and J are subsets of $\{0, 1, \cdots, 2n\}$, $\text{card } I \le n$, $\text{card } J \le n$, $I \cap J = \varnothing$, and $V(I, J) = \{x_i \mid i \in I\} \cup \{-x_j \mid j \in J\}$. Show that:

(i) $\mathscr{C}$ is a $(2n - 1)$-complex;

(ii) $0 \notin \text{set } \mathscr{C}$;

(iii) the radial projection from 0 establishes a homeomorphism between set $\mathscr{C}$ and the $(2n - 1)$-dimensional unit sphere S^{2n-1}.

Formulate and prove the analogous result for R^{2n+1}.

27. Let P be a d-polytope in R^d which is not a d-simplex.

(i) Prove that there exists a nonsingular permissible projective image P' of P with the following properties:

(a) P' is the intersection of the $f = f_{d-1}(P')$ closed halfspaces $H_1, \cdots, H_f$;

(b) the intersection $\bigcap_{i=1}^{f-1} H_i$ is a bounded set.

(ii) Characterize the facets of P which may correspond to bd H_f.

(iii) Show that for a suitable P', already the intersection of some $d + 1$ of the halfspaces H_i is bounded.

(iv) Show that if P has more than $2d$ facets, then it is possible to choose $P' = P$.

28. Let P be the convex hull of d segments having a common point relatively interior to each of them, and such that none of the segments is contained in the affine hull of the union of the other segments. Show that P is projectively equivalent to Q^d.

29. Let $V_1, \cdots, V_d$ be linearly independent points in R^d, let

$$T^d = \mathrm{conv}\{0, V_1, \cdots, V_d\},$$

and let K be a polytope $K = \mathrm{conv}\{0, V_1, \cdots, V_d, W_1, \cdots, W_n\}$, where $W_i \in \mathrm{int\,cone}_0 T^d$ for all $i = 1, \cdots, n$. Prove: For every $\lambda > 1$ there exists a projective transformation P_λ permissible for K, such that $P_\lambda(0) = 0$, $P(V_j) = V_j$ for $j = 1, \cdots, d$, and $T^d \subset P_\lambda K \subset \lambda T^d$.

30. Let a polytope P (and its combinatorial type) be called *projectively unique* provided every polytope P' combinatorially equivalent to P is projectively equivalent to P. Show:

(i) The cartesian product $T^p \times T^r$ of two simplices is projectively unique.

(ii) The combinatorial type dual to a projectively unique combinatorial type is itself projectively unique.

(iii) The combinatorial types of all 3-polytopes with at most 9 edges are projectively unique.

(iv) No 3-polytope with 10 or more edges is projectively unique. (The only proof of this fact known to the author uses Steinitz's theorem 13.1.1; it would be of interest to find a more direct and elementary proof.)

No characterization of projectively unique d-polytopes, $d \geq 4$, is known. Shephard [12] has established the existence of a projectively unique polytope with 7 vertices, 17 edges, 18 2-faces and 8 facets; however, there exist 4-polytopes having the same numbers of faces of all dimensions which are not projectively unique (for example, the 4-polytope D^* of figure 10.4.1).

31. Let P be a d-polytope and let $V \in \mathrm{vert}\,P$. We shall say that P is *pyramidal* at V provided $\mathrm{conv}(\mathrm{vert}\,P \sim \{V\}) \in \mathscr{F}(K)$. Establish the following results:

(i) Let P be a polytope pyramidal at V, let $F = \mathrm{conv}(\mathrm{vert}\,P \sim \{V\})$, and let $W \in$ vert F. Then F is pyramidal at W if and only if P is pyramidal at W.

(ii) A d-polytope P is pyramidal at r different vertices if and only if P is an r-fold d-pyramid.

32. Prove that simplicial polytopes are 'stable' in the following sense: If $P \in \mathscr{P}_s^d$ there exists an $\varepsilon = \varepsilon(P)$ with the property: If P' is any polytope with $f_0(P) = f_0(P')$ such that for each vertex V of P there is a vertex V' of P' with $\rho(V, V') < \varepsilon$, then P' is combinatorially equivalent to P.

33. Let P^d be a self-dual d-polytope. Prove that the $(d + 1)$-pyramid with basis P^d is self-dual. Prove also the self-duality of the $(d + 1)$-polytope obtained as the union of a $(d + 1)$-pyramid over P with a $(d + 1)$-prism over P, the prism and the pyramid intersecting in P.

4.9   Additional notes and comments

Polytope theory is alive and well—this is demonstrated by its wealth of interesting examples and constructions. Here are notes on some more.

0/1-Polytopes.
$0/1$-Polytopes are convex hulls of subsets of $\{0,1\}^d$. These polytopes have
gained enormous importance in view of combinatorial optimization. In particular, one is interested in the description of (large classes of) facets of such
polytopes, for specific families of incidence vectors, for use in cutting plane
approaches to $0/1$-integer programming; this is the object of the field of polyhedral combinatorics, as surveyed succinctly in Schrijver [b].

Some specific subclasses of $0/1$-polytopes have received special attention.
We refer to a few prototypical expositions that also demonstrate the setting
and use of such special classes in optimization: to Cook et al. [a] for matching polytopes and variations, to Grötschel–Padberg [a] for travelling salesman
polytopes, and to Deza–Laurent [a] for cut polytopes.

Ziegler [b] provides a survey of general $0/1$-polytopes. A recent break-
through is by Bárány–Pór [a], who showed that the maximal number of facets
of a d-dimensional $0/1$-polytope grows super-exponentially in d: Certain random $0/1$-polytopes have more than $c^{d\log d}$ facets, for a constant $c > 1$.

Hypersimplices.
The polytopes K_k^d of exercise 4.8.16 appear in Coxeter [1, Ch. VIII, esp. p. 163],
who refers to Stott [2]. They are interesting $0/1$-polytopes of type $(2, d-2)$.
Due to prominent work by Gel'fand–Goresky–MacPherson–Serganova [a] and
Gel'fand–Kapranov–Zelevinsky [a] they also became known as *hypersimplices*.
See also De Loera–Sturmfels–Thomas [a].

k-Simplicial h-simple polytopes.
The polytopes of type (k, h) with $k, h \geq 2$ are very interesting, They are, however, hard to construct (see also the comments in section 5.6, and in section 9.9
on exercise 9.7.7), and their properties, such as their f-vectors, are not well
understood. The first interesting case is that of $h = k = 2$, where the 4-simplex,
the hypersimplex K_2^4 and its dual, and the regular 24-cell (the polytope M^4
of exercise 4.8.17) provide examples. Recently, Eppstein–Kuperberg–Ziegler
[a] provided a construction (using hyperbolic geometry) of infinitely many 4-
polytopes of type $(2,2)$; see also the notes in section 5.6. In contrast, it is not
clear whether d-polytopes of type $(4,4)$ other than simplices exist for arbitrarily large d—compare this to exercise 4.8.18.

Cubical polytopes.
Interesting examples of cubical polytopes arose from several different directions. Cubical zonotopes correspond to generic linear hyperplane arrangements; see the notes in section 18.4. The combinatorial types of cubical d-polytopes with at most 2^{d+1} vertices were completely classified by Blind–Blind [b]. Joswig–Ziegler [a] constructed *neighborly cubical polytopes*: cubical d-polytopes with the $(\lceil \frac{d}{2} \rceil - 1)$-skeleton of an n-cube for all $n > d \geq 2$ (see the notes in section 12.4).

Free sums of polytopes.
The generalized bipyramids $P^d = \text{conv}(P^s \cup P^t)$ of exercise 4.8.4 have later been called *free sums* of polytopes (Henk et al. [a], Kalai [e]), and denoted $P^d = P^s \oplus P^t$. The construction is dual to forming products.

Cyclic polytopes.
For even d, $C(d+2,d)$ is a generalized bipyramid; indeed, with exercise 4.8.24, $C(2n+2,2n)$ is equivalent to the generalized bipyramid T_n^{2n} of exercise 4.8.5. For odd d, the cyclic polytope $C(v,d)$ is combinatorially equivalent to a different type of modified bipyramid $\text{conv}(I \cup C(v-1,d-1))$, where the interval I meets the affine hull of the cyclic polytope $C(v-1,d-1)$ in a single point that is in the relative interior of I but is a vertex of $C(v-1,d-1)$. This modified bipyramid construction is dual to forming "wedges" over facets; for this construction we refer to Klee–Walkup [a] and to Holt–Klee [a].

For the trigonometric moment curve of exercise 4.8.23 see Ziegler [a, Ex. 2.21(ii)]. The trigonometric coordinates provide realizations of the cyclic polytopes of full symmetry, including a canonical "center".

Computing explicit examples.
From the data of exercise 4.8.6, `polymake` together with `javaview` "automatically" yield rotating color 3D output, which in b/w print looks as follows:

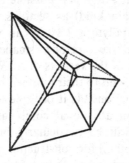

CHAPTER 5

Fundamental Properties and Constructions

Despite the simplicity of the notion of a polytope, our understanding of what properties a polytope may, or may not, have is severely hampered by the difficulty of producing polytopes having certain desired features. One example of such a difficulty will help to focus the problem. Let us consider 4-polytopes with 8 vertices; with some patience, combining the different types of polytopes considered in Chapter 4, the reader will probably find 4-polytopes with 8 vertices and 16 edges, or with any number of edges between 18 and 28. Now he may ask himself whether it is possible to complete the list by a polytope with 17 edges. But how is one to start looking for such a polytope?

The answer to this particular query is not too hard to come by (see section 10.4), but there are many similar questions to which we still do not have answers. In particular, there is no easy and direct prescription as to what tack to take for the solution of a given problem. The best we can do is to investigate some more or less general properties of polytopes in the hope that one or another of them may be useful in solving specific problems.

This chapter is devoted to a presentation of some such properties, and of certain techniques for a 'planned' construction of polytopes. Thus the spirit of this chapter is quite different from that of chapter 3; there we mainly had one polytope and we were concerned about the features it exhibits. The principal questions to be discussed in the present chapter are: To what extent is it possible to obtain all polytopes (or all polytopes of a certain kind) as sections, or as projections, of some 'standard' types of polytopes? How can we relate the structure of a polytope P with the structure of a 'smaller' polytope, obtained as the convex hull of some of the vertices of P? What happens to the faces of different dimensions, and to their number, if we replace a given polytope by a 'sufficiently near' one? Is it possible to devise an algorithmic procedure for the determination of all combinatorial types of polytopes?

Two techniques will be particularly important in the sequel: the 'beneath-beyond' method for construction of new polytopes from given ones (section 5.2), and the Gale-transforms and Gale-diagrams which frequently allow a reformulation of a problem in more readily tractable terms (section 5.4). As will become obvious in the following chapters, many of the recently obtained results use one of these methods.

5.1 Representations of Polytopes as Sections or Projections

In the present section we shall discuss a number of results on representations of polytopes as sections, or projections, of other polytopes. (A *section* of a polytope P is the intersection of P with some flat; a *projection* of P is the transform TP of P under a (singular) affine map T.) The first result is very simple, and should convey the flavor of this type of results; the following group of results discusses to what extent variants and generalizations have been considered in the literature. The last part of the section deals with results obtained recently by M. A. Perles; they are important because of the method used, as well as because of their applicability to various other problems.

We shall prove only theorem 1: for proofs of the other results the reader may use the hints provided, or consult the original papers. We shall use the term *facet* to denote the $(d - 1)$-dimensional faces of a d-polytope. The first result is:

1. *Every d-polytope with $f \geq d + 1$ facets is a section of an $(f - 1)$-simplex.*

PROOF Any permissible transformation of a section of a polytope may be extended to a permissible projective transformation of the polytope (the reader should prove this, using exercise 2.2.7), and every permissible projective image of a simplex is itself a simplex. In order to prove theorem 1, it is therefore enough to establish that every d-polytope with $f \geq d + 1$ facets is a permissible projective image of a section of an $(f - 1)$-simplex.

Since a section of a section of a polytope is itself a section of the polytope, the last assertion shall be proved if we establish that every d-polytope P with $f > d + 1$ facets is a permissible projective image of a section of a $(d + 1)$-polytope P^+ with f facets.

In order to prove this we note that P is not a d-simplex and therefore, by exercise 4.8.27, there exists a permissible, regular projective image P_0 of P with the following property: If $H = \operatorname{aff} P_0$ is the d-flat spanned by P_0, and if $H_1, \cdots, H_f$ are closed halfspaces of H such that $P_0 = \bigcap_{1 \leq i \leq f} H_i$, then $P' = \bigcap_{1 \leq i \leq f-1} H_i$ is a d-polytope. Let A be a point outside H and let

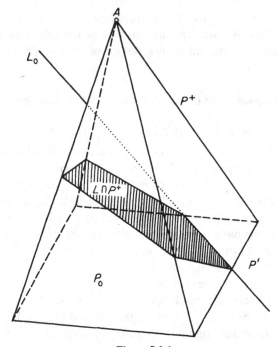

Figure 5.1.1

P^+ be the pyramid with apex A and basis P' (see figure 5.1.1). Then P^+ is a $(d + 1)$-polytope with f facets. Denoting by L_0 the boundary of H_f in H, let L be the d-flat determined by L_0 and some relatively interior point of $\operatorname{conv}(\{A\} \cup P_0)$. Then clearly $P^+ \cap L$ is a projection from A of P_0. This completes the proof of theorem 1.

The following result is dual to theorem 1; the reader is invited to find a direct proof for it.

2. *Every d-polytope with $v \geq d + 1$ vertices is a (parallel) projection of a $(v - 1)$-simplex.*

Using prisms or bipyramids, respectively, instead of the pyramids used in the proof of theorem 1, the following (mutually dual) theorems may be proved:

3. *Every centrally symmetric d-polytope with $2f \geq 2d$ facets is a (central) section of the f-cube.*

4. *Every centrally symmetric d-polytope with $2v \geq 2d$ vertices is a (parallel) projection of the v-dimensional octahedron.*

The above results have probably been known for a long time. The first written appearance of theorem 2 seems to be in Motzkin [1]. Theorem 1 appears in Davis [1, 2], theorem 3 in Klee [3]. The qualitative aspects of theorems 1 to 4 were found also by Naumann [1]. Naumann's proof is much more involved, but his results are also somewhat stronger; for example, he establishes

3*. *Every d-polytope with f facets is obtainable as a section of a* $(2^d \cdot (d + 1) \cdot f)$*-dimensional (regular) cube.*

Results of this type have been used in the determination of the 'projection constants' of certain Minkowski spaces (see Grünbaum [4]).

Another related problem, which also arose from questions in the geometry of Banach spaces, is due to J. Lindenstrauss. Its solution was found by Klee [6] who proved the following results:

5. *If P is a centrally symmetric d-polytope, $d \geq 2$, there is a 2-dimensional section of P, and an orthogonal projection of P into a 2-dimensional plane, each of which has at least 2d vertices.*

5*. *If P is a d-polytope, $d \geq 2$, there is an orthogonal projection of P into a 2-dimensional plane, and, through each point of int P, a 2-dimensional section of P, each of which has at least $d + 1$ vertices.*

It is easily seen that for $d = 3$, if P in theorem 5 is neither a cube nor an octahedron, a section and a projection exist which have at least 8 vertices. It would be interesting to find extensions of this observation to higher dimensions, and to find analogues of theorems 5 and 5* for higher-dimensional sections and projections of P.

For other related results see Croft [2].

In the opposite direction one may inquire whether there exists a (central) k-polytope such that among its (central) d-dimensional sections occur affine images of all members of certain families of d-polytopes. A problem of this type was first raised by S. Mazur in *The Scottish Book* (Ulam [1]). Various results were obtained by Bessaga[1], Grünbaum [2], Melzak [1, 2], Naumann [1, 2], and Shephard [3]. Most far-reaching are the results of Klee [3]; among them we mention only the following:

6. *The least integer k such that some k-polytope has, for every d-polytope P with at most f facets, a d-dimensional section affinely equivalent to P, satisfies*

$$\frac{d}{d+1}(f + 1) \leq k \leq f - 1.$$

Another line of research should be mentioned here. Let P be a compact convex set in R^d; as shown by Süss–Kneser [1], if every (parallel) projection of P into 2-dimensional planes is a polygon with at most four vertices, then P is the convex hull of four points. Bol [1] proved that if every projection is a polygon with at most five vertices, then P is the convex hull of at most six points. Bol [1] also obtained:

7. *If every projection of a compact convex set $P \subset R^3$ into 2-dimensional planes is a polygon, then P is a polytope. Moreover, if each such projection of P has at most n vertices, $n \geq 5$, then P is either a pentagon, or each 2-face of P has at most $2n - 6$ vertices.*

These investigations were generalized in many directions by Klee [2]; among analogous results dealing with various notions of polyhedral sets he obtained:

8. *A bounded convex subset K of E^d is a polytope if and only if, for some j with $2 \leq j < d$, all the projections of K into j-dimensional spaces are polytopes.*

9. *If K is a bounded convex subset of E^d and if $p \in \text{int } K$ then K is a polytope if and only if every section of K by a 2-dimensional plane containing p is a polygon.*

For a related result (in case $d = 3$) see Valentine [1], p. 142.

The proof of theorem 1 shows that a d-polytope P with f facets may be obtained in many different ways as the intersection an $(f - 1)$-simplex and a d-flat. (See also exercise 3.) Therefore it may be asked whether this representability still holds if additional restrictions are imposed on the flats and simplices considered. The remaining part of the present section deals with some interesting results in this direction. The results and their proofs have recently been communicated to the author by M. A. Perles.

10. *Let $P \subset R^d$ be a d-polytope with at most f facets, let T be an $(f - 1)$-simplex in R^{f-1}, and let $p \in \text{int } T$. Then there exists a d-flat L in R^{f-1} such that $p \in L$ and $T \cap L$ is affinely equivalent to P.*

The following are the main steps of the proof; the details are left to the reader.

(i) No generality is lost in assuming that $p = 0$, the origin of R^{f-1}, and that P has exactly f facets.

(ii) The flat L required by the theorem exists if and only if there exist a point $q \in \text{int } P$ and a *linear* transformation A from R^d into R^{f-1}, such that $A(-q + P) = T \cap AR^d$.

(iii) For a given $q \in \operatorname{int} P$, the above A exists if and only if there exists a linear transformation A^* from R^{f-1} onto R^d such that A^* maps T^*, the polar of T, onto $(-q + P)^*$, the polar of $-q + P$. (A^* is the adjoint of A; see exercises 3.4.12 and 3.4.13.)

(iv) Let F_i, $i = 1, 2, \cdots, f$, be the facets of P; then $\hat{F}_i = y_i$ are the vertices of P^* (see section 3.4). Let $y_i(q) = (-q + F_i)\hat{}$ be the vertices of $(-q + P)^*$; then $y_i(q) = y_i/(1 - \langle q, y_i \rangle)$ (see exercise 3.4.4). Let x_i, $i = 1, \cdots, f$, be the vertices of T^*; there exists a unique f-tuple of numbers λ_i, $1 \leq i \leq f$, such that $\sum_{i=1}^{f} \lambda_i x_i = 0$, $\lambda_i > 0$ for $i = 1, 2, \cdots, f$, and $\sum_{i=1}^{f} \lambda_i = 1$.

A linear transformation A^* from R^{f-1} to R^d maps T^* onto $(-q + P)^*$ if and only if (possibly after a suitable permutation of the indices i) $A^* x_i = y_i(q)$ for $i = 1, \cdots, f$. A necessary and sufficient condition for the existence of such an A^* is $\sum_{i=1}^{f} \lambda_i y_i(q) = 0$.

(v) So far, the question of existence or non-existence of a d-flat L as required in theorem 10 has been reduced to the question whether there exists a point $q \in \operatorname{int} P$ such that $\sum_{i=1}^{f} \lambda_i y_i(q) = 0$.

For an arbitrary $q \in \operatorname{int} P$ we define

$$M(q) = \{(\alpha_1, \cdots, \alpha_f) \in R^f \mid \sum_{i=1}^{f} \alpha_i y_i(q) = 0\}.$$

It follows (see (iv) above) that

$$M(q) = \{(\alpha_1(1 - \langle q, y_1 \rangle), \cdots, \alpha_f(1 - \langle q, y_f \rangle)) \mid (\alpha_1, \cdots \alpha_f) \in M(0)\},$$

and it is also easy to see that

$$\bigcup_{q \in \operatorname{int} P} M(q) = \{(\alpha_1(1 - \beta_1), \cdots, \alpha_f(1 - \beta_f)) \mid (\alpha_1, \cdots, \alpha_f) \in M(0),$$

$$(\beta_1, \cdots, \beta_f) \in M(0)^{\perp}, \qquad \beta_i < 1 \quad \text{for} \quad i = 1, \cdots, f\},$$

where $M(0)^{\perp}$ denotes the orthogonal complement of $M(0)$ in R^f. Let

$$G = \bigcup_{q \in \operatorname{int} P} M(q) \cap \{(\gamma_1, \cdots, \gamma_f) \in R^f \mid \sum_{i=0}^{f} \gamma_i = 1,$$

$$\gamma_i > 0 \qquad \text{for} \quad i = 1, \cdots, f\}.$$

Then

$$G = \{(\alpha_1(1 - \beta_1), \cdots, \alpha_f(1 - \beta_f)) \mid (\alpha_1, \cdots, \alpha_f) \in M(0),$$

$$(\beta_1, \cdots, \beta_f) \in M(0)^{\perp}, \sum_{i=1}^{f} \alpha_i = 1, \ \alpha_i > 0 \ \text{and} \ \beta_i < 1 \ \text{for} \ i = 1, \cdots, f\}.$$

By the above, the assertion of theorem 10 is true if and only if $(\lambda_1, \cdots, \lambda_f) \in G$. Consider the function

$$\varphi(\alpha_1, \cdots, \alpha_f, \beta_1, \cdots, \beta_f) = \sum_{i=1}^{f} \lambda_i \log(\alpha_i(1 - \beta_i)) \quad \text{for} \quad (\alpha_1, \cdots, \alpha_f) \in M(0),$$

$$\alpha_i \geq 0, \sum_{i=1}^{f} \alpha_i = 1, (\beta_1, \cdots, \beta_f) \in M(0)^{\perp}, \beta_i \leq 1,$$

where $\log 0 = -\infty$. Since φ is an upper semicontinuous function defined on a compact set, it attains a finite maximum ω at some point $(\alpha'_1, \cdots, \alpha'_f, \beta'_1, \cdots, \beta'_f)$ with $\alpha'_i > 0, \beta'_i < 1$. ω is the maximum value of the function $\sum_{i=1}^{f} \lambda_i \log \gamma_i$ for $(\gamma_1, \cdots, \gamma_f) \in G$. Assume, without loss of generality, that $\beta'_i = 0$ for $1 \leq i \leq f$. (The point $(\alpha'_1(1 - \beta'_1), \cdots, \alpha'_f(1 - \beta'_f))$ lies in $M(q)$ for some $q \in \text{int } P$, and we may replace P by $-q + P$, if $q \neq 0$.)

Fix a point $(\beta_1, \cdots, \beta_f) \in M(0)^{\perp}$, and define

$$g(t) = \varphi(\alpha'_1, \cdots, \alpha'_f, \beta_1 t, \cdots, \beta_f t).$$

Then $g'(0) = 0$, since $g(t)$ attains a maximum at $t = 0$. But an easy calculation shows that $g'(0) = -\sum_{i=1}^{f} \lambda_i \beta_i$. It follows that $(\lambda_1, \cdots, \lambda_f) \in M(0)^{\perp\perp} = M(0)$, and, by assumption, $\sum_{i=1}^{f} \lambda_i = 1$ and $\lambda_i > 0$ for $1 \leq i \leq f$, hence $(\lambda_1, \cdots, \lambda_f) \in G$.

The reader may verify, using a similar orthogonality argument, that $(\alpha'_1, \cdots, \alpha'_f) = (\lambda_1, \cdots, \lambda_f)$, and that the point $(\lambda_1, \cdots, \lambda_f)$ cannot belong to $M(q)$ for more than one $q \in \text{int } P$.

Theorem 10, and the method used in its proof, have many analogues, variants, and applications in other proofs; some of them will be mentioned in the exercises. Here we shall dwell on one of them only.

It is well known that the notion of convexity may be defined in vector spaces over any ordered field, and that many of the usual properties of convex sets remain valid in such more general settings (though some definitions, and some proofs, have to be modified; see, for example, Weyl [1], Motzkin [2], Klee [10]). While other exceptions are known (see section 5.5), the simplest known result on convex polytopes in R^d which fails in vectors spaces over ordered fields is probably theorem 10. Indeed, we have

11. *In vector-spaces over the field of rational numbers, theorem* 10 *is not true.*

For an example establishing theorem 11 see exercise 5.1.6.

Exercises

1. Show that it is impossible to strengthen theorems 1–4 by decreasing the dimension of the simplex (or cube, or octahedron).

2. Let P be a polytope, L_1 and L_2 two flats of the same dimension. Show that $L_1 \cap P$ and $L_2 \cap P$ are combinatorially equivalent provided L_1 and L_2 have 0-dimensional intersections with the same faces of P.

3. Supply the details of the following proof of theorem 1. Let $P \subset R^d$ be a d-polytope with f facets, $0 \in \text{int } P$, and let $v_1, \cdots, v_f$ be the vertices of the polar P^* of P. Let $\alpha_1, \cdots, \alpha_f$ be positive numbers such that

$$\sum_{i=1}^{f} \alpha_i v_i = 0 \quad \text{and} \quad \sum_{i=1}^{f} \alpha_i = 1,$$

and define for $x \in R^d$

$$Ax = (\alpha_1(1 - \langle v_1, x \rangle), \cdots, \alpha_f(1 - \langle v_f, x \rangle)) \in R^f.$$

Show that

(i) $AR^d \subset \{(y_1, \cdots, y_f) \in R^f \mid \sum_{i=1}^{f} y_i = 1\}$;

(ii) A is one-to-one;

(iii) $AK = AR^d \cap T$, where T is the $(f - 1)$-simplex

$$T = \{(y_1, \cdots, y_f) \in R^f \mid \sum_{i=1}^{f} y_i = 1, \quad y_1 \geq 0 \quad \text{for} \quad i = 1, \cdots, f\}.$$

4. Show that for given P, T, and p in theorem 10 there exist at most f! different d-flats L satisfying the conditions of theorem 10, with $f = f_{d-1}(P)$. Their exact number is equal to f! divided by the order of the group of affine automorphisms of P.

5. Let $P \subset R^d$ be a centrally symmetric d-polytope with $2f$ facets, such that $P = -P$, and let $C^f \subset R^f$ be the f-cube. Using a method analogous to parts (ii), (iii) of the proof of theorem 10, show that there exists a d-dimensional subspace L of R^f, such that $L \cap C^f$ is linearly equivalent to P.

6. Let $T \subset R^d$ be a 3-simplex with vertices at points having rational coordinates; let p be the centroid of T, and let $P \subset R^2$ be the quadrangle conv$\{(1, 1), (1, -1), (-1, 3), (-1, -3)\}$. Show that P is not affinely equivalent to any set of the type $T \cap L$, where $p \in L$ and L is a 2-flat $L = \{x \in R^3 \mid \langle x, y \rangle = \alpha\}$ with α and the coordinates of y all rational.

5.2   The Inductive Construction of Polytopes

In various problems it is important to know whether a given polytope may be changed in such a way that the new polytope has some desired properties. We shall encounter examples of such situations in chapters 6 and 7. The simplest case, which shall be discussed in the present section, is that in which the new polytope has at most one new vertex.

Let $P \subset R^d$ be a d-polytope, H a hyperplane such that $H \cap \text{int } P = \varnothing$, and let $V \in R^d$. We shall say that V is *beneath* H, or *beyond* H, (with respect to P) provided V belongs to the open halfspace determined by H which contains int P, or does not meet P, respectively. If $V \in R^d$ and F is a facet of the d-polytope $P \subset R^d$, we shall say that V is *beneath* F or *beyond* F provided V is beneath or beyond aff F, respectively.

Various ideas and constructions related to the 'beneath–beyond' concepts have appeared in the literature since Euler's times, mostly in the (dual) variant of 'cutting off' vertices or larger parts of a given polytope. (See, for example, Brückner [1, 2, 3], Steinitz–Rademacher [1].) Nevertheless it seems that the systematic use of these notions, in particular for higher-dimensional polytopes, is rather new. The terminology was used first in Grünbaum [12]. We shall see important applications of theorem 1 in chapters 6 and 7, and in other sections of the book.

The relation between the facial structure of a polytope and that of the convex hull of its union with one additional point is determined by the following theorem.

1. *Let P and P^* be two d-polytopes in R^d, and let V be a vertex of P^*, $V \notin P$, such that $P^* = \text{conv}(\{V\} \cup P)$. Then*

   (i) *a face F of P is a face of P^* if and only if there exists a facet F' of P such that $F \subset F'$ and V is beneath F';*

   (ii) *if F is a face of P then $F^* = \text{conv}(\{V\} \cup F)$ is a face of P^* if and only if*
either   (*) *$V \in \text{aff } F$;*
or   (**) *among the facets of P containing F there is at least one such that V is beneath it and at least one such that V is beyond it.*

   Moreover, each face of P^* is of one and only one of those types.

PROOF   It is obvious that a face of P^* is either a face of P, or the convex hull of the union of V with some face of P. It is equally obvious that a facet F of P is a facet of P^* if and only if V is beneath F with respect to either (and therefore both) P or P^*. Therefore, if F_0 is a face of P contained in the facet F of P, and if V is beneath F, then F_0 is a face of P^*. On the other hand, if F_0^* is a face of P^* such that $V \notin F_0^*$ then F_0^* is a face of P

and, since F_0^* is the intersection of all the facets of P^* which contain F_0^*, there exists a facet F^* of P^* for which $F_0^* \subset F^*$ and $V \notin F^*$. Clearly V is beneath F^* and the proof of part (i) is completed.

Let now F be a face of P and $F^* = \text{conv}(\{V\} \cup F)$ a face of P^*. Then clearly $F = P \cap \text{aff } F^*$. We consider the intersection of the whole configuration with the 2-dimensional plane E determined by the three points V, x_0 and y_0, where $x_0 \in \text{relint } F$, and $y_0 \in \text{int } P$. Then $P_0 = E \cap P$ is a 2-polytope, i.e. a polygon (see figures 5.2.1 and 5.2.2). The line

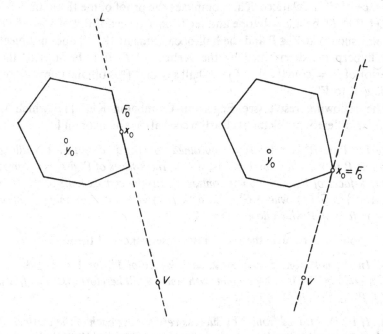

Figure 5.2.1   Figure 5.2.2

$L = \text{aff}\{x_0, V\}$ is the intersection of E with aff F^*, and $F_0 = L \cap P_0$ is either a vertex of P_0 or an edge of P_0. In the last case, $V \in \text{aff } F_0 \subset \text{aff } F$, and we have (*). If F_0 is a vertex of P_0, V is (in E) beneath one and beyond the other of the edges of P_0 incident with F_0. Denoting by F_1 and F_2 any pair of facets of P containing those two edges, it follows that V is (in R^d) beneath one and beyond the other of F_1 and F_2, and we have (**).

There remains to be shown that if F is a face of P satisfying the conditions (*) or (**), then $F^* = \text{conv}(\{V\} \cup F)$ is a face of F^*. This is entirely trivial in case of (*). Let therefore F be a face of P such that $F \subset F_1 \cap F_2$, where F_1 and F_2 are facets of P, V being beyond F_1 and beneath F_2. Let $H_i = \text{aff } F_i$, and let H_0 be a hyperplane such that $H_0 \cap P = F$. Rotating H_i slightly about $H_i \cap H_0$ towards H_0, new hyperplanes H_1^* and H_2^* are obtained such that $H_i^* \cap P = F$, while V is beyond H_1^* and beneath H_2^*. But then the hyperplane $H^* = \text{aff}(\{V\} \cup (H_1^* \cap H_2^*))$ contains V and satisfies $H^* \cap P = F$. Thus

$$H^* \cap P^* = H^* \cap \text{conv}(\{V\} \cup P) = \text{conv}(\{V\} \cup F) = F^*$$

is a face of P^*, as claimed. This completes the proof of the theorem.

Let $P \subset R^d$ be a d-polytope and let V be a vertex of P. Let $V' \in R^d$ be a point such that $V' \notin P$ and the half-open segment $]V, V']$ does not meet any hyperplane determined by the vertices of P. If V belongs to the interior of $P' = \text{conv}(\{V'\} \cup P)$ we shall say that P' is obtained from P by *pulling V to V'*.

The following result (see Eggleston–Grünbaum–Klee [1]) which is useful in different problems, is clearly a special case of theorem 1.

2. *Let $P' \subset R^d$ be a d-polytope obtained from the d-polytope P by pulling $V \in \text{vert } P$ to V'. Then, for $1 \le k \le d - 1$, the k-faces of P' are as follows: (i) the k-faces of P which do not contain V; (ii) the convex hulls of the type $\text{conv}(\{V'\} \cup G^{k-1})$, where G^{k-1} is a $(k - 1)$-face not containing V of a $(d - 1)$-face of P which does contain V.*

We note in particular the following consequences of theorem 2.

3. *In the notation of theorem 2, each k-face of P', for $1 \le k \le d - 1$, which contains V' is a k-pyramid with apex V'. Therefore $f_0(P') = f_0(P)$, and $f_i(P') \ge f_i(P)$ for $1 \le i \le d - 1$.*

4. *If P^* is obtained from P by successively pulling each of the vertices of P, then P^* is a simplicial d-polytope satisfying $f_0(P^*) = f_0(P)$, $f_i(P^*) \ge f_i(P)$ for $1 \le i \le d - 1$. Moreover, if some j-face of P is not a j-simplex, then strict inequality holds for all i with $j - 1 \le i \le d - 1$.*

We shall use theorem 4 in chapter 10. The idea to apply suitable 'perturbations' of the vertices in order to reduce questions about the maximal possible number of faces of arbitrary d-polytopes (having a given number of vertices) to the corresponding problems dealing with simplicial polytopes only, has been frequently used. The case corresponding to $i = d - 1$ in theorem 4 was employed (without proof) by Gale [5]. Klee [13] gave a justification of Gale's assumption, for all i, by a process called 'pushing' (see exercise 16). As far as applications to extremal problems are concerned, 'pulling' and 'pushing' seem to have the same value; the properties of the 'pulling' procedure, and its generalization contained in theorem 1, seem to be more easy to establish, and to be intuitively simpler.

A consequence of theorem 4 may be succintly formulated using the following notion.

Let $\mathscr{P}'$ and $\mathscr{P}''$ be two families of d-polytopes in R^d. We shall say that $\mathscr{P}''$ is *dense* in $\mathscr{P}'$ provided for every $\varepsilon > 0$ and every $P' \in \mathscr{P}'$ there exists a $P'' \in \mathscr{P}''$ such that the Hausdorff distance $\rho(P', P'')$ is less than ε. Theorem 4 obviously implies:

5. *The family $\mathscr{P}_s^d$ of all simplicial d-polytopes is dense in the family $\mathscr{P}^d$ of all d-polytopes.*

Exercises

1. Find projectively equivalent d-polytopes P_1 and P_2 such that among the d-polytopes of the form conv($\{V_1\} \cup P_1$) occur types not obtainable in the form conv($\{V_2\} \cup P_2$), where V_1 and V_2 are arbitrary points.

2. Determine all the combinatorial types of 4-polytopes with 6 vertices, and construct their Schlegel diagrams.

3. In the notation of section 3.3, determine the set of points x_0 which may serve as centers of projection for the formation of Schlegel diagrams of a given polytope P.

4. Show that there are $[\frac{1}{2}d]$ different combinatorial types of simplicial d-polytopes with $d + 2$ vertices.

5. Formulate and prove the result dual to theorem 1.

6. If P and P^* are related as in theorem 1, or in its dual, what is the relation between their Schlegel diagrams?

7. Determine under what conditions does $f_k(P') = f_k(P)$ hold in theorem 3 for a given k.

8. Let $P \subset R^d$ be a d-polytope and let V be a point of R^d which is beyond (or on) the facet F of P and beneath all those facets of P which intersect F in a $(d - 2)$-face. Show that V is beneath all the facets of P different from F.

9. Let $P \subset R^d$ be a d-polytope, F a facet of P, and X, Y points such that F is a facet both of conv($\{X\} \cup P$) and of conv($\{Y\} \cup P$). Show that F is a facet of conv($\{X, Y\} \cup P$). Does this statement remain true if F is a face of lower dimension?

10. Let $P \subset R^d$ be a d-polytope and let X be a point of R^d, $X \notin P$.

Prove that the set $\cup F_i$, where F_i ranges over all the facets of P such that X is beyond the facet, is homeomorphic to the $(d-1)$-dimensional ball B^{d-1}.

. 11. Let $P \subset R^d$ be a d-polytope and let $X_1, X_2 \notin P$. If the segment $\mathrm{conv}\{X_1, X_2\}$ is an edge of $\mathrm{conv}(\{X_1, X_2\} \cup P)$ show that there exist facets F_1 and F_2 of P such that X_i is beyond or on F_i, $i = 1, 2$, and $\dim(F_1 \cap F_2) \geq d - 2$.

12. Let $P \subset R^d$ be a d-polytope with facets $F_1, \cdots, F_f$, where $f = f_{d-1}(P)$. For $i = 1, \cdots, f$, let X_i be a point which is beyond or on F_i, and beneath all the other facets of P. Let $P' = \mathrm{conv}(\{X_1, \cdots, X_f\} \cup P)$. Then $\mathrm{conv}\{X_i, X_j\} \cap \mathrm{int}\, P \neq \varnothing$ for all $1 \leq i < j \leq f$ if and only if each k-face of P is a k-face of P', for all $k = 0, 1, \cdots, d - 2$.

13. It may be conjectured that whenever $k + h \leq d$, the family $\mathscr{P}^d(k, h)$ (see section 4.5) is dense in $\mathscr{P}^d$. Theorem 5 is clearly a particular case of this conjecture. Prove the following additional partial results:

(i) The family of all simple d-polytopes is dense in $\mathscr{P}^d$.

(ii) $\mathscr{P}^4(2, 2)$ is dense in $\mathscr{P}^4$. (This conjecture of D. W. Walkup was recently established by Shephard [7].)

14. Prove that the family $\mathscr{P}^d_c$ of all cubical d-polytopes is dense in $\mathscr{P}^d$. (This is a result due to Shephard [9], who also proved a number of other interesting results of a similar type. The proof uses exercise 4.8.29 and theorem 5.)

15. Even for $d = 3$ it is not known whether the family of all self-dual polytopes is dense in $\mathscr{P}^d$.

16. Following Klee [13] we shall say that a d-polytope P' is obtained from the d-polytope $P \subset R^d$ by *pushing the vertex V of P to the position V'* provided V' is a point of P such that the half-open interval $]V, V']$ does not intersect any $(d-1)$-hyperplane determined by vertices of P, and provided P' is the convex hull of the set consisting of V' and of the vertices of P different from V.

Find the analogue of theorem 2 if 'pulling' is replaced by 'pushing'.

Show that theorems 3 and 4 remain valid if 'pulling' is replaced by 'pushing'.

17. Let $C \subset R^d$ be a d-polytope with facets $G_1, \cdots, G_t$, where $t = f_{d-1}(C)$. For $1 \leq i \leq s$, with $1 \leq s \leq t$, let z_i be a point which lies beyond G_i but beneath all other facets of C. Assume also that all the segments $[z_i, z_j]$, $1 \leq i < j \leq s$, intersect $\mathrm{int}\, C$. For $1 \leq i \leq s$, let P_i be a d-polytope which has a facet F_i, such that F_i is projectively equivalent to G_i. Prove:

(i) There exist nonsingular projective transformations T_i, $1 \leq i \leq s$, with T_i permissible for F_i, such that $T_iF_i = G_i$ and $T_iP_i \subset \mathrm{conv}(G_i \cup \{z_i\})$.

(ii) If T_i are such transformations, then the set $K = C \cup \overset{s}{\underset{i=1}{\cup}} T_iP_i$ is a d-polytope, and

$$\mathcal{F}(K) = (\{K\} \cup \mathcal{F}(C) \cup \overset{s}{\underset{i=1}{\cup}} \mathcal{F}(T_iP_i)) \sim \{G_1, \cdots, G_s, C\}.$$

(This construction is frequently used; K is often said to arise by *adjoining* copies of $P_1, \cdots, P_s$ along their facets $F_1, \cdots, F_s$ to the corresponding facets $G_1, \cdots, G_s$ of C.)

5.3   Lower Semicontinuity of the Functions $f_k(P)$

We saw in theorem 5.2.4 that every d-polytope P may be approximated (in the sense of the Hausdorff metric) arbitrarily closely by polytopes P' such that $f_k(P') \geq f_k(P)$ for all k. Following Eggleston–Grünbaum–Klee [1] we shall supplement this by showing that each of the functions $f_k(P)$ is lower semicontinuous in P.

1. *Let P be a polytope in R^d. Then there exists an $\varepsilon = \varepsilon(P) > 0$ such that every polytope $P' \subset R^d$ for which $\rho(P', P) < \varepsilon$ satisfies $f_k(P') \geq f_k(P)$ for all k.*

PROOF   Let $\{P_i \mid i = 1, 2, \cdots\}$ be any sequence of polytopes converging to P. It is clearly enough to show that $\lim_i \sup\{f_k(P_i)\} \geq f_k(P)$. Now, if the sequence $\{f_0(P_i) \mid i = 1, 2, \cdots\}$ is unbounded, then $\{f_k(P_i) \mid i = 1, 2, \cdots\}$ is unbounded for each k $(0 \leq k < \dim P)$ and there is nothing to prove. Thus we may assume that $\{f_0(P_i) \mid i = 1, 2, \cdots\}$ is a bounded sequence. Passing, if necessary, to subsequences we may without loss of generality assume that:

(i) all P_i have the same dimension d', where $\dim P \leq d' \leq d$;

(ii) $f_k(P_i) = f_k$ for all $k = 0, 1, \cdots, d' - 1$ and $i = 1, 2, \cdots$;

(iii) for every $k = 0, 1, \cdots, d' - 1$, the k-faces of P_i may be denoted by $P_i^k(j), j = 1, 2, \cdots, f_k$, in such a manner that, for every k and j, the sequence $\{P_i^k(j) \mid i = 1, 2, \cdots\}$ be convergent to a polytope $P_0^k(j)$, of dimension at most k.

Clearly, $\mathrm{bd}\, P \subset \underset{\substack{0 \leq k \leq d'-1 \\ 1 \leq j \leq f_k}}{\cup} P_0^k(j)$. Let s satisfying $0 \leq s < \dim P$, be fixed; let Q denote the union of all the sets $P_0^k(j)$ of dimension less than s. Then Q obviously contains no s-dimensional subset.

Let now $\delta > 0$ be sufficiently small to ensure that for each s-face F of P, the 3δ-neighborhood of the union of all the s-faces of P different from F does not contain F. Then there is an s-dimensional (relatively open) subset F_0 of F which is at distance greater than 2δ from the union of the s-faces of P different from F. By the above, F_0 is not contained in Q; let $x(F) \in F_0 \sim Q$. Then $x(F)$ belongs to some $P_0^{k_0}(j_0)$ which is s-dimensional and contained in F. It follows that there exists $i(F)$ such that for all $i \geq i(F)$ the set $P_i^{k_0}(j_0)$ is contained in the δ-neighborhood of $P_0^{k_0}(j_0)$, and therefore in the δ-neighborhood of F. By Exercise 3.3.5, the union of the s-faces of $P_i^{k_0}(j_0)$ also converges to $P_0^{k_0}(j_0)$. Since $x(F) \in P_0^{k_0}(j_0)$, for all i (greater than or equal to a suitable $i'(F) \geq i(F)$) there exists an s-face of $P_i^{k_0}(j_0)$ which contains $x(F)$ in its δ-neighborhood. Let this face (or one of them) be denoted by $K_i(F)$. We shall show that if the s-faces F_1 and F_2 of P are different then $K_i(F_1)$ is different from $K_i(F_2)$ for all $i \geq i(P) = \max\{i'(F) \mid F$ an s-face of $P\}$. Indeed, assuming $K_i(F_1) = K_i(F_2)$ it follows that $K_i(F_2)$ is in the δ-neighborhood of F_1; but $x(F_2)$ is in the δ-neighborhood of $K_i(F_2)$. Hence $x(F_2)$ is in the 2δ-neighborhood of F_1—contradicting the definition of $x(F_2)$. Thus for $i \geq i(P)$ to each s-face F of P there corresponds an s-face $K_i(F)$ of P_i, to different s-faces of P corresponding different s-faces of P_i. Therefore $f_s(P_i) \geq f_s(P)$ and the proof of the theorem is completed.

We shall see some applications of theorem 1 in chapter 10.

Without proof we mention an extension of theorem 1 to certain complexes, due to Eggleston–Grünbaum–Klee [1].

A complex $\mathcal{K}$ is said to have property $A(s)$ provided for each convex subset C of set $\mathcal{K}$ such that $C = \text{relint } C$, and for each face F of $\mathcal{K}$ with $\dim F = s \leq \dim C$, the assumption $C \cap \text{relint } F \neq \emptyset$ implies $C \subset F$.

The result may be formulated as:

2. *Let $\{\mathcal{K}_i\}$ be a sequence of complexes in R^d such that the sequence $\{\text{set } \mathcal{K}_i\}$ converges to set $\mathcal{K}$ for a complex $\mathcal{K}$ which has property $A(s)$. Then $\liminf_i f_s(\mathcal{K}_i) \geq f_s(\mathcal{K})$ if at least one of the following conditions is satisfied:*

(i) *The sequence $\{f_0(\mathcal{K}_i)\}$ is bounded;*

(ii) *for each point x belonging to some s-face of $\mathcal{K}$ there is a sequence $\{x_i\}$ with limit x, such that x_i belongs to an s-face of $\mathcal{K}_i$.*

It should be noted that a satisfactory characterization of complexes which possess the lower semicontinuity property is still lacking.

Exercises

1. Let $P \subset R^d$ be a d-polytope. Prove that there exists an $\varepsilon = \varepsilon(P) > 0$ such that if $P' \subset R^d$ is any polytope satisfying $\rho(P, P') < \varepsilon$ and $f_i(P') = f_i(P)$ for $i = 0, 1, \cdots, d - 2$, then P' is combinatorially equivalent with P.

Show by examples that it is impossible to drop the requirement $f_{d-2}(P') = f_{d-2}(P)$ in this result. (An acquaintance with chapter 8 is desirable in solving this exercise.)

2. Prove that for every polytope P the complex $\mathscr{C}(P)$ has property $A(s)$ for every s.

3. Find examples of complexes $\mathscr{K}$ which do not have property $A(s)$, such that
   (i) theorem 2 holds for $\mathscr{K}$;
   (ii) theorem 2 does not hold for $\mathscr{K}$.

5.4 Gale-Transforms and Gale-Diagrams

The present section is devoted to the exposition of a powerful technique, applicable to various problems involving polytopes. The method of *Gale-transforms* and *Gale-diagrams*, to be discussed now, is rather algebraic in character. Various special aspects of it, or of related ideas, may be found in different papers, such as Motzkin [2,4,6], and in particular Gale [3,4,5]. (Compare also the proof of theorem 5.1.10.) Our exposition follows a private communication from M. A. Perles. All the new results obtained by this method and presented in different parts of the book (sections 5.5, 6.3, etc.) as well as the very useful notion of *coface* (see below), are due to Perles. The reader will find it well worth his while to become familiar with the concepts of Gale-transforms and Gale-diagrams, since for many of the results obtained through them no alternative proofs have been found so far. It is very likely that the method will yield many additional results.

We turn first to a description of the Gale-transform and its properties.

Let $X = (x_1, \cdots, x_n)$ be an n-tuple of points in R^d, with dim aff $X = r$. The set $D(X) \subset R^n$ of *affine dependences* of X consists of all points $a = (\alpha_1, \cdots, \alpha_n) \in R^n$ such that

$$\sum_{i=1}^n \alpha_i x_i = 0$$

and

$$\sum_{i=1}^n \alpha_i = 0.$$

It is obvious that $D(X)$ is a linear subspace of R^n, and that for each $x \in R^d$ we have $D(x + X) = D(X)$. More generally, it is easily checked that for $X, Y \subset R^d$ we have $D(X) = D(Y)$ if and only if X and Y are affinely equivalent with x_i corresponding to y_i, i.e., if $y_i = Ax_i$ for some nonsingular affine transformation A and for $i = 1, \cdots, n$. Let $x_i = (x_{i,1}, \cdots, x_{i,d})$ for $i = 1, \cdots, n$; we shall consider the n by $d + 1$ matrix

$$D_0 = \begin{pmatrix} x_{1,1} & x_{1,2} & \cdots & x_{1,d} & 1 \\ x_{2,1} & x_{2,2} & \cdots & x_{2,d} & 1 \\ \cdot & \cdot & \cdots & \cdot & \cdot \\ x_{n,1} & x_{n,2} & \cdots & x_{n,d} & 1 \end{pmatrix}.$$

The rank of D_0 clearly equals $r + 1$; therefore, among the columns $x^{(1)}, \cdots, x^{(d+1)}$ of D_0, considered as vectors in R^n, there are $r + 1$ linearly independent ones (the column $x^{(d+1)}$ may be assumed to be one of them). Hence the subspace $M(X) = \lim\{x^{(1)}, \cdots, x^{(d+1)}\}$ of R^n has dimension $r + 1$. Its orthogonal complement $M(X)^\perp = \{a \in R^n \mid \langle a, y \rangle = 0$ for all $y \in M(X)\}$ clearly coincides with $D(X)$. It follows that

$$\dim D(X) = \dim M(X)^\perp = n - \dim M(X) = n - r - 1.$$

Let the $n - r - 1$ vectors $a^{(1)}, \cdots, a^{(n-r-1)}$ of R^n form a basis of $D(X)$, and let D_1 be the n by $n - r - 1$ matrix having columns $a^{(1)}, \cdots, a^{(n-r-1)}$,

$$D_1 = \begin{pmatrix} \alpha_{1,1} & \alpha_{1,1} & \cdots & \alpha_{1,n-r-1} \\ \alpha_{2,1} & \alpha_{2,2} & \cdots & \alpha_{2,n-r-1} \\ \cdot & \cdot & \cdots & \cdot \\ \alpha_{n,1} & \alpha_{n,2} & \cdots & \alpha_{n,n-r-1} \end{pmatrix}.$$

The *rows* of D_1 may be considered as vectors in R^{n-r-1}; we shall denote the jth row by $\bar{x}_j = (\alpha_{j,1}, \alpha_{j,2}, \cdots, \alpha_{j,n-r-1})$, for $j = 1, \cdots, n$.

The final result of the above construction is the assignment of a point $\bar{x}_j \in R^{n-r-1}$ to each point $x_j \in X \subset R^d$ (or rather, to each $j \in \{1, \cdots, n\}$), where $d \geq r$, $n = \text{card } X$, and $r = \dim \text{aff } X$. The n-tuple $\bar{X} = (\bar{x}_1, \cdots, \bar{x}_n) \subset R^{n-r-1}$ is the *Gale-transform* of X. It should be observed that the n-tuple $\bar{X} \subset R^{n-r-1}$, which linearly spans R^{n-r-1}, does not necessarily consist of n *different* points, even if the n points of X are different. Hence the *set* $\bar{X}$ may consist of less than n points, some of the points $z \in \bar{X}$ having a *multiplicity* greater than 1 and equal to the number of points $x_i \in X$ (or, more precisely, to the number of $i \in \{1, \cdots, n\}$) such that $\bar{x}_i = z$. The Gale-transform $\bar{X}$ obviously depends on a factor which has no geometric significance for X (namely the choice of the basis $a^{(j)}$ in $M(X)^\perp$), and it would be more proper to call it *a* Gale-transform of X. Nevertheless, many geometric properties of X have as counterparts meaningful geometric properties of the Gale-transform $\overline{X}$. The facts listed below deal with some such properties; the reader is invited to supply their proofs, which require only the basic results of linear algebra.

(i) $\sum_{i=1}^{n} \bar{x}_i = 0$, $0 \in \text{relint conv } \overline{X}$, and $\overline{X}$ linearly (and even positively) spans R^{n-r-1}.

(ii) If the n-tuple $\overline{X'} = \overline{x_i'}, \cdots, \overline{x_n'}$ is the Gale-transform of X obtained by using a basis of $D(X)$ different from $a^{(1)}, \cdots, a^{(n-r-1)}$, then $\overline{X'}$ and $\overline{X}$ are linearly equivalent. Conversely, whenever A is a regular linear transformation of R^{n-r-1} into itself, the n-tuple $A\bar{x}_1, \cdots, A\bar{x}_n$ may be obtained as the Gale-transform of X by a suitable choice of basis in $D(X)$.

(iii) If $M^\perp$ is an $(n-r-1)$-dimensional subspace of R^n, $d \geq r$, orthogonal to the vector $(1, 1, \cdots, 1) \in R^n$, there exists an n-tuple $X \subset R^d$ such that $\dim \text{aff } X = r$ and $D(X) = M^\perp$. Moreover, if $Z = (z_1, \cdots, z_n)$ is an n-tuple of (not necessarily different) points in R^{n-r-1} such that $\sum_{i=1}^{n} z_i = 0$ and $\dim \text{lin } Z = n - r - 1$, there exists an n-tuple $X \subset R^d$ with $\dim \text{aff } X = r$ such that $\overline{X} = Z$.

(iv) Let $X \subset R^d$; if $J = \{i_1, \cdots, i_m\} \subset \{1, \cdots, n\} = N$ we shall write $X(J)$ for $(x_{i_1}, \cdots, x_{i_m})$ and similarly $\overline{X}(J) = (\bar{x}_{i_1}, \cdots, \bar{x}_{i_m})$, assuming of course that $i_1 < i_2 < \cdots < i_m$. If $\dim \text{aff } X = r$, then $\overline{X(J)}$ (which is an m-tuple in R^{m-q-1}, where $q = \dim \text{aff } X(J)$) is linearly equivalent to the m-tuple obtained from $\overline{X}(J)$ by orthogonal projection onto the subspace of R^{n-r-1} which is orthogonal to $\text{lin } \overline{X}(N \sim J) \subset R^{n-r-1}$.

(v) The n points of X are in general position in R^d if and only if the n-tuple $\overline{X}$ consists of n points in linearly general position in R^{n-d-1} (i.e., no $(n-d-2)$-dimensional subspace contains $n-d-1$ of them).

(vi) Let P be a nonsingular projective transformation of R^d into itself, permissible for $X = (x_1, \cdots, x_n) \subset R^d$ (i.e. $Px = (Ax + b)/(\langle c, x \rangle + \delta)$, and $\langle c, x_i \rangle + \delta \neq 0$ for $x_i \in X$). Let $Y = (Px_1, \cdots, Px_n)$. Then $\overline{Y}$ is linearly equivalent to the n-tuple $((\langle c, x_1 \rangle + \delta)\bar{x}_1, \cdots, (\langle c, x_n \rangle + \delta)\bar{x}_n)$. Conversely, if $X, Y \subset R^d$ are two n-tuples such that there exist nonzero numbers $\lambda_1, \cdots, \lambda_n$ with the property $\bar{y}_i = \lambda_i \bar{x}_i$ for $i = 1, \cdots, n$, then there exist $c \in R^d$ and δ such that $\lambda_i = \langle c, x_i \rangle + \delta$ for $i = 1, \cdots, n$, and a linear transformation A and vector $b \in R^d$ such that

$$Px = \frac{Ax + b}{\langle c, x \rangle + \delta}$$

is a regular projective transformation permissible for X, satisfying $y_i = Px_i$ for $i = 1, \cdots, n$. Moreover, P is permissible for $\text{conv } X$ if and only if $\lambda_i > 0$ for $i = 1, \cdots, n$, (or $\lambda_i < 0$ for $i = 1, \cdots, n$).

Let $X \subset R^d$ and let $Y = X(J) \subset X$ for some $J \subset N$. We shall say that Y is a *face* of X provided $\text{conv}(X \sim Y) \cap \text{aff } Y = \emptyset$ (compare section 2.4 and theorem 2.6.1; here, of course, $X \sim Y$ means $X(N \sim J)$). Clearly, if X consists of n points in general position, or if $X = \text{vert } P$ where P is a polytope, then $Y \subset X$ is a face of X if and only if conv Y is a face of conv X. We shall say that $Y \subset X$ is a *coface* of X provided $X \sim Y$ is a face of X, i.e. if and only if conv $Y \cap \text{aff}(X \sim Y) = \emptyset$.

The notion of coface permits an easy translation of some geometric properties of X into properties of $\overline{X}$, due to the following result:

1. $Y = X(J) \subset X$ *is a coface of* X *if and only if either* $Y = \emptyset$ *or* $0 \in \text{relint conv } \overline{X}(J)$.

PROOF Assume that $Y = X(J) \subset X \subset R^d$ is not a coface of X; then $Y \neq \emptyset$. Without loss of generality assume also that $\dim \text{aff } X = d$. The $\text{conv } Y \cap \text{aff}(X \sim Y) \neq \emptyset$, hence there exists an n-vector $b = (\beta_1, \cdots, \beta_n)$ such that $\sum_{i=1}^{n} \beta_i x_i = 0$, $\sum_{i=1}^{n} \beta_i = 0$, $\sum_{i \in J} \beta_i = 1$, and $\beta_i \geq 0$ whenever $i \in J$. Since $b \in D(X)$ there exist $\gamma_1, \cdots, \gamma_{n-d-1}$ such that $b = \sum_{j=1}^{n-d-1} \gamma_j a^{(j)}$. In other words, denoting

$$c = (\gamma_1, \cdots, \gamma_{n-d-1}) \in R^{n-d-1}$$

we have $\beta_i = \langle c, \bar{x}_i \rangle$ for $i = 1, \cdots, n$. Thus $\langle c, x_i \rangle \geq 0$ for every i such that $x_i \in Y$, with strict inequality for some of those i. Hence there exists a hyperplane H separating 0 and conv $\overline{X}(J)$, with $\overline{X}(J) \not\subset H$. By the separation theorem 2.2.2, this implies $0 \notin \text{relint conv } \overline{X}(J)$. Since all the steps of the above argument are reversible, this completes the proof of theorem 1.

The following results are immediate consequences of theorem 1 and the properties of Gale-transforms mentioned above; their proofs are left to the reader.

2. *If* $X \subset R$ *and* $\text{aff } X = R^d$ *then* $X = \text{vert } P$ *for some d-polytope* P *with n vertices if and only if either* (i) $\bar{x} = 0$ *for all* $x \in X$ *(and P is a d-simplex), or* (ii) *every open halfspace* H^+ *of* R^{n-d-1}, *such that* $0 \in \text{bd } H^+$, *satisfies* $\text{card } \{i \mid \bar{x}_i \in H^+\} \geq 2$.

3. *If* $0 \in \overline{X}$ *then* conv X *is pyramidal at every* $x_i \in X$ *such that* $\bar{x}_i = 0$. *Conversely, if* conv X *is pyramidal at* x_i *and if* $x_i \neq x_j$ *whenever* $i \neq j$, *then* $\bar{x}_i = 0$.

4. *Let* $P \subset R^d$ *be a d-polytope, and* $V = (v_1, \cdots, v_n) = \text{vert } P$. P *is simplicial if and only if* $\dim \text{conv } \overline{V}(J) = \dim \text{conv } \overline{V}$ *for every nonempty coface* $V(J) \subset V$.

5. *Let P and P' be d-polytopes in R^d and let $V = (v_1, \cdots, v_n) = $ vert P and $V' = (v'_1, \cdots, v'_n) = $ vert P'. The polytopes P and P' are combinatorially equivalent under a mapping φ of $\mathscr{F}(P)$ onto $\mathscr{F}(P')$ such that $v'_{\vartheta(i)} = \varphi v_i$ for $i = 1, \cdots, n$ and a permutation ϑ of $1, 2, \cdots, n$, if and only if*

(*) *for every $J \subset \{1, \cdots, n\}$, the condition $0 \in$ relint conv $\overline{V}(J)$ is equivalent to $0 \in$ relint conv $\overline{V}'(\vartheta(J))$, where $\vartheta(J) = \{\vartheta(j) \mid j \in J\}$.*

The Gale-transforms $\overline{V}$ and $\overline{V}'$ of two sets V and V' shall be called *isomorphic* provided (*) holds. Thus theorem 5 amounts to saying that two polytopes P and P' are combinatorially equivalent if and only if the Gale-transforms of the n-tuples of their vertices are isomorphic.

6. *Let P, P' be d-polytopes in R^d, and let $(v_1, \cdots, v_n) = $ vert P and $(v'_1, \cdots, v'_n) = $ vert P'. Then P and P' are affinely equivalent by an affine transformation A of R^d onto itself such that $v'_i = Av_i, i = 1, \cdots, n$, if and only if there exists a nonsingular linear transformation B of R^{n-d-1} onto itself such that $\overline{v}'_i = B\overline{v}_i$ for $i = 1, \cdots, n$.*

7. *Let P and P' be d-polytopes in R^d and let $(v_1, \cdots, v_n) = $ vert P and $(v'_1, \cdots, v'_n) = $ vert P'. Then P and P' are projectively equivalent by a projective transformation A permissible for P and such that $v'_i = Av_i$, $i = 1, \cdots, n$, if and only if there exist positive reals $\lambda_1, \cdots, \lambda_n$, and a regular linear transformation B of R^{n-d-1} onto itself, such that $\overline{v}'_i = \lambda_i B\overline{v}_i$ for all $i = 1, \cdots, n$.*

Let $X = (x_1, \cdots, x_n)$ and aff $X = R^d$. For any Gale-transform $\overline{X}$ of X we define the *Gale-diagram* $\hat{X}$ of X by $\hat{X} = (\hat{x}_1, \cdots, \hat{x}_n)$, where

$$\hat{x}_i = 0 \quad \text{if } \overline{x}_i = 0,$$
$$\hat{x}_i = \frac{\overline{x}_i}{\|\overline{x}_i\|} \text{ if } \overline{x}_i \neq 0,$$

and $\|x\|$ is the (Euclidean) length of the vector x.

Clearly $\hat{X}$ is a subset of $\{0\} \cup S^{n-d-2}$, where S^k denotes the k-sphere, i.e. the boundary of the unit ball of R^{k+1} (with center at 0).

Isomorphism of two Gale-diagrams is defined by the same condition which was used in the definition of isomorphic Gale-transforms. The reader may verify that theorems 1, 2, 3, 4, 5, and 7 are valid also if 'Gale-diagrams' are substituted for 'Gale-transforms' throughout. It should be noted that the Gale-diagram $\hat{X}$ of a set X coincides with the Gale-transform $\overline{Y}$ of some Y if and only if $\sum_{i=1}^{n} \hat{x}_i = 0$; in this case Y is projectively equivalent to X under a mapping permissible for conv X. Another important property is: Whenever $Z = (z_1, \cdots, z_n)$ is an n-tuple of (not necessarily different) points of $\{0\} \cup S^{n-d-2}$ such that aff $Z = R^{n-d-1}$ and $0 \in$ int conv Z, there exists $X = (x_1, \cdots, x_n) \subset R^d$ such that aff $X = R^d$ and $z_i = \hat{x}_i$ for $i = 1, \cdots, n$.

An additional property of Gale-diagrams is

8. *Let* $X = (x_1, \cdots, x_n)$ *be an n-tuple of points in* R^d, *and let* $\hat{X} = (\hat{x}_1, \cdots, \hat{x}_n)$ *be a Gale-diagram of* X. *Let* H *be a hyperplane which strictly separates* x_n *from* $x_1, \cdots, x_{n-1}$; *let* $y_i = H \cap [x_i, x_n]$ *for* $1 \le i \le n - 1$, *and* $Y = (y_1, \cdots, y_{n-1})$. *Then the* $(n - 1)$-*tuple* $\hat{Y} = (\hat{x}_1, \cdots, \hat{x}_{n-1})$ *is a Gale-diagram of* Y. *In particular, if* $X = $ vert P *for a polytope* P, *and if all* $[x_i, x_n], 1 \le i \le n - 1$, *are edges of* P, *then* $\hat{Y}$ *is a Gale-diagram of the vertex figure* $H \cap P$ *of* P *at* x_n.

5.5   Existence of Combinatorial Types*

The combinatorial structure of a given d-polytope P is obviously determined by the *scheme* of P. The *scheme* of P is an enumeration of the vertices of P, of the 1-faces of P, of the 2-faces, $\cdots$, of the $(d - 1)$-faces of P, where each face is designated by the subset of vert P contained in it. If the schemes of two polytopes are given, it is obviously possible to decide whether the two polytopes are isomorphic (or whether they are dual) to each other. Unless the number of vertices is very small, the actual carrying out of the task may be rather time-consuming; however, there is no question of principle involved.

As a consequence, if we are presented with a finite set of polytopes it is possible to find those among them which are of the same combinatorial type. It may seem that this fact, together with theorem 5.2.1 which determines all the polytopes obtainable as convex hulls of a given polytope and one additional point, are sufficient to furnish an *enumeration* of combinatorial types of d-polytopes. By this we mean a procedure which yields an inductive determination of all $c(k, d)$ combinatorial types of d-polytopes which have a given number k of vertices. However, from the result of exercise 5.2.1 it follows that it may be necessary to use different representatives of a given combinatorial type in order to obtain all the polytopes having one vertex more which are obtainable from polytopes of the given combinatorial type. Therefore it is not possible to carry out the inductive determination of all the combinatorial types in the fashion suggested above.

---

* The author is indebted to Professor M. O. Rabin for helpful discussions on the subject of this section.

This naturally leads to the question whether there is *any* algorithm which would yield all the different combinatorial types of polytopes. The answer is affirmative but—at least at present—the proof uses a theorem of Tarski on the decidability of first-order sentences in the field of real numbers. In order to avail ourselves this rather heavy tool, we start by introducing the notion of an *abstract scheme*.

An *abstract scheme* is a family $\mathscr{V}$ of nonempty subsets of a set $V = \{v_1, v_2, \cdots, v_k\}$, such that $V \notin \mathscr{V}$ but each singleton $\{v_i\}$, $1 \leq i \leq k$, belongs to $\mathscr{V}$. Clearly, the scheme of a d-polytope P is an abstract scheme provided $V = \text{vert } P$ and the family $\mathscr{V}$ consists of the sets vert F for all proper faces F of P. We shall say that an abstract scheme is *realized* by the d-polytope P provided it is isomorphic (in the obvious sense) to the scheme of P.

The key step in the enumeration of d-polytopes is:

1. *There exists an algorithm for deciding whether there exists a d-polytope which realizes a given abstract scheme.*

From this there results:

2. *The enumeration problem for d-polytopes is solvable, i.e. there exists an algorithm for the determination of all the different combinatorial types of d-polytopes with k vertices.*

Assuming, for the moment, the assertion 1, the proof of theorem 2 is immediate. Clearly all abstract schemes with card $V = k$ are easily determinable. By theorem 1, those abstract schemes which are realizable by d-polytopes may be singled out. Finally, as mentioned above, a single representative of each combinatorial type may be chosen.

In order to prove theorem 1, we start by recalling some definitions from mathematical logic. A *statement in elementary algebra* is any expression constructed according to the usual rules and involving only the symbols $+$, $-$, $.$, $=$, $<$, $(,)$, $[,]$, 0, 1, $\vee$ (disjunction), & (conjunction), $\sim$ (negation), $\forall$ (universal quantifier), $\exists$ (existential quantifier), and real variables. The quantifiers $\forall$ and $\exists$ act only upon the real variables.

The part of Tarski's theorem (Tarski [1], Seidenberg [1], Cohen [1]) which we need may be formulated as follows:

Every statement in elementary algebra containing no free variables (i.e. such that each variable is bound either by $\forall$ or by $\exists$) is effectively decidable, that is, there exists an algorithm for deciding whether any such statement is true or false.

Now, given an abstract scheme the question of its realizability by a

d-polytope may be put in the following form, in which Tarski's theorem becomes applicable.

The scheme will be realizable by a d-polytope if and only if it is possible to find reals $x_{i,j}$, where $1 \leq i \leq k = \text{card } V$ and $1 \leq j \leq d$, such that the following statements are equivalent for every nonempty set $W \subset V$:

(i) $W \in \mathscr{V}$.

(ii) There exist reals y_j and c, where $1 \leq j \leq d$, such that $\sum_{j=1}^{d} y_j^2 > 0$ and

$$\sum_{j=1}^{d} x_{i,j} y_j \begin{cases} = c & \text{if } v_i \in W \\ > c & \text{if } v_1 \notin W. \end{cases}$$

The above, obviously, expresses the quest for vertices such that appropriate sets of them form proper faces of the polytope, while other sets do not form faces. In this formulation, Tarski's theorem shows that the problem is effectively decidable, and the proof of theorem 1 is completed.

It would be rather interesting to have an elementary proof of theorem 2. In case of 3-polytopes such a proof may be derived from Steinitz's theorem (see chapter 13). An elementary proof of theorem 2 would also be interesting in connection with the solution of the following problem of V. Klee:

Is every combinatorial type of polytopes rational, that is, does every combinatorial type of polytopes have representatives all vertices of which have rational coordinates in a suitable system of Cartesian coordinates?

In other words, instead of dealing with the real d-dimensional space, one could consider polytopes in the *rational d-dimensional space*. Though some of the proofs would have to be changed, many of the results on polytopes contained in the preceding sections remain valid in the rational space (see, however, theorem 5.1.11). But Tarski's theorem does not apply to the field of rational numbers and therefore it does not lead to the solution of the enumeration problem for rational polytopes.

As we shall see in chapter 13, a theorem of Steinitz leads to an affirmative solution of Klee's problem for 3-polytopes. For higher dimensions the enumeration problem for rational polytopes is still unsolved; however, a negative solution to Klee's problem in sufficiently high dimensions has recently been obtained by M. A. Perles (see theorem 4 below).

By a slight modification of the proof of theorem 2 it is possible to show that the different combinatorial types of d-complexes with a given number of vertices are effectively determinable. Regarding complexes, however, even the case of 2-complexes in R^3 of Klee's problem is still unsolved.

An interesting side-light is shed on the above problems by the following observation on configurations. A *configuration* is a finite set of points and lines in a projective plane, with prescribed incidence relations. Since any two points in a projective plane are on a line, we shall shorten the description of configurations by indicating only those lines which are incident to at least three points of the configuration. (For the related notion of *arrangements* see chapter 18.)

3. *There exist configurations in the real projective plane which are not realizable in the rational projective plane (or in any rational projective space).* *

A very simple configuration $\mathscr{C}$ with this property consists of 9 points and 9 lines.† Let the points of $\mathscr{C}$ be A, B, C, D, E, F, G, H, I, and let the following sets of more than two points (and only those sets) be collinear: ABEF, ADG, AHI, BCH, BGI, CEG, CFI, DEI, DFH. The realizability of the configuration $\mathscr{C}$ in the real projective plane is easily established on hand of figure 5.5.1, which is derived in an obvious way from the regular pentagon. On the other hand, the reader can easily verify that *every* realization of $\mathscr{C}$ in the real plane is projectively equivalent to the configuration of figure 5.5.1 in one of the two ways: either as indicated in figure 5.5.1, or as indicated in figure 5.5.2. The cross-ratio (A, B; E, F) is

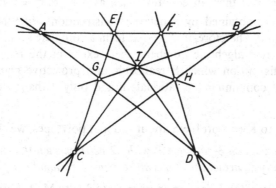

Figure 5.5.1

* For a similar difference between the projective geometries over the real and the rational fields, see the notion of 'accessible points' in Coxeter [6, p. 126].

† It may be conjectured that no configuration of less than 9 points has this property.

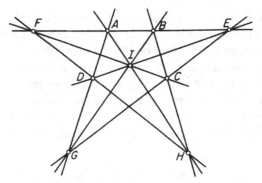

Figure 5.5.2

$\frac{1}{2}(3 - \sqrt{5})$ in the first case, and $\frac{1}{2}(3 + \sqrt{5})$ in the second. Therefore $\mathscr{C}$ may not be realized by points having rational coordinates—hence $\mathscr{C}$ is not realizable in the rational projective plane.

As a matter of fact,* if F is an ordered field such that every configuration realizable in the real projective plane is realizable also in the projective plane over F, then F contains a subfield isomorphic to the field of all real algebraic numbers. The proof is very simple on observing that if on a line L a projective system of coordinates is introduced by specifying points 0, 1, and ∞, for any point x on L it is possible to construct the point $\sum_{i=0}^{n} \alpha_i x^i \in L$ for each choice of rational numbers α_i. To 'construct' means to draw appropriate lines in the plane, i.e. to specify a suitable configuration. But then an equation such as $\sum_{i=0}^{n} \alpha_i x^i = 0$ means that a suitable line, obtained by a definite construction which starts at x, intersects L in 0. Therefore, if $x_1, \cdots, x_k$ form a complete set of conjugate, nonrational, real algebraic numbers, there exists in the real projective plane a configuration which is realizable in a projective plane over an ordered field containing the rationals if and only if that field contains $x_1, \cdots, x_k$.

Returning to Klee's problem about rational polytopes, we shall show:

4. *There exists an 8-polytope P with 12 vertices such that no polytope combinatorially equivalent to P has all vertices at rational points.*

Theorem 4 and the following proof of it are due to M. A. Perles (private communication). It is easily seen that the construction of P could be modified to yield polytopes of any dimension $d \geq 8$ having any number $v \geq d + 4$ of vertices, which are not rationally realizable.

* For this remark the author is indebted to Professor H. A. Heilbronn.

We shall first describe the construction of P. We consider the configuration $\mathscr{C}$ of 9 points shown in figure 5.5.1, and assume its plane to be $\{(x_1, x_2, x_3) \in R^3 \mid x_3 = 1\}$. Using the 9 points of this configuration, we form the 12-tuple

$$\hat{V} = \left(\frac{A}{\|A\|}, \frac{B}{\|B\|}, \frac{C}{\|C\|}, \frac{D}{\|D\|}, \frac{E}{\|E\|}, \frac{F}{\|F\|}, \frac{G}{\|G\|}, \frac{H}{\|H\|}, -\frac{F}{\|F\|}, -\frac{G}{\|G\|}, \right.$$
$$\left. -\frac{H}{\|H\|}, -\frac{I}{\|I\|} \right).$$

It is easily checked that $\hat{V}$ satisfies the conditions (see section 5.4) sufficient for it to be the Gale-diagram of the 12-tuple of vertices of some 8-polytope P with 12 vertices $v_1, \cdots, v_{12}$. Let now P' be an 8-polytope combinatorially equivalent to P, let $V' = \text{vert } P' = (v'_1, \cdots, v'_{12})$, and let v'_i correspond to v_i for $i = 1, \cdots, 12$. Considering the Gale-diagram $\hat{V}'$ of V' we note that the assumed isomorphism between $\hat{V}$ and $\hat{V}'$ implies that a subset of $\hat{V}$ is on a line, or in a plane, through the origin if and only if the corresponding subset of $\hat{V}'$ has the same property. (The reader should check that this property results from the manner in which plus or minus signs were assigned to the points of the configuration $\mathscr{C}$ used in defining $\hat{V}$.) Therefore, the 12 points of $\hat{V}'$ are on 9 lines $L_1, \cdots, L_9$ through the origin 0. The intersections $A', \cdots, I'$ of the lines L_i with a suitable plane L (such that $0 \notin L$) determine 9 points which yield a configuration equivalent to $\mathscr{C}$. If it were possible for P' to have all vertices at rational points, the Gale-diagram $\hat{V}'$ could be chosen so that the lines L_i have rational direction coefficients. If L is then chosen in a similar manner, all the points $A', \cdots, I'$ would be rational—which is impossible in view of theorem 3.

This completes the proof of theorem 4.

In a certain sense, the result of theorem 4 is best possible: As we shall see in exercise 6.5.3, every d-polytope with at most $d + 3$ vertices is rationally realizable. On the other hand, the dimension 8 is probably too high—but no lower-dimensional, rationally not realizable polytopes are known.

Exercises

1. Prove the effective determinability of the (number of) different combinatorial types of d-complexes with a given number of vertices.

2. Prove the assertion made about the realizations of the configuration $\mathscr{C}$ of figure 5.5.1.

3. Show that the polytope P constructed in the proof of theorem 4 is projectively unique (see exercise 4.8.30), but that P has a facet which is not projectively unique. (Hint: Consider the coface $\left(\dfrac{F}{\|F\|}, -\dfrac{F}{\|F\|}\right)$.)

4. Show that the configuration of 11 points and 11 lines in figure 5.5.3 (the 'projective construction' of $\sqrt{2}$) is not rationally realizable. Deduce from it the existence of an 11-polytope with 15 vertices which is projectively unique and not rationally realizable.

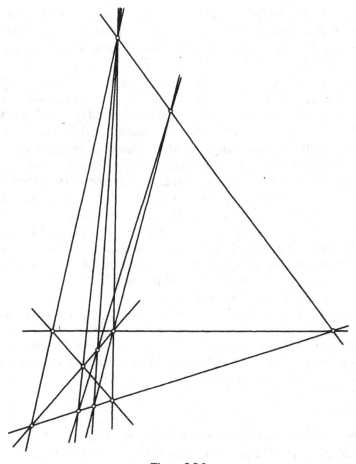

Figure 5.5.3

5.6 Additional notes and comments

Sections and projections.

The representation of a polytope as a simplex or cube intersected with an affine subspace is one of the basic steps in the construction and formulation of linear programs, the transformation to (various kinds of) "standard form". The key idea for this is the introduction of *slack variables* (see Dantzig [1, Sect. 4-5]), which here appears in thin disguise in exercise 5.1.3. The "is a section" in theorem 5.1.1 may be interpreted as "is projectively equivalent to", or more restrictively as "is affinely equivalent to".

Page 73 gives a glimpse of the now very powerful connections to functional analysis (Banach space theory). We refer to Ball [a] and Matoušek [b, Chap. 14] for introductions. See also Giannopoulos–Milman [a].

Gale-diagrams.

On Perles' theory of Gale-diagrams, for which this chapter is the original published source, Grünbaum wrote: "The reader will find it well worth his while to become familiar with the concepts of Gale-transforms and Gale-diagrams, since for many of the results obtained through them no alternative proofs have been found so far. It is very likely that the method will yield many additional results." It did! Among them are:

o Gale-diagrams can be formulated in terms of linear programming duality (the Farkas lemma). They were interpreted in terms of oriented matroid duality by McMullen [f] (see also Ziegler [a, Lect. 6]).

o Mnëv's [a] [b] universality theorem for d-polytopes with $d+4$ vertices is a vast generalization and extension of Perles' example of a non-rational 8-polytope with 12 vertices: *For every semi-algebraic variety V defined over the integers, there is a d-polytope with $d+4$ vertices whose realization space is "stably equivalent" to V.* Later, Richter-Gebert [b] [c] (with a different method) achieved a stunning universality theorem for 4-polytopes with arbitrarily many vertices. (See also Richter-Gebert–Ziegler [a] and Günzel [a].)

o Bokowski and Sturmfels [d] formalized "affine Gale-diagrams" as a different way to handle and visualize the ("linear") Gale-diagrams explained here; these would augment figure 5.5.1 by the signs of the vector configuration $\hat{V}$ that is derived from it in the proof of theorem 5.5.4.

No non-rational polytope with fewer than 12 vertices has been constructed yet. However, Richter-Gebert's results include the construction of non-rational 4-polytopes (with roughly 30 vertices). An alternative construction method, the "Lawrence construction", (Billera–Munson [a], Björner et al. [a, Sect. 9.3]), produces non-rational $(n+2)$-dimensional polytopes with $2n$ vertices from non-rational configurations of $n \geq 9$ points in the plane (as in figure 5.5.2).

Enumeration of combinatorial types.

Lindström [a] has shown that every combinatorial type of d-polytope is realizable in the d-dimensional vector space over the field of algebraic numbers.

Enumerating combinatorial types by Tarski's procedure for solving polynomial inequalities is impractical. The same holds for Collins' [a] "cylindrical algebraic decomposition". Nevertheless, one has enumerated and classified

o 4-polytopes with 8 vertices (1294 types: Altshuler–Steinberg [b] [c]),
o simplicial 4-polytopes with 9 vertices (1142 types: Altshuler–Bokowski–Steinberg [a]; see also Engel [a]),
o neighborly 6-polytopes with 10 vertices (37 types: Bokowski–Shemer [a]),
o and partially the neighborly 4-polytopes with 10 vertices (exactly 431 types: Altshuler [a], Bokowski–Sturmfels [a], and Bokowski–Garms [a]).

The general approach towards such results is explained in Bokowski–Sturmfels [b]. The most successful strategy has three essential steps:

1. Enumerate the combinatorial types in a larger class (e. g., combinatorial or shellable spheres).
2. For each type enumerate the compatible oriented matroids, i. e., orientation data for a hypothetical realization of the vertices (it may be that none exist).
3. Find coordinates for these, or prove that none exist. (For this, there are "final polynomial" proofs, according to Bokowski–Richter–Sturmfels [a]; the special "biquadratic final polynomials" of Bokowski–Richter–Gebert [a] can be found efficiently by linear programming.)

Decidability of the existence of combinatorial types of rational polytopes is open, related to Hilbert's tenth problem: Is it decidable whether a given rational polynomial in several variables has a rational solution? (See Sturmfels [b].)

Algorithmic aspects.

Both the inductive construction of polytopes with the "beneath-beyond" method and the perturbation via "pulling" and "pushing" of vertices (pages 80–82), are essential for the algorithmic treatment of polytopes. For example, beneath-beyond approaches appear in convex hull algorithms; see de Berg et al. [a], or Brönniman [a]. Pulling and pushing are key tools for constructing of triangulations (of the polytope, and/or of the polytope boundary); see Lee [b].

2-Simple 2-simplicial polytopes.

Exercise 5.2.13(ii) was not established by Shephard [7]. In view of the difficulty in constructing 4-polytopes of type $(2,2)$, as discussed in section 4.9, the Walkup conjecture seems daring; it is still open. However, Problem 5.2.15 has a positive answer: It may be derived from theorem 5.2.5 using "connected sums" (see Ziegler [a, Example 8.41]) of the form $P \# P^*$, for simplicial P.

CHAPTER 6

Polytopes with Few Vertices

The aim of the present chapter is a discussion, as complete as possible at present, of polytopes with 'few' vertices. In section 6.1 we deal with d-polytopes having $d + 2$ vertices; Sections 6.2 and 6.3 present two different approaches to the classification of d-polytopes with $d + 3$ vertices, while section 6.4 deals with a remarkable phenomenon concerning centrally symmetric polytopes.

6.1 d-Polytopes with $d + 2$ Vertices

We start with a discussion of simplicial d-polytopes with $d + 2$ vertices. Each $d + 1$ vertices of such a polytope P must be affinely independent, since otherwise P would be a pyramid on a $(d - 1)$-basis different from a simplex. Let v be a vertex of P. Then the remaining $d + 1$ vertices of P determine a d-simplex T^d, and $P = \text{conv}(T^d \cup \{v\})$. The vertex v is beyond a certain number k of facets* of T^d, where $1 \le k \le d - 1$. Since all k-tuples of facets of T^d are combinatorially equivalent, all polytopes P for which v is beyond k facets of T^d are equivalent. Let their combinatorial type, as well as any polytope of that type, be denoted by T^d_k.

We shall first show that the types T^d_k and T^d_{d-k} coincide. Let P be a simplicial d-polytope with $d + 2$ vertices $\{v_0, v_1, \cdots, v_{d+1}\}$. Since each $d + 1$ points v_i are affinely independent there exists, by Radon's theorem 2.3.6, a unique k with $1 \le k \le [\frac{1}{2}d]$, and reals β_i, γ_j for $0 \le i \le k < j \le d + 1$, such that (possibly after a permutation of the indices) we have

$$\sum_{0 \le i \le k} \beta_i v_i = \sum_{k+1 \le j \le d+1} \gamma_j v_j,$$

where

$$\sum_{0 \le i \le k} \beta_i = \sum_{k+1 \le j \le d+1} \gamma_j = 1$$

and $\beta_i > 0, \gamma_j > 0$ for all i, j.

---
* The reader is reminded that a 'facet' is a $(d - 1)$-face of a d-polytope.

97

In other words, it is possible to split the vertices of P into two groups V_1 and V_2, containing $k + 1$ respectively $d - k + 1$ of the vertices, in such a way that the simplices conv V_1 and conv V_2 intersect in one single point 0, relatively interior to both simplices.

Let us consider the facets of P. Each facet is determined by the two vertices of P which do not belong to it. These two vertices may not belong to the same V_i since then the point 0 would be in the facet although it belongs to the interior of P. Therefore each facet is the convex hull of some k points of V_1 and some $d - k$ points of V_2. On the other hand, for every choice of k points of V_1 and of $d - k$ points of V_2, the convex hull of their union is a facet F of P since the remaining two vertices of P are both in that open halfspace determined by aff F which contains 0.

Consequently, each $v_i \in V_1$ is beneath $d - k + 1$ facets of P, while each $v_j \in V_2$ is beneath $k + 1$ facets of P. If T_m is the d-simplex determined by the vertices of P different from v_m, by theorem 5.2.1 those facets of P for which v_m is beneath them are also facets of T_m. Therefore $v_i \in V_1$ is beyond $d + 1 - (d - k + 1) = k$ facets of T_i, while $v_j \in V_2$ is beyond $d + 1 - (k + 1) = d - k$ facets of T_j. In other words, P is of the combinatorial type T_k^d and also of the type T_{d-k}^d, and therefore the types T_k^d and T_{d-k}^d coincide. This establishes

1. *There exist $[\frac{1}{2}d]$ different combinatorial types of simplicial d-polytopes with $d + 2$ vertices. A polytope T_k^d of the kth type, $k = 1, 2, \cdots, [\frac{1}{2}d]$, is obtained as the convex hull of the union of a d-simplex T^d with a point which is beyond k facets of T^d. A polytope of the same type results if the point is beyond $d - k$ facets of T^d. The polytope $T_{[\frac{1}{2}d]}^d$ is combinatorially equivalent to the cyclic polytope $C(d + 2, d)$.*

In order to prove the last assertion, it is enough to note that if $x_0, x_1, \cdots, x_d, x_{d+1}$ are the vertices of $C(d + 2, d)$ arranged according to their order on the moment curve, Gale's evenness condition (theorem 4.7.2) implies that any one-to-one correspondence between the points of V_1 (respectively V_2) and the x_i's with odd (respectively even) i establishes a combinatorial equivalence between $T_{[\frac{1}{2}d]}^d$ and $C(d + 2, d)$.

The proof of theorem 1 shows also that $f_{d-1}(T_k^d) = (k + 1)(d - k + 1)$. This shall be generalized in the next theorem.

2. *For $0 \leq m \leq d - 1$, the number of m-faces of T_k^d is*

$$f_m(T_k^d) = \binom{d + 2}{d - m + 1} - \binom{k + 1}{d - m + 1} - \binom{d - k + 1}{d - m + 1}.$$

PROOF The total number of $(m + 1)$-tuples of vertices of T_k^d is $\binom{d + 2}{m + 1} = \binom{d + 2}{d + 1 - m}$. A given $(m + 1)$-tuple V^* of vertices of T_k^d determines an m-face of T_k^d if and only if there is a facet F of T_k^d which contains the $(m + 1)$-tuple V^*. Such a facet F exists, by the above, if and only if neither V_1 nor V_2 is contained in V^*. In other words, counting the number of m-faces of T_k^d, from the total of $\binom{d + 2}{m + 1}$ $(m + 1)$-tuples, we have to exclude those which contain V_1 and those which contain V_2. The number of the former is $\binom{d - k + 1}{m + 1 - (k + 1)} = \binom{d - k + 1}{d - m + 1}$, while the number of $(m + 1)$-tuples of the latter type is

$$\binom{k + 1}{m + 1 - (d - k + 1)} = \binom{k + 1}{d - m + 1}.$$

This completes the proof of theorem 2.

Because of the use we shall make below, it is worthwhile to amend the expressions for $f_m(T_k^d)$ given in theorem 2 so as to remain valid for all m. As easily checked, the formula yields the correct values $f_m = 1$ for $m = -1$, $f_m = 0$ for $m < -1$ and $m > d + 1$; the values for $m = d$ and $m = d + 1$ are too small by 1. Therefore an expression valid for all m is

$$f_m(T_k^d) = \binom{d + 2}{d - m + 1} - \binom{k + 1}{d - m + 1} - \binom{d - k + 1}{d - m + 1} + \delta_{d,m} + \delta_{d+1,m}.$$

We turn now to the problem of finding the polytopes T_k^d which have a maximal or a minimal number of m-faces, $m = 1, 2, \cdots, d - 1$. Since for $k < [\tfrac{1}{2}d]$ we have

$$\binom{k + 2}{d - m + 1} - \binom{k + 1}{d - m + 1} = \binom{k + 1}{d - m} \leq \binom{d - k}{d - m}$$

$$= \binom{d - k + 1}{d - m + 1} - \binom{d - k}{d - m + 1}$$

$$= \binom{d - k + 1}{d - m + 1} - \binom{d - (k + 1) + 1}{d - m + 1},$$

with strict inequality if and only if $k \leq m$, it follows that

(i) for $m < [\tfrac{1}{2}d]$

$$f_m(T_1^d) < f_m(T_2^d) < \cdots < f_m(T_m^d) < f_m(T_{m+1}^d) = f_m(T_{m+2}^d) = \cdots = f_m(T_{[\frac{1}{2}d]}^d);$$

(ii) for $m \geq [\frac{1}{2}d]$

$$f_m(T_1^d) < f_m(T_2^d) < \cdots < f_m(T_{[\frac{1}{2}d]}^d).$$

Hence we have

3. *For every simplicial d-polytope P with d + 2 vertices, and for every* $m = 1, 2, \cdots, d - 1$.

$$f_m(T_1^d) \leq f_m(P) \leq f_m(T_{[\frac{1}{2}d]}^d).$$

If $f_m(P) = f_m(T_1^d)$ *for some m with* $1 \leq m \leq d - 1$, *then* $P = T_1^d$. *If* $f_m(P) = f_m(T_{[\frac{1}{2}d]}^d)$ *for some m then: if* $[\frac{1}{2}d] - 1 \leq m \leq d - 1$, $P = T_{[\frac{1}{2}d]}^d$; *if* $1 \leq m < [\frac{1}{2}d] - 1$, $P = T_k^d$ *for some k satisfying* $m + 1 \leq k \leq [\frac{1}{2}d]$.

Let now P be a d-polytope with $d + 2$ vertices which is not simplicial. Then all but one of the vertices of P are contained in a hyperplane, and P is a d-pyramid having as basis a $(d - 1)$-polytope P_1 with $d + 1 = (d - 1) + 2$ vertices. If P_1 is a simplicial $(d - 1)$-polytope then $P_1 = T_k^{d-1}$ for some k with $1 \leq k \leq \frac{1}{2}(d - 1)$; if P_1 is not simplicial then it is a $(d - 1)$-pyramid with $(d - 2)$-basis P_2 which has $d = (d - 2) + 2$ vertices. Proceeding by induction we obtain

4. *Each d-polytope P with d + 2 vertices is, for suitable r and k with* $0 \leq r \leq d - 2$ *and* $1 \leq k \leq [\frac{1}{2}(d - r)]$, *an r-fold d-pyramid with* $(d - r)$-*dimensional basis* T_k^{d-r}. *Denoting such P by* $T_k^{d,r}$ *we have*

$$f_m(T_k^{d,r}) = \binom{d + 2}{d - m + 1} - \binom{k + r + 1}{d - m + 1} - \binom{d - k + 1}{d - m + 1} + \binom{r + 1}{d - m + 1},$$

for all $m = 0, 1, \cdots, d - 1$. *There are* $[\frac{1}{4}d^2]$ *different combinatorial types of d-polytopes with d + 2 vertices.*

PROOF Clearly $T_k^{d,0} = T_k^d$. In order to establish the expressions given for $f_m(T_k^{d,r})$, it is enough to combine the values given above for $f_m(T_k^d)$ (in the form valid for all m) with theorem 4.2.2. Thus

$$f_m(T_k^{d,r}) = \sum_i \binom{r}{i} f_{m-i}(T_k^{d-r})$$

$$= \sum_i \binom{r}{i} \left\{ \binom{d - r + 2}{m - i + 1} - \binom{k + 1}{m - i + k - d + r} \right.$$

$$\left. - \binom{d - r - k + 1}{m - i - k} + \delta_{d-r,m-i} + \delta_{d-r+1,m-i} \right\}$$

$$= \binom{d+2}{m+1} - \binom{k+r+1}{m+k-d+r} - \binom{d-k+1}{m-k}$$

$$+ \binom{r}{r+m-d} + \binom{r}{r+m-d-1},$$

which equals to the expression given above. Since $T_k^{d,r}$ is not of the same combinatorial type as $T_{k^*}^{d^*,r^*}$ unless $d = d^*$, $r = r^*$ and $k = k^*$, the number $[\frac{1}{4}d^2]$ of different combinatorial types follows easily from the inequalities $0 \leq r \leq d - 2$ and $1 \leq k \leq [\frac{1}{2}(d - r)]$.

From theorem 4 it follows that
(i) for $r \geq 0$ and $1 < k \leq [\frac{1}{2}(d - r)]$,

$$f_m(T_{k-1}^{d,r}) \leq f_m(T_k^{d,r}),$$

with strict inequality if and only if $k \leq m + 1$;
(ii) for $r > 0$ and $1 \leq k \leq [\frac{1}{2}(d - r)]$,

$$f_m(T_k^{d,r}) \leq f_m(T_k^{d,r-1}),$$

with strict inequality if and only if $d \leq m + k + r$.
Hence we have

5. *For every d-polytope P with $d + 2$ vertices and for every m with $1 \leq m \leq d - 1$,*

$$f_m(P) \geq f_m(T_1^{d,d-2})$$

and

$$f_m(P) \leq f_m(T_{[\frac{1}{2}d]}^{d,0}) = f_m(T_{[\frac{1}{2}d]}^d).$$

For any m, $1 \leq m \leq d - 1$, equality holds in the first relation if and only if $P = T_1^{d,d-2}$. For any m satisfying $[\frac{1}{2}d] - 1 \leq m \leq d - 1$, equality holds in the second relation if and only if $P = T_{[\frac{1}{2}d]}^{d,0} = T_{[\frac{1}{2}d]}^d$.

Combining theorems 3 and 5, an easy computation yields the following theorem.

6. *If r and k are such that $T_k^{d,r}$ has maximal possible number of m-faces, then $r = 0$ and $k = [\frac{1}{2}d]$ provided at least one of the following conditions is satisfied:*
    (i) *d is even and $m \geq [\frac{1}{2}d] - 1$;*
    (ii) *$T_k^{d,r}$ is simplicial (i.e., $r = 0$) and $m \geq [\frac{1}{2}d] - 1$;*
    (iii) *$m \geq [\frac{1}{2}d]$.*

6.2 d-Polytopes with $d + 3$ Vertices

The structure of d-polytopes with $d + 3$ vertices is much more complicated than that of d-polytopes with $d + 2$ vertices, and our knowledge of it is very recent. In the present section we shall restrict our attention to simplicial d-polytopes with $d + 3$ vertices and we shall present here a method of describing them which will enable us to solve questions about maximal, or minimal, numbers of faces of different dimensions. A different method of investigation will be used in section 6.3 to determine the possible combinatorial types of d-polytopes with $d + 3$ vertices, and to solve some additional problems.

Let $V = \{v_0, v_1, \cdots, v_{d+2}\}$ be the vertices of a simplicial d-polytope $P \subset R^d$. Without loss of generality we may assume that every $d + 1$ points of V are affinely independent. Let $V' = V \sim \{v_{d+2}\} = \{v_0, \cdots, v_{d+1}\}$ and $P' = \text{conv } V'$. Then there exists a unique decomposition $V' = X \cup Y$, $X \cap Y = \varnothing$ such that conv $X \cap$ conv Y is a single point, which we take as the origin 0. Let us denote $X = \{x_0, \cdots, x_s\}$, $Y = \{y_0, \cdots, y_t\}$; then $s \geq 1$, $t \geq 1$, and $s + t = d$. Clearly, $P^* = \text{conv } X$ and $P^{**} = \text{conv } Y$ are simplices of dimensions s respectively t. Each facet F of P' is the convex hull of the union of a facet of P^* and a facet of P^{**}. Therefore the facets of P' may be labeled by a pair of integers (p, q), where $0 \leq p \leq s, 0 \leq q \leq t$, in such a way that $F(p, q) = \text{conv}(V' \sim \{x_p, y_q\})$.

Let $H^* = \text{aff } X$ and $H^{**} = \text{aff } Y$; then $R^d = H^* + H^{**}, H^* \cap H^{**} = \{0\}$. Since X respectively Y are affine bases of H^* respectively H^{**}, there is a unique relation of the form

$$0 = \sum_{i=0}^{s} \lambda_i^* x_i = \sum_{j=0}^{t} \mu_j^* y_j \quad \text{where} \quad \sum_{i=0}^{s} \lambda_i^* = \sum_{j=0}^{t} \mu_j^* = 1$$

and $\lambda_i^* > 0$ and $\mu_j^* > 0$ for all i and j.

Also, each $z \in R^d$ has a unique representation

$$z = \sum_{i=0}^{s} \lambda_i x_i + \sum_{j=0}^{t} \mu_j y_j \quad \text{with} \quad \sum_{i=0}^{s} \lambda_i = \sum_{j=0}^{t} \mu_j = 1.$$

Since $0 \in \text{int } P'$, z will be beyond a facet $F(p, q)$ of P' if and only if for some κ, $0 < \kappa < 1$, we have $\kappa z \in \text{aff } F(p, q)$. Using the above representation of 0 it follows that

$$z = \sum_{\substack{0 \leq i \leq s \\ i \neq p}} \left(\lambda_i - \frac{\lambda_p}{\lambda_p^*} \lambda_i^* \right) x_i + \sum_{\substack{0 \leq j \leq t \\ j \neq q}} \left(\mu_j - \frac{\mu_q}{\mu_q^*} \mu_j^* \right) y_j,$$

and therefore the condition $\lambda z \in \text{aff } F(p,q)$, i.e. $z \in (1/\kappa) \text{aff } F(p,q)$, becomes

$$\frac{1}{\kappa} = \sum_{\substack{0 \leq i \leq s \\ i \neq p}} \left(\lambda_i - \frac{\lambda_p}{\lambda_p^*} \lambda_i^* \right) + \sum_{\substack{0 \leq j \leq t \\ j \neq q}} \left(\mu_j - \frac{\mu_q}{\mu_q^*} \mu_j^* \right) = 2 - \left(\frac{\lambda_p}{\lambda_p^*} + \frac{\mu_q}{\mu_q^*} \right).$$

Since $0 < \kappa < 1$ if and only if $1/\kappa > 1$, a necessary and sufficient condition for z to be beyond $F(p,q)$ is $\lambda_p/\lambda_p^* + \mu_q/\mu_q^* < 1$. In other words, z is beneath $F(p,q)$ if and only if $\lambda_p/\lambda_p^* + \mu_q/\mu_q^* > 1$.

Let us assume now that $z = v_{d+2}$, and that the labeling of X and Y is such that $\lambda_0/\lambda_0^* > \lambda_1/\lambda_1^* > \lambda_2/\lambda_2^* > \cdots > \lambda_s/\lambda_s^*$ and $\mu_0/\mu_0^* > \mu_1/\mu_1^* > \mu_2/\mu_2^* > \cdots > \mu_t/\mu_t^*$. Then, clearly, if v_{d+2} is beneath $F(p,q)$ it is also beneath every $F(p',q')$ with $p' \leq p, q' \leq q$.

We shall represent the facial structure of P' by a *diagram* in a $(p;q)$-plane, consisting of the lattice points $(p;q)$ with $0 \leq p \leq s, 0 \leq q \leq t$, the facet $F(p,q)$ of P' being represented by the point $(p;q)$. For the polytope $P = \text{conv}(P' \cup \{v_{d+2}\})$ we shall use the following representation in the $(p;q)$-diagram of P': a point $(p;q)$ of the diagram shall be *marked*, e.g. by a star, if and only if v_{d+2} is beneath $F(p,q)$. For given P', this *star-diagram* of P is clearly determined by P in a unique way. From our conventions it follows that if $(p;q)$ is starred so are all $(p';q')$ with $0 \leq p' \leq p$ and $0 \leq q' \leq q$. Thus the general appearance of a star-diagram is as illustrated in figure 6.2.1.*

From the developments so far it is not clear whether every star-diagram satisfying the above conditions is indeed the star-diagram of a polytope. The answer to this query is affirmative (see theorem 6.3.4), but we do not need it for the purpose of the present discussion. Our aim here is to determine the changes in the facial structure of P accompanying the addition or deletion of certain stars in the star-diagram. Though it would require a certain amount of technicalities we could (without reference to polytopes) define 'faces' of various dimensions of star-diagrams, investigate changes in their number, and determine the extremal values. To solve the extremal problem for polytopes we would only have to show that the star-diagrams with an extremal number of 'faces' are indeed star-diagrams of polytopes. Though not invoking theorem 6.3.4 (the proof of which is independent of the present considerations), we shall refrain from this

---

* Note that for every polytope P the points $(0;0)$, $(0;1)$ and $(1;0)$ of the star-diagram of P are necessarily starred, while $(s;t)$ is necessarily without a star.

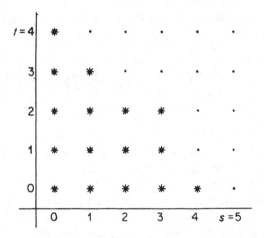

Figure 6.2.1

complication; nevertheless, we shall explicitly indicate polytopes corresponding to the extremal values of f_k.

Using theorem 5.2.1, we shall investigate the change of $f_k(P)$ resulting
from the addition of a star to a point $(b; c)$ of the star-diagram. According
to the definition of star-diagrams the assumption that $(b; c)$ may have a
star as well as be without it implies that $(b; c)$ is a point of the star diagram
such that all points $(p; c)$ and $(b; c)$ with $p < b, q < c$, have stars, and those
with $p > b$ or $q > 0$ do not have stars. If an $(m + r + 1)$-face F of P' has
vertices $\{x_{i_0}, \cdots, x_{i_m}; y_{j_0}, \cdots, y_{j_r}\}$, then F is contained in all the facets
$F(p, q)$ of P' such that $p \notin \{i_0, \cdots, i_m\}$ and $q \notin \{j_0, \cdots, j_r\}$. Now, if
$F = \text{conv}\{x_{i_0}, \cdots, x_{i_m}; y_{j_0}, \cdots, y_{j_r}\}$ is a face of P' with the property that
v_{d+2} is beyond all the facets of P' containing F if $(b; c)$ is not starred, but
v_{d+2} becomes beneath at least one facet (namely $F(b, c)$) of P' containing
F if $(b; c)$ is starred, then necessarily

$$b \notin \{i_0, i_1, \cdots, i_m\} \supset \{0, 1, \cdots, (b - 1)\}$$

and

$$c \notin \{j_0, j_1, \cdots, j_r\} \supset \{0, 1, \cdots, (c - 1)\}.$$

Therefore, putting $m + r + 1 = k$, it follows that there are

$$\binom{s - b}{m + 1 - b}\binom{t - c}{r + 1 - c}$$

such k-faces with given m and r, and thus there are altogether

$$\sum_{m+r+1=k} \binom{s-b}{m+1-b}\binom{t-c}{r+1-c}$$

k-faces of P' which are faces of P if $(b;c)$ is starred, and are not faces of P if $(b;c)$ is not starred.

Unless $m = s - 1$ and $r = t - 1$, the same k-faces of P' will serve as bases of pyramidal $(k + 1)$-faces of P with apex v_{d+2} if $(b;c)$ is starred, and not be bases of such faces if $(b;c)$ is not starred. On the other hand, a similar counting argument shows that the starring of $(b;c)$ will disqualify

$$\sum_{m+r+1=k} \binom{b}{m+1-(s-b)}\binom{c}{r+1-(t-c)}$$

such k-faces of P' to serve as bases of $(k + 1)$-faces of P containing v_{d+2}, while they are such bases if $(b;c)$ is not starred.

Therefore, the increase $\Delta_k(b,c)$ in the number of k-faces of P which results from the starring of $(b;c)$ equals

$$\Delta_k(b,c) = \sum_{m+r+1=k} \binom{s-b}{m+1-b}\binom{t-c}{r+1-c}$$

$$+ \sum_{m+r+1=k-1} \binom{s-b}{m+1-b}\binom{t-c}{r+1-c}$$

$$- \sum_{m+r+1=k-1} \binom{b}{m+1-(s-b)}\binom{c}{r+1-(t-c)}$$

$$= \sum_m \binom{s-b}{m+1-b}\binom{t-c}{k-m-c} + \sum_m \binom{s-b}{m+1-b}$$

$$\times \binom{t-c}{k-1-m-c} - \sum_m \binom{b}{s-m-1}\binom{c}{t-k+m+1}$$

$$= \binom{s+t-b-c}{k-b-c+1} + \binom{s-t-b-c}{k-b-c} - \binom{b+c}{s+t-k}$$

$$= \binom{s+t+1-(b+c)}{s+t-k} - \binom{b+c}{s+t-k}.$$

Hence

$\Delta_k(b, c) > 0$ if and only if $b + c \leq k + 1$ and $2(b + c) < d + 1$;

$\Delta_k(b, c) < 0$ if and only if $b + c \geq d - k$ and $2(b + c) > d + 1$;

$\Delta_k(b, c) = 0$ if and only if either $k + 1 < b + c < d - k$, or $2(b + c) = d + 1$.

Taking into account that $b \leq s$ and $c \leq t$, it follows that $f_k(P)$ will be maximal provided $s = [\frac{1}{2}d]$, $t = d - [\frac{1}{2}d] = [\frac{1}{2}(d + 1)]$, and all the points $(b; c)$ with $b + c < \frac{1}{2}(d + 1)$ are starred. That such a star-diagram may be realized by a d-polytope is shown by the cyclic d-polytope $C(d + 3, d)$ with $d + 3$ vertices. Using Gale's 'evenness condition' (section 4.7) it is easy to see that this is indeed the star-diagram of $C(d + 3, d)$; for even $d = 2n$ the $d + 3$ points of the moment curve should appear in the order

$$x_0, y_n, x_1, y_{n-1}, x_2, \cdots, y_1, x_n, y_0, z,$$

while for odd $d = 2n + 1$ they should be preceded by y_{n+1}.

This proves

1. *For every* k, $1 \leq k \leq d - 1$, *and every simplicial d-polytope P with $d + 3$ vertices,*

$$f_k(P) \leq f_k(C(d + 3, d)).$$

Taking into account the lower semicontinuity of $f_k(P)$ as a function of P (see theorem 5.3.1) this implies

2. *For every* k, $1 \leq k \leq d - 1$, *and every d-polytope P with $d + 3$ vertices,*

$$f_k(P) \leq f_k(C(d + 3, d)).$$

The proof of theorem 1 may be strengthened in case $d = 2n$ is even. In this case

$\Delta_k(b, c) > 0$ if and only if $b + c \leq k + 1$ and $b + c \leq n$;

$\Delta_k(b, c) < 0$ if and only if $b + c \geq 2n - k$ and $b + c \geq n + 1$;

$\Delta_k(b, c) = 0$ if and only if $k + 1 < b + c < 2n - k$.

If, moreover, $k \geq n - 1$ then these criteria simplify and yield:

$\Delta_k(b, c) > 0$ if and only if $b + c \leq n$;

$\Delta_k(b, c) < 0$ if and only if $b + c > n$.

Thus in case $d = 2n$ and $k \geq n - 1$ the star-diagram maximizing $f_k(P)$ is uniquely determined by the condition that $s = t = n$ and all points $(b; c)$ with $b + c \leq n$ are starred. Hence

3. *For even* $d = 2n$, *if* P *is a simplicial* d-*polytope with* $d + 3$ *vertices such that* $f_k(P) = f_k(C(d + 3, d))$ *for some* k *satisfying* $n - 1 \leq k \leq 2n - 1$, *then* P *is combinatorially equivalent to* $C(d + 3, d)$.

Our next aim is to strengthen theorem 3 by showing that it remains valid for all d-polytopes with $d + 3$ vertices. Let P be a nonsimplicial d-polytope with $d + 3$ vertices; we shall derive a contradiction from the assumption that $f_k(P) = f_k(C(d + 3, d))$ for some k satisfying $n - 1 \leq k \leq 2n - 1$. By theorem 5.3.1, without loss of generality we may assume that P has only one nonsimplicial facet F, such that $f_0(F) = d + 1$ and F is a simplicial $(d - 1)$-polytope. Let V be a vertex of F and let Q be the convex hull of the vertices of P different from V. Using the notation from pages 102 and 103, let

$$X \cup Y = \text{vert } Q, X \cap Y = \varnothing, \text{conv } X \cap \text{conv } Y = \{0\},$$

and let $V = \sum_{i=1}^{s} \lambda_i x_i + \sum_{j=1}^{t} \mu_j y_j$, with $\sum_{i=1}^{s} \lambda_i = \sum_{j=1}^{t} \mu_j = 1$. Without loss of generality we may assume that $\lambda_i/\lambda_i^* \neq \lambda_k/\lambda_k^*$ and $\mu_j/\mu_j^* \neq \mu_k/\mu_k^*$ for $i \neq k \neq j$. Let L be a straight line such that $L \cap P = \{V\}$, and such that L is not contained in the affine hull of any proper face of P. Let V^+, V^- be points of L strictly separated by V, and let $P^+ = \text{conv}(Q \cup \{V^+\})$ and $P^- = \text{conv}(Q \cup \{V^-\})$. By the lower semicontinuity of $f_k(P)$ we shall have $f_k(P) \leq f_k(P^+)$ and $f_k(P) \leq f_k(P^-)$ for all V^+, V^- sufficiently near to V. Assuming, without loss of generality that Q is simplicial it follows that P^+ and P^- are simplicial. Since the difference between P^+ and P^- is that, relatively to Q, V^+ and V^- differ in their position (beneath or beyond) with respect to some facets of Q, by theorem 3 it is not possible that both P^+ and P^- have the maximal number of k-faces, for some k with $n - 1 \leq k \leq 2n - 1$. But $f_k(P) \leq \min\{f_k(P^+), f_k(P^-)\}$ and thus we established

4. *For even* $d = 2n$, *if* P *is a* d-*polytope with* $d + 3$ *vertices such that* $f_k(P) = f_k(C(d + 3, d))$ *for some* k *satisfying* $n - 1 \leq k \leq 2n - 1$, *then* P *is combinatorially equivalent to* $C(d + 3, d)$.

This clearly implies the case $v = d + 3$ of theorem 7.2.3.

In order to find the minimal value of $f_k(P)$ for simplicial P we note that

$$\Delta_k(b, c) + \Delta_k(b', c') = 0$$

whenever $(b + c) + (b' + c') = d + 1$. It follows easily that (for given s and t) $f_k(P)$ is minimal if and only if either only $(0; 0)$ is starred or if all

points except $(s; t)$ are starred. Since the first case does not yield a polytope with $d + 3$ vertices, the polytope P minimizing $f_k(P)$ for given s and
t is the convex hull of T_s^d with a point which is beyond one and only one
facet of T_s^d. Since in this situation

$$f_k(P) = f_k(T_s^d) + \binom{d}{k} - \delta_{k, d-1},$$

$f_k(P)$ will be minimal (for variable s, t) if and only if T_s^d minimizes $f_k(T_s^d)$.
By theorem 6.1.3 this happens if and only if $s = 1$. Hence

  5. *For every simplicial d-polytope P with $d + 3$ vertices, and for each k
satisfying $1 \le k \le d - 1$,*

$$f_k(P) \ge \binom{d+2}{k+1} + \binom{d}{k} - \binom{d}{k-1} - 2\delta_{k, d-1}.$$

*Equality holds for some k, $1 \le k \le d - 1$, if and only if P is combinatorially
equivalent to the convex hull of T_1^d with a point which is beyond one and only
one facet of T_1^d.*

6.3 Gale-Diagrams of Polytopes with Few Vertices

In the present section we shall see how the Gale-diagrams, discussed in
section 5.4, may be used to solve problems about d-polytopes with $d + 2$
or with $d + 3$ vertices. The new results of this section are due to M. A.
Perles (private communication).

  Let, first, P be a d-polytope with $d + 2$ vertices, and let $V = \text{vert } P$.
The Gale-transform $\bar{V}$ is a $(d + 2)$-tuple of points in $R^1 = R$ (since in
this case $n = d + 2$ and thus $n - d - 1 = 1$). The Gale-diagram $\hat{V}$ is
contained in the 3-point set $\{0, 1, -1\} \subset R$, those points having multiplicities m_0, m_1, m_{-1} assigned in such a way that $m_0 \ge 0, m_1 \ge 2, m_{-1} \ge 2$,
and $m_0 + m_1 + m_{-1} = d + 2$ (see theorems 5.4.2 and 5.4.3). If P and P'
are two such polytopes, (m_0, m_1, m_{-1}) and (m_0', m_1', m_{-1}') being the associated multiplicities, then P and P' are combinatorially equivalent if and only
if either $(m_0, m_1, m_{-1}) = (m_0', m_1', m_{-1}')$ or $(m_0, m_1, m_{-1}) = (m_0', m_{-1}', m_1')$.
P is simplicial if and only if $m_0 = 0$. It is rather easy to deduce from these
observations all the results of section 6.1; this task is left to the reader as
a useful exercise.

We turn now to the much more interesting discussion of d-polytopes P with $d + 3$ vertices. Their Gale-diagrams are contained in the set $C^+ = \{0\} \cup C$, where C denotes the unit circle centered at the origin 0 of R^2.

For ease of explanation and pictorial representation, when drawing a Gale-diagram $\hat{V}$ of $V = \text{vert } P$, we shall show the circle C as well as all the *diameters of $\hat{V}$*, that is, diameters of C which have at least one endpoint in $\hat{V}$. Points of $\hat{V}$ shall be shown in the illustrations by small circles, and if a point of $\hat{V}$ has multiplicity greater than 1, its multiplicity will be marked near the point.

The reader is invited to verify (using theorems 5.4.1 and 5.4.5) that the seven 3-polytopes with 6 vertices, shown by their Schlegel-diagrams in figure 6.3.1, have Gale-diagrams isomorphic to those shown beneath the Schlegel diagrams. (The letters in the diagrams should help the identification.)

Using theorem 5.4.5 it is easy to verify that if $\hat{V}$, $\hat{V}'$ are Gale-diagrams of two d-polytopes P, P' with $d + 3$ vertices each, and if the only difference between $\hat{V}$ and $\hat{V}'$ is in the position of one of the diameters—its position in $\hat{V}'$ being obtained by rotating the corresponding diameter in $\hat{V}$ through an angle sufficiently small not to meet any other diameter— then $\hat{V}$ and $\hat{V}'$ are isomorphic (and P and P' are combinatorially equivalent). For example, the Gale-diagrams in the last row of figure 6.3.1 are isomorphic to those above them.

By a repeated application of this remark we see that each combinatorial type of d-polytopes with $d + 3$ vertices has representatives for which the consecutive diameters of its Gale-diagram are equidistant. We shall call such Gale-diagrams *standard* diagrams.

Another change which may be performed on a Gale-diagram $\hat{V}$ and results in an isomorphic Gale-diagram is as follows:
If D_1 and D_2 are consecutive diameters of the Gale-diagram $\hat{V}$, each of which has only one endpoint in $\hat{V}$, and these two points of $\hat{V}$ are not separated by any other diameter of $\hat{V}$, we may omit $D_2 \cap \hat{V}$ and D_2 if we simultaneously increase the multiplicity of $D_1 \cap \hat{V}$ by the multiplicity of $D_2 \cap \hat{V}$. For example, the first four Gale-diagrams in figure 6.3.2 are isomorphic (but the fifth is not isomorphic to any of them). Clearly, the change opposite to the one just described also yields a Gale-diagram isomorphic to the given one. It follows that each combinatorial type of d-polytopes with $d + 3$ vertices may be represented by Gale-diagrams which are either *contracted*, or else *distended*—the first meaning it has the least possible number of diameters among all isomorphic diagrams,

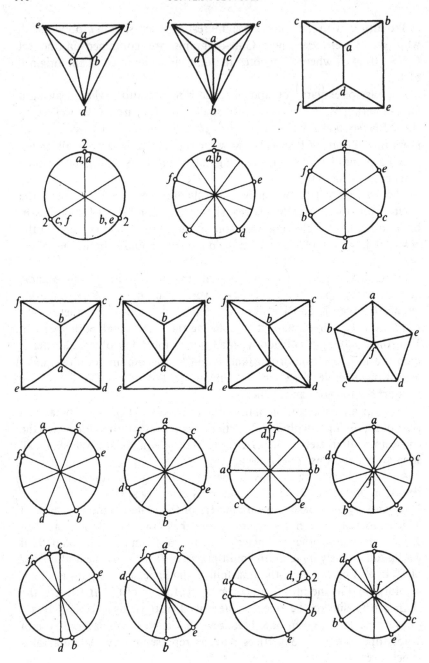

Figure 6.3.1

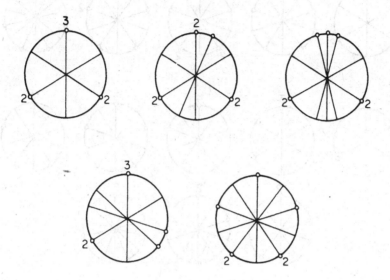

Figure 6.3.2

the latter indicating the largest possible number of diameters.

Using the above notions and the facts mentioned in section 5.4, it is easy to establish the following result.

1. *Two d-polytopes with d + 3 vertices are combinatorially equivalent if and only if the contracted (or else, the distended) standard forms of their Gale-diagrams are orthogonally equivalent (i.e. isomorphic under an orthogonal linear transformation of* R^2 *onto itself).*

Theorem 1 clearly enables one to determine, with relatively little effort, all the combinatorial types of d-polytopes with $d + 3$ vertices. This task is particularly simple for simplicial polytopes; in this case $0 \notin \hat{V}$, and no diameter of $\hat{V}$ has both endpoints in $\hat{V}$. Therefore the contracted Gale diagram has an odd number (≥ 3) of diameters, the points of $\hat{V}$ being situated on alternate endpoints of the diameters. The different possible contracted standard Gale-diagrams for $d = 4, 5$, and 6 are shown in figure 6.3.3.

Counting the distended standard Gale-diagrams, Perles established the following general formula for the number $c_s(d + 3, d)$ of different combinatorial types of simplicial d-polytopes with $d + 3$ vertices.

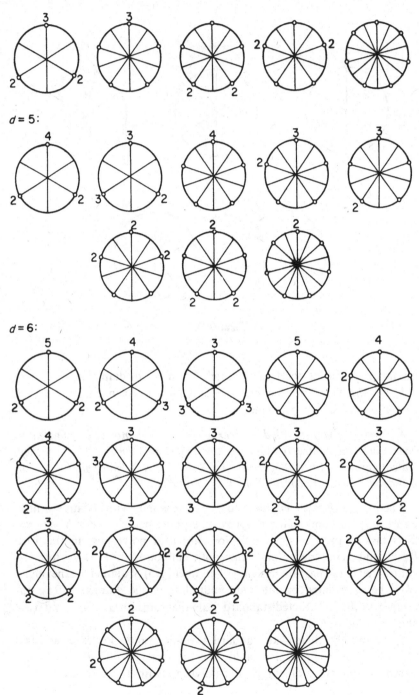

Figure 6.3.3. Standard contracted Gale diagrams of simplicial
d-polytopes with $d + 3$ vertices, for $d = 4, 5, 6$

2. Let $d + 3 = 2^{\alpha_0} p_1^{\alpha_1} \cdots p_k^{\alpha_k}$, where the p_i's are distinct odd primes, $\alpha_0 \geq 0$, $\alpha_i \geq 1$ for $i = 1, \cdots, k$. Then $c_s(d + 3, d)$ equals

$$2^{[d/2]} - \left[\frac{d+4}{2}\right] + 2^{-\alpha_0 - 2} \sum_{\substack{\gamma_1, \cdots, \gamma_k \\ 0 \leq \gamma_1 \leq \alpha_1, \cdots, 0 \leq \gamma_k \leq \alpha_k}} \left(\prod_{j=1}^{k} p_j^{-\gamma_j} \prod_{j \in \{i \mid \gamma_i < \alpha_i\}} \frac{p_j^{-1}}{p_j} \cdot 2^g\right)$$

where $g = 2^{\alpha_0} \prod_{j=1}^{k} p_j^{\gamma_j}$.

A simpler form of this formula, using Euler's φ-function, is

$$c_s(d + 3, d) = 2^{[d/2]} - \left[\frac{d+4}{2}\right] + \frac{1}{4(d+3)} \sum_{h \text{ odd divisor of } d+3} \varphi(h) \cdot 2^{(d+3)/h}.$$

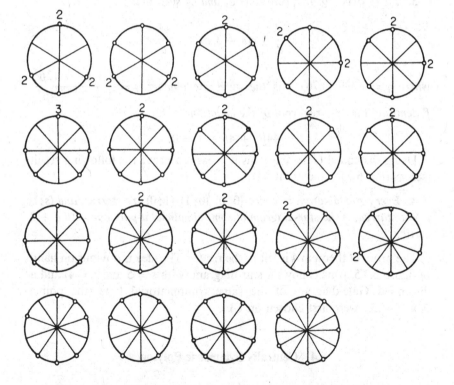

Figure 6.3.4. Standard contracted Gale diagrams of nonsimplicial, nonpyramidal 4-polytopes with 7 vertices

(Euler's φ-function is defined by

$$\varphi(h) = h \prod_{p \text{ prime divisor of } h} \left(1 - \frac{1}{p}\right).$$

For small values of d, the values of $c_s(d + 3, d)$ may be found in table 1.

Similarly, for small d it is not hard to determine *all* the different combinatorial types of d-polytopes with $d + 3$ vertices. Contracted standard Gale-diagrams of the 19 combinatorial types of non-simplicial 4-polytopes with 7 vertices which are not 4-pyramids over 3-polytopes with 6 vertices, are shown in figure 6.3.4. The numbers $c(d + 3, d)$ of different combinatorial types of d-polytopes with $d + 3$ vertices (reproduced for $d \leq 6$ in table 2) have been determined by Perles for $d = 4, 5, 6$. No general formula for $c(d + 3, d)$ has been found so far, but Perles established

3. *There exist positive constants c_1 and c_2 such that*

$$c_1 \frac{\gamma^d}{d} \leq c(d + 3, d) \leq c_2 \frac{\gamma^d}{d},$$

where $\gamma = 2.83928676 \cdots$ is the algebraic number $\gamma = \dfrac{2 + 6\beta + 12\beta^2}{1 + 2\beta - 4\beta^2}$, β denoting the only real root of the equation

$$44\beta^3 + 4\beta - 1 = 0.$$

Using distended Gale-diagrams it is easy to prove the following result (see section 6.2 for the notation):

4. *Every star-diagram in which $(0;0)$, $(0;1)$, $(1;0)$ are starred, and $(s;t)$ is not starred, is the star-diagram of some simplicial d-polytope with $d + 3$ vertices.*

The proof of theorem 4 is left to the reader. The idea is obvious on hand of figure 6.3.5, which shows a star-diagram (with $s = 5$ and $t = 4$) and a distended Gale-diagram of the same combinatorial type (the points $\hat{z}, \hat{x}_0, \cdots, \hat{x}_5$ were first chosen on C).

6.4   Centrally Symmetric Polytopes

A polytope $P \in \mathscr{P}^d$ is *centrally symmetric* provided $-P$ is a translate of P. In the sequel we shall usually make the tacit assumption that the *center*

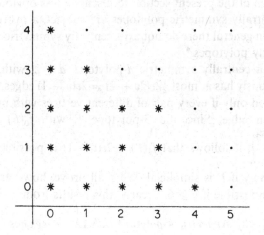

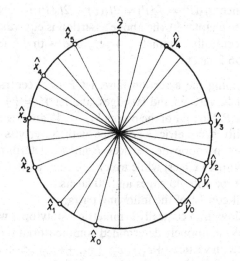

Figure 6.3.5

of the centrally symmetric polytope P is at the origin 0, i.e. that $P = -P$. The centrally symmetric d-polytopes form an interesting and important subclass $\mathscr{P}^d_*$ of $\mathscr{P}^d$.

The requirement of central symmetry imposes certain natural and obvious restrictions on the polytopes and on the numbers $f_k(P)$ of their k-faces. The most obvious property is the evenness of $f_k(P)$ for all $P \in \mathscr{P}^d_*$ and $0 \le k \le d - 1$.

It is the aim of the present section to discuss a less obvious restriction to which centrally symmetric polytopes are subject, a restriction which implies that in general there do not exist centrally symmetric analogues of the neighborly polytopes.*

Let P be a centrally symmetric d-polytope, $d \geq 2$, with $2v$ vertices. Then P obviously has at most $\frac{1}{2}2v(2v-2) = 2v(v-1)$ edges, this number being achieved only if every pair of different vertices with non-zero sum determine an edge. Since no 3-polytope F with $f_0(F) > 4$ satisfies $f_1(F) = \begin{pmatrix} f_0(F) \\ 2 \end{pmatrix}$, it follows that $f_1(P) = 2v(v-1)$ is possible for $P \in \mathscr{P}_*^4$, $f_0(P) = 2v$, only if P is simplicial. We shall prove, however, that this is altogether impossible if $v \geq 6$. Clearly, this results from

1. *No centrally symmetric 4-polytope with* 12 *vertices has* 60 *edges.*

Note that since $2f_3(P) = f_2(P) = 2f_1(P) - 2f_0(P)$ for every simplicial 4-polytope P (see chapter 9), the above assertion is equivalent to the non-existence of a centrally symmetric 4-polytope with 12 vertices and 48 facets, or with 96 2-faces.

PROOF Assuming the assertion false, let P be a centrally symmetric 4-polytope with 12 vertices and 60 edges, and let $\mathscr{V} = \{\pm V_i \mid 1 \leq i \leq 6\}$ be the vertices of P. No four of the points $V_1, \cdots, V_6$ are linearly dependent, since no centrally symmetric 3-polytope with 8 vertices has 24 edges. Every four pairs of opposite vertices from $\mathscr{V}$ determine therefore a 4-octahedron (with 24 edges) while every five such pairs form a centrally symmetric polytope with 40 edges and 30 facets.

The proof will consist of the following parts:

(i) We shall show that centrally symmetric 4-polytopes with 10 vertices and 40 edges have a uniquely determined combinatorial type, exemplified by the polytope with vertices $\pm e_1, \pm e_2, \pm e_3, \pm e_4, \pm(e_1 + e_2 + e_3 + e_4)$, where the e_i's are unit vectors in direction of the coordinate axes in a Cartesian coordinate system.

(ii) By a counting argument we shall show that two of the pairs from $\mathscr{V}$ must be in one of two definite configurations with respect to the 4-octa-hedron determined by the remaining four pairs.

(iii) We shall complete the proof by showing that neither of the two special configurations has all the desired properties.

---

*Neighborly polytopes are generalizations of the cyclic polytopes; we shall discuss them in chapter 7 and in section 9.6.

(i) Let K be a centrally symmetric simplicial 4-polytope with 10 vertices. Using, if necessary, an appropriate affine transform, we may assume that the vertices of K are $\pm e_1$, $\pm e_2$, $\pm e_3$, $\pm e_4$, and $\pm e = \pm(\alpha, \beta, \gamma, \delta)$, where the e_i denote unit vectors as above. Without loss of generality we assume that $\alpha > \beta > \gamma > \delta > 0$; clearly also $\alpha + \beta + \gamma + \delta > 1$.

Let Q be the 4-octahedron $\text{conv}\{\pm e_1, \pm e_2, \pm e_3, \pm e_4\}$. The 16 facets of Q are of five types:

    I. One facet, denoted by $(+ + + +)$, which has vertices e_1, e_2, e_3, e_4.
   II. Four facets, one of which is denoted by $(+ + + -)$ and has vertices e_1, e_2, e_3, e_4, the other three, defined analogously, being $(+ + - +), (+ - + +), (- + + +)$.
  III. Six facets, designated by the rather obvious notation $(+ + - -)$, $(+ - + -), (+ - - +), (- + + -), (- + - +), (- - + +)$.
  IV. Four facets $(+ - - -), (- + - -), (- - + -), (- - - +)$.
   V. One facet $(- - - -)$.

We note, first, that e is not beyond any facet of Q of types IV or V; indeed, if e were beyond $(+ - - -)$, for example, then $\alpha - \beta - \gamma - \delta > 1$; since $\alpha, \beta, \gamma, \delta$ are positive, it would follow that $\alpha \pm \beta \pm \gamma \pm \delta > 1$ for every choice of signs. But this would imply that e is beyond all facets of Q which contain e_1, and therefore e_1 would not be a vertex of K.

Hence e is beyond $(+ + + +)$ and possibly beyond some facets of types II and III.

Our assumptions about the magnitudes of $\alpha, \beta, \gamma, \delta$, imply: If e is beyond a certain facet of type II, it is also beyond those facets of type II which have the $-$ sign more to the right. Thus if e is beyond $(+ + - +)$ this implies that e is beyond $(+ + + -)$.

Similarly, if e is beyond a facet of type III, it is beyond the facets having the $-$ signs more to the right, and also beyond those having a $+$ sign instead of a $-$ sign.

Note that e is beneath $(- - + +)$ and $(- + - +)$, and that e is not beyond both $(- + + -)$ and $(+ - - +)$.

It follows that only the following possibilities arise. It may be verified that all the cases are indeed realizable by centrally symmetric 4-polytopes.

If e is beyond $(- + + -)$ then K has 26 facets and 36 edges:

$(+ - - +)$	22	32
$(+ - - +)$ and $(- + + +)$	24	34
$(+ - + -)$	24	34
$(+ - + -)$ and $(- + + +)$	26	36
$(+ + - -)$	24	34
$(+ + - -)$ and $(+ - + +)$	26	36

$(+ + - -)$ and $(- + + +)$	28	38
$(- + + +)$	30	40
$(+ - + +)$	28	38
$(+ + - +)$	26	36
$(+ + + -)$	24	34
$(+ + + +)$	22	32

It follows that the maximal possible number of edges, or facets, is attained if and only if e is beyond all the facets of types I and II, and beneath all the facets of types III, IV, and V. An equivalent description of this case is to say that e is exactly beyond a certain facet (its 'central' facet) and the four facets incident with it in its four 2-faces.

(ii) Let us now return to the polytope $P \in \mathscr{P}_*^4$ with 6 pairs of vertices and 48 facets. Taking any four pairs of opposite vertices and considering the 4-octahedron Q they determine, we may consider P as being obtained by taking first the convex hull K of Q with one of the remaining pairs, and then the convex hull of K with the last pair. Since K must have the maximal possible number of edges, it must be of the type described above. Therefore, considering the two additional pairs we see that their mutual relationship relative to Q may be described by specifying whether their 'central' facets

(a) are the same;
(b) have a common 2-face;
(c) have a common edge.

In case (a) six of the facets of Q are facets of P; in case (b) no facet of Q is a facet of P, while in case (c) four facets of Q are facets of P. Among the $\binom{6}{4} = 15$ possible choices of Q by quadruples of pairs of vertices of P, let a_1, a_2, a_3 denote the number of occurrences of the cases (a), (b), (c). Then, since each facet of P will be counted precisely once, we have $a_1 + a_2 + a_3 = 15$ and $6a_1 + 4a_3 = 48$. Each solution of this system satisfies either $a_1 > 0$ or $a_3 > 0$; thus the nonexistence of P will be established if we prove that in either of the two cases some edge which should occur in P does not occur.

(iii) If $a_1 > 0$, let V_5 and V_6 be beyond the same facets of $Q = \text{conv}\{\pm V_i | 1 \leq i \leq 4\}$. Then, by exercise 5.2.11, the six facets beneath which are $\pm V_5$ and $\pm V_6$, separate the boundary of Q, and also the boundary of P, into two simply connected parts, one of which contains V_5 and V_6 in its interior, while $-V_5$ and $-V_6$ are contained in the interior of the other. Therefore P cannot contain the edges $(V_5, -V_6)$, and $(V_6, -V_5)$.

In order to complete the proof in case $a_3 > 0$, we need a simple lemma. Its proof is immediate on considering the projection of the 4-space onto the plane determined by $0, X, Y$, which carries aff F_0 into a point. Clearly, the lemma is generalizable to higher dimensions.

LEMMA   *Let Q be a 4-polytope, 0 a point of int Q, F_1 and F_2 two facets of Q such that $F_0 = F_1 \cap F_2$ is a 2-face of Q. Let X be a point beyond F_2 such that the segment $0X$ intersects F_1, and let Y be a point beyond F_2 and beneath F_1. Then X is beyond the hyperplane $\mathrm{aff}(F_0 \cup \{Y\})$.*

Assuming now that $a_3 > 0$, let $Q = \mathrm{conv}\{\pm V_i \mid 1 \le i \le 4\}$, let $\mathrm{conv}\{V_1, V_2, V_3, V_4\}$ be the 'central' facet for V_5, and let $\mathrm{conv}\{V_1, -V_2, V_3, -V_4\}$ be the 'central' facet for V_6. The edge (V_1, V_3) of $K = \mathrm{conv}\{V_i \mid 1 \le i \le 5\}$ is contained in the following three facets of K: $\mathrm{conv}\{V_1, -V_2, V_3, -V_4\}$, $\mathrm{conv}\{V_1, -V_2, V_3, V_5\}$, $\mathrm{conv}\{V_1, V_3, -V_4, V_5\}$. Applying the lemma, with $V_6 = X$, $V_5 = Y$, $\mathrm{conv}\{V_1, -V_2, V_3, -V_4\} = F_1$, and either $\mathrm{conv}\{V_1, -V_2, V_3, V_4\}$ or $\mathrm{conv}\{V_1, V_2, V_3, -V_4\}$ as F_2, we see that V_6 is beyond the facets $\mathrm{conv}\{V_1, -V_2, V_3, V_5\}$ and $\mathrm{conv}\{V_1, V_3, -V_4, V_5\}$ of K. Since V_6 is also beyond $\mathrm{conv}\{V_1, -V_2, V_3, -V_4\}$, it follows that V_6 is beyond *all* the facets of K which contain (V_1, V_3). Therefore $P = \mathrm{conv}(K \cup \{V_6, -V_6\})$ does not contain the edge (V_1, V_3).

This completes the proof of theorem 1.

6.5   Exercises

1. Derive the results of section 6.1 directly from Radon's theorem, without using the notions of 'beyond' and 'beneath'.

2. Show that $f_{d-1}(T^d_{[\frac{1}{2}d]}) = [\frac{1}{4}(d+2)^2]$.

3. Using Gale transforms prove the following result of M. A. Perles; it complements theorem 5.5.4 and provides a partial affirmative solution to Klee's problem about rational polytopes (see section 5.5). Let $P \subset R^d$ be a d-polytope with $d + 3$ vertices $x_1, \cdots, x_{d+3}$, and let $\varepsilon > 0$; then there exists a d-polytope $P' \subset R^d$ with vertices $x'_1, \cdots, x'_{d+3}$, which has the following properties:

   (i) $\|x_i - x'_i\| < \varepsilon$ for $i = 1, \cdots, d+3$.

   (ii) P and P' are combinatorially equivalent under the mapping $\varphi x_i = x'_i, i = 1, \cdots, d+3$.

   (iii) x'_i has rational coordinates for $i = 1, \cdots, d+3$.

(iv) P and P' have the same affine structure (that is, for every $X \subset \text{vert } P$, $\dim \text{aff } X = \dim \text{aff}\{\varphi x_i \mid x_i \in X\}$.

4. It would be interesting to have a complete classification of centrally symmetric d-polytopes with few vertices. Here 'few' means $2d + 2$, $2d + 4$, etc. While the discussion of the case $f_0 = 2d + 2$ is probably quite simple (though it seems that it has not been performed so far), the result of section 6.4 indicates that the case $f_0 = 2d + 4$ is likely to be rather complicated. Probably a suitable variant of the Gale diagram technique could be useful in this connection.

5. Prove the following result of M. A. Perles: If P is a projectively unique d-polytope with $d + 3$ vertices, then its contracted standard Gale diagram has one of the forms indicated in figure 6.5.1. (The conditions on the multiplicities m_i are listed beneath each diagram.)

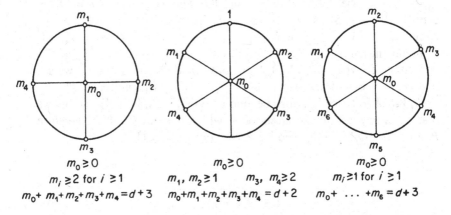

$$m_0 \geqslant 0$$
$$m_i \geqslant 2 \text{ for } i \geqslant 1$$
$$m_0 + m_1 + m_2 + m_3 + m_4 = d + 3$$

$$m_0 \geqslant 0$$
$$m_1, m_2 \geqslant 1 \qquad m_3, m_4 \geqslant 2$$
$$m_0 + m_1 + m_2 + m_3 + m_4 = d + 2$$

$$m_0 \geqslant 0$$
$$m_i \geqslant 1 \text{ for } i \geqslant 1$$
$$m_0 + \ldots + m_6 = d + 3$$

Figure 6.5.1

6. Use Gale diagrams and property (iv) from page 87 to establish the following result of M. A. Perles: If $d + 3$ points of R^d are in general position (i.e. no $d + 1$ are contained in a hyperplane) then some $d + 2$ of them are the vertices of a polytope combinatorially equivalent to $C(d + 2, d)$. (This is a strengthening of the result of exercise 2.4.12.)

7. (M. A. Perles) Let P be a d-polytope with at most $d + 3$ vertices, and let G be the group of combinatorial automorphisms of P (i.e. lattice automorphisms of $\mathscr{F}(P)$). Then there exists a polytope P', combinatorially equivalent to P, such that each element of the group of combinatorial automorphisms of P' (which is isomorphic to G) is induced by a *symmetry* of P' (i.e. an orthogonal transformation of P' onto itself).

6.6 Remarks

The classification of simplicial d-polytopes with $d + 2$ vertices (or of their duals) is well known. Thus, a discussion is given in Schoute [1] and in Sommerville [2], showing that there are $[\frac{1}{2}d]$ different combinatorial types of such polytopes. The treatment of the general d-polytopes with $d + 2$ vertices seems not to have been published before, although traces of it may be found in various papers (see, for example, Grünbaum–Motzkin [2]; compare also section 2.3).

The notion of the star-diagram of a simplicial d-polytope with $d + 3$ vertices seems to be new. In proving the case $k = d - 1$ of theorem 6.2.1, Gale [5] develops a correspondence between the facets of such polytopes and cyclic triangles in an oriented complete graph with $d + 3$ nodes. Gale [4,5] also established, by methods different from those used here, the cases $m = d - 1$ of theorem 6.1.3 ($d = 2n$) and $k = d - 1$ of theorem 6.2.3.

The results of section 6.3 are due to M. A. Perles (private communication).

Our knowledge of the numbers $c_s(v, d)$ or $c(v, d)$ of different combinatorial types of simplicial, or of all, d-polytopes with v vertices is very limited; no general results are known for $v \geq d + 4$. Correcting an earlier attempt of Brückner [3], Grünbaum–Sreedharan [1] have recently shown that $c_s(8, 4) = 37$. The available information on $c(v, d)$ and $c_s(v, d)$ is reproduced in tables 1 and 2.

6.7   Additional notes and comments

The number of d-polytopes with d + 3 vertices.
There is interest in both explicit formulas and the asymptotic growth of the numbers $c(d+3,d)$, $c_s(d+3,d)$ and $c_n(d+3,d)$ of all, resp. all simplicial, resp. all simplicial neighborly d-polytopes with $d+3$ vertices.

o For $c_s(d+3,d)$ a formula without proof appears as theorem 6.3.2 in this book—the proof is left to the reader (that is, to you) as an exercise. You should also try to derive the asymptotics from it.

o For $c(d+3,d)$ a formula was given by Lloyd [a], but that one is not correct: For $d \geq 4$ it does not produce the (correct) values determined by Perles, see page 424. However, McMullen has remarked that Lloyd's formula produces the correct asymptotics, as announced in theorem 6.3.3.

o For even d all neighborly d-polytopes with $d+3$ vertices are cyclic (in particular, simplicial) by theorem 7.2.3; thus $c_n(d+3,d) = 1$ for even d. For odd d Altshuler–McMullen [a] derived a formula for $c_n(d+3,d)$. McMullen [e] furthermore derived a formula for the number of (not necessarily simplicial) neighborly d-polytopes with $d+3$ vertices.

The numbers of types of spheres and of polytopes.
A surprising discovery was that there are "by far more spheres than polytopes": In a wide range of parameters (with $n \geq d+4$, $d \geq 4$), the fraction of polytopally realizable types among the combinatorial types of simplicial $(d-1)$-spheres with n vertices is extremely small.

There are few polytopes: Using upper bounds on the topological complexity of real algebraic varieties (by Milnor, Thom, Warren, Oleinik and Petrovsky), Goodman–Pollack [a] gave unexpectedly low upper bounds on the number of simplicial d-polytopes. After improvements by Alon [a], we know

$$\left(\frac{n-d}{d}\right)^{nd/4} \leq c_s(n,d) \leq c(n,d) \leq \left(\frac{n}{d}\right)^{d^2 n(1+o(1))}$$

for the numbers of (simplicial) polytopes, where the $o(1)$ denotes a term that goes to zero as n/d becomes large.

There are many spheres: A simple (but elegant) construction by Kalai [b] provides "many simplicial spheres" as boundaries of shellable balls in the boundary complexes of cyclic polytopes, and hence the lower bound

$$2^{\frac{1}{(n-d)(d+1)} \binom{n-\lceil(d+2)/2\rceil}{\lceil(d+1)/2\rceil}}$$

for the number of simplicial $(d-1)$-spheres with n vertices. Combined with the Goodman–Pollack result, this implies for example that for all fixed $d > 4$,

POLYTOPES WITH FEW VERTICES

for sufficiently large $n \gg d$, most simplicial spheres are not polytopal. This also holds for $d = 4$, according to Pfeifle–Ziegler [a], while Pfeifle [b] has shown that all of Kalai's 3-spheres are polytopal.

"Centrally symmetric neighborly" polytopes.
A centrally symmetric d-polytope, $d \geq 4$, with $2v \gg 2d$ vertices cannot be *centrally symmetric 2-neighborly* in the sense that each vertex is adjacent to all other vertices except for its opposite vertex: It must have many fewer edges than the required number $2v(v - 1)$. Conceptual proofs for this were given by Burton [a] and by Schneider [a].

In contrast, centrally symmetric 2-neighborly simplicial 3-spheres with $2v$ vertices exist for all $v \geq 4$ according to Jockusch [a]; the first examples were given by Grünbaum [c] in 1969. According to a conjecture of Lutz [a] there should even be centrally symmetric neighborly $(d - 1)$-spheres with a vertex-transitive group action for all even d and $n = 2v \geq 2d$.

Strengthening theorem 6.4.1, Pfeifle [a] showed that there is not even a centrally symmetric neighborly fan in R^d with $2d + 4$ rays. In particular, there is no "star convex" neighborly centrally symmetric 4-polytope with 12 vertices.

Central diagrams.
As suggested in exercise 6.5.4, "central diagrams" for dealing with centrally symmetric polytopes with few vertices were developed by McMullen and Shephard [a].

However, the power of these is quite limited, at least for classification purposes. While centrally symmetric d-polytopes with $2d$ vertices are affinely equivalent to the d-dimensional cross-polytope, a complete classification of the centrally symmetric d-polytopes with $2d + 2$ vertices is out of reach. This can be seen as follows: We may assume that the first $2d$ vectors are $\{\pm e_1, \ldots, \pm e_d\}$, spanning a d-dimensional cross polytope. Now we take one extra point, and try to classify the collections of facets of the cross polytope it "sees". Dually, this means that we are trying to classify the vertex sets of a d-cube $[-1, +1]^d$ that can be cut off by a hyperplane. There are interesting results for $d \leq 5$ (see Emamy-K.–Lazarte [a] and Emamy-K.–Caiseda [a]), but the problem for general d is well-known to be intractable: It appears, for example, in the disguise of a classification of knapsack polytopes (see e. g. Weismantel [a]), threshold gates (Muroga [a], Håstad [a]), etc.

These fundamental problems vindicate the rather cumbersome, explicit arguments that Grünbaum uses here to classify the centrally-symmetric 4-polytopes with 12 vertices.

CHAPTER 7

Neighborly Polytopes

The neighborly polytopes, rediscovered barely a decade ago as a rather freakish family of polytopes, have in recent years gained importance and received much attention due to their connection with certain extremal problems (see section 10.1). In the present chapter we discuss the main properties of neighborly polytopes.

7.1 Definition and General Properties

The present chapter is devoted to the study of an interesting and important family of polytopes, the neighborly polytopes. We have already met examples of neighborly polytopes—the cyclic polytopes $C(v, d)$ discussed in section 4.7.

Let k be a positive integer. A d-polytope P shall be called k-*neighborly* provided every k-membered subset V of the set vert P of vertices of P determines a proper face $F = \operatorname{conv} V$ of P such that $V = \operatorname{vert} F$.

Thus for $d \geq 1$ every d-polytope is 1-neighborly; every d-simplex is k-neighborly for each k satisfying $1 \leq k \leq d$. Clearly, no d-polytope is k-neighborly for $k > d$.

The example of the cyclic polytopes $C(v, d)$ shows that for every d and every $v > d$ there exist d-polytopes with v vertices which are $[\frac{1}{2}d]$-neighborly.

In order to gain more insight into the structure of neighborly polytopes, we start with some simple observations.

1. *If P is a k-neighborly d-polytope, $1 \leq k \leq d$, then every k vertices of P are affinely independent.*

Indeed, let $V = \{v_1, \cdots, v_k\}$ be a k-membered subset of vert P. Assuming V to be affinely dependent, let $v_k \in \operatorname{aff}\{v_1, \cdots, v_{k-1}\}$. We take any $w \in \operatorname{vert} P$, $w \notin V$, and consider the k-membered set $W = \{v_1, \cdots, v_{k-1}, w\} \subset \operatorname{vert} P$. Since P is k-neighborly, $F = \operatorname{conv} W$ is a face of P such that $W = \operatorname{vert} F$. Now, for each hyperplane H such that $F = H \cap P$ we have $v_k \in \operatorname{aff}\{v_1, \cdots, v_{k-1}\} \subset \operatorname{aff} W \subset H$; thus $v_k \in F$. Since v_k is a vertex of P, v_k is a vertex of any face of P which contains it. Hence we reach the contradiction $v_k \in \text{vert } F = W = \{v_1, \cdots, v_{k-1}, w\} \not\ni v_k$, which establishes our assertion 1.

As a consequence of theorem 1 we see that the convex hull of every k members of vert P is a $(k-1)$-face of P; more precisely, it is a $(k-1)$-simplex. On the other hand, since each $(k-1)$-face of P contains at least k vertices, it follows that every $(k-1)$-face of P is a $(k-1)$-simplex. Hence we have

2. *If P is a k-neighborly d-polytope, and if $1 \le k^* \le k$, then P is k^*-neighborly.*

3. *If P is a k-neighborly d-polytope and V is a subset of vert P, card $V > k$, then conv V is k-neighborly.*

Our next aim is to show that no d-polytope different from a simplex is 'more neighborly' that the cyclic polytopes. More precisely, we have

4. *If P is a k-neighborly d-polytope and $k > [\frac{1}{2}d]$ then $f_0(P) = d + 1$; that is, P is a d-simplex.*

Indeed, assuming $f_0(P) > d + 1$, let $V \subset \text{vert } P$ contain $d + 2$ vertices of P. By Radon's theorem 2.3.6 there exist sets W and Z such that $W \cup Z = V$, $W \cap Z = \varnothing$, and

(*) $\qquad\qquad\qquad \text{conv } W \cap \text{conv } Z \ne \varnothing.$

Without loss of generality we may assume that card $W \le \lfloor (d + 2)/2 \rfloor = [d/2] + 1 \le k$. Because of (*) every supporting hyperplane H of P which contains W must have a non-empty intersection with conv Z, and therefore $H \cap \text{vert } Z \ne \varnothing$. Therefore vert conv W properly contains W, in contradiction to theorem 2, the fact that card $W \le k$, and the k-neighborliness of P.

7.2 $[\frac{1}{2}d]$-Neighborly d-Polytopes

The most interesting neighborly polytopes are the $[\frac{1}{2}d]$-neighborly d-polytopes. The family of all $[\frac{1}{2}d]$-neighborly d-polytopes shall be denoted by $\mathcal{N}^d$, and the subfamily of $\mathcal{N}^d$ consisting of all simplicial members of $\mathcal{N}^d$ shall be denoted by $\mathcal{N}_s^d$. The cyclic polytopes show that $\mathcal{N}_s^d \ne \varnothing$. In the remaining sections of the present chapter, and in the following chapters, we shall simplify our terminology by referring to members of $\mathcal{N}^d$ as 'neighborly d-polytopes'.

The present section and section 9.6 are devoted to a more detailed study of neighborly d-polytopes.

As a strengthening of the results of the preceding section we have

1. *For even $d = 2n$ every neighborly d-polytope is simplicial; that is, $\mathcal{N}^d = \mathcal{N}^d_s$.*

The proof is very simple. The theorem being obvious for $d = 2$, we assume $d > 2$. Let F be a $(d - 1)$-face of $P \in \mathcal{N}^d$; if F were not a simplex, then $f_0(F) > d$. By theorem 3 of the preceding section F is n-neighborly; since $\dim F = d - 1 = 2n - 1$ and $n > n - 1 = [\frac{1}{2} \dim F]$, the inequality $f_0(F) > d$ is in contradiction to theorem 4 of the preceding section.

There is no analogue to theorem 1 for odd $d = 2n + 1$. Indeed, every $(2n + 1)$-pyramid with an n-neighborly $2n$-polytope as basis is in $\mathcal{N}^{2n+1}$ but—unless the basis is a $2n$-simplex—is clearly not simplicial. Thus, $\mathcal{N}^{2n+1}$ contains $\mathcal{N}^{2n+1}_s$ as a proper subfamily.

Additional information is available for the families $\mathcal{N}^d_s$ of simplicial neighborly d-polytopes. From the definition of neighborly d-polytopes it follows at once that if $P \in \mathcal{N}^d$ has $v = f_0(P)$ vertices, then $f_k(P) = \binom{v}{k+1}$ for each k in the range $0 \le k \le [\frac{1}{2}d] - 1$. Now if P is simplicial, that is $P \in \mathcal{N}^d_s$, then theorem 9.5.1 implies

2. *For $P \in \mathcal{N}^d_s$ the numbers $f_{[\frac{1}{2}d]}(P), \cdots, f_{d-1}(P)$ (as well as $f_1(P), \cdots, f_{[\frac{1}{2}d]-1}(P)$) are functions of $f_0(P)$ and d only.*

Explicit expressions for $f_k(P)$, where $P \in \mathcal{N}^d_s$ and $1 \le k \le d - 1$, in terms of f_0 and d, will be derived in section 9.6.

We mention here the following result of Gale [4], which is a special case of theorems 6.1.6 and 6.2.4.

3. *If $P \in \mathcal{N}^{2n} = \mathcal{N}^{2n}_s$ and if $v = f_0(P) \le 2n + 3$, then P is combinatorially equivalent to the cyclic polytope $C(v, 2n)$.*

Motzkin [4] (see also Gale [4]) claimed that theorem 3 holds without the restriction $v \le 2n + 3$. The results 1 and 2 above seem to lend additional support to this assertion. However, we shall see that Gale's theorem 3 is in a sense the maximal valid range of Motzkin's assertion. Indeed, we have

4. *There exists a neighborly 4-polytope with 8 vertices which is not combinatorially equivalent to the cyclic polytope $C(8, 4)$ with 8 vertices.*

NEIGHBORLY POLYTOPES 125

The proof of theorem 4 will be furnished by constructing a neighborly 4-polytope N_8 with 8 vertices and by showing that N_8 is not combinatorially equivalent to $C(8, 4)$.

In order to define N_8 let A, B, C, D, E, F, G be the seven vertices of the cyclic polytope $C(7, 4)$, arranged according to increasing values of t on the moment curve (t, t^2, t^3, t^4) (see section 4.7 for the definitions and facts used here). Then the 14 facets of $C(7,4)$ are:* (A,B,C,D), (A,B,C,G), (A,B,D,E), (A,B,E,F), (A,B,F,G), (A,C,D,G), (A,D,E,G), (A,E,F,G), (B,C,D,E), (B,C,E,F), (B,C,F,G), (C,D,E,F), (C,D,F,G), (D,E,F,G). The edge (A,G) of $C(7,4)$ is obviously incident with the following five facets of $C(7,4)$: (A,B,C,G), (A,B,F,G), (A,C,D,G), (A,D,E,G), (A,E,F,G). In the terminology introduced in section 5.2, let X be a point beyond (A,B,C,G) sufficiently near the centroid of (A,B,C,G) (that is, beneath all the other facets of $C(7, 4)$); let Y denote the midpoint of (A,G). Then each point of the ray $\{\lambda X + (1 - \lambda)Y \mid \lambda < 0\}$ is beneath (A,B,C,G) and beyond the other four facets of $C(7, 4)$ which contain (A,G). Let $Z = \lambda X + (1 - \lambda)Y$, for a $\lambda < 0$ sufficiently near to 0, so that Z is beneath all the facets of $C(7, 4)$ which are not incident with (A,G). Defining N_8 as the convex hull of $\{Z\} \cup C(7, 4)$ it is easily verified (with the help of theorem 5.2.1) that the 20 facets of N_8 are: (A,B,C,D), (A,B,C,G), (A,B,D,E), (A,B,E,F), (A,B,F,Z), (A,B,G,Z), (A,C,D,Z), (A,C,G,Z), (A,D,E,Z), (A,E,F,Z), (B,C,D,E), (B,C,E,F), (B,C,F,G), (B,F,G,Z), (C,D,E,F), (C,D,F,G), (C,D,G,Z), (D,E,F,G), (D,E,G,Z), (E,F,G,Z). It is now obvious that N_8 is 2-neighborly, and that each of the edges (B,C), (B,F), (C,G), (F,G) of N_8 is incident with exactly five facets of N_8.

This completes the proof of the assertion that N_8 is not combinatorially equivalent to $C(8, 4)$, since an edge of $C(8, 4)$ can be incident with either 3, or 4, or 6 facets of $C(8, 4)$, but never with exactly five facets.

It may be remarked that if the point X, used in the construction of N_8, had been chosen near the centroid of (A,C,D,G), we would have obtained another neighborly 4-polytope N_8^* with eight vertices, such that N_8^* is not combinatorially equivalent to either $C(8, 4)$ or N_8.

It is immediate that the construction used in the proof of theorem 4 may be applied also to 4-polytopes with more than 8 vertices, as well as in higher dimensions.

7.3 Exercises

1. Show that every member of $\mathcal{N}^{2n+1}$ is quasi-simplicial.
2. Show that in case of odd d the analogues of theorems 7.2.3 and 7.2.4 are:

* Throughout this section a face with vertices $A, B, \cdots$ shall be denoted $(A, B, \cdots)$.

If $P \in \mathcal{N}_s^d$ and $f_0(P) = d + 2$ then P is combinatorially equivalent to $C(d + 2, d)$. There exist polytopes $P \in \mathcal{N}^d$ with $f_0(P) = d + 2$, as well as $P \in \mathcal{N}_s^d$ with $f_0(P) = d + 3$, which are not combinatorially equivalent with cyclic polytopes.

3. Prove that the vertices of every member of $\mathcal{N}^{2n}$ are in general position in R^{2n}.

4. (Derry [2]) For every set A of $d + 3$ points in general position in R^d there exists a projective transformation T permissible for A such that TA is the set of vertices of the d-polytope conv TA, which is neighborly. Every two such polytopes obtained from a given set A are projectively equivalent.

5. Let us call a finite set $A \subset R^d$ *k-neighborly* [*k-almost-neighborly*] provided conv $A \cdot$ is a k-neighborly d-polytope and $A = $ vert conv A [for every k-membered subset B of A, conv $B \subset$ rel bd conv A]. Prove the following Helly-type theorems (compare exercise 2.4.11):

(i) $A \subset R^d$ is k-neighborly if and only if every $B \subset A$ such that card $B = d + 2$ is k-neighborly.

(ii) $A \subset R^d$ is k-almost-neighborly if and only if every $B \subset A$ such that card $B \leq 2d + k$ is k-almost-neighborly.

The notion of k-almost-neighborly sets was considered (under the name *k-convex sets*) by Motzkin [6]. The 'Helly-number' $2d + k$ of (ii) could probably be lowered to $2d + 1$ (compare exercise 2.4.11). The assertion in Motzkin [6] that (ii) remains true if $2d + k$ is replaced by $d + 2$ is easily shown to be false for each $d \geq 2$.

6. Prove the following statements analogous to exercise 2.4.13:

(i) Given integers d and v, with $2 \leq d < v$, there exists an integer $m(d, v)$ with the following property: Whenever $A \subset R^d$ consists of $m(d, v)$ or more points in general position, there exists $B \subset A$ such that card $B = v$, $B = $ vert conv B, and conv B is a neighborly d-polytope. (Hint: Use exercises 6.5.6 and 7.3.5(i), and Ramsey's theorem.)

(ii) Given integers k, d, v with $2 \leq 2k \leq d < v$, there exists an integer $m'(d, k, v)$ with the following property: Whenever $A \subset R^d$ satisfies card $A \geq m'(d, k, v)$ and dim aff $A = d$, there exists $B \subset A$ such that card $B = v$, dim conv $B = d$, and B is k-almost-neighborly. (Hint: First establish the existence of $m'(d, k, 2d + k)$, using an inductive argument involving (i) and the remark on page 4; then apply the result of exercise 7.3.5(ii) and Ramsey's theorem, with $q = 2d + k$, $p_1 = v$, and $p_2 = m'(d, k, 2d + k)$.

7. Prove that a d-polytope P with v vertices is k-neighborly if and only if each open hemisphere of S^{v-d-2} contains at least $k + 1$ points of the Gale diagram of P.

7.4 Remarks

The first to discuss cyclic polytopes and to realize their neighborliness was
Carathéodory [1,2]. His results on cyclic polytopes (and, in particular, their
existence) were forgotten for a long time. Steinitz [6] mentions the existence
of (duals of) neighborly 4-polytopes referring to Brückner [3] (see below),
but is unaware of Carathéodory's work on them although he quotes
Carathéodory's paper [2] on the very next page. Sz.-Nagy [1] noted the
existence of 2-neighborly d-polytopes with up to $[\frac{3}{2}d]$ vertices, and asked
whether this is the greatest possible number. The existence of 2-neighborly
polytopes, in particular such that have all vertices on a sphere, invalidates
the results of Chabauty [1,2] and partly of Chabauty [3]. Unaware of
Carathéodory's paper, Gale [2,3] and later Motzkin [4] rediscovered the
neighborly polytopes. Gale [2,3], inspired by H. Kuhn's accidental
discovery of a 2-neighborly 11-polytope with 24 vertices, established the
existence of k-neighborly $(2k)$-polytopes for every positive k. In an abstract,
Motzkin [4] described the generation of neighborly polytopes as convex
hulls of finite subsets of 'strictly comonotone curves'. Neighborly polytopes seem to have been discussed in Fieldhouse's thesis [1] which,
however, was not published and was not accessible to the author. The
above exposition of properties of neighborly polytopes follows mainly
Gale [4], as elaborated in mimeographed lecture-notes of Klee [9].

   In a certain sense, however, the history of neighborly polytopes begins
even before Carathéodory. In a paper published in 1909, Brückner [3]
clearly had in mind the duals of 2-neighborly 4-polytopes. (Even earlier
in the booklet Brückner [1], the 2-neighborly 4-polytopes with 7 or less
vertices were discovered; Brückner's examples were reproduced in
Schoute [1] and Sommerville [2]. However, Brückner's discussion deals
not with 4-polytopes, but with Schlegel-diagrams, or, more precisely,
with 3-diagrams. In view of theorems 11.5.1 and 11.5.2 Brückner's work
has only heuristic value as far as 4-polytopes are concerned. As shown by
theorem 11.5.2, one of Brückner's [3] diagrams is indeed not a Schlegel-
diagram of any 4 polytope.

   If the mere discussion of appropriate 3-diagrams (or similar concepts)
is counted as occurrence of neighborly polytopes, even Brückner [3] was
not the first discoverer. In 1905, in connection with the spatial analogue
of the 4-color problem for planar maps, Tietze [1] constructed, for every
$n \geq 1$, a partition of the Euclidean 3-space into n (unbounded) convex
sets, each two of which have a 2-dimensional intersection. In all fairness
to Brückner, however, it must be said that although it is possible to modify

Tietze's construction so as to obtain a partition of a 3-polytope into 3-polytopes, each two of which have an intersection which is a face of both—this did not occur to Tietze in [1], and doing it takes some effort. Tietze's paper [1], and the whole problem of 'neighboring polytopes' in R^3 was forgotten for many years. (For a stimulating discussion see Tietze [2]). In 1947 Besicovitch [1] gave a different construction, which was later extended to R^d by Rado [1] and Eggleston [1]. The results of these papers can be summarized as follows (compare Danzer–Grünbaum–Klee [1], p. 151, where some related problems are also discussed).

1. *If R^d contains a family of m polytopes such that each i of them have a $(d - i + 1)$-dimensional intersection for every i with $1 \le i \le d$, then $m \le d + 2$. There exists such families containing $d + 2$ members.*

2. *There exist in R^d infinite families of closed convex sets such that each i of the sets have a $(d - i + 1)$-dimensional intersection for all i with $1 \le i \le [\frac{1}{2}(d + 1)]$. No families in R^d have this property for all i with $1 \le i \le 1 + [\frac{1}{2}(d + 1)]$.*

Clearly theorems 1 and 2 are related to the results on (duals of) neighborly polytopes although neither of them implies the other (compare section 11.5).

Let a family $\mathscr{P}$ of d-polytopes in R^d be called *neighborly* provided each two members of $\mathscr{P}$ have a $(d - 1)$-dimensional intersection. While it follows from theorem 2 that arbitrarily large families of this type exist for each $d \ge 3$, there are many open problems if the members of $\mathscr{P}$ are subject to additional requirements.

One variant, due to Bagemihl [1], is to restrict $\mathscr{P}$ to a family of d-simplices. Clearly, if $d = 2$, such a family contains at most 4 members. For $d = 3$, Bagemihl [1] showed that the maximal number $m(3)$ of members of $\mathscr{P}$ satisfies $8 \le m(3) \le 17$ and conjectured that $m(3) = 8$, while Baston [1] proved that $m(3) \le 9$. It would be interesting to know whether, for general d, the correspondingly defined $m(d)$ satisfies $m(d) = 2^d$.

In another variant (see Grünbaum [6]) the members of $\mathscr{P}$ are required to be translates of the same d-polytope P. The maximal possible number of members of $\mathscr{P}$ is in this case 3 for $d = 2$, and 5 for $d = 3$. In general, it equals to the maximal possible number of points in a *strictly antipodal* subset of R^d (see Grünbaum [6]); thus $\mathscr{P}$ may contain $2d - 1$ members (Danzer–Grünbaum [1]), and this is probably the correct upper bound. The bound $2d - 1$ (in R^d) may be achieved even if the d-polytope P is assumed to be centrally symmetric.

Still another variant arises if the members of $\mathscr{P}$ are assumed to be centrally symmetric. For $d = 2$ it is clear that the maximal possible number of members of $\mathscr{P}$ is 4; however, already the case $d = 3$ is still open.

The number of similar problems may easily be increased, but their solutions seem to be rather elusive. For additional problems, results, and references in this general area see Danzer–Grünbaum–Klee [1], and section 6 of Grünbaum [10].

Returning to the subject of k-neighborly d-polytopes we note that the proof of theorem 7.1.4 yields also the following stronger result:

3. *If P is a k-neighborly d-polytope, $d \geq 2k$, then all $(2k - 1)$-faces of P are simplices.*

M. A. Perles raised (oral communication) the question whether there exist d-polytopes P such that P is k-neighborly, while the dual of P is h-neighborly. Using the above theorem 3 it follows at once that if such a P exists then $d \geq 2k + 2h - 2$. Thus the first interesting case would be to establish or disprove the existence of 2-neighborly 6-polytopes which have 2-neighborly duals. Using theorem 6.1.4 it may be shown that no 6-polytope with 8 vertices (or 8 facets) has this property, but the problem is open for polytopes of dimension 6 or more having more vertices.

Despite their rather recent discovery, there exists a widespread feeling (see for example Gale [3]) that neighborly d-polytopes are rather common —in some sense—among all d-polytopes. Exercise 6.5.6 may be considered as a confirmation of this belief in a special case. Another expression of this attitude is a recently posed question of V. Klee (private communication), the affirmative answer to which is the content of exercise 7.3.4(i). Still another aspect is contained in a recent assertion of Motzkin [6], the main part of which appears in exercise 7.3.4 (ii).

7.5 Additional notes and comments

Examples.
Explicit examples of (even-dimensional) neighborly polytopes are still not easy
to come by. However, research about neighborly polytopes, to a large extent triggered by the Grünbaum's book, has produced some substantial new
insights, including important new construction methods.

The number of types of neighborly 4-polytopes with n vertices is

n	5	6	7	8	9	10
# types	1	1	1	3	23	431

where the analysis by Grünbaum–Sreedharan [a] showed that with 8 vertices
there are only the three types $C(8,4)$, N_8, and N_8^* described on page 125; the
enumerations for $n = 9$ and $n = 10$ are due to Altshuler–Steinberg [a], resp.
Altshuler [a], Bokowski–Sturmfels [a] and Bokowski–Garms [a]. For dimension 6 we similarly have

n	7	8	9	10
# types	1	1	1	37

where the result for $n \leq 9$ follows from theorem 7.2.3, while the non-trivial
result for $n = 10$ is due to Bokowski–Shemer [a].

Construction techniques.
Shemer's fundamental paper [a] introduced a "sewing" construction that allows one to add vertices to neighborly polytopes. Exploiting this technique,
Shemer produced super-polynomially many different combinatorial types of
neighborly d-polytopes with n vertices, for every fixed even $d \geq 4$:

$$c_n(n,d) \geq \tfrac{1}{2}((\tfrac{d}{2}-1)\lfloor \tfrac{n-2}{d+1} \rfloor)!$$

For $n = d+4$, he derived a super-polynomial lower bound of

$$c_n(d+4,d) \geq \frac{(d+2)!}{24 \cdot 2^{d/2}(\tfrac{d}{2}+2)!} \sim \frac{1}{\sqrt{18}}\left(\frac{d}{e}\right)^{d/2}.$$

These lower bounds are remarkably close to the best known upper bounds
on the numbers $c(n,d) \geq c_s(n,d) \geq c_n(n,d)$ of combinatorial types of all d-
polytopes with n vertices; compare the notes in section 6.7.

See also Barnette [g], whose "facet splitting technique" produces duals of
neighborly polytopes.

Random polytopes.
The question "What is the probability that a random (simplicial) polytope is neighborly?" is not well-posed, since the answer heavily depends on the model of random polytopes. (See, e. g., Buchta–Müller–Tichy [a] and Bárány [b].) A model with few cyclic, but many neighborly polytopes, was studied by Bokowski–Richter-Gebert–Schindler [a].

Gale-diagrams.
A characterization of neighborly polytopes in terms of their Gale-diagrams is hidden in exercise 7.3.7; compare Sturmfels [c]. Note that "Gale's Lemma", which appeared in Bárány's [a] solution to the Kneser problem, is equivalent to the existence of neighborly polytopes with n vertices, for all $n > d$.

Cyclic polytopes.
For cyclic polytopes of even dimension, Sturmfels [a] proved that in every realization the vertices do lie on a curve of order d. This was based on a result of Shemer [a] that all even-dimensional neighborly polytopes are *rigid* in the sense that the combinatorial type of P also determines the combinatorial types of all subpolytopes. A connection between cyclic polytopes and the theory of totally positive matrices was established in Sturmfels [e].

Some problems posed on pages 128–129.
Zaks [b] proved $m(3) = 8$, and Perles [a] showed $m(d) < 2^{d+1}$; see also Aigner–Ziegler [a, Chap. 12]. Zaks [a] showed that for every $d \geq 3$ there are arbitrarily large neighborly families of centrally symmetric convex d-polytopes in R^d. See Erickson [a] for a strong extension to congruent polytopes.

Grünbaum's conjecture on "strictly antipodal sets" is wrong: A probabilistic argument of Erdős–Füredi [a] showed that these can be even exponentially large; see also Aigner–Ziegler [a, Chap. 13].

Three open problems.
Shemer [a, p. 314] asked whether, for even d, every neighborly d-polytope with n vertices can be extended to a neighborly d-polytope with $n + 1$ vertices.

Does every combinatorial type of simplicial polytope occur as a quotient (iterated vertex figure) of an even-dimensional neighborly polytope? This is a question by Perles (see Sturmfels [d]). Kortenkamp [a] gave a neat proof for d-polytopes with at most $d + 4$ vertices, by a Gale-diagram construction.

A problem posed on page 129: Is there *any* polytope other than a simplex that is 2-neighborly and dual 2-neighborly? One can show that there are no such d-polytopes with less than $d + 4$ vertices or facets.

CHAPTER 8

Euler's Relation

In the preceding chapters we became acquainted with methods of generating convex polytopes and with the relations existing among a polytope and its faces or other points of the space. Our interest was mostly centered on positional relationships between a polytope (or its faces) and some other given geometric object.

In the present chapter we turn to a different field of investigation: we try to find meaningful variations of the simple-minded question 'What polytopes do exist', and we endeavor to find at least partial answers to some of those problems. One of the simplest questions that may be asked in this connection is: What d-tuples of numbers can occur as the numbers of vertices, edges, $\cdots$, $(d - 1)$-faces of convex d-polytopes?

The present chapter, and some of the following ones, are devoted to an exposition of the presently known partial answers. In this chapter we shall deal with the result known as 'Euler's formula' and some of its ramifications; chapter 9 will deal with analogues of Euler's formula valid for the subclass of simplicial polytopes and for some other special classes of polytopes.

8.1 Euler's Theorem

We start by recalling from chapter 3 the $f_k(P)$ notation.

Let P be a d-polytope, and let k be an integer such that $0 \leq k \leq d - 1$. We shall denote by $f_k(P)$ the number of k-faces of P; when no confusion as to the polytope in question is likely to arise, we shall sometimes write f_k instead of $f_k(P)$. We find it convenient to use the symbol $f_k(P)$ also for the improper faces $\emptyset$ and P of P; in other words, we put $f_{-1}(P) = 1$ and $f_d(P) = 1$. Where convenient, we shall as well use the notation $f_k(P) = 0$ for $k > d$ and for $k < -1$.

With each d-polytope P we shall associate the d-dimensional vector $f(P) = (f_0(P), f_1(P), \cdots, f_{d-1}(P))$, which will be called the f-vector of P. For any family $\mathscr{P}$ of d-polytopes we shall denote by $f(\mathscr{P})$ the set $f(\mathscr{P}) = \{f(P) \mid P \in \mathscr{P}\}$.

The problem mentioned above is, therefore, to determine $f(\mathscr{P}^d)$, the set of all f-vectors of d-polytopes. For dimensions $d \geq 4$ this problem is, however, far from being solved; for $d = 3$ the complete solution is rather easy, and it will be given in section 10.3. In the present chapter we shall solve, for general d, a much easier problem: the determination of the affine hull of $f(\mathscr{P}^d)$. We formulate the result as *Euler's theorem*:

1. *The affine hull of the f-vectors of all members of the family $\mathscr{P}^d$ of all d-polytopes is given by*

$$\text{aff } f(\mathscr{P}^d) = \{f = (f_0, \cdots, f_{d-1}) \mid \sum_{i=0}^{d-1} (-1)^i f_i = 1 - (-1)^d\}.$$

The relation $\sum_{i=0}^{d-1} (-1)^i f_i(P) = 1 - (-1)^d$, which by the theorem holds for every d-polytope P, is known as Euler's equation. Using the extended symbols $f_d(P)$ and $f_{-1}(P)$, Euler's equation may be written in the equivalent but more symmetric forms

$$\sum_{i=0}^{d} (-1)^i f_i(P) = 1$$

or

$$\sum_{i=-1}^{d} (-1)^i f_i(P) = 0.$$

Before proceeding to prove the theorem, we shall consider a few special cases.

(1) Euler's equation obviously holds for $d = 1$, as well as for $d = 2$ (though the inductive proof which will be given below applies already to the latter case.) For 3-polytopes, Euler's equation $f_0 - f_1 + f_2 = 2$ is not obvious any more, although very simple proofs are known (see below).

(2) Next we consider a d-prismoid P (see chapter 4), that is, a d-polytope which is the convex hull of two polytopes P_{-1} and P_1, each of dimension $d - 1$ at most, and such that the intersection $(P_{-1} \cup P_1) \cap H_{-1} \cap H_1$ is empty, where H_i is a hyperplane containing $P_i, i = \pm 1$. Let H_0 be a hyperplane belonging to the pencil determined by H_{-1} and H_1, and passing through an interior point of P, and let $P_0 = P \cap H_0$. (See figure 8.1.1 where the case $d = 3$ is illustrated.) For a d-prismoid P the following assertions obviously hold:

(i) $f_0(P) = f_0(P_{-1}) + f_0(P_1)$

(ii) for $1 \leq k \leq d - 1$, a k-face F of P is either a face of either P_{-1} or P_1, or it has vertices in both P_{-1} and P_1. In the latter case, there corresponds to the k-face F of P in a biunique fashion a $(k - 1)$-face $F \cap H_0$ of P_0. Thus, for $1 \leq k \leq d - 1$, we have

$$f_k(P) = f_k(P_{-1}) + f_k(P_1) + f_{k-1}(P_0).$$

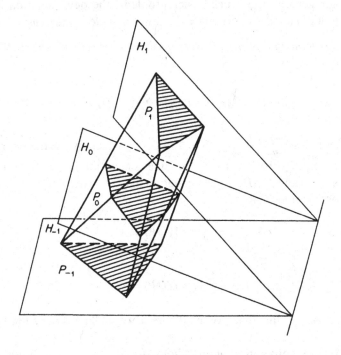

Figure 8.1.1

Therefore,

$$\sum_{k=0}^{d-1} (-1)^k f_k(P) = \sum_{k=0}^{d-1} (-1)^k (f_k(P_{-1}) + f_k(P_1)) + \sum_{k=0}^{d-2} (-1)^{k+1} f_k(P_0).$$

If the validity of Euler's formula is assumed for the $(d - 1)$-polytope P_0 and for the at most $(d - 1)$-dimensional polytopes P_{-1} and P_1, it follows that

$$\sum_{k=0}^{d-1} (-1)^k f_k(P) = 2 - (1 - (-1)^{d-1}) = 1 - (-1)^d;$$

in other words, the d-polytope P also satisfies Euler's equation.

(3) The validity of Euler's relation for any d-simplex can be established by induction from the reasoning in (2) (since P_{-1} and P_0 may be taken as $(d-1)$-simplices, and P_1 as a point). In this case, however, a direct verification is also possible since for a d-simplex P we have

$$f_k(P) = \binom{d+1}{k+1}, \quad k = 0, \cdots, d-1,$$

and since

$$\sum_{k=0}^{d-1} (-1)^k \binom{d+1}{k+1} = -(1-1)^{d+1} + 1 + (-1)^{d+1}\binom{d+1}{d+1} = 1 - (-1)^d.$$

(4) Still preparatory to the proof of the general case of Euler's relation, we consider the following situation.

Let P be a d-polytope (in d-space) and let H_0 be a hyperplane meeting the interior of P and containing exactly one vertex V_0 of P. Let H^+ and H^- be the two closed half-spaces determined by H_0, and let $P_0 = P \cap H_0$, $P_1 = P \cap H^+$ and $P_2 = P \cap H^-$. (See figure 8.1.2 illustrating the case $d = 3$). We note the following relations between the f-vectors of P_0, P_1, and P_2:

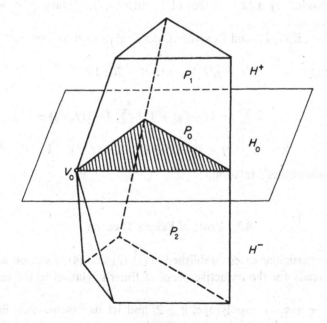

Figure 8.1.2

(i) $f_0(P) = f_0(P_1) + f_0(P_2) - 2f_0(P_0) + 1$

since, except for V_0, the vertices of P are exactly those vertices of P_1 or P_2 which do not belong to P_0;

(ii) $f_1(P) = f_1(P_1) + f_1(P_2) - 2f_1(P_0) - f_0(P_0) + 1$

since an edge of P is either an edge of P_1 or of P_2 not contained in P_0, or it is divided by a vertex of P_0 different from V_0 into an edge of P_1 and an edge of P_2;

(iii) $f_2(P) = f_2(P_1) + f_2(P_2) - 2f_2(P_0) - f_1(P_0)$

$$\vdots$$

$$f_{d-2}(P) = f_{d-2}(P_1) + f_{d-2}(P_2) - 2f_{d-2}(P_0) - f_{d-3}(P_0)$$

since, for $2 \le k \le d - 2$, a k-face of P is either a face of P_1 or of P_2 but not of P_0, or it is divided by a $(k - 1)$-face of P_0 into a k-face of P_1 and k-face of P_2;

(iv) $f_{d-1}(P) = f_{d-1}(P_1) - 1 + f_{d-1}(P_2) - 1 - f_{d-2}(P_0)$

since each $(d - 1)$-face of P is either a face of P_1 or of P_2 different from P_0, or it is divided by a $(d - 2)$-face of P_0 into a $(d - 1)$-face of P_1 and a $(d - 1)$-face of P_2.

Therefore, if P_0, P_1, and P_2 satisfy Euler's relation, it follows that

$$\sum_{k=0}^{d-1} (-1)^k f_k(P) = \sum_{k=0}^{d-1} (-1)^k (f_k(P_1) + f_k(P_2)) - 2(-1)^{d-1}$$

$$- 2 \sum_{k=0}^{d-2} (-1)^k f_k(P_0) + 1 + \sum_{k=0}^{d-2} (-1)^k f_k(P_0) - 1$$

$$= 2(1 - (-1)^d) + 2(-1)^d - (1 - (-1)^{d-1}) = 1 - (-1)^d,$$

i.e. P satisfies Euler's relation.

8.2 Proof of Euler's Theorem

Using the particular cases established in (1), (2), and (4) of section 8.1, we are now ready for the inductive proof of Euler's relation in the general case.

Let P be a given d-polytope, $d \ge 2$, and let us assume that Euler's equation is already established for all polytopes of dimension less than d.

Let H be a hyperplane such that no translate of H contains two or more vertices of P. (See figure 8.2.1 for an illustration of the case $d = 3$.) Let $H_1, H_2, \cdots, H_v$ (where $v = f_0(P)$) be the hyperplanes parallel to H, each containing one vertex of P, the notation being arranged in such a way that H_j separates H_i from H_k for $i < j < k$. For $1 \le i \le v - 1$, let K_i be the part of K between H_i and H_{i+1}. Then each K_i is a d-prismoid, as considered above. The inductive assumption implies, therefore, the validity of Euler's formula for each d-prismoid K_i. Let $K^{(j)} = \overset{j}{\underset{i=1}{\cup}} K_i$, for $j = 1$, $2, \cdots, v - 1$. Then, for $j = 1, 2, \cdots, v - 2$, the conditions assumed in (4) for P, H_0, P_1 and P_2, are satisfied by $K^{(j+1)}$, H_{j+1}, $K^{(j)}$, and K_{j+1}. Since $K^{(1)} = K_1$, and all K_i satisfy Euler's formula, it follows successively that $K^{(2)}, K^{(3)}, \cdots, K^{(v-1)}$ satisfy Euler's formula. But $K^{(v-1)} = P$, and thus the validity of Euler's relation for P is established.

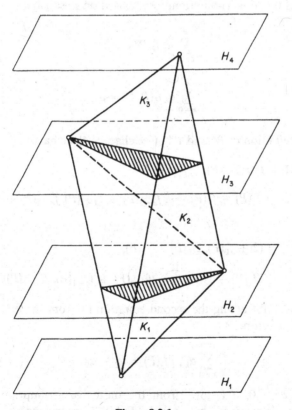

Figure 8.2.1

Till now we have established one half of the theorem, namely the fact that $f(\mathscr{P}^d)$, and therefore also its affine hull aff $f(\mathscr{P}^d)$, are contained in the 'Euler hyperplane' $E^d = \{f = (f_0, \cdots, f_{d-1}) \mid \sum_{i=0}^{d-1} (-1)^i f_i = 1 - (-1)^d\}$. In order to complete the proof of the theorem we have to show that aff $f(\mathscr{P}^d)$ coincides with E^d. In other words, we have to show that every linear equation among $f_0(P), \cdots, f_{d-1}(P)$, which is satisfied by all d-polytopes P, is a multiple of Euler's equation. While it is not hard to describe a finite family $\mathscr{P}_0$ of d-polytopes such that the affine hull of $f(\mathscr{P}_0)$ is $(d-1)$-dimensional and therefore necessarily coincides with E^d, a much shorter proof can be given by induction. Indeed, the assertion aff $f(\mathscr{P}^d) = E^d$ is obvious for $d = 1$ or $d = 2$; assuming its truth established for a certain d, we consider aff $f(\mathscr{P}^{d+1})$.

Let $\sum_{k=0}^{d} \alpha_k f_k = \beta$ be the equation of a hyperplane containing aff $f(\mathscr{P}^{d+1})$. For every d-polytope $P \in \mathscr{P}^d$, the $(d+1)$-pyramid P^* with base P, and the $(d+1)$-bipyramid P^{**} based on P, satisfy

(*)

$$\sum_{k=0}^{d} \alpha_k f_k(P^*) = \beta$$

$$\sum_{k=0}^{d} \alpha_k f_k(P^{**}) = \beta$$

By the definition of P^* and P^{**} (see chapter 4) we have

$$f(P^*) = (1 + f_0(P), f_0(P) + f_1(P),$$

$$f_1(P) + f_2(P), \cdots, f_{d-2}(P) + f_{d-1}(P), f_{d-1}(P) + 1)$$

and

$$f(P^{**}) = (2 + f_0(P), 2f_0(P) + f_1(P),$$

$$2f_1(P) + f_2(P), \cdots, 2f_{d-2}(P) + f_{d-1}(P), 2f_{d-1}(P)).$$

Therefore, subtracting the second equation (*) from the first we have, for every d-polytope P,

(**)

$$\sum_{k=0}^{d-1} \alpha_{k+1} f_k(P) = \alpha_d - \alpha_0.$$

Since $\sum \alpha_k^2 > 0$, it follows from the inductive assumption that the hyperplane determined by (**) is Euler's hyperplane E^d, that is,

$\alpha_k = (-1)^k \alpha_0$ for $0 \le k \le d$. Considering the $(d + 1)$-simplex we see that $\beta = (1 - (-1)^{d+1})\alpha_0$; in other words, the equation $\sum_{k=0}^{d} \alpha_k f_k = \beta$ is a multiple of Euler's equation $\sum_{k=0}^{d} (-1)^k f_k = 1 - (-1)^{d+1}$. This completes the proof of the theorem.

In section 9.3 we shall see that the Euler hyperplane is spanned even by the much smaller family of quasi-simplicial polytopes.

8.3 A Generalization of Euler's Relation

Euler's formula is self-dual in the sense that the passage from a d-polytope $P \subset R^d$ to the dual d-polytope P^*, and the use of the relations $f_i(P) = f_{d-1-i}(P^*)$, yield again only Euler's formula for P^*. Nevertheless, a variant of Euler's formula, which we shall have occasion to use in chapter 9, is obtainable by an application of duality.

Let P be a d-polytope and let F be an m-face of P. Let $h_k(F)$ denote the number of k-faces of P which contain F if $k \ge m$, or the number of k-faces of P contained in F if $k \le m$. Thus for $k \le m$ we have $h_k(F) = f_k(F)$.

With this notation we have the following easy generalization of Euler's formula:

1. *Let P be a d-polytope and F a k-face of P, $-1 \le k \le d - 1$. Then*

$$\sum_{j=k}^{d-1} (-1)^j h_j(F) = (-1)^{d-1}.$$

In order to prove this result we consider a d-polytope P^* dual to P. To each j-face of P there corresponds a unique $(d - 1 - j)$-face of P^*, the correspondence reversing inclusion-relations. Therefore, if F^* is the $(d - 1 - k)$-face of P^* corresponding to F, for $j \ge k$ we have $h_j(F) = h_{d-1-j}(F^*) = f_{d-1-j}(F^*)$. Using Euler's relation for the $(d-1-k)$-polytope F^* we obtain

$$\sum_{j=k}^{d-1} (-1)^j h_j(F) = \sum_{j=k}^{d-1} (-1)^j f_{d-1-j}(F^*)$$

$$= (-1)^{d-1} \sum_{i=0}^{d-1-k} (-1)^i f_i(F^*) = (-1)^{d-1},$$

as claimed.

8.4 The Euler Characteristic of Complexes

Our next aim is to obtain analogs of Euler's relation which are valid for certain complexes (see section 3.3). We start with some definitions.

Let $\mathscr{C}$ be a d-complex; for $0 \le k \le d$ we define $f_k(\mathscr{C})$ as the number of k-dimensional members of $\mathscr{C}$. Clearly this is consonant with our earlier notation for polytopes, since $f_k(\mathscr{C}(P)) = f_k(P)$ for $0 \le k \le d$, and $f_k(\mathscr{B}(P)) = f_k(P)$ for $0 \le k \le d - 1$ whenever P is a d-polytope.

The *Euler characteristic* $\chi(\mathscr{C})$ of a d-complex $\mathscr{C}$ is defined by

$$\chi(\mathscr{C}) = \sum_{k=0}^{d} (-1)^k f_k(\mathscr{C}).$$

Using this notation we see that Euler's relation for a d-polytope P may be written in the two equivalent forms

$$\chi(\mathscr{B}(P)) = 1 - (-1)^d$$

or

$$\chi(\mathscr{C}(P)) = 1.$$

Thus in two special cases—for complexes of type $\mathscr{B}(P)$, as well as for those of type $\mathscr{C}(P)$—the Euler characteristic of a complex depends only on the dimension of the complex. We shall show that stars in $\mathscr{B}(P)$ have a similar property.

1. *If F is a proper face of a polytope P then* $\chi(\mathrm{st}(F\,;\mathscr{B}(P))) = 1$.

For an elementary proof of theorem 1, it is convenient to establish a more general result, the formulation of which requires some additional concepts.

Let $\mathscr{C}$ be a d-complex and let $0 \le k \le d$. The *k-skeleton* $\mathrm{skel}_k\mathscr{C}$ is the subcomplex of $\mathscr{C}$ formed by all the members of $\mathscr{C}$ which have dimension at most k. Thus $\mathrm{skel}_0\mathscr{C}$ is the set of vertices of $\mathscr{C}$, while $\mathrm{skel}_d\mathscr{C} = \mathscr{C}$. For a polytope P, $\mathrm{skel}_1\mathscr{C}(P) = \mathrm{skel}_1\mathscr{B}(P)$ is the *graph* determined by the vertices and edges of P.

Let P be a d-polytope, F an m-face of P, and $0 \le m \le k \le d - 1$. The *k-star* $\mathrm{st}_k(F\,;P)$ of F in P is defined by

$$\mathrm{st}_k(F\,;P) = \mathrm{st}(F\,;\mathrm{skel}_k\mathscr{B}(P)).$$

We shall establish

2. *If F is an m-face of a d-polytope P and if $0 \le m \le k \le d - 1$, then* $\chi(\mathrm{st}_k(F\,;P)) = 1$.

Since $\text{st}_{d-1}(F; P) = \text{st}(F; \mathscr{B}(P))$, the above theorem 1 is the special case $k = d - 1$ of theorem 2.

We shall prove theorem 2 by induction on d and k. The assertion is obvious if $d = 1$ or $d = 2$. It is also true for arbitrary d provided $k = m$, since then $\text{st}_k(F; P) = \mathscr{C}(F)$ and by Euler's relation $\chi(\mathscr{C}(F)) = 1$. Thus we may assume $k > m$. Let us denote by F_i^k, where i belongs some index-set $I(k)$, the different k-dimensional members $\text{st}_k(F; P)$. If F^* is a proper face of some F_i^k then either (i) F^* is not a face of any F_j^k with $j \neq i$; or (ii) F^* is an element of $\text{st}_{k-1}(F; P)$.

Indeed, if F^* is a face of both F_i^k and F_j^k, $i \neq j$, then $F^* \subset F_i^k \cap F_j^k = F^{**}$ which is a face of P of dimension at most $k - 1$; since F is a face of F^{**} it follows that $F^{**} \in \text{st}_{k-1}(F; P)$ and thus also $F^* \in \text{st}_{k-1}(F; P)$. Therefore

$$f_n(\text{st}_k(F; P)) = f_n(\text{st}_{k-1}(F; P)) + \sum_{i \in I(k)} f_n(\mathscr{C}(F_i^k) \sim \text{st}_{k-1}(F; P))$$

for $n = 0, 1, \cdots, k$, where $f_k(\text{st}_{k-1}(F; P)) = 0$. Now $\mathscr{C}(F_i^k) \sim \text{st}_{k-1}(F; P)$ $= \mathscr{C}(F_i^k) \sim \text{st}_{k-1}(F; F_i^k)$ and it follows that $f_n(\mathscr{C}(F_i^k) \sim \text{st}_{k-1}(F; P))$ $= f_n(F_i^k) - f_n(\text{st}_{k-1}(F; F_i^k))$. The inductive assumption and Euler's relation imply therefore

$$\chi(\text{st}_k(F; P)) = \sum_{n=0}^{k} (-1)^n f_n(\text{st}_k(F; P))$$

$$= \sum_{n=0}^{k-1} (-1)^n f_n(\text{st}_{k-1}(F; P)) + \sum_{i \in I(k)} \sum_{n=0}^{k} (-1)^n f_n(F_i^k)$$

$$- \sum_{i \in I(k)} \sum_{n=0}^{k-1} (-1)^n f_n(\text{st}_{k-1}(F; F_i^k))$$

$$= 1 + \text{card } I(k) - \text{card } I(k) = 1.$$

This completes the proof of theorem 2.

8.5 Exercises

1. Show that the Euler hyperplane E^d is spanned by the f-vectors of centrally symmetric d-polytopes.

2. If P is a polyhedral set we denote by $f_k^0(P)$ the number of bounded k-faces of P, by $f_k^\infty(P)$ the number of unbounded k-faces of P, and by $f_k(P) = f_k^0(P) + f_k^\infty(P)$ the total number of k-faces of P. Using the results

of sections 2.4 and 2.5, derive from Euler's theorem the following relations, valid for d-dimensional, line-free, unbounded polyhedral sets P:

$$\text{(i)} \quad \sum_{i=0}^{d-1} (-1)^i f_i^0(P) = 1$$

$$\text{(ii)} \quad \sum_{i=1}^{d} (-1)^{i+1} f_i^\infty(P) = 1$$

$$\text{(iii)} \quad \sum_{i=0}^{d} (-1)^i f_i(P) = 0$$

$$\text{(iv)} \quad \sum_{i=0}^{d} (-1)^i (f_i(P) + f_{i+1}^\infty(P)) = 1$$

Obviously, (i) and (iv) are valid even if P is bounded, while (iii) is valid whenever P is a polyhedral set which is not the vector sum of a polytope and a linear variety.

Using theorem 2.5.4. (ii) may be generalized to

$$\sum_{i=k+1}^{d} (-1)^{i+1} f_i^\infty(P) = (-1)^k ,$$

where P is any unbounded, d-dimensional polyhedral set which is not the vector sum of a polytope and a linear variety, and k is the maximal dimension of affine varieties contained in P.

3. Show that the Euler hyperplane E^d is spanned already by the f-vectors of d-polytopes with at most $d + 2$ vertices.

4. (Sommerville [1]). Show that the f-vector of every r-fold d-pyramid satisfies the equations

$$\sum_{j=0}^{d-1-t} (-1)^j \binom{d-1-j}{t} f_j = \binom{d}{t} - (-1)^{d-t}$$

for each $t = 0, 1, \cdots, r$.

5. Use theorem 8.3.1 to show that if P is a simplicial d-polytope and if F is a k-face of P, $-1 \le k \le d - 1$, then

$$\chi(\text{link}(F\,;\mathscr{B}(P))) = 1 + (-1)^{d-k} .$$

6. Show that the boundary complex $\mathscr{B}(P)$ of every d-polytope P is an *Eulerian $(d - 1)$-manifold* (Lefschetz [1], Klee [11]). This means that for every k-face F of P, $-1 \leq k \leq d - 1$, the equation

$$\chi(\text{link}(F\,;\mathscr{B}(P))) = 1 + (-1)^{d-k}$$

holds.

8.6 Remarks

Euler's equation for 3-polytopes, $f_0 - f_1 + f_2 = 2$, was discovered by Euler [1,2] in 1752. It has been hailed as 'the first important event in topology' (Alexandroff–Hopf [1], p. 1), and as 'the first landmark' in the theory of polytopes (Klee [18]). Enquiries into its range of validity have sparked many papers during the first half of the nineteenth century, and were actually among the first to turn the spotlight on the notion of convexity. For a history of these early endeavors see in particular Brückner [2], Zacharias [1], Steinitz [6]. It is interesting that 'Euler's theorem' was known to Descartes about a hundred years earlier; however, his manuscript was lost and forgotten, and the present knowledge of it stems from a partial copy found in 1860 among the papers of Leibnitz (see Zacharias [1], Steinitz–Rademacher [1], p. 9). Possibly even more remarkable is the absence of any result related to Euler's theorem from the writings of the ancient Greek mathematicians.

With the emergence of higher-dimensional geometry about the middle of the nineteenth century came Schäfli's [1] discovery of Euler's relation for polytopes of dimensions exceeding 3. Though Schläfli's discovery was made in 1852, it was published in full only in 1902, and probably had little influence on other workers in this area. The 'public discovery' of Euler's relation in higher dimensions came in the early 1880's, almost simultaneously by many geometers (Stringham [1], Forchhammer [1], Hoppe [1], Durège [1], Rudel [1]). Unfortunately, all the proofs published by these authors were deficient at one point of their reasoning: it was assumed, without any attempt at justifying this assumption, that it is possible to build up the boundary complex $\mathscr{B}(P)$ of a d-polytope P by successively adding facets of P (and their faces) in such a fashion that (except for the first and last step) the newly added facet intersects the already present $(d - 1)$-complex in a 'simply-connected' (i.e. homotopically trivial) $(d - 2)$-complex. Though this assumption is possibly valid, its truth is certainly not obvious even for $d = 3$. In case $d = 3$ the assumption may be established by an argument involving the Jordan curve theorem. Using Jordan's theorem in higher dimensions, or some other high-powered result, it is conceivable that one could establish the general validity of the above assumption—it seems, however, that this has not been done so far. Moreover, an example of M. E. Rudin [1] casts serious doubts on the validity of the assumption for $d \geq 4$. Curiously enough, geometers have for a long time ignored this flaw in the first attempts to prove Euler's theorem: essentially the same proof is presented in both Schoute's [1] and Sommerville's [2] books. On the other hand, starting with an attempt by Poincaré [1] in 1893, topologists were considering the Euler characteristic of manifolds, and its relation to the Betti numbers of the manifolds. Some flaws in Poincaré's paper [1] were corrected by him in 1899 (Poincaré [2]) and this seems to be the first real proof of Euler's relation in higher dimensions. A more modern version of Poincaré's proof, in which the use of homology and algebraic apparatus was purposely kept minimal, is given in chapter IX of Coxeter [1]. In a more circuitous way, Euler's theorem is proved also in Alexandroff–Hopf [1].

In all those proofs, the elementary-geometric character of Euler's theorem is completely lost. It was only with Hadwiger's paper [2] in 1955 that the question of an elementary proof of Euler's theorem for polytopes was considered. Another relatively elementary proof was given by Klee [5] in 1963. In both papers, however, algebraic overtones—extraneous to Euler's formula for polytopes—are introduced through the consideration of the *Konvexring* (in Hadwiger [2]) and lattices (in Klee [5]). The elementary proof given in the preceding pages is probably not the simplest possible, but it is completely elementary and does not exceed the framework of convex polytopes.

The Euler characteristic may be (and usually is) defined for topological complexes. Clearly, this is a situation much more general than the one considered in the present chapter; the methods of proof are, correspondingly, less elementary. On the other hand, the topological approach yields not only Euler's formula for d-polytopes, but at the same time the analogous relations for spherical polytopes, topological subdivisions of the d-sphere or the projective d-space, etc. For details and proofs the reader is referred to topology texts such as Alexandroff–Hopf [1].

The second part of theorem 8.1.1, namely the assertion that aff $f(\mathscr{P}^d)$ has dimension $d - 1$, is quite recent. For $d = 3$ it was established by Steinitz [1] (see section 10.3), but for $d \geq 4$ it appears first in Höhn [1] in 1953. Höhn's proof coincides with exercise 8.5.3.

8.7   Additional notes and comments

Sweeps.

The method of proof that Grünbaum employs for the proof of Euler's formula 8.1.1 (also known as the Euler–Poincaré formula) may in modern computational geometry terms be seen as a *sweep*: A hyperplane that in the beginning does not intersect the polytope is moved in parallel across the polytope in such a way that it never hits two vertices at the same time.

For a simple, but striking, application of this point of view, see Seidel's [a] proof of the asymptotic upper bound theorem.

Sweeps form a crucial technique of computational geometry; see e.g. de Berg et al. [a, Chap. 2]. Furthermore, they are of fundamental importance for the algebraic geometry view on polytopes: Any generic sweep for a rational polytope corresponds to a (linear) Morse function for the associated compact toric variety.

Polytopes are shellable.

Bruggesser–Mani [a] proved that all polytopes are shellable. The proof is by now a classic; it may be found also, e. g., in Ziegler [a, Sect. 8.2]. In brief, Bruggesser and Mani's *line shellings* are generated as follows: Consider a directed line in general position through the interior of a d-polytope P. Imagine a rocket that starts at the point where the line leaves the polytope, and take the facets of P in the order in which the rocket passes through the facet hyperplanes, so that the facets become "visible" from the rocket; the rocket "passes though infinity" and then, coming from the other side, again the facets are taken in the order of passage through their hyperplanes, that is, in the order in which they disappear from the horizon.

With hindsight, it is ironic that via polarization Grünbaum's proof already solves the problem about the existence of a shelling, as posed in section 8.6. Here is an explicit translation table:

primal	dual
polytope P	polar polytope P^*
hyperplane H	point x
parallel family of hyperplanes H	points x on a line
(parallel hyperplane) sweep	point moves on a line
H does not intersect P	x is in the interior of P^*
H hits two vertices of P	x lies on two facet hyperplanes of P^*

Thus polarization of Grünbaum's proof leads us to consider a line in general position through the interior of P^*, letting a point move on this line, starting in the interior of P^*, and to take the facets of P^* in the order in which they

move through the facet hyperplanes of P^*: This is exactly the construction of a
Bruggesser–Mani line shelling of P^*.

Eulerian lattices.

One of the many reformulations of Euler's theorem is that the face lattice L of
any non-empty polytope P has the same number of elements of even and of odd
rank. Any interval $[G, F] \subset L$, for faces $G \subset F$ of P, is again the face lattice
of a polytope (namely, of an iterated vertex figure of F); thus it follows that
the odd/even property is also true for all non-trivial intervals of L. Following
Klee [11], one may thus consider the combinatorial model of *Eulerian lattices*,
i. e., finite graded lattices such that every non-trivial interval has the odd/even
property; equivalently (see Stanley [h, Chap. 3]), the Möbius function of such
a lattice is given by $\mu(G, F) = (-1)^{\ell(G,F)}$, where $\ell(G, F) = \dim F - \dim G$ is
the *length* of the interval $[G, F]$.

This generalization provides an entirely combinatorial model that includes
the face lattices of the regular cell decompositions of spheres and of odd-
dimensional manifolds that have the *intersection property* (i. e., the intersection
of any two cells of the complex is a cell of the complex, which may be empty).
While there has been considerable effort to understand Eulerian posets and
their f-vectors (a recent survey is Stanley [f]), Eulerian lattices have not yet
received enough attention. See also Eppstein–Kuperberg–Ziegler [a], where
regular cell complexes with the intersection property are called "strongly regular".

The vector $(10, 20, 20, 10)$ is an example of an f-vector of an Eulerian lattice
(corresponding to the boundary complex of two disjoint 4-simplices) that is
not the f-vector of a polytope (since every 4-polytope with $f_1 = 2f_0$ and $f_2 =
2f_3$ is simple and simplicial, i. e., it is a 4-simplex having $(5, 10, 10, 5)$ as its
f-vector). A "connected" example of this type is $(13, 78, 195, 260, 195, 65)$,
corresponding to a simplicial 5-dimensional manifold that was constructed by
Kühnel [a].

CHAPTER 9

Analogues of Euler's Relation

In chapter 8 we have seen that the affine hull of the set $f(\mathscr{P}^d)$ of f-vectors of all d-polytopes is the Euler hyperplane. In the present chapter we shall be interested in finding the affine hulls of sets $f(\mathscr{P})$, for certain families $\mathscr{P}$ of d-polytopes.

In principle, this task consists of two parts, in which quite different techniques find application. The first step is the establishment of various linear equations satisfied by $f(P)$ for all $P \in \mathscr{P}$. It is based on a combinatorial procedure useful in many fields, the 'double counting of incidences', and it has rather little geometric content. The final result obtained by this method is an affine variety which contains $f(\mathscr{P})$. However, there is *a priori* not much reason to assume that this variety is indeed the affine hull of the vectors $f(\mathscr{P})$. For families $\mathscr{P}$ which interest us we shall show, by the use of appropriate geometric constructions, that this coincidence does happen. This is the second part of the method, and it naturally varies from family to family.

9.1   The Incidence Equation

The method of 'double counting of incidences' has various formulations, the simplest of which is probably expressed by the commutative law for the multiplication of positive integers. The formulation useful for our purposes involves a somewhat specialized set-up, which we shall first define abstractly. Applications to a number of specific problems will be found in the following sections.

Let us assume that we are given a positive integer d and, for each i satisfying $0 \le i \le d$, a nonempty family $\mathscr{F}_i$ of objects which we shall call *i-objects*. One illustration of such a situation, to which we shall refer repeatedly, is given by any d-polytope or d-complex, where the *i*-faces may be considered as *i*-objects. If $0 \le i, j \le d$, an *i*-object $F_i \in \mathscr{F}_i$ and a *j*-object $F_j \in \mathscr{F}_j$ may be *incident*, or they may fail to be incident; the only assumptions we wish to make on this relationship being that each object

143

be incident to itself, and that the incidence-relation be symmetric. It is convenient to introduce a function $\varphi(F', F'')$ defined for all $F', F'' \in \bigcup\limits_{0 \le i \le d} \mathscr{F}_i$

. by

$$\varphi(F', F'') = \begin{cases} 1 \text{ if } F' \text{ and } F'' \text{ are incident,} \\ 0 \text{ if } F' \text{ and } F'' \text{ are not incident.} \end{cases}$$

Our assumptions imply that $\varphi(F, F) = 1$ for every F and that $\varphi(F', F'') = \varphi(F'', F')$ for all F' and F''. In the illustration mentioned above one could interpret 'incidence' of two faces to have the usual meaning, i.e. that one of the faces is contained in the other.

For given families $\mathscr{F}_i$ of objects, let $v_i(F)$ denote the number of i-objects incident to a given object F; thus $v_i(F) = \sum_{F_i \in \mathscr{F}_i} \varphi(F_i; F)$. We shall be interested in families which satisfy the following conditions:

There exists an m, $1 \le m \le d$, such that

(i) There exist real numbers $\alpha_0, \cdots, \alpha_m$, and β_m such that each $F \in \mathscr{F}_m$ satisfies

$$\sum_{i=0}^{m} \alpha_i v_i(F) = \beta_m.$$

(ii) For every i, $0 \le i \le m$, there exists a constant $\gamma_{i,m}$ such that $v_m(F_i) = \gamma_{i,m}$ for every $F_i \in \mathscr{F}_i$.

In the illustration, where the faces of a polytope are the objects, an example of an equation as required in condition (i) is provided by Euler's relation or by one of its variants (such as the relations of section 8.4). Conditions of type (ii) are satisfied, for example, by the faces of simple polytopes (see exercise 4.8.12).

Let μ_i denote the number of i-objects (i.e. the number of elements of $\mathscr{F}_i$). We shall prove

1. *If $\mathscr{F}_i$ are families satisfying the above assumptions, then the* incidence equation

(*) $$\beta_m \mu_m = \sum_{0 \le i \le m} \alpha_i \gamma_{i,m} \mu_i$$

holds.

PROOF   For a fixed m, we consider the number

$$I = \sum_{\substack{F_m \in \mathscr{F}_m, F_i \in \mathscr{F}_i \\ 0 \le i \le m}} \alpha_i \varphi_i(F_i, F_m).$$

The number I is computable in two ways. First, keeping F_m fixed, we have

$$I = \sum_{\substack{F_m \in \mathscr{F}_m}} \sum_{\substack{F_i \in \mathscr{F}_i \\ 0 \leq i \leq m}} \alpha_i \varphi(F_i, F_m) = \sum_{F_m \in \mathscr{F}_m} \sum_{0 \leq i \leq m} \alpha_i \nu_i(F_m).$$

Using condition (i) it follows that

$$I = \sum_{F_m \in \mathscr{F}_m} \beta_m = \beta_m \mu_m.$$

On the other hand, keeping F_i fixed and using condition (ii), we have

$$I = \sum_{0 \leq i \leq m} \alpha_i \sum_{F_i \in \mathscr{F}_i} \sum_{F_m \in \mathscr{F}_m} \varphi_i(F_i, F_m) = \sum_{0 \leq i \leq m} \alpha_i \sum_{F_i \in \mathscr{F}_i} \gamma_{i,m}$$

$$= \sum_{0 \leq i \leq m} \alpha_i \gamma_{i,m} \mu_i.$$

This completes the proof of theorem 1.

It should be noted that the setting for 'double counting of incidences' used above is by no means unique, or the most general possible. For example, an easy generalization would result by allowing $\varphi(F', F'')$ to be any real-valued function symmetric in its variables. Also, instead of computing the number I, some other magnitude could be considered. We shall meet examples of such variants of the method in sections 10.1 and 13.1. Another variant may be found in Sommerville [1]. The approach used in the present section is closely related to that used by Klee [11].

It is also obvious that similar counting arguments may be applied to spherical polytopes, to complexes or topological complexes, and in many other situations. For an example of such a procedure compare section 18.1.

9.2 The Dehn–Sommerville Equations

In chapter 8 we have seen that Euler's equation

$$\sum_{i=0}^{d-1} (-1)^i f_i(P) = 1 - (-1)^d$$

is the only linear relation satisfied by the f-vectors $f(P)$ of all d-polytopes P. However, the f-vectors of simplicial d-polytopes satisfy additional linear equations, which may be obtained as special cases of the incidence equation of theorem 9.1.1.* In fact, denoting by $\mathscr{P}_s^d$ the family of all simplicial d-polytopes, we shall establish the following result:

---

* Naturally, the equations for simplicial polytopes may as well be derived independently, using the method applied in section 9.1.

1. *The dimension of the affine hull* $A^{(d)} = \text{aff} f(\mathscr{P}^d_s)$ *of the f-vectors of all simplicial d-polytopes is* $[\frac{1}{2}d]$. *The affine variety* $A^{(d)}$ *is the intersection of the hyperplanes* E^d_k, $-1 \le k \le d - 2$, *which are determined by the equations*

$$(E^d_k) \qquad \sum_{j=k}^{d-1} (-1)^j \binom{j+1}{k+1} f_j = (-1)^{d-1} f_k.$$

We shall call (E^d_k) the *Dehn–Sommerville equations*. The equation (E^d_{-1}) obviously coincides with Euler's equation.

The system of d equations (E^d_{-1}), (E^d_0), $\cdots$, (E^d_{d-2}) is not independent; some of its independent subsystems shall be determined below.

PROOF   We shall first show that the f-vector of every $P \in \mathscr{P}^d_s$ satisfies the Dehn–Sommerville equations. To this effect we note that for $0 \le i \le d$, the family of all $(d - i - 1)$-faces of P may be considered as a family of i-objects (in the sense used in section 9.1), the incidence relationship being that customary for faces. Putting $i = d - j - 1$ and $m = d - k - 1$, we have (using $h_j(F) = v_{d-j-1}(F) = v_i(F)$) from the equation of theorem 8.3.1

$$(-1)^{d-1} = \sum_{j=k}^{d-1} (-1)^j h_j = \sum_{j=k}^{d-1} (-1)^j v_{d-j-1} = \sum_{i=0}^{d-k-1} (-1)^{d-i-1} v_i$$

$$= \sum_{i=0}^{m} (-1)^{d-i-1} v_i,$$

i.e.

$$\sum_{i=0}^{m} (-1)^i v_i = 1.$$

Therefore our 'objects' satisfy condition (i) of section 9.1 whenever $-1 \le k \le d - 2$ (i.e. $1 \le m \le d$) if we put $\alpha_i = (-1)^i$ and $\beta_m = 1$. On the other hand, if $-1 \le k \le j \le d - 1$, then each j-face is incident with $\binom{j+1}{k+1}$ k-faces, P being a simplicial polytope. Putting again $i = d - j - 1$ and $m = d - k - 1$, it follows that condition (ii) of section 9.1 is satisfied for all $1 \le m \le d$, with $\gamma_{i,m} = \binom{d-i}{d-m}$ Hence the incidence equation of theorem 9.1.1 holds, and reduces to

$$\mu_m = \sum_{i=0}^{m} (-1)^i \binom{d-i}{d-m} \mu_i$$

for $1 \le m \le d$. Introducing the numbers of faces $f_j = \mu_{d-j-1} = \mu_i$ we obtain

$$\sum_{j=k}^{d-1} (-1)^{d-j-1} \binom{j+1}{k+1} f_j = f_k$$

for $-1 \le k \le d-2$. But except for a factor $(-1)^{d-1}$ those are exactly the Dehn–Sommerville equations we set out to prove.

Our next aim is to find the dimension of the intersection of the hyperplanes E_k^d for $-1 \le k \le d-2$. Clearly some of the equations (E_k^d) are dependent; for example, equation (E_{d-2}^d) is equivalent to (E_{d-3}^d) whenever $d \ge 2$. Our intention is to show that $\dim A^{(d)} \le [\frac{1}{2}d]$ by exhibiting $d - [\frac{1}{2}d] = [\frac{1}{2}(d+1)]$ independent equations (E_k^d). The simplest way of doing this is the following (later we shall indicate other systems of $[\frac{1}{2}(d+1)]$ independent linear equations which determine $A^{(d)}$ and which are more convenient for certain purposes):

If $d = 2n$ is even, then the $n = d - n$ equations $(E_0^d), (E_2^d), (E_4^d), \cdots, (E_{d-2}^d)$ are independent. Indeed, f_{2j} occurs only in the first $j+1$ equations for $j = 0, 1, \cdots, n-1$. The same applies to the n equations $(E_{-1}^d), (E_1^d), (E_3^d), \cdots, (E_{d-3}^d)$.

If $d = 2n + 1$ is odd, then the $n + 1 = d - n$ equations $(E_{-1}^d), (E_1^d), (E_3^d), \cdots, (E_{d-2}^d)$ are independent. Indeed, (E_{-1}^d) is the only non-homogenous equation, and for $j = 1, 2, \cdots, n$, terms involving f_{2j-1} appear only in those equations (E_k^d) for which $k \le 2j - 1$. Another independent system is formed by the $n + 1$ equations $(E_{-1}^d), (E_0^d), (E_2^d), (E_4^d), \cdots, (E_{d-3}^d)$.

In order to complete the proof of theorem 1 we shall now show that $\dim A^d \ge [\frac{1}{2}d]$. For this purpose we shall use the cyclic d-polytopes (see section 4.7 for their definition and the simple properties used here). Let $C(v, d)$ denote a cyclic d-polytope with $v \ge d + 1$ vertices, and let

$$n = [\tfrac{1}{2}d]; \quad \text{then } f_k(C(v, d)) = \binom{v}{k+1} \quad \text{for } 0 \le k \le n-1. \text{ Our assertion}$$

that the dimension of $A^{(d)}$ is at least n shall be established provided we prove that $f(\mathscr{P}_s^d)$ contains $n + 1$ affinely independent points. But the $n + 1$ f-vectors $f(C(v, d)), f(C(v+1)), \cdots, f(C(v+n, d))$ are affinely independent for any $v \ge d + 1$. Indeed, already the n-dimensional vectors formed by their first n coordinates are affinely independent. In order to see this we have only to consider the determinant

$$D(v, n) = \begin{vmatrix} 1 & \binom{v}{1} & \binom{v}{2} & \cdots & \binom{v}{n} \\ 1 & \binom{v+1}{1} & \binom{v+1}{2} & \cdots & \binom{v+1}{n} \\ & & \cdots & & \\ 1 & \binom{v+n}{1} & \binom{v+n}{2} & \cdots & \binom{v+n}{n} \end{vmatrix}.$$

By repeatedly subtracting each row from the following one it is immediate that

$$D(v, n) = \begin{vmatrix} 1 & \binom{v}{1} & \binom{v}{2} & \cdots & \binom{v}{n} \\ 0 & 1 & \binom{v}{1} & \cdots & \binom{v}{n-1} \\ 0 & 0 & 1 & \cdots & \binom{v}{n-2} \\ & & \cdots & & \\ 0 & 0 & 0 & 0 \cdots & 1 \end{vmatrix} = 1,$$

which establishes our assertion.

This completes the proof of theorem 1.

Theorem 1 and its proof determine the affine hull $A^{(d)}$ of $f(\mathscr{P}_s^d)$ by specifying (independent) linear equations satisfied by all the f-vectors $f(P)$, $P \in \mathscr{P}_s^d$, as well as by giving examples of affine bases of $A^{(d)}$. We shall supplement those results by finding another system of equations which determine $A^{(d)}$, as well as another affine basis for $A^{(d)}$. The new forms have occasionally certain computational advantages (see, e.g., section 9.5).

First, following Sommerville [1], we shall obtain a new system of equations, equivalent to the equations (E_k^d).

2. *The $[\frac{1}{2}d]$-dimensional affine variety $A^{(d)} = \mathrm{aff}\, f(\mathscr{P}_s^d)$ is determined by the equations*

$$(E_k^{d*}) \qquad \sum_{i=-1}^{k-1} (-1)^{d+i} \binom{d-i-1}{d-k} f_i = \sum_{i=-1}^{d-k-1} (-1)^i \binom{d-i-1}{k} f_i,$$

where $0 \le k \le \frac{1}{2}(d-1)$.

PROOF Let k and m be integers satisfying $0 \leq m + 1 \leq k \leq \frac{1}{2}d$. Multiplying equation (E_m^d) by $(-1)^{m-1}\begin{pmatrix} d - m - 1 \\ d - k \end{pmatrix}$ and adding, we obtain

$$\sum_{m=-1}^{k-1} (-1)^{d+m} \begin{pmatrix} d - m - 1 \\ d - k \end{pmatrix} f_m$$

$$= \sum_{m=-1}^{k-1} (-1)^{m-1} \begin{pmatrix} d - m - 1 \\ d - k \end{pmatrix} \sum_{i=m}^{d-1} (-1)^i \begin{pmatrix} i + 1 \\ m + 1 \end{pmatrix} f_i$$

$$= \sum_{i=-1}^{d-1} (-1)^i \left\{ \sum_{m=-1}^{k-1} (-1)^{m+1} \begin{pmatrix} d - m - 1 \\ d - k \end{pmatrix} \begin{pmatrix} i + 1 \\ m + 1 \end{pmatrix} \right\} f_i$$

$$= \sum_{i=-1}^{d-1} (-1)^i \left\{ \sum_{j=0}^{k} (-1)^j \begin{pmatrix} d - j \\ k - j \end{pmatrix} \begin{pmatrix} i + 1 \\ j \end{pmatrix} \right\} f_i \qquad *$$

$$= \sum_{i=-1}^{d-1} (-1)^i \begin{pmatrix} d - i - 1 \\ k \end{pmatrix} f_i = \sum_{i=-1}^{d-k-1} (-1)^i \begin{pmatrix} d - i - 1 \\ k \end{pmatrix} f_i .$$

Therefore each $f(P)$, for $P \in \mathcal{P}_s^d$, satisfies (E_k^{d*}).

On the other hand, since f_{d-p} occurs, for $1 \leq p \leq [(d + 1)/2]$, only in the first p equations (E_k^{d*}), it is clear that the equations are independent. The intersection of the hyperplanes determined by them is therefore of dimension at most $d - [(d + 1)/2] = [d/2]$. Therefore the equations (E_k^{d*}), $0 \leq k \leq [(d - 1)/2]$, determine $A^{(d)}$, as claimed.

Our next aim is to obtain an affine basis for $A^{(d)} = \text{aff} f(\mathcal{P}_s^d)$, different from the basis determined by the f-vectors of cyclic polytopes, or the basis mentioned in exercise 9.7.1. All those bases have the disadvantage that approximately one half of the coordinates of their elements are not readily determined; moreover, even when this task is completed (in the case of cyclic polytopes, we shall accomplish this in section 9.6), the resulting expressions are awkward and unmanageable.

---

* We used the fact that, for $0 \leq c \leq a$, $\sum_{i=0}^{c} (-1)^i \begin{pmatrix} a - i \\ c - i \end{pmatrix} \begin{pmatrix} b \\ i \end{pmatrix} = \begin{pmatrix} a - b \\ c \end{pmatrix}$. In order to establish this identity, let its left hand side be denoted by $R(a, b, c)$. Then obviously $R(a, b, c) = R(a - 1, b, c - 1) + R(a - 1, b, c)$, while $\begin{pmatrix} a - b \\ c \end{pmatrix}$ satisfies the analogous relation. Therefore our identity shall be established by induction if we prove it in the cases $R(a, b, 0)$ and $R(a, b, a)$. But this is easy since $R(a, b, 0) = (-1)^0 \begin{pmatrix} a - 0 \\ 0 \end{pmatrix} \begin{pmatrix} b \\ 0 \end{pmatrix} = 1 = \begin{pmatrix} a - b \\ 0 \end{pmatrix}$, and $R(a, b, a) = \sum_{i=0}^{a} (-1)^i \begin{pmatrix} b \\ i \end{pmatrix} = (-1)^a \begin{pmatrix} b - 1 \\ a \end{pmatrix} = \begin{pmatrix} a - b \\ a \end{pmatrix}$.

The basis of $A^{(d)}$ we are going to determine now is much simpler, but it has an aesthetic drawback : its members are not f-vectors of simplicial d-polytopes. With minor changes, the exposition below follows Klee [11].

3. *If $d = 2n$ is even, then $A^{(d)}$ is the affine hull of the $n + 1$ vectors $h^{(k)}$, $0 \leq k \leq n$, where $h^{(k)} = (h_0^{(k)}, h_1^{(k)}, \cdots, h_{2n-1}^{(k)})$ with*

$$h_i^{(k)} = \binom{k}{1 + i - k} \quad for \quad 0 \leq i \leq 2n - 1, 0 \leq k \leq n.$$

PROOF The affine independence of the $n + 1$ vectors $h^{(k)}$ is obvious. Thus we only have to show that each of the vectors $h^{(k)}$ satisfies the Dehn–Sommerville equations (E_m^d). Indeed, we have*

$$\sum_{i=m}^{d-1} (-1)^i \binom{i+1}{m+1} h_i^{(k)} = \sum_{i=m}^{d-1} (-1)^i \binom{i+1}{m+1} \binom{k}{1+i-k}$$

$$= \sum_{j=m-k+1}^{d-k} (-1)^{k+1+j} \binom{j+k}{m+1} \binom{k}{j}$$

$$= \sum_{j=0}^{k} (-1)^{k+1+j} \binom{j+k}{m+1} \binom{k}{j}$$

$$= -\binom{k}{m+1-k} = (-1)^{d-1} h_m^{(k)},$$

as claimed.

* We note the simple formula $\sum_{i=0}^{k} (-1)^i \binom{n}{i} = (-1)^k \binom{n-1}{k}$, which holds for all integers k, n, and which is easily verified by induction. Using this we derive

$$\sum_{i=0}^{k} (-1)^i \binom{n}{i} \binom{i}{t} = \binom{n}{t} \sum_{i=0}^{k} (-1)^i \binom{n-t}{i-t} = \binom{n}{t} \sum_{j=0}^{k-t} (-1)^{j-t} \binom{n-t}{j} = (-1)^k \binom{n}{t} \binom{n-t-1}{k-t},$$

which is also valid for all integers k, n, t. Let now $R(n, m, t) = \sum_{i=0}^{n} (-1)^i \binom{n}{i} \binom{i+m}{t}$, where n, m, t are nonnegative integers. Clearly $R(n, m, t) + R(n, m, t + 1) = R(n, m + 1, t + 1)$. We shall prove by induction that $R(n, m, t) = (-1)^n \binom{m}{t-n}$. Since the binomial coefficients have the same addition formula as R, we have to establish the equality only for $m = 0$ and for $t = 0$. Let $n > 0$; by the above, $R(n, 0, t) = (-1)^n \binom{n}{t} \binom{n-t-1}{n-t} = (-1)^n \delta_{n,t} = (-1)^n \binom{0}{t-n}$, while $R(n, m, 0) = 0 = (-1)^n \binom{m}{0-n}$. This proves the identity, a particular case of which was used in the text above, for $n > 0$; its validity for $n = 0$ is obvious.

4. *If $d = 2n + 1$ is odd, then $A^{(d)}$ is the affine hull of the $n + 1$ vectors $g^{(k)}, 0 \le k \le n$, where $g^{(k)} = (g_0^{(k)}, g_1^{(k)}, \cdots, g_{2n}^{(k)})$ with*

$$g_i^{(k)} = \binom{1 + k}{1 + i - k} + \binom{k}{i - k} + 2\delta_{i,0}(1 - \delta_{k,0}),$$

for $0 \le k \le n, 0 \le i \le 2n$.

($\delta_{i,j}$ is Kronecker's δ and equals 1 if $i = j$ and zero otherwise).

PROOF   The affine independence of the $n + 1$ vectors $g^{(k)}$ is again obvious. The fact that the $g^{(k)}$'s satisfy the Dehn–Sommerville equations is also not hard to prove, although the computations are slightly more involved than in the case of even d. Indeed if

$$e_m^{(k)} = \sum_{i=m}^{d-1} (-1)^i \binom{i + 1}{m + 1} g_i^{(k)},$$

it is easily checked that $e_{-1}^{(0)} = 1$, $e_0^{(0)} = 2 = g_0^{(0)}$, and $e_m^{(0)} = 0$ for $m > 0$ as required. If $k > 0$ we have*

$$e_0^{(k)} = 2 + \sum_{i=0}^{d-1} (-1)^i \binom{i + 1}{1} \left[\binom{k + 1}{1 + i - k} + \binom{k}{i - k} \right]$$

$$= 2 + (-1)^{k+1} \sum_{j=0}^{k+1} (-1)^j \binom{j + k}{1} \binom{k + 1}{j}$$

$$+ (-1)^k \sum_{j=0}^{k} (-1)^j \binom{j + k + 1}{1} \binom{k}{j}$$

$$= 2 + \binom{k}{-k} + \binom{k + 1}{1 - k} = 2 + \delta_{k,1} = (-1)^{2n} g_0^{(k)}.$$

On the other hand, for $k > 0$ and $m > 0$, we have by an analogous computation,

$$e_m^{(k)} = \binom{k}{m - k} + \binom{k + 1}{m + 1 - k} = (-1)^{2n} g_m^{(k)}.$$

This completes the proof of theorem 4.

* The identity used was proved in the footnote on p. 150.

The Dehn–Sommerville equations (E_k^d) of theorem 9.2.1 easily generalize to certain complexes. We recall that a simplicial $(d - 1)$-complex $\mathscr{C}$ is called an *Eulerian* $(d - 1)$-*manifold* (Klee [11]) provided for each k-dimensional simplex $F \in \mathscr{C}, 0 \leq k \leq d - 1$, we have

$$\chi(\text{link}(F\,;\mathscr{C})) = 1 + (-1)^{d-k},$$

where χ denotes the Euler characteristic. Then we have (Klee [11]):

5. *The f-vector of every Eulerian $(d - 1)$-manifold $\mathscr{C}$ satisfies the Dehn–Sommerville equations (E_k^d), for every k such that $0 \leq k \leq d - 2$.*

The proof is practically a repetition of the corresponding part of the proof of theorem 9.2.1. The only difference is that the assumption $\chi(\text{link}(F\,;\mathscr{C})) = 1 + (-1)^{d-k}$ has to be used in order to derive the equation $\sum_{j=k}^{d-1} (-1)^j h_j(F) = (-1)^{d-1}$, which was in the case of polytopes given by theorem 8.3.1. The derivation is immediate, since

$$h_j(F) = f_{j-1-k}(\text{link}(F, \mathscr{C}))$$

for every simplicial complex $\mathscr{C}$.

Another difference between theorems 1 and 5 is that $k = -1$ is excluded in the latter. Indeed, instead of Euler's equation (E_{-1}^d) we have for Eulerian $(d - 1)$-manifolds

$$\sum_{i=0}^{d-1} (-1)^i f_i(\mathscr{C}) = \chi(\mathscr{C}).$$

However $\chi(\mathscr{C})$ is not arbitrary; indeed we have (Alexandroff–Hopf [1], chapter XIV; Klee [11]).

6. *If $\mathscr{C}$ is an Eulerian $(d - 1)$-manifold and if $d = 2n$ is even, then $\chi(\mathscr{C}) = 0$.*

The proof is very simple. The remark (page 147) that for $d = 2n$ Euler's equation (E_{-1}^d) is a linear combination of the equations $(E_0^d), (E_2^d), \cdots, (E_{d-2}^d)$ implies, in view of theorem 5, that $\sum_{i=0}^{d-1} (-1)^i f_i(\mathscr{C}) = 0$. But $\chi(\mathscr{C})$ is equal to the expression at left, hence the assertion of the theorem.

It should be noted that the proof of the equation (E_k^d) for a fixed k (in theorems 1 and 5) uses only relations among faces of dimensions $\geq k$. This observation yields the first part of the following result (Wall [1]), the second part of which is also easily established.

7. *If $\mathcal{K}$ is a simplicial $(d-1)$-complex such that for each simplex T in $\mathcal{K}$, $\chi(\text{link}(T, \mathcal{K})) = 1 + (-1)^{d-j}$ provided $\dim T = j \geq k$, then $\mathcal{K}$ satisfies the equations (E_i^d) for all $i \geq k$. Moreover, for such a $\mathcal{K}$ these are the only essential equations, in the sense that every linear relation among the f_i's valid for $\mathcal{K}$ and all its stelar subdivisions is dependent on those equations.*

9.3 Quasi-Simplicial Polytopes

On comparing theorem 9.2.1 with Euler's theorem 8.1.1 the difference between the dimension $[\frac{1}{2}d]$ of the affine hull $A^{(d)}$ of the f-vectors of simplicial d-polytopes, and the dimension $d-1$ of the affine hull E^d of the f-vectors of all d-polytopes, strikes the eye. One is tempted to guess that a gradual increase in the dimension of the affine hull will occur if the family of simplicial d-polytopes is gradually enlarged. One possibility in that direction would be to consider all d-polytopes P such that, for a certain $k \leq d-1$, all the k-faces of P (and therefore also all lower-dimensional faces) are simplices. Somewhat unexpectedly one finds that the transition is not gradual: Already the family of quasi-simplicial d-polytopes $\mathscr{P}_q^d$ (see section 4.5), corresponding to $k = d-2$, has a set of f-vectors whose affine hull is of dimension $d-1$, i.e. coincides with the Euler hyperplane E^d.

The present section is devoted to a proof of this result. The theorem and the idea of its proof were communicated to the author by M. Perles.

1. *The dimension of the affine hull $A_q^{(d)} = \text{aff} f(\mathscr{P}_q^d)$ of the f-vectors of quasi-simplicial d-polytopes is $d-1$. The affine variety $A_q^{(d)}$ coincides with the Euler hyperplane E^d.*

PROOF Simplicial d-polytopes are quasi-simplicial, and so are d-pyramids whose bases are simplicial $(d-1)$-polytopes. We shall show that the affine hull of the f-vectors of these particular quasi-simplicial d-polytopes already has dimension $d-1$.

As we saw in section 4.2, if P is a simplicial $(d-1)$-polytope and P^* is a d-pyramid with basis P, then $f_i(P^*) = f_i(P) + f_{i-1}(P)$ for $0 \leq i \leq d-1$. Thus $f(P^*)$ is the affine image of $f(P)$ under the affine transformation T (of the $(d-1)$-space into the d-space) determined by $T(f_0, f_1, \cdots, f_{d-2}) = (1 + f_0, f_0 + f_1, f_1 + f_2, \cdots, f_{d-3} + f_{d-2}, f_{d-2}, f_{d-2} + 1)$. Since affine transformations commute with the operation of taking affine hulls, the affine hull of the set of f-vectors of d-pyramids with simplicial $(d-1)$-bases coincides with $T(A^{(d-1)}) = T(\text{aff} f(\mathscr{P}_s^{d-1}))$. We shall prove the theorem by showing that the affine hull of $A^{(d)} \cup T(A^{(d-1)})$ is $(d-1)$-dimensional. This will be accomplished by taking suitable affine bases of $A^{(d)}$ and of $A^{(d-1)}$, and by showing that the affine hull of the union of the basis of $A^{(d)}$ with the image under T of the basis of $A^{(d-1)}$ contains d affinely independent vectors. As bases of $A^{(d)}$ and $A^{(d-1)}$ we shall use the bases $h^{(i)}$ and $g^{(i)}$ determined in theorems 9.2.3 and 9.2.4.

First, let $d = 2n + 1$ be odd. We denote by $g^{(0)}, \cdots, g^{(n)}$ the basis of $A^{(d)}$, and by $h^{(0)}, \cdots, h^{(n)}$ the basis of $A^{(d-1)}$. The vectors $T(h^{(i)})$, for $0 \le i \le n-1$, and $2T(h^{(i)}) - g^{(i)}$, for $0 \le i \le n$, clearly belong to $A_q^{(d)}$. This $(2n+1)$-membered set of vectors is affinely independent. Indeed, already the vectors formed by their first $2n$ components are affinely independent. To see this we consider the determinant of order $2n + 1$ formed by a column of 1's followed by the first $2n$ coordinates of our vectors. If the vectors are taken in the order $2T(h^{(0)}) - g^{(0)}$, $T(h^{(0)})$, $2T(h^{(1)}) - g^{(1)}$, $T(h^{(1)}), \cdots, T(h^{(n-1)})$, $2T(h^{(n)}) - g^{(n)}$, the determinant becomes

$$\begin{vmatrix}
1 & 0 & 0 & 0 & 0 & 0 & & \cdots & & & 0 \\
1 & 1 & 0 & 0 & 0 & 0 & & \cdots & & & 0 \\
1 & 1 & 1 & 0 & 0 & 0 & & \cdots & & & 0 \\
1 & 2 & 2 & 1 & 0 & 0 & & \cdots & & & 0 \\
1 & 0 & 1 & 2 & 1 & 0 & & \cdots & & & 0 \\
1 & 1 & 1 & 3 & 3 & 1 & & \cdots & & & 0 \\
1 & 0 & 0 & 1 & 3 & 3 & & \cdots & & & 0 \\
\cdot & \cdot & \cdot & \cdot & \cdot & \cdot & & \cdot & & & \\
1 & 1 & 0 & \cdots & 0 & \binom{n}{0} & \binom{n}{1} & \binom{n}{2} & \cdots & \binom{n}{n} & 0 \\
1 & 0 & 0 & \cdots & 0 & 0 & \binom{n}{0} & \binom{n}{1} & \cdots & \binom{n}{n-1} & \binom{n}{n}
\end{vmatrix} = 1.$$

Thus, the vectors are affinely independent, and the dimension of $A_q^{(d)}$ is at least (and hence exactly) $2n = d - 1$.

Next, let $d = 2n$ be even. Proceeding as above let $h^{(0)}, \cdots, h^{(n)}$ be the affine basis of $A^{(d)}$ and $g^{(0)}, \cdots, g^{(n-1)}$ that of $A^{(d-1)} = A^{(2n-1)}$. Then the $2n$ vectors $h^{(0)}$, $2h^{(1)} - T(g^{(0)})$, $h^{(1)}$, $2h^{(2)} - T(g^{(1)})$, $h^{(2)}, \cdots, h^{(n-1)}$, $2h^{(n)} - T(g^{(n-1)})$, which belong to $A_q^{(d)}$, are affinely independent. Indeed, forming the determinant analogous to the one used above, we find

$$
\begin{vmatrix}
1 & 0 & 0 & 0 & 0 & 0 & 0 & 0 & \cdots & & & & & 0 \\
1 & -1 & 0 & 0 & 0 & 0 & 0 & 0 & \cdots & & & & & 0 \\
1 & 1 & 1 & 0 & 0 & 0 & 0 & 0 & \cdots & & & & & 0 \\
1 & -4 & -4 & -1 & 0 & 0 & 0 & 0 & \cdots & & & & & 0 \\
1 & 0 & 1 & 2 & 1 & 0 & 0 & 0 & \cdots & & & & & 0 \\
1 & -3 & -3 & -3 & -3 & -1 & 0 & 0 & \cdots & & & & & 0 \\
1 & 0 & 0 & 1 & 3 & 3 & 1 & 0 & \cdots & & & & & 0 \\
\multicolumn{14}{c}{\cdots\cdots\cdots\cdots\cdots\cdots\cdots\cdots\cdots\cdots} \\
1 & 0 & 0 & 0 \cdots 0 & \binom{n-1}{0} & \binom{n-1}{1} & \binom{n-1}{2} & \cdots & \binom{n-1}{n-1} & & & & & 0 \\
1 & -3 & -2 & 0 \cdots 0 & -\binom{n}{0} & -\binom{n}{1} & -\binom{n}{2} & \cdots & -\binom{n}{n-1} & & & -\binom{n}{n} \\
\end{vmatrix}
$$

$$= (-1)^n \neq 0.$$

Thus the dimension of $A_q^{(d)}$ is $d-1$ in all cases, and the proof of the theorem is completed.

9.4 Cubical Polytopes

The 'incidence equation' of theorem 9.1.1, which was used in section 9.2 to derive the Dehn–Sommerville equations for the f-vectors of simplicial polytopes, may with equal success be applied to other families of polytopes.

In the present section we shall apply this method—in a way closely similar to that used above—in order to determine the affine hull of the f-vectors of cubical d-polytopes. We recall from section 4.6 that a d-polytope is called cubical provided each of its facets is a combinatorial $(d-1)$-cube. The class of all cubical d-polytopes shall be denoted by $\mathscr{P}_c^d$.

Intuitively, it seems reasonable to expect that there are 'fewer' cubical d-polytopes than simplicial ones and that, consequently, the dimension of the affine hull of their f-vectors is smaller than the dimension of $A^{(d)} = \operatorname{aff} f(\mathscr{P}_s^d)$.

However, we have the following result:

1. *The dimension of the affine hull* $A_c^{(d)} = \text{aff} f(\mathscr{P}_c^d)$ *of the f-vectors of cubical d-polytopes is* $[\tfrac{1}{2}d]$. *The affine variety* $A_c^{(d)}$ *is the intersection of the hyperplanes determined by the equations*

$$\sum_{j=k}^{d-1} (-1)^j 2^{j-k} \binom{j}{k} f_j = (-1)^{d-1} f_k, \qquad \text{for} \quad 0 \le k \le d - 2,$$

with the Euler hyperplane

$$\sum_{j=0}^{d-1} (-1)^j f_j = 1 - (-1)^d.$$

The proof of the theorem is analogous to the proof of theorem 9.2.1. In the notation we employed there, the only change consists in a different value of $\gamma_{i,m}$. Indeed for every $p \in \mathscr{P}_c^d$ and for $-1 \le k \le j \le d - 1$, each j-face of P is incident with $2^{j-k}\binom{j}{k}$ k-faces, and therefore

$$\gamma_{i,m} = 2^{m-i} \binom{d-i-1}{d-m-1}.$$

This yields the equations of theorem 9.4.1. Since this system of equations (including Euler's) obviously contains $[\tfrac{1}{2}(d+1)]$ independent equations, it follows that the dimension of $A_c^{(d)}$ is at most $d - [\tfrac{1}{2}(d+1)] = [\tfrac{1}{2}d]$.

The following example of a family of cubical d-polytopes shall complete the proof by establishing that $A_c^{(d)}$ has dimension at least $[\tfrac{1}{2}d]$, and is therefore exactly $[\tfrac{1}{2}d]$-dimensional.

In the notation of section 4.6, let us consider the cuboids C_k^d for $0 \le k \le [\tfrac{1}{2}d]$. We recall that for $i + k \le d - 1$, and thus in particular for $0 \le i < k \le [\tfrac{1}{2}d]$, we have

$$f_i(C_k^d) = \sum_{j=0}^{k} (-1)^j \binom{k}{j} 2^{d+k-i-2j} \binom{d-j}{i}.$$

Let $n = [\tfrac{1}{2}d]$. We shall show that the vectors $f(C_k^d)$, $0 \le k \le n$, are affinely independent. This will be accomplished by proving the affine independence of the vectors formed by the first n coordinates of the vectors $f(C_k^d)$. The last assertion is equivalent to the nonvanishing of the determinant

$$D(d) = \begin{vmatrix} 1 & \sum_j (-1)^j \binom{0}{j}\binom{d-j}{0}2^{d-2j} & \sum_j (-1)^j \binom{0}{j}\binom{d-j}{1}2^{d-1-2j} & \cdots \; \sum_j (-1)^j \binom{0}{j}\binom{d-j}{n-1}2^{d-(n-1)-2j} \\[2ex] 1 & \sum_j (-1)^j \binom{1}{j}\binom{d-j}{0}2^{d+1-2j} & \sum_j (-1)^j \binom{1}{j}\binom{d-j}{1}2^{d+1-1-2j} & \cdots \; \sum_j (-1)^j \binom{1}{j}\binom{d-j}{n-1}2^{d+1-(n-1)-2j} \\[2ex] \cdots & & & \cdots \\[2ex] 1 & \sum_j (-1)^j \binom{n}{j}\binom{d-j}{0}2^{d+n-2j} & \sum_j (-1)^j \binom{n}{j}\binom{d-j}{1}2^{d+n-1-2j} & \cdots \; \sum_j (-1)^j \binom{n}{j}\binom{d-j}{n-1}2^{d+n-(n-1)-2j} \end{vmatrix}$$

In order to evaluate $D(d)$ we proceed as follows. First, subtracting each row multiplied by 2 from the following, we find that $D(d)$ equals

$$(-1)^n \begin{vmatrix} 1 & \sum_j (-1)^j \binom{0}{j}\binom{d-j}{0}2^{d-2j} & \sum_j (-1)^j \binom{0}{j}\binom{d-j}{1}2^{d-1-2j} & \cdots \; \sum_j (-1)^j \binom{0}{j}\binom{d-j}{n-1}2^{d-(n-1)-2j} \\[2ex] 1 & \sum_j (-1)^j \binom{0}{j}\binom{d-1-j}{0}2^{d-1-2j} & \sum_j (-1)^j \binom{0}{j}\binom{d-1-j}{1}2^{d-2-2j} & \cdots \; \sum_j (-1)^j \binom{0}{j}\binom{d-1-j}{n-1}2^{d-1-(n-1)-2j} \\[2ex] \cdots & & & \cdots \\[2ex] 1 & \sum_j (-1)^j \binom{n-1}{j}\binom{d-1-j}{0}2^{d-2+n-2j} & \sum_j (-1)^j \binom{n-1}{j}\binom{d-1-j}{1}2^{d-2+n-1-2j} & \cdots \; \sum_j (-1)^j \binom{n-1}{j}\binom{d-1-j}{n-1}2^{d-2+n-(n-1)-2j} \end{vmatrix}$$

Repeating this procedure successively with the last $n, n-1, \cdots, 2$ rows we obtain

$$D(d) = (-1)^{n(n-1)/2} \begin{vmatrix} 1 & \sum_j (-1)^j \binom{0}{j}\binom{d-j}{0} 2^{d-2j} & \sum_j (-1)^j \binom{0}{j}\binom{d-j}{1} 2^{d-1-2j} & \cdots & \sum_j (-1)^j \binom{0}{j}\binom{d-j}{n-1} 2^{d-(n-1)-2j} \\ 1 & \sum_j (-1)^j \binom{0}{j}\binom{d-1-j}{0} 2^{d-1-2j} & \sum_j (-1)^j \binom{0}{j}\binom{d-1-j}{1} 2^{d-2-2j} & \cdots & \sum_j (-1)^j \binom{0}{j}\binom{d-1-j}{n-1} 2^{d-1-(n-1)-2j} \\ \cdot & \cdot \quad \cdot \quad \cdot \quad \cdot & \cdot \quad \cdot \quad \cdot \quad \cdot & & \cdot \quad \cdot \quad \cdot \\ 1 & \sum_j (-1)^j \binom{0}{j}\binom{d-n-j}{0} 2^{d-n-2j} & \sum_j (-1)^j \binom{0}{j}\binom{d-n-j}{1} 2^{d-n-1-2j} & \cdots & \sum_j (-1)^j \binom{0}{j}\binom{d-n-j}{n-1} 2^{d-n-(n-1)-2j} \end{vmatrix}$$

Thus $D(d) = (-1)^{\frac{1}{2}n(n+1)}D(d, n)$, where, for $1 \le 2k \le d$,

$$D(d, k) = \begin{vmatrix} 1 & 2^d \binom{d}{0} & 2^{d-1}\binom{d}{1} & \cdots & 2^{d-(k-1)}\binom{d}{k-1} \\ 1 & 2^{d-1}\binom{d-1}{0} & 2^{d-2}\binom{d-1}{1} & \cdots & 2^{d-1-(k-1)}\binom{d-1}{k-1} \\ \cdot & \cdot \quad \cdot \quad \cdot & \cdot \quad \cdot \quad \cdot & \cdots & \cdot \quad \cdot \quad \cdot \\ 1 & 2^{d-k}\binom{d-k}{0} & 2^{d-k-1}\binom{d-k}{1} & \cdots & 2^{d-k-(k-1)}\binom{d-k}{k-1} \end{vmatrix}$$

Considering $D(d, k)$, we subtract from each row the following one multiplied by 2 and obtain

$$D(d, k) = \begin{vmatrix} -1 & 0 & 2^{d-1}\binom{d-1}{0} & \cdots & 2^{d-(k-1)}\binom{d-1}{k-2} \\ -1 & 0 & 2^{d-2}\binom{d-2}{0} & \cdots & 2^{d-1-(k-1)}\binom{d-2}{k-2} \\ \cdot & \cdot & \cdot & \cdots & \cdot \\ -1 & 0 & 2^{d-k}\binom{d-k}{0} & \cdots & 2^{d-(k-1)-(k-1)}\binom{d-k}{k-2} \\ 1 & 2^{d-k}\binom{d-k}{0} & 2^{d-k-1}\binom{d-k}{1} & \cdots & 2^{d-k-(k-1)}\binom{d-k}{k-1} \end{vmatrix}.$$

Developing $D(d, k)$ with respect to the elements of the second column we obtain $D(d, k) = (-1)^k 2^{d-k} D(d-1, k-1)$. Thus, by induction,

$$D(d, k) = (-1)^{\sum_{i=2}^{k} i} 2^{(d-k)(k-1)} D(d-(k-1), 1)$$

$$= (-1)^{\frac{1}{2}(k-1)(k+2)} 2^{(d-k)(k-1)} \begin{vmatrix} 1 & 2^{d-k+1} \\ 1 & 2^{d-k} \end{vmatrix}$$

$$= (-1)^{\frac{1}{2}k(k+1)} 2^{k(d-k)}.$$

It follows that $D(d) = 2^{n(d-n)} \neq 0$, as claimed.

This completes the proof of the affine independence of the f-vectors of $C_0^d, C_1^d, \cdots, C_n^d$, and with it the proof of theorem 1.

A set of independent equations determining $A_c^{(d)}$, simpler than the equations of theorem 1, is given by the following result:

2. *The flat $A_c^{(d)}$ is the intersection of the Euler hyperplane with the hyperplanes determined by the equations*

$$\sum_{j=k}^{d-1} (-1)^j \binom{j}{k} f_j = 0$$

for $k \equiv d \pmod 2$ and $1 \leq k \leq d - 2$.

PROOF   Let the left-hand side of the equation be denoted by a_k. Using the equations of theorem 1 we have

$$(-1)^{d-1}a_k = \sum_{j=k}^{d-1}(-1)^j\binom{j}{k}(-1)^{d-1}f_j$$

$$= \sum_{j=k}^{d-1}\sum_{i=j}^{d-1}(-1)^{j+i}\binom{j}{k}\binom{i}{j}2^{i-j}f_i$$

$$= \sum_{j=k}^{d-1}\sum_{i=j}^{d-1}(-1)^{i-j}2^{i-j}\binom{i-k}{i-j}\binom{i}{k}f_i.$$

Changing the order of summation, introducing $t = i - j$, and noting that

$$\sum_{t=0}^{i-k}(-1)^t 2^t\binom{i-k}{t} = (1-2)^{i-k} = (-1)^{i-k},$$

the last expression may be transformed in the following manner:

$$(-1)^{d-1}a_k = \sum_{i=k}^{d-1}\sum_{t=0}^{i-k}(-1)^t 2^t\binom{i-k}{t}\binom{i}{k}f_i$$

$$= \sum_{i=k}^{d-1}(-1)^{i-k}\binom{i}{k}f_i = (-1)^k a_k.$$

Since $k \equiv d \pmod 2$ this yields $a_k = -a_k$, i.e. $a_k = 0$ as claimed.

9.5   Solutions of the Dehn–Sommerville Equations

The present section is a continuation of section 9.2. We are again considering simplicial d-polytopes and our aim is to 'solve' the system of equations determining $A^{(d)} = \operatorname{aff} f(\mathscr{P}_s^d)$. In a certain sense we already solved this problem by finding affine bases (for instance, those on pp. 147, 150) for $A^{(d)}$. However, the geometric background of the problem makes another type of solution desirable. The algebraic solution by bases of $A^{(d)}$ yields, naturally, all the d-tuples $(f_0, \cdots, f_{d-1})$ belonging to $A^{(d)}$. On the other hand, our interest is centered on the proper subset $f(\mathscr{P}_s^d)$ of $A^{(d)}$. The complete determination of $f(\mathscr{P}_s^d)$ (and of $f(\mathscr{P}^d)$) is the final aim, and our preoccupation with $A^{(d)}$ is only a first step in this direction.

Though the present state of knowledge is so inadequate that we are unable to give a complete characterization even for $f(\mathscr{P}_s^4)$ (see chapter 10), there is a possibility of solving the equations determining $A^{(d)}$ in a way which does yield additional information. We may solve the equations in such a manner that certain of the numbers f_i are expressed in terms of the remaining numbers f_i. Obviously, there is a large number of possible choices of the 'independent' f_i's. The particular form of solution established in the following lines is made desirable by its applications in later chapters. Using again the notation $n = [d/2]$ and $f_{-1} = 1$ we have the following theorem.

1. *Every vector* $f = (f_0, \cdots, f_{d-1}) \in A^{(d)}$ *satisfies the relations*

$$f_{n+t} = \sum_{i=-1}^{d-n-2} \left\{ \sum_{k=0}^{d-n-1} (-1)^{k+i+1} \binom{k}{d-n-1-t}\binom{d-1-i}{d-k} \right\} f_i$$

$$+ \sum_{i=-1}^{n-1} (-1)^{n+i+1} \binom{d-1-i}{d-n-1-t}\binom{n+t-i-1}{t} f_i,$$

for $0 \le t \le d-n-1$.

It is easily seen that for even $d = 2n$ the terms corresponding to $i = -1$ cancel out, while for odd $d = 2n+1$ they combine to

$$(-1)^n 2 \binom{d}{n-t}\binom{n+t}{n}.$$

For special values of t the above formulae may be simplified. Thus we have:
For $d = 2n$,

$$f_n = \sum_{i=0}^{n-1} (-1)^{n+i+1} \frac{i+1}{n+1} \binom{2n-i}{n} f_i,$$

$$f_{2n-1} = \sum_{i=0}^{n-1} (-1)^{n+i+1} \frac{i+1}{n} \binom{2n-2-i}{n-1} f_i,$$

while for $d = 2n+1$

$$f_n = (-1)^n 2 \binom{2n+1}{n} + \sum_{i=0}^{n-1} (-1)^{n+i+1} \binom{2n+1-i}{n+1} f_i,$$

$$f_{2n} = (-1)^n 2 \binom{2n}{n} + 2 \sum_{i=0}^{n-1} (-1)^{n+i+1} \binom{2n-1-i}{n} f_i.$$

In order to prove theorem 1 we start from the equations E^{d*} (page 148). Multiplying equation (E_k^{d*}) by $(-1)^{k+s}\binom{k}{s}$, where s is some fixed non-negative integer, $s \leq d - n - 1$, adding the resulting equations, and changing the order of summation, we obtain

$$\sum_{i=-1}^{d-n-2}\left\{\sum_{k=0}^{d-n-1}(-1)^{d+i+k+s}\binom{k}{s}\binom{d-i-1}{d-k}\right\}f_i$$

$$= \sum_{i=-1}^{d-1}\left\{\sum_{k=0}^{d-n-1}(-1)^{i+k+s}\binom{k}{s}\binom{d-i-1}{k}\right\}f_i.$$

On the right hand side, we consider separately the summands corresponding to $i = -1, 0, 1, \cdots, n-1$, and those corresponding to $i = n, \cdots, d-1$; in the second sum the effective range of k is only up to $d - i - 1$. Therefore, using the relation*

$$\sum_{i=0}^{c}(-1)^i\binom{a}{i}\binom{i}{b} = (-1)^c\binom{a}{b}\binom{a-b-1}{c-b},$$

the right hand side is transformed into

$$\sum_{i=-1}^{n-1}(-1)^{d+n+s+i+1}\binom{d-i-1}{s}\binom{d-i-s-2}{d-n-s-1}f_i$$

$$+ \sum_{i=n}^{d-1}(-1)^{d+s+1}\binom{d-i-1}{s}\binom{d-i-s-2}{d-i-s-1}f_i.$$

But $\binom{d-i-s-2}{d-i-s-1} = \delta_{0,d-i-s-1}$; therefore the second sum reduces to $(-1)^{d+s+1}f_{d-s-1}$. Putting $n + t = d - s - 1$ we finally obtain the equations of theorem 1.

9.6   The f-Vectors of Neighborly d-Polytopes

It was already pointed out in theorem 7.2.2 that for a simplicial neighborly d-polytope P the dimension d and the number of vertices $v = f_0(P)$ are sufficient to determine the complete f-vector $f(P)$. The interest in the numbers $f_k(P)$ stems in part from certain extremal properties of simplicial

* See the footnote on page 149.

neighborly d-polytopes, which we shall discuss in chapter 10. In the present section our aim is to find expressions for $f_k(P)$ simpler than those obtained from theorem 9.5.1 by inserting $f_k(P) = \begin{pmatrix} v \\ k+1 \end{pmatrix}$ for $0 \le k < [\tfrac{1}{2}d]$.

We define $f_k(v, d)$ to mean $f_k(P)$ where $P \in \mathcal{N}_s^d$ is any simplicial neighborly d-polytope with $f_0(P) = v$ vertices. By theorem 9.5.1, $f_k(v, d)$ is properly defined, that is, it does not depend on the particular $P \in \mathcal{N}_s^d$ chosen. In this notation we may formulate the result as follows.

1. *Let $n = [\tfrac{1}{2}d] \ge 1$ and let $0 \le k \le d - 1$. Then*

$$f_k(v, d) = \frac{v - (d - 2n)(v - k - 2)}{v - k - 1} \sum_{j \ge 0}^{n} \begin{pmatrix} v - 1 - j \\ k + 1 - j \end{pmatrix} \begin{pmatrix} v - k - 1 \\ 2j - k - 1 + d - 2n \end{pmatrix}.$$

In order to prove the theorem we shall first show that the $f_k(v, d)$ as defined in the theorem satisfy $f_k(v, d) = \begin{pmatrix} v \\ k+1 \end{pmatrix}$ for $0 \le k < n$; this relation is characteristic for neighborly d-polytopes. The proof will be completed by showing that the numbers $f_k(v, d)$ of the theorem satisfy the Dehn–Sommerville equations (theorem 9.2.1).

Let $0 \le k \le n - 1$; introducing into the formulae for $f_k(v, d)$ the new index of summation $i = k + 1 - j$, we obtain

$f_k(v, d)$

$$= \frac{v - (d - 2n)(v - k - 2)}{v - k - 1} \sum_{i \ge 0} \begin{pmatrix} v - k - 2 + i \\ i \end{pmatrix} \begin{pmatrix} v - k - 1 \\ k + 1 + d - 2n - 2i \end{pmatrix}.$$

Using the relation (*)

$$\sum_{i \ge 0} \begin{pmatrix} a + i \\ a \end{pmatrix} \begin{pmatrix} a + 1 \\ h + 1 - 2i \end{pmatrix} = \begin{pmatrix} a + h + 1 \\ a \end{pmatrix},$$

the sum on the right can be evaluated in closed form and we obtain

$$f_k(v, d) = \frac{v - (d - 2n)(v - k - 2)}{v - k - 1} \begin{pmatrix} v + d - 2n - 1 \\ v - k - 2 \end{pmatrix}.$$

Therefore, for even $d = 2n$ we have

$$f_k(v, 2n) = \frac{v}{v - k - 1} \begin{pmatrix} v - 1 \\ v - k - 2 \end{pmatrix} = \begin{pmatrix} v \\ v - k - 1 \end{pmatrix} = \begin{pmatrix} v \\ k + 1 \end{pmatrix},$$

* For sake of continuity of argument, we defer the proof till page 167.

while for odd $d = 2n + 1$ we have

$$f_k(v, 2n + 1) = \frac{k + 2}{v - k - 1} \binom{v}{v - k - 2} = \binom{v}{v - k - 1} = \binom{v}{k + 1}.$$

Our first assertion is thus established for all d.

We turn now to the second part of the proof, and we remark, first, that it is sufficient to show that the numbers $f_k(v, d)$ of the theorem satisfy a maximal independent subsystem of the Dehn–Sommerville equations. As noted on p. 147, in case $d = 2n$ such a system is formed by the n equations $E^d_{-1}, E^d_1, E^d_3, \cdots, E^d_{d-3}$, while for $d = 2n + 1$ we may take the $n + 1$ equations $E^d_{-1}, E^d_0, E^d_2, E^d_4, E^d_{d-3}$.

Considering the case of even $d = 2n$, we thus have to show that

$$\sum_{k=2r}^{2n-1} (-1)^k \binom{k+1}{2r} \frac{v}{v-k-1} \sum_{j=0}^{n} \binom{v-1-j}{k+1-j}\binom{v-k-1}{2j-k-1} = 0$$

holds for $r = 0, 1, 2, \cdots, n - 1$. Using

$$\binom{v-1-j}{k+1-j}\binom{v-k-1}{2j-k-1} = \frac{v-k-1}{j}\binom{v-j-1}{j-1}\binom{j}{k+1-j},$$

this is equivalent to establishing

$$\sum_{j=0}^{n} \left\{ \sum_{k=2r}^{2n-1} (-1)^k \binom{k+1}{2r}\binom{j}{k+1-j} \right\} \frac{v}{j}\binom{v-j-1}{j-1} = 0.$$

The last equation, however, is immediate since the effective range of k is only from $2r$ to $2j - 1$ and since (*)

$$\sum_{k=2r}^{2j-1} (-1)^k \binom{k+1}{2r}\binom{j}{k+1-j} = 0.$$

In case of odd $d = 2n + 1$ we shall first show that the numbers $f_k(v, 2n + 1)$ satisfy Euler's equation E^d (that is, E^d_{-1}). Then, in analogy to the above, we have

$$\sum_{k=0}^{2n} (-1)^k \frac{k+2}{v-k-1} \sum_{j=0}^{n} \binom{v-1-j}{k+1-j}\binom{v-k-1}{2j-k}$$

$$= \sum_{j=0}^{n} \left\{ \sum_{k=0}^{2n} (-1)^k (k+2) \binom{j+1}{k+1-j} \right\} \frac{1}{j+1}\binom{v-j-1}{j}.$$

* See page 167.

Now, for $j = 0$ the range of k is reduced to the single value $k = 0$, and the contribution to the sum is 2; for $0 < j \le n$ the value of the sum on k is zero since, introducing $i = 2j - k$, we have

$$\sum_{k=0}^{2n} (-1)^k(k+2)\binom{j+1}{k+1-j} = 2(j+1)\sum_{i=0}^{j+1}(-1)^i\binom{j+1}{i}$$

$$- \sum_{i=1}^{j+1}(-1)^i i\binom{j+1}{i}$$

$$= (j+1)\left\{2\sum_{i=0}^{j+1}(-1)^i\binom{j+1}{i} + \sum_{i=0}^{j}(-1)^i\binom{j}{i}\right\} = 0.$$

Thus, Euler's equation is satisfied.

In order to complete the proof we have to show that, for $r = 0, 1, \cdots, n-1$ the expression

$$\sum_{k=2r+1}^{2n} (-1)^k\binom{k+1}{2r+1}\frac{k+2}{v-k-1}\sum_{j=0}^{n}\binom{v-1-j}{k+1-j}\binom{v-k-1}{2j-k}$$

vanishes. Transforming as above we see that this sum equals to

$$\sum_{j=0}^{n}\left\{\sum_{k=2r+1}^{2n}(-1)^k\binom{k+1}{2r+1}(k+2)\binom{j+1}{k+1-j}\right\}\frac{1}{j+1}\binom{v-j-1}{j};$$

But this sum is easily seen to equal zero since the sum on k is 0 for each j. Indeed, introducing $i = k+1-j$ we have

$$\sum_{k=2r+1}^{2n} (-1)^k\binom{k+1}{2r+1}(k+2)\binom{j+1}{k+1-j}$$

$$= \sum_{i=2r-j+2}^{j+1}(-1)^{i+j-1}\binom{i+j}{2r+1}(i+j+1)\binom{j+1}{i}$$

$$= 2(-1)^{j+1}(r+1)\sum_{i=2r-j+2}^{j+1}(-1)^i\binom{i+j+1}{2r+2}\binom{j+1}{i}.$$

But

$$\binom{i+j+1}{2r+2}\binom{j+1}{i} = 0 \quad \text{for} \quad i < 0 \quad \text{or} \quad i < 2r-j+1.$$

Therefore our sum equals

$$2(-1)^{j+1}(r+1)\left\{\sum_{i=0}^{j+1}(-1)^i\binom{i+j+1}{2r+2}\binom{j+1}{i}+(-1)^j\binom{j+1}{2r-j+1}\right\}.$$

Using the relation (see footnote on page 150)

$$\sum_{i=0}^{a}(-1)^i\binom{a}{i}\binom{i+a}{b}=(-1)^a\binom{a}{b-a}$$

we finally obtain

$$2(-1)^{j+1}(r+1)\left\{(-1)^{j+1}\binom{j+1}{2r+2-(j+1)}+(-1)^j\binom{j+1}{2r-j+1}\right\}=0,$$

as claimed.

This completes the proof of theorem 1.

Note that the above expressions for $f_k(v, d)$ can be put into the following form which is occasionally more convenient (see, for example, the second part of the proof of theorem 1):

(*) $$f_k(v, 2n) = \sum_{j=1}^{n}\frac{v}{j}\binom{v-j-1}{j-1}\binom{j}{k-j+1}$$

and

$$f_k(v, 2n+1) = \sum_{j=0}^{n}\frac{k+2}{j+1}\binom{v-j-1}{j}\binom{j+1}{k-j+1}$$

for $0 \le k \le d-1$.

Either from these equations, or from those in theorem 1, it is easy to deduce

2. *For* $0 \le k < 2n+1 < v$

$$f_k(v, 2n+1) = \frac{k+2}{v+1}\cdot f_{k+1}(v+1, 2n+2).$$

Equation (*) appears without proof in Motzkin [4]; the same abstract contains also a relation similar to theorem 2, but marred by misprints.

We now give proofs for the identities used above. First, let

$$F(x, y) = \sum_{i \geq 0} \binom{x + i}{x}\binom{x + 1}{y + 1 - 2i};$$

we shall show that $F(x, y) = \binom{x + y + 1}{x}$. Indeed,

$$F(x, 0) = \binom{x + 0}{x}\binom{x + 1}{0 - 2 \cdot 0 + 1} = x + 1 = \binom{x + 0 + 1}{x},$$

and

$$F(0, y) = \sum_{i \geq 0} \binom{i}{0}\binom{1}{y - 2i + 1} = \left[\binom{1}{0} \text{ or } \binom{1}{1}\right] = 1 = \binom{0 + y + 1}{0}.$$

Thus our assertion shall be established by induction if we show that
$F(x + 1, y) + F(x, y + 1) = F(x + 1, y + 1)$. But

$F(x + 1, y) + F(x, y + 1)$

$$= \sum_{i \geq 0}\left[\binom{x + 1 + i}{i}\binom{x + 2}{y - 2i + 1} + \binom{x + i}{i}\binom{x + 1}{y - 2i + 2}\right]$$

$$= \sum_{i \geq 0}\left[\binom{x + i + 1}{i}\binom{x + 1}{y - 2i + 1} + \binom{x + i}{i - 1}\binom{x + 1}{y - 2i + 2}\right.$$

$$\left. + \binom{x + i}{i}\binom{x + 1}{y - 2i + 2}\right]$$

$$= \sum_{i \geq 0}\left[\binom{x + i + 1}{i}\binom{x + 1}{y - 2i + 1} + \binom{x + i + 1}{i}\binom{x + 1}{y - 2i + 2}\right]$$

$$= \sum_{i \geq 0}\binom{x + i + 1}{i}\binom{x + 2}{y - 2i + 2} = F(x + 1, y + 1),$$

as claimed.

Next, we consider the expression

$$S(n, m, k) = \sum_{i = 0}^{n} (-1)^i \binom{n}{i}\binom{i + m}{k}.$$

Clearly $S(n, m, k) = S(n, m - 1, k - 1) + S(n, m - 1, k)$; therefore, by induction,

$$S(n, m, k) = \sum_{i = 0}^{r} \binom{r}{i} S(n, m - r, k - i).$$

Also, by the footnote on page 150,

$$S(n, 0, k) = \sum_{i=0}^{n} (-1)^i \binom{n}{i}\binom{i}{k} = (-1)^n \binom{n}{k}\binom{n-k-1}{n-k} = (-1)^n \delta_{n,k} \, .$$

Therefore, with $r = m$, we have

$$S(n, m, k) = \sum_{i=0}^{m} \binom{m}{i} S(n, 0, k-i) = \sum_{i=0}^{m} \binom{m}{i}(-1)^n \delta_{n,k-i} = (-1)^n \binom{m}{k-n}.$$

In particular, for $n = m$ we obtain

(*) $$\sum_{i=0}^{n} (-1)^i \binom{n}{i}\binom{i+n}{k} = (-1)^n \binom{n}{k-n}.$$

Using (*) we can evaluate

$$\sum_{i=2r-1}^{2j-1} (-1)^i \binom{i+1}{2r}\binom{j}{i+1-j} = (-1)^{j-1} \sum_{t=0}^{j} (-1)^t \binom{t+j}{2r}\binom{j}{t}$$

$$= -\binom{j}{2r-j}.$$

Therefore

$$\sum_{i=2r}^{2j-1} (-1)^i \binom{i+1}{2r}\binom{j}{i+1-j} = 0,$$

as claimed on page 164.

9.7 Exercises

1. Prove that dim $A^{(d)} \geq [\tfrac{1}{2}d] = n$ by showing the affine independence of the f-vectors of the $n+1$ d-polytopes T_r^d, $0 \leq r \leq n$, considered in exercise 4.8.5 and in section 6.1.

2. Let $f_k(v, d)$ denote the number of k-faces of a cyclic d-polytope $C(v, d)$ with v vertices (see section 4.7). Using the Dehn–Sommerville equations, and $f_k(v, d) = \binom{v}{k+1}$ for $0 \leq k \leq [\tfrac{1}{2}d] - 1$, show that

$$f_k(d+2, d) = \binom{2n+2}{k+1} - 2\binom{n+1}{k-n} \qquad \text{for } d = 2n,$$

$$f_k(d+2, d) = \binom{2n+3}{k+1} - \binom{n+1}{k-n-1} - \binom{n+2}{k-n} \qquad \text{for } d = 2n+1,$$

$$f_{d-1}(v, d) = \frac{v}{n}\binom{v-n-1}{n-1} = \binom{v-n}{n} + \binom{v-n-1}{n-1} \qquad \text{for } d = 2n,$$

$$f_{d-1}(v, d) = 2\binom{v-n-1}{n} \qquad \text{for } d = 2n+1.$$

3. Show that for $k = d - 1$ theorem 9.6.1 reduces to theorem 4.7.3.

4. Determine the Dehn–Sommerville equations for simple d-polytopes by each of the following methods:

(i) by duality, from theorem 9.2.1;

(ii) directly from the 'incidence equation' of theorem 9.1.1;

(iii) using 'double counting of incidences.'

5. Show that the affine hull of the set of f-vectors of all centrally symmetric simplicial d-polytopes coincides with $A^{(d)}$. (Hint: Use inductively bipyramids and Kleetopes (see section 11.4) over them, starting—for $d = 3$—with the octahedron and the Kleetope over it.)

6. Show that the affine hull of the set of f-vectors of all self-dual d-polytopes has dimension $[\frac{1}{2}d]$. (Hint: Use the polytopes defined in exercise 4.8.33).

7. As in section 4.5, let $\mathscr{P}^d(k, h)$ denote the family of all d-polytopes of type (k, h).

(i) Show that the affine hull of the f-vectors of all members of $\mathscr{P}^4(2, 2)$ is determined by the equations $f_0 = f_3$ and $f_1 = f_2$. (Hint: Find the equations by counting in different ways the number of incidences of edges and 2-faces; to find three affinely independent f-vectors, consider the simplex, the polytope mentioned in exercise 4.8.15, and the regular (Platonic) 4-polytope with 24 vertices (see Coxeter [1]).)

(ii) Show that the f-vectors of polytopes in $\mathscr{P}^d(k, d - k)$ satisfy the equation $(k + 1)f_k = (d - k + 1)f_{k-1}$.

(iii) It may be conjectured that the dimension of the affine hull of the set of f-vectors of all polytopes in $\mathscr{P}^d(k, h)$ is

(a) $\quad 0 \qquad$ if $k + h > d$

(b) $\quad [\frac{1}{2}d] \quad$ if $k + h = d$ and $k = 1$ or $h = 1$

(c) $\quad d - 2$ if $k + h = d$ and $k \neq 1 \neq h$

(d) $\quad d - 1$ if $k + h < d$.

Assertion (a) is obvious (see exercise 4.8.11), while (b) is theorem 9.2.1 and its dual. (i) above and theorem 9.3.1 establish (c) and (d) for $d \leq 4$. M. A. Perles and G. C. Shephard (private communication) proved (c) for $k = 2$ and $d \leq 7$. They also proved the existence, for each $d \geq 4$, of infinitely many d-polytopes of type $(2, d - 2)$, and for $d \geq 5$, the existence of polytopes in $\mathscr{P}^d(3, d - 3)$ different from the d-simplex.

8. Let $A^{(d)}$ and $A_s^{(d)}$ denote the affine hull of the set of f-vectors of all simplicial respectively simple d-polytopes. Show that $\text{aff}(A^{(d)} \cup A_s^{(d)})$ is the Euler hyperplane in R^d for $d \leq 4$ and for $d = 6$, but that

$$\delta(d) = \dim \text{aff}(A^{(d)} \cup A_s^{(d)}) < d - 1$$

for $d = 5$ and for $d \geq 7$.

It would be interesting to determine $\delta(d)$ for all d.

9. Show that the equations of theorem 9.4.1 (including Euler's equation) can be solved for $f_n, \cdots, f_{d-1}$ in terms of $f_0, \cdots, f_{n-1}$ (where $n = [\frac{1}{2}d]$).

10. For $0 \leq k \leq d$, let the vector $w^{(k,d)} = (w_0^{(k,d)}, \cdots, w_{d-1}^{(k,d)})$ be defined by

$$w_i^{(k,d)} = \begin{cases} \dbinom{k+1}{i+1} & \text{for } 0 \leq i \leq k - 1 \\ 0 & \text{for } k \leq i \leq d - 1. \end{cases}$$

Prove that the set $W^{(d)} = \{w^{(k,d)} \mid 0 \leq k \leq d, k \equiv d \pmod 2\}$ is an affine basis of $A^{(d)}$.

9.8   Remarks

The history of the Dehn–Sommerville equations is rather interesting. The 4-dimensional case is rather trivial; it was known already to Brückner [1,3]. Dehn [1] proved in 1905 that the f-vectors of simplicial d-polytopes satisfy 2 independent linear relations for $d = 4$, and 3 such relations for $d = 5$. He also conjectured that for general d there are $[\frac{1}{2}d] + 1$ such relations.* The complete systems (E_m^d) and (E_m^{d*}) were obtained in 1927 by Sommerville [1], in a manner very similar to the one used here. There seems to be no additional mention of the Dehn–

---

* This is obviously a typographical error since already the cases $d = 2$ and $d = 4$ contradict it; Dehn certainly had the correct $[\frac{1}{2}(d + 1)]$ in mind.

Sommerville equations in the literature* till Fieldhouse's thesis [1] in 1961. Independently of all previous work Klee [11] rediscovered in 1963 the Dehn–Sommerville equations, in the more general formulation for manifolds and 'incidence systems' (see below).

It is interesting to note that though Sommerville [1] mentions the fact that $[\frac{1}{2}(d + 1)]$ of the equations (E_m^d) are independent, he does not consider the question of the completeness of this system—that is, the question about the dimension of $A^{(d)} = \mathrm{aff}\, f(\mathscr{P}_s^d)$. Klee [11] shows by a family of examples (from which the above exercise 9.7.1 is derived) that the dimension of the *linear* hull of $f(\mathscr{P}_s^d)$ is at least $[\frac{1}{2}(d + 1)]$.

Theorem 9.5.1 was first obtained by Fieldhouse [1] (the formulas reproduced in the summary Fieldhouse [2] are not correct for odd d). Klee [11,13] found the expressions for f_{d-1} in terms of $f_0, \cdots, f_{n-1}$ (where $n = [\frac{1}{2}d]$), and noted that in the expression for $f_k, k \geq n$, in terms of $f_0, \cdots, f_{n-1}$, the coefficient of f_{n-1} is positive. Recently Riordan [1] obtained various recurrences and interrelations for the coefficients appearing in those equations, as well as expressions for the f_i's of even index in terms of those of odd index, and vice versa.

Klee [11] is certainly the most important paper on the Dehn–Sommerville equations, not only because of the generality of the results obtained but also since it is the first, necessary, stepping-stone for Klee's [11] treatment of the problem of maximizing f_{d-1} given f_0 (see section 10.1). The method used by Klee [13] is similar to that used in the present exposition; he obtained most of the results of section 9.2.

In a far-reaching generalization of theorems 9.2.1 and 9.3.1, and exercises 8.5.1 and 9.7.5, M. Perles [2] recently established the following theorem:

Let G be any finite group of linear transformations of R^d, $d \geq 2$. Denote by $\mathscr{P}_s^d(G)$ respectively $\mathscr{P}_q^d(G)$ the family of all simplicial (respectively quasi-simplicial) d-polytopes P such that P is mapped onto itself by each transformation in G. Then $\dim \mathrm{aff}\, f(\mathscr{P}_s^d(G)) = [\frac{1}{2}d]$, and $\dim \mathrm{aff}\, f(\mathscr{P}_q^d(G)) = d - 1$.

---

*It is remarkable that even Sommerville himself fails to mention the equations in his book [2].

9.9   Additional notes and comments

The h-vector.
In his seminal 1970 paper in which he also established the upper bound theorem (see sections 10.1 and the notes in 10.6), McMullen [c] made extensive use of the h-vector $h(P) = (h_0, h_1, \ldots, h_d)$ of a simplicial d-polytope. Its components are given by

$$h_k := \sum_{i=0}^{k} (-1)^{k-i} \binom{d-i}{d-k} f_{i-1},$$

where we write $f(P) = (f_0, f_1, \ldots, f_{d-1})$ for the f-vector, and agree that $f_{-1} = 1$. The h-vector is linearly equivalent to the f-vector, via

$$f_{i-1} = \sum_{k=0}^{i} \binom{d-k}{d-i} h_k.$$

In terms of the h-vector, the Dehn–Sommerville equations have the strikingly simple form (already observed by Sommerville [1])

$$h_k = h_{d-k} \qquad \text{for } 0 \le k \le d.$$

The h-vector of a neighborly d-polytope with n vertices is given by

$$h_k(C(n,d)) = \binom{n-d-1+k}{k} \qquad \text{for } 0 \le k \le [\tfrac{d}{2}],$$

so the upper bound theorem is reduced to showing that $h_k(P) \le \binom{n-d-1+k}{k}$ holds for all (simplicial) d-polytopes P with n vertices.

The components h_k of the h-vector of a simplicial polytope have a number of important interpretations:
o h_k counts the number of shelling steps (i.e., facets) in an arbitrary shelling for which the unique minimal "new" face has exactly k vertices.
o Dually, h_k counts the number of vertices in which an arbitrary sweep hyperplane for P^* encounters exactly k new edges.
(In particular, reversal of the shelling resp. the sweep direction immediately establishes the Dehn–Sommerville equations. Also, the combinatorial interpretation yields that $h_k \ge 0$ holds for all k.)
o h_k is the dimension of the k-graded part of the face ring (see also the notes in section 10.6) of the boundary complex of P, modulo a homogeneous system of parameters of degree 1.
(This is still true if we consider any triangulated sphere in place of the boundary of a polytope P, and this leads to Stanley's [a] [g] proof of the upper bound theorem for spheres.)

o h_k is the rank of the $2k$-th singular homology group of the toric variety associated with a (rational, centered) simplicial polytope P.

(This allows one to apply the hard Lefschetz theorem for the toric variety, and thus establishes additional restrictions on the components of the h-vector; this is the key to Stanley's [b] proof of the necessity of the g-theorem, which characterizes the f-vectors of simplicial polytopes. See Ewald [a], Stanley [c], and the notes in section 10.6.)

Dehn–Sommerville with symmetries.
See Barvinok [a] for an equivariant generalization of the Dehn–Sommerville equations.

Binomial sums.
There has been a lot of progress in the derivation and manipulation of binomial identities such as those appearing in this chapter. Three valuable and enjoyable references for this are Graham–Knuth–Patashnik [a], Wilf [a], and Petkovšek–Wilf–Zeilberger [a].

Cubical h-vectors.
There are two different, competing notions of an h-vector for cubical polytopes, one by Stanley [d] as a special case of his "toric h-vector for general polytopes" (see also Chan [a]), and a simpler combinatorial one due to Adin [a]. Both versions yield non-negative vectors that can be computed from a shelling, and they satisfy cubical Dehn–Sommerville equations of the form $h_k^{cub} = h_{d-k}^{cub}$.

On exercise 9.7.7.
As noted in section 4.9, our ability to construct polytopes of type (k,h) is very limited. Despite the remark on page 170, no family of infinitely many polytopes of type $(2, d - 2)$ for any fixed $d > 4$ is known. However, we have the quite remarkable Wythoff construction, as explained by Coxeter [1, §§5.7, 11.6–11.8]. It produces uniform polytopes as the convex hulls of special orbits of finite reflection groups; thus it provides the Gosset–Elte polytopes r_{st} of dimension $d = r + s + t + 1$, for parameters $r \geq 1$, $s \geq t \geq 0$ with $\frac{1}{r+1} + \frac{1}{s+1} + \frac{1}{t+1} > 1$. (See Coxeter [a] for a slick proof of this finiteness condition.) P. McMullen has observed that r_{st} has the type $(r + 2, s + t - 1)$.

One may work out that this construction includes as special cases the half-cubes $h\gamma_d = 1_{d-3,1}$, which are dual to the examples of type $(3, d - 3)$ that appear in exercise 4.8.18. Furthermore, the construction produces the remarkable 8-dimensional polytope 2_{41} of type $(4,4)$, related to the Coxeter group E_8, with 2160 vertices and 17520 facets.

CHAPTER 10

Extremal Problems Concerning Numbers of Faces

In chapters 8 and 9 we obtained information on the f-vectors of d-polytopes, and of some special classes of d-polytopes, by determining the affine hull of all the f-vectors of polytopes in the class considered. In other words, we were concerned with the linear relations satisfied by the f-vectors of all polytopes belonging to a certain class. Chapter 10 reports on the present knowledge about non-linear relations existing between the components of f-vectors of certain classes of polytopes. As will be seen, results in this direction are still rather incomplete—despite the considerable interest and efforts devoted to some aspects of these problems.

10.1 Upper Bounds for f_i, $i \geq 1$, in Terms of f_0

We have seen in section 5.2 that the process of 'pulling' the vertices of a d-polytope P leaves $f_0(P)$ unchanged while $f_i(P)$, for $1 \leq i \leq d - 1$, is either unchanged or increased. If the pulling is performed, in turn, with all the vertices of P, a simplicial d-polytope P' is obtained such that $f_0(P') = f_0(P)$, and $f_i(P') \geq f_i(P)$ for $1 \leq i \leq d - 1$, with $f_m(P') = f_m(P)$ implying that all k-faces of P are simplices, for all $k \leq m$. In other words, we have

1. *For all k, d, v with $1 \leq k < d < v$,*

$$\max\{f_k(P) \mid P \in \mathscr{P}^d, f_0(P) = v\} = \max\{f_k(P) \mid P \in \mathscr{P}^d_s, f_0(P) = v\}.$$

Moreover, if $P \in \mathscr{P}^d$, $f_0(P) = v$, and if $f_{d-1}(P)$ has maximal possible value, then $P \in \mathscr{P}^d_s$.

Therefore, when endeavoring to determine upper bounds for $f_i(P)$ in terms of $f_0(P)$, we may without loss of generality restrict our attention to simplicial polytopes P. The reduction of the problem to a question on simplicial polytopes is important because it enables us to use the Dehn–Sommerville equations from chapter 9.

172

We shall denote by $\mu_k(v, d)$ the maximal possible number of k-faces of d-polytopes P with $f_0(P) = v$. In particular, $\mu(v, d)$ shall occasionally be used instead of $\mu_{d-1}(v, d)$.

Before stating and proving the general results, we shall examine in some detail the simplest cases $d = 3, 4, 5$.

(1) If $d = 3$ we have the following special cases of the equations from section 9.5 for the f-vectors of polytopes in $\mathscr{P}_s^3$ (see table 3):

$$f_1 = 3f_0 - 6,$$

$$f_2 = 2f_0 - 4.$$

Therefore, for every $P \in \mathscr{P}^3$,

$$f_1(P) \leq 3f_0(P) - 6 = \mu_1(f_0, 3),$$

$$f_2(P) \leq 2f_0(P) - 4 = \mu_2(f_0, 3).$$

By duality there follow also the relations

$$f_1(P) \leq 3f_2(P) - 6 = \mu_1(f_2, 3),$$

$$f_0(P) \leq 2f_2(P) - 4 = \mu_2(f_2, 3).$$

Note that either inequality of each pair is a consequence of the other inequality and of Euler's equation $f_0 - f_1 + f_2 = 2$.

(2) If $d = 4$ then obviously $f_1 \leq \binom{f_0}{2}$ with equality only for neighborly 4-polytopes. The equations of section 9.5 yield for each $P \in \mathscr{P}_s^4$

$$f_2 = 2f_1 - 2f_0 \leq 2\binom{f_0}{2} - 2f_0,$$

$$f_3 = f_1 - f_0 \leq \binom{f_0}{2} - f_0.$$

Therefore, the f-vector of each 4-polytope satisfies

$$f_1 \leq \tfrac{1}{2}f_0(f_0 - 1) = \mu_1(f_0, 4),$$

$$f_2 \leq f_0(f_0 - 3) = \mu_2(f_0, 4),$$

$$f_3 \leq \tfrac{1}{2}f_0(f_0 - 3) = \mu_3(f_0, 4),$$

with equality in any of the relations characterizing the polytope as neighborly, and implying equality in all inequalities.

(3) Similarly, if $d = 5$ we have, for $P \in \mathscr{P}_s^5$,

$$f_2 = 4f_1 - 10f_0 + 20$$
$$f_3 = 5f_1 - 15f_0 + 30$$
$$f_4 = 2f_1 - 6f_0 + 12.$$

Since again $f_1 \leq \binom{f_0}{2}$, the f-vector of each 5-polytope P satisfies

$$f_1 \leq \tfrac{1}{2}f_0(f_0 - 1) = \mu_1(f_0, 5),$$
$$f_2 \leq 2f_0(f_0 - 6) + 20 = \mu_2(f_0, 5),$$
$$f_3 \leq \tfrac{5}{2}f_0(f_0 - 7) + 30 = \mu_3(f_0, 5),$$
$$f_4 \leq f_0(f_0 - 7) + 12 = \mu_4(f_0, 5).$$

Equality in any of the relations implies that P is a 2-neighborly 5-polytope. If P is a simplicial 5-polytope then equality in any of the relations implies equality in all of them.

Already for $d = 6$ such a simple approach does not work. The Dehn–Sommerville equations yield

$$f_3 = 5f_0 - 5f_1 + 3f_2,$$
$$f_4 = 6f_0 - 6f_1 + 3f_2,$$
$$f_5 = 2f_0 - 2f_1 + f_2.$$

and the obvious inequalities $f_1 \leq \binom{f_0}{2}, f_2 \leq \binom{f_0}{3}$, are not sufficient to derive best-possible upper bounds for f_k, $k = 3, 4, 5$.

Our next aim is a set of inequalities generalizing $f_r \leq \binom{f_0}{r+1}$, which shall, in conjunction with the Dehn–Sommerville equations, yield the most important part of the known values of $\mu_k(v, d)$.

2. *For every simplicial polytope P, and for all nonnegative k and $r, k \leq r$, we have*

$$\binom{r+1}{k} f_r \leq \binom{f_0 + k - 1 - r}{k} f_{r-k}$$

independent of the dimension of P. Moreover, equality holds if and only if P is an $(r + 1)$-neighborly polytope.

PROOF We shall estimate in two ways the number $g_{r,r-k}(P)$ of incidences of an r-face of P with an $(r - k)$-face of P. On the one hand, each r-face of P is an r-simplex and has, therefore, $\begin{pmatrix} r + 1 \\ r - k + 1 \end{pmatrix} = \begin{pmatrix} r + 1 \\ k \end{pmatrix}$ $(r - k)$-faces; hence $g_{r,r-k}(P) = \begin{pmatrix} r + 1 \\ k \end{pmatrix} f_r$. On the other hand, theorem 3.1.8 implies that if an $(r - k)$-face F^{r-k} is incident with an r-face F^r there exists a $(k - 1)$-face F^{k-1} such that

$$F^{k-1} \cap F^{r-k} = \varnothing \quad \text{and} \quad F^r = P \cap \text{aff}\,(F^{k-1} \cup F^{r-k}).$$

Thus the number of incidences of an F^{r-k} with r-faces of P does not exceed the number of $(k - 1)$-faces of P which are disjoint from F^{r-k}. Clearly, the last number is not greater than $\begin{pmatrix} v \\ k \end{pmatrix}$, where $v = f_0 - (r - k + 1)$ is the number of vertices of P not incident with F^{r-k}. Therefore,

$$g_{r,r-k}(P) \le \begin{pmatrix} v \\ k \end{pmatrix} f_{r-k},$$

and the proof of theorem 2 is completed.

Let $f_k(v, d)$ denote the number of k-faces of any cyclic d-polytope with v vertices; as shown in chapter 9, $f_k(v, d)$ is also the number of k-faces of every simplicial neighborly d-polytope with v vertices. Obviously $\mu_k(v, d) \ge f_k(v, d)$. We shall prove

3. *The relation $\mu_k(v, d) = f_k(v, d)$ holds at least in the following cases:*
   (i) *for every k, $1 \le k \le d - 1$, provided v is large enough;*
   (ii) *for $k = d - 1$ provided d is even and $v \ge [\frac{1}{2}d]^2 - 1$ or d is odd and $v \ge [\frac{1}{2}(d + 1)]^2 - 2$:*
   (iii) *for $k = [\frac{1}{2}d]$ provided $d = 2n$ and $v \ge \frac{1}{2}(n^2 + 3n - 4)$ or $d = 2n + 1$ and $v \ge \frac{1}{2}(n^2 + 5n - 2)$:*
   (iv) *for every k, $1 \le k \le d - 1$, provided $v \le d + 3$.*
   (v) *for every v and k, $1 \le k < d < v$, provided $d \le 8$.*

We shall prove the various assertions of theorem 3 one after another.
(i) By theorem 9.5.1 every $P \in \mathscr{P}_s^d$ satisfies, for each k with

$$n = [d/2] \le k \le d - 1,$$

an equation of the type

$$f_k(P) = \binom{d-n}{k-n} f_{n-1}(P) + \sum_{i=-1}^{n-2} \alpha_i f_i(P),$$

where the α_i's are numerical coefficients depending only on k and d (the values of the α_i's are given in theorem 9.5.1). In order to obtain an upper bound for $f_k(P)$ in terms of $f_0(P) = v$, we note that

$$\frac{\binom{n}{i+1}}{\binom{v-i-1}{n-i-1}} f_{n-1}(P) \le f_i(P) \le \binom{v}{i+1},$$

where the lower bound results from theorem 2 above. (Note that equality holds throughout if and only if P is a simplicial n-neighborly d-polytope with v vertices). Therefore

$$f_k(P) \le \left\{ \binom{d-n}{k-n} + \sum{}^* \alpha_i \frac{\binom{n}{i+1}}{\binom{v-i-1}{n-i-1}} \right\} f_{n-1}(P) + \sum{}^{**} \alpha_i \binom{v}{i+1},$$

where the summation in $\sum^*$ extends over all i such that $\alpha_i < 0$, and in $\sum^{**}$ over all i with $\alpha_i \ge 0$. The expression in brackets is nonnegative for v large enough and therefore for such v

$$f_k(P) \le \left\{ \binom{d-n}{k-n} + \sum{}^* \alpha_i \frac{\binom{n}{i+1}}{\binom{v-i-1}{n-i-1}} \right\} \binom{v}{n} + \sum{}^{**} \alpha_i \binom{v}{i+1}.$$

Since all the inequalities become equations if P is an n-neighborly d-polytope, this completes the proof of the assertion (i).

Clearly, taking the values of α_i from theorem 9.5.1, explicit lower bounds on v (in terms of k and d) may be found from the above proof. However, the slightly more involved method used below in the proof of (ii) yields better lower bounds for v.

(ii) The proof follows a pattern similar to the above but is more elaborate in order to obtain a best-possible result.

We first consider the case of even $d = 2n$; for any $P \in \mathscr{P}_s^d$, by theorem 9.5.1 and the inequalities in theorem 2 above we have

$$f_{2n-1} = \frac{1}{n} \sum_{i=0}^{n-1} (-1)^{n+i+1}(i+1) \binom{2n-2-i}{n-1} f_i$$

$$= \frac{1}{n} \sum_{j=0}^{n-1} (-1)^j(n-j) \binom{n-1+j}{n-1} f_{n-1-j}$$

$$= \frac{1}{n} \sum_{k=0}^{[\frac{1}{2}(n-1)]} \left\{ (n-2k) \binom{n-1+2k}{n-1} f_{n-1-2k} \right.$$

$$\left. - (n-2k-1) \binom{n+2k}{n-1} f_{n-2-2k} \right\}$$

$$\leq \frac{1}{n} \sum_{k=0}^{[\frac{1}{2}(n-1)]} \left\{ (n-2k) \binom{n-1+2k}{n-1} \right.$$

$$\left. - \binom{n+2k}{n-1} \frac{(n-2k)(n-2k-1)}{f_0-(n-1-2k)} \right\} f_{n-1-2k}$$

$$= \sum_{k=0}^{[\frac{1}{2}(n-1)]} \left\{ 1 - \frac{(n-2k-1)(n+2k)}{(2k+1)(f_0-(n-1-2k))} \right\} \frac{(n-2k)}{n}$$

$$\times \binom{n-1+2k}{n-1} f_{n-1-2k}$$

$$= \sum_{k=0}^{[\frac{1}{2}(n-1)]} \beta_k \cdot \frac{n-2k}{n} \binom{n-1+2k}{n-1} f_{n-1-2k}.$$

Now, provided all β_k are nonnegative, an upper bound for f_{2n-1} is obtained from this inequality by using $f_{n-1-2k} \leq \binom{f_0}{n-2k}$. The resulting upper bound is best possible since in the case of n-neighborly d-polytopes all inequalities used become equations. As to the sign of β_k, it is obviously the same as the sign of

$$(2k+1)(f_0-(n-1-2k)) - (n-2k-1)(n+2k)$$

$$= (2k+1)f_0 - (n-2k-1)(n+4k+1)$$

$$= (2k+1)\left(f_0 - \frac{n(n-1)}{2k+1} - n - 1 + 2(2k+1) \right).$$

The value of the last expression clearly decreases with decreasing k. Therefore all β_k will be nonnegative provided $\beta_0 \geq 0$; in this case the condition becomes $f_0 - n^2 + n - n - 1 + 2 \geq 0$, i.e. $f_0 \geq n^2 - 1 = [\frac{1}{2}d]^2 - 1$ as claimed.

We omit the completely analogous computations for the case of odd $d = 2n + 1$, which yield the condition $f_0 \geq n^2 + 2n - 1 = [\frac{1}{2}(d + 1)]^2 - 2$.

(iii) The proof in this case parallels the proof in case (ii) (using the equations from page 161) and is omitted.

(iv) In case of d-polytopes with $d + 2$ vertices, our assertion follows immediately from theorem 6.1.3 or 6.1.5, and the observation that $C(d + 2, d) = T^d_{[\frac{1}{2}d]} = T^{d,0}_{[\frac{1}{2}d]}$.

In case of d-polytopes with $d + 3$ vertices, the assertion is contained in theorem 6.2.1.

Before proceeding to prove the last assertion of theorem 3 we shall quote, without proof, a recent result of Kruskal [2].

Let r and k be two positive integers. The k-canonical representation of r is the representation of r in the form

$$r = \binom{r_1}{k} + \binom{r_2}{k - 1} + \cdots + \binom{r_i}{k - i + 1},$$

where $r_1 = \max\left\{ s \mid r \geq \binom{s}{k} \right\}$, and in general, r_p is defined provided

$$r > \binom{r_1}{k} + \cdots + \binom{r_{p-1}}{k - p + 2},$$

and equals

$$r_p = \max\left\{ s \mid r \geq \binom{r_1}{k} + \cdots + \binom{r_{p-1}}{k - p + 2} + \binom{s}{k - p + 1} \right\}.$$

For given r, k, and j we define

$$r^{(j|k)} = \binom{r_1}{j} + \binom{r_2}{j - 1} + \cdots + \binom{r_i}{j - i + 1},$$

where $r_1, \cdots, r_i$ are defined by the k-canonical representation of r.

Let $\mathscr{C}$ denote a simplicial complex. We define

$$\kappa(r; k; j) = \max\{ f_j(\mathscr{C}) \mid f_k(\mathscr{C}) = r \} \qquad \text{if} \quad j > k$$

and

$$\kappa(r;k;j) = \min\{f_j(\mathscr{C}) \mid f_k(\mathscr{C}) = r\} \qquad \text{if} \quad j < k.$$

The result of Kruskal [2] may be formulated as follows:

4. *For every* r, k, j *the relation*

$$\kappa(r;k;j) = r^{(j+1|k+1)}$$

holds.

It is worth noting that Kruskal's theorem is a purely combinatorial result, not depending on the geometric interpretation given to it here.

We shall use Kruskal's theorem in order to prove that $\mu_k(v, 7) = f_k(v, 7)$. Considering the case $k = 6$, the Dehn–Sommerville equations show that every simplicial 7-polytope satisfies $f_6 = 2(f_2 - 4f_1 + 10f_0 - 20)$. We shall show that $f_6 \leq f_6(f_0, 7)$.

For $f_0 \leq 10$ this assertion is contained in part (iv) of theorem 3; hence we may assume that $f_0 \geq 11$. Since $f_6(11, 7) = 70$, we need consider only 7-polytopes with $f_6 \geq 70$. The 7-canonical representation of 70 being

$$70 = \binom{9}{7} + \binom{8}{6} + \binom{6}{5},$$ theorem 4 implies that for these polytopes

$$f_2 \geq 70^{(3|7)} = \binom{9}{3} + \binom{8}{2} + \binom{6}{1} = 118.$$ Hence, for all polytopes con-

sidered, the 3-canonical representation of f_2 will be of one of the types

$$f_2 = \binom{a}{3} + \binom{b}{2} + \binom{c}{1}$$

or

$$f_2 = \binom{a}{3} + \binom{b}{2}$$

or

$$f_2 = \binom{a}{3},$$

where $a \geq 9$. If the representations of f_2 is of the third type then, by theorem 4, $f_1 \geq a^{(2|3)} = \binom{a}{2}$ and $f_0 \geq a$. Since for $a \geq 9$ we have

$\binom{a}{3} - 4\binom{a}{2} \leq \binom{a+1}{3} - 4\binom{a+1}{2}$ it follows that

$$f_6 \leq 2\left\{\binom{a}{3} - 4\binom{a}{2} + 10f_0 - 20\right\} \leq 2\left\{\binom{f_0}{3} - 4\binom{f_0}{2} + 10f_0 - 20\right\}$$

$$= f_6(f_0, 7),$$

and the assertion is established.

Thus we may assume that the 3-canonical representation of f_2 is of the first or second type, and consequently $a < f_0$. From the definition of canonical representation it follows that $a > b > c$. Now, for fixed f_0, f_6 clearly increases or decreases with the appropriate change in $f_2 - 4f_1$. By theorem 4,

$$f_2 - 4f_1 \leq f_2 - 4f_2^{\{2|3\}} = \left\{\binom{a}{3} + \binom{b}{2} + \binom{c}{1}\right\} - 4\left\{\binom{a}{2} + \binom{b}{1} + \binom{c}{0}\right\},$$

the binomial coefficients involving c being possibly absent. If c does occur, increasing it to its limit $c = b - 1$ clearly increases the right hand side of this relation. Thus we have only to estimate from above the expressions

$$\binom{a}{3} + \binom{b}{2} + \binom{b-1}{1} - 4\left\{\binom{a}{2} + \binom{b}{1} + 1\right\}$$

$$< \binom{a}{3} + \binom{b+1}{2} - 4\left\{\binom{a}{2} + \binom{b+1}{1}\right\}.$$

and

$$\binom{a}{3} + \binom{b}{2} - 4\left\{\binom{a}{2} + \binom{b}{1}\right\}.$$

In other words, it is enough to find an appropriate upper bound for the last expression, under the assumption $b \leq a$. We claim that $\binom{a+1}{3} - 4\binom{a+1}{2}$ is such a bound. This assertion is equivalent to

$$\binom{a}{2} - \binom{b}{2} = \binom{a+1}{3} - \binom{a}{3} - \binom{b}{2} \geq 4\left\{\binom{a+1}{2} - \binom{a}{2} - \binom{b}{1}\right\}$$

$$= 4(a - b),$$

that is, to

$$a(a - 1) - b(b - 1) \geq 8(a - b)$$

or $a^2 - b^2 \geq 9(a - b)$. But since $a \geq b$ this holds if and only if $a + b \geq 9$, which follows from the assumption $a \geq 9$. Thus

$$f_2 - 4f_1 \leq \binom{a + 1}{2} - 4\binom{a + 1}{2} \leq \binom{f_0}{3} - 4\binom{f_0}{2},$$

the last inequality being a consequence of $f_0 > a \geq 9$. Finally $f_6 \leq f_6(f_0, 7)$, and hence $\mu_6(v, 7) \leq f_6(v, 7)$. The equation $\mu_k(v, 7) = f_k(v, 7)$ for all k may be proved either by an analogous argument, or deduced from the case $k = 6$ by using the relations

$$2f_5 = 7f_6$$

$$2f_4 = 9f_6 + 4f_1 - 12f_0 + 24$$

$$2f_3 = 5f_6 + 10f_1 - 30f_0 + 60,$$

which follow from the Dehn–Sommerville equations. This completes the proof of the case $d = 7$ of part (v) of theorem 3.

Since the cases $d \leq 5$ have been established at the beginning of the present section and the case $d = 6$ results on combining parts (ii), (iii) and (iv) of theorem 3 (together with $f_4 = 3f_5$), we shall now consider the case of 8-dimensional polytopes. The proof of this case follows from the case $d = 7$ in view of the following lemma (which also yields at once a proof for $d = 6$ from the case $d = 5$),

5. *For all $k < d < v$,*

$$\mu_k(v, d) \leq \frac{v}{k + 1}\mu_{k-1}(v - 1, d - 1).$$

Indeed, let P be a simplicial d-polytope and let P_i, $1 \leq i \leq v = f_0(P)$, be the polytopes obtained as sections of P by hyperplanes, each of which strictly separates one of the vertices of P from all the other vertices. Clearly

$$f_k(P) = \frac{1}{k + 1}\sum_{i=1}^{v} f_{k-1}(P_i) \qquad \text{for all} \quad k = 0, 1, \cdots, d - 1.$$

Also, for each i, $f_0(P_i) \leq f_0(P) - 1 = v - 1$. Therefore

$$f_j(P_i) \leq \mu_j(v - 1; d - 1)$$

and hence

$$f_k(P) \leq \frac{v}{k+1}\mu_{k-1}(v-1, d-1),$$

which establishes lemma 5.

Now, if d is even and if $\mu_{k-1}(v-1, d-1) = f_{k-1}(v-1, d-1)$, then lemma 5 and theorem 9.6.2 imply

$$f_k(v, d) \leq \mu_k(v, d) \leq \frac{v}{k+1}\mu_{k-1}(v-1, d-1)$$

$$= \frac{v}{k+1}f_{k-1}(v-1, d-1) = f_k(v, d),$$

i.e. $f_k(v, d) = \mu_k(v, d)$.

This completes the proof of part (v) of theorem 3, which is therefore established in all its parts.

Remarks

The values of $\mu_k(v, 3)$ and $\mu_k(v, 4)$ have been known for a relatively long time (see, for example, Brückner [1, 3], Saaty [1]. The 'upper bound conjecture' $\mu_k(v, d) = f_k(v, d)$ was first announced in an abstract of Motzkin [4]. (The formulation of Motzkin [4] is categorical, but since no detailed exposition appeared in the intervening years it seems reasonable to refer to the statement as to a conjecture. See also above, sections 7.2 and 9.6.) A proof of the case $k = d - 1$ of the conjecture was announced by Jacobs–Schell [1]; however, the proof seems to have been incorrect. The first positive step was made by Fieldhouse [1, 2]. Using the easily established cases with $d \leq 5$ and the above lemma 5, he proved $\mu_k(v, 6) = f_k(v, 6)$. Next came Gale's [5] proof that $\mu_{d-1}(v, d) = f_{d-1}(v, d)$ for all $v \leq d + 3$. The most important contribution is Klee's paper [13], in which parts (i) and (ii) of theorem 3 are established. Using the analogues of the Dehn–Sommerville equations for Eulerian $(d - 1)$-manifolds $\mathscr{C}$ of Euler characteristic $\chi(\mathscr{C})$ (Klee [11]; see section 9.2), Klee [13] extended parts (i) and (ii) of theorem 3, to all complexes of this type and conjectured the validity of the 'upper bound conjecture' for all such manifolds.

The case $k = 1$ of theorem 2 is also due to Klee [13].

Rather obviously, the case $k = d - 1$ of the 'upper bound conjecture' is most important and interesting. The 'gap' $d + 4 \leq v \leq [\frac{1}{2}d]^2 - 2$ for even d, and $d + 4 \leq v \leq [\frac{1}{2}(d + 1)] - 3$ for odd d, of values of v for which the 'upper bound conjecture' is still open is rather frustrating. The first unsolved cases, $13 \leq v \leq 22$ for $d = 9$ and $14 \leq v \leq 23$ for $d = 10$, seem not to be decidable by the methods developed till now.

The last part of theorem 1 may probably be strengthened as follows:

If $P \in \mathscr{P}^d$ and $f_k(P) = \mu_k(f_0(P), d)$ for some $k \geq [\frac{1}{2}(d - 1)]$, then $P \in \mathscr{P}^d_s$.

This problem was first noted by Klee [13] who also observed that the solution is affirmative if $k = d - 1$, $k = d - 2$, or $d = 2k + 2$. The last case is a consequence of theorem 7.2.1. The simplest undecided cases are $d = 5, k = 2$ and $d = 6, k = 3$.

10.2   Lower Bounds for f_i, $i \geq 1$, in Terms of f_0

As a counterpart to the upper bounds for the numbers $f_k(P)$, $P \in \mathscr{P}^d$ or $P \in \mathscr{P}^d_s$, in terms of $f_0(P)$, which were given in the preceding section, we shall now present the known results on the corresponding lower bounds.

The present problem differs from the previous one in an important aspect: There is, for the time being, no reasonable and general conjecture as to the d-polytopes P which minimize $f_k(P)$ or as to the minimal values themselves. As a consequence our knowledge is much more restricted in case of the lower bounds than in case of the upper bounds.

We are interested in the two functions $\varphi_k(v, d)$ and $\varphi_k^*(v, d)$ defined by

$$\varphi_k(v, d) = \inf\{ f_k(P) \mid P \in \mathscr{P}^d \quad \textbf{and} \quad f_0(P) = v \}$$

and

$$\varphi_k^*(v, d) = \inf\{ f_k(P) \mid P \in \mathscr{P}^d_s \quad \textbf{and} \quad f_0(P) = v \}.$$

Let us define, for $v \geq d + 1$,

$$\phi_k^*(v, d) = \binom{d}{k} v - \binom{d + 1}{k + 1} k \quad \text{for} \quad 1 \leq k \leq d - 2$$

$$\phi_{d-1}^*(v, d) = (d - 1)v - (d + 1)(d - 2).$$

It has been repeatedly conjectured that $\varphi_k^*(v, d) = \phi_k^*(v, d)$ for all $v, d,$ and k. For the rather interesting history of this 'lower bound conjecture' see below.

We shall prove the rather weak result

1. *The relation*

$$\varphi_k^*(v, d) = \phi_k^*(v, d), \qquad k = 1, 2, \cdots, d - 1$$

holds provided

either   (i) $d \le 3$;

or   (ii) $v \le d + 3$.

PROOF Assertion (i) is completely trivial for $d = 2$. In case $d = 3$ we have $f_1 = 3f_0 - 6$ and $f_2 = 2f_0 - 4$, hence $\varphi_k^*(v, 3) = \phi_k^*(v, 3)$ as claimed.

As to assertion (ii), in case $v = d + 2$ the proof follows from theorem 6.1.3 since $\phi_k^*(d + 2, d) = f_k(T_1^d)$. In case $v = d + 3$ our claim follows from theorem 6.2.5 since

$$\phi_k^*(d + 3, d) = \binom{d + 2}{k + 1} + \binom{d}{k} - \binom{d}{k - 1} - 2\delta_{k, d-1}.$$

This completes the proof of theorem 1.

Let us define, for $d < v \le 2d$,

$$\phi_k(v, d) = \binom{d + 1}{k + 1} + \sum_{1 \le i \le v - d - 1} \binom{d - i}{k}$$

$$= \binom{d + 1}{k + 1} + \binom{d}{k + 1} - \binom{2d + 1 - v}{k + 1}.$$

It may be conjectured that $\varphi_k(v, d) = \phi_k(v, d)$ for all k and $d + 1 \le v \le 2d$. However, we are able to prove only the weaker result.

2. (i) *For all k, r, and s such that $1 \le k \le d - 1$, $1 \le r \le \min\{4, d\}$, and $s > r$, we have*

$$\varphi_k(d + s, d) \ge \varphi_k(d + r, d) = \phi_k(d + r, d).$$

(ii) $\varphi_1(v, 2) = v$ for all $v \ge 3$.

(iii) $\varphi_1(v, 3) = [\frac{1}{2}(3v + 1)]$

       $\varphi_2(v, 3) = [\frac{1}{2}(v + 5)]$ for all $v \ge 4$.

PROOF Assertion (ii) is trivial. If P is a 3-polytope then

$$2f_1(P) \ge 3f_0(P) + \varepsilon,$$

where $\varepsilon = 0$ if $f_0(P)$ is even, and $\varepsilon = 1$ if $f_0(P)$ is odd. In any case

$$f_1(P) \geq [\tfrac{1}{2}(3f_0(P) + 1)],$$

which establishes the first assertion of (iii); the second then results from Euler's theorem.

The proof of (i) is much longer. The case $r = 1$ is trivial, and the case $r = 2$ follows at once from theorem 6.1.5. Let now P^d denote the $(d - 3)$-fold d-pyramid with a 3-dimensional three-sided prism as basis. Then an easy computation yields

$$f_k(P^d) = \binom{d + 1}{k + 1} + \binom{d - 1}{k} + \binom{d - 2}{k} = \phi_k(d + 3, d).$$

Thus $\varphi_k(d + 3, d) \leq \phi_k(d + 3, d)$ and we have to show that the reversed inequality holds. Let $d > 3$ and let P be a d-polytope with $d + 3$ vertices.
If P is a d-pyramid with $(d - 1)$-basis F then $f_0(F) = d + 2$ and

$$f_k(P) = f_k(F) + f_{k-1}(F) \geq \varphi_k(d + 2, d - 1) + \varphi_{k-1}(d + 2, d - 1)$$

$$= \phi_k(d + 2, d - 1) + \phi_{k-1}(d + 2, d - 1) = \phi_k(d + 3, d).$$

If P is not a d-pyramid but some facet F of P has $d + 1$ vertices, then there are vertices V_1 and V_2 of P not belonging to F, and a facet F' of P which contains V_1 but not V_2. Each k-face of P is of one and only one of the following types:
(a) a k-face of F;
(b) a k-face containing V_1 of F';
(c) a k-face containing V_2.
The number of k-faces of type (a) is

$$f_k(F) \geq \varphi_k(d + 1, d - 1) = \phi_k(d + 1, d - 1).$$

Those of type (b) may be counted by considering a section of F' by a $(d - 1)$-hyperplane H strictly separating V_1 from the other vertices of F'. The required number of k-faces is then

$$f_{k-1}(F' \cap H) \geq \varphi_{k-1}((d - 2) + 1, d - 2) = \phi_{k-1}(d - 1, d - 2).$$

Similarly, by considering a section of P by a hyperplane which strictly separates V_2 from the other vertices of P, the number of k-faces of type (c) is found to be at least $\varphi_{k-1}((d - 1) + 1, d - 1) = \phi_{k-1}(d, d - 1)$. Therefore

$$f_k(P) \geq \phi_k(d + 1, d - 1) + \phi_{k-1}(d - 1, d - 2) + \phi_{k-1}(d, d - 1)$$

$$= \phi_k(d + 3, d).$$

The remaining possibility is that P is simplicial. But then

$$f_k(P) \geq \varphi_k^*(d + 3, d) > \phi_k(d + 3, d).$$

Thus $\varphi_k(d + 3, d) = \phi_k(d + 3, d)$, as claimed.

By theorem 6.1.2 we have $f(T_1^4) = (6, 14, 16, 8)$. Thus if T^* is a 4-polytope dual to T_1^4, we have

$$\varphi_k(8, 4) \leq f_k(T^*) = \phi_k(8, 4) \qquad \text{for} \quad k = 1, 2, 3.$$

Similarly, for $d > 4$ the $(d - 4)$-fold d-pyramids with 4-dimensional basis T^* show that $\varphi_k(d + 4, d) \leq \phi_k(d + 4, d)$ for all $k = 1, 2, \cdots, d - 1$.

Thus we have only to show that every d-polytope P with $f_0(P) \geq d + 4$ satisfies $f_k(P) \geq \phi_k(d + 4, d)$.

Again we have to distinguish a number of possibilities.

(a) P is a d-pyramid over a $(d - 1)$-dimensional basis with at least $d + 3 = (d - 1) + 4$ vertices. Then, by induction,

$$f_k(P) \geq \phi_k(d + 3, d - 1) + \phi_{k-1}(d + 3, d - 1) = \phi_k(d + 4, d).$$

(b) There is a facet of P with $d + 2 = (d - 1) + 3$ vertices; in analogy to the above, we have

$$f_k(P) \geq \phi_k(d + 2, d - 1) + \phi_{k-1}(d, d - 1) + \phi_{k-1}(d - 1, d - 2)$$
$$= \phi_k(d + 4, d).$$

(c) Each facet of P has at most $d + 1 = (d - 1) + 2$ vertices, but there exists a $(d - 2)$-face F with $d = (d - 2) + 2$ vertices. Then, using exercise 3.1.2 and the fact that at least two vertices of P belong to neither of the two facets of P which contain F, we have

$$f_k(P) \geq \phi_k(d, d - 2) + 2\phi_{k-1}(d, d - 2) + \phi_{k-1}(d, d - 1)$$
$$+ \phi_{k-1}(d - 1, d - 2) \geq \phi_k(d + 4, d).$$

(d) Each facet of P is a simplicial $(d - 1)$-polytope and there exist two facets, F_1 and F_2, with $d + 1 = (d - 1) + 2$ vertices each, intersecting in a $(d - 2)$-simplex. Then there exists a vertex of P not contained in $F_1 \cup F_2$, and therefore

$$f_k(P) \geq 2\phi_k(d + 1, d - 1) - \binom{d - 1}{k + 1} + \phi_{k-1}(d, d - 1) \geq \phi_k(d + 4, d).$$

(e) Each facet of P is a simplicial $(d - 1)$-polytope, and there exist facets F_1 and F_2 such that $f_0(F_1) = d + 1$, $f_0(F_2) = d$, and $F_1 \cap F_2$ is a $(d - 2)$-simplex. Then there exist two vertices of P not contained in $F_1 \cup F_2$ and therefore

$$f_k(P) \geq \phi_k(d + 1, d - 1) + \binom{d}{k + 1} - \binom{d - 1}{k + 1} + \phi_{k-1}(d, d - 1)$$

$$+ \phi_{k-1}(d - 1, d - 2) \geq \phi_k(d + 4, d).$$

(f) P is a simplicial d-polytope. Without loss of generality we may assume that every $d + 1$ vertices of P are affinely independent. Let V be a vertex of P, let P^* be the convex hull of the vertices of P different from V, and let L be a line through V which does not intersect P^* nor any $(d - 2)$-flat determined by vertices of P^*. Let V' be the point of L nearest to V which is on a $(d - 1)$-dimensional hyperplane determined by a facet of P^*. Then for each $V'' \in L$ which is between V and V', the d-polytope $P'' = \text{conv}(P^* \cup \{V''\})$ is of the same combinatorial type as P. But when V'' approaches V', the polytope P'' tends to $P' = \text{conv}(P^* \cup \{V'\})$, and therefore, by theorem 5.3.1, $f_k(P') \leq f_k(P'') = f_k(P)$. Now $f_0(P') = f_0(P)$, but P' is not simplicial and hence, by the above,

$$f_k(P) \geq f_k(P') \geq \phi_k(d + 4, d).$$

This completes the proof of theorem 2.

Denoting, as before, by $f_k(r, d)$ the number of k-faces of the cyclic polytope $C(r, d)$, theorem 10.1.3 yields by duality

3. *The relation*

$$\varphi_{d-1}(v, d) = \min\{r \mid f_{d-1}(r, d) \geq v\}$$

holds provided either

(i) $d \leq 8$;

or   (ii) v *is sufficiently large relative to* d.

Remarks

Clearly, if the 'upper bound conjecture' (section 10.1) is true then the relation in theorem 3 holds for all d and v.

The simplest nontrivial consequence of theorem 2 is the fact that no 3-polytope has exactly 7 edges; this was known already to Euler (see, for example, Steinitz–Rademacher [1], Cairns [3]). The similar 'gap' in the possible numbers of edges in higher dimensions was discussed by Buck [2]. The case $r = 2$ of theorem 2 was established by Klee [13]. Theorem 2 clearly implies that for sufficiently large d more 'gaps' occur; thus for $d = 6$, $f_1 = 21$ for $f_0 = 7$, $26 \leq f_1 \leq 28$ for $f_0 = 8$, and $f_1 \geq 30$ for $f_0 \geq 9$; for $d \geq 11$ three gaps occur in the series of possible values of f_1.

The only serious attempts at proving the 'lower bound conjecture' were made in case $d = 4$. Then the conjecture reduces to the simple (and naturally equivalent) statements: Every simplicial 4-polytope with v vertices has at least $4v - 10$ edges and at least $3v - 10$ facets. The first attempt to establish these bounds (in the dual formulation for simple 4-polytopes) was made by Brückner [3]. Unfortunately, his arguments are not valid (see Steinitz [6]); equally invalid are the arguments of Fieldhouse [2]. Another invalid 'proof' was found by the author (fortunately very few copies of the mimeographed preprint survived); an analysis of this attempt led to the construction of the 3-diagram which is not the Schlegel diagram of any 4-polytope (see theorem 11.5.1). It seems that new methods will have to be developed in order to deal with the 'lower bound conjecture'. At present, it is one of the more challenging open problems about polytopes.

A different approach to problems related to the 'lower bound conjecture' for simplicial polytopes was recently discovered by Klee [21]. For convenience of exposition, we shall first reformulate part of the 'lower bound conjecture' in the dual form:

Does every simple d-polytope with n facets have at least $(d - 1)n - (d - 2)(d + 1)$ vertices?

Generalizing the concept of simple d-polytopes we shall say that a polyhedral set P of dimension d is a *d-polyhedron* provided P is line-free (i.e. vert $P \neq \emptyset$) and simple (i.e. each vertex of P is incident to exactly d facets of P).

Using this definition it is possible to inquire about the analogue of the 'lower bound conjecture' for d-polyhedra. Rather surprisingly, this problem is easier to solve than the original one; indeed, the result of Klee [21] is:

4. *Every d-polyhedron with n facets has at least $n - d + 1$ vertices.*

The existence of d-polyhedra having n facets and $n - d + 1$ vertices is easy to establish. The ingenious idea used by Klee [21] in the proof of theorem 4 is the reduction of the problem concerning d-polyhedra to a problem about graphs; in this setting, an inductive proof becomes easy since there is no need to worry about the polyhedral structure (which is much more intricate than the graph-structure).

It is to be expected that a similar technique may solve the analogous problems dealing with the minimal number of k-faces for $k \geq 1$. Conceivably, the proper approach to the 'lower bound conjecture' for simple polytopes will turn out to be via d-polyhedra having specified numbers of bounded and of unbounded facets.

10.3   The Sets $f(\mathscr{P}^3)$ and $f(\mathscr{P}_s^3)$

The aim of the present section is to give a complete description of the sets $f(\mathscr{P}^3)$ and $f(\mathscr{P}_s^3)$.

For simplicial 3-polytopes the Dehn–Sommerville equations reduce to $f_0 - f_1 + f_2 = 2$, $2f_1 = 3f_2$. These equations may be re-written in the form

(*) $f_1 = 3f_0 - 6 \qquad f_2 = 2f_0 - 4.$

Clearly, we have also

$$f_0 \geq 4.$$

This obviously proves one half of the following theorem:

1. $f(\mathscr{P}_s^3) = \{(f_0, 3f_0 - 6, 2f_0 - 4) \mid f_0 \geq 4\}.$

In order to complete the proof we have to show that for each $f_0 \geq 4$ there exists a $P \in \mathscr{P}_s^3$ with $f(P) = (f_0, 3f_0 - 6, 2f_0 - 4)$. If $f_0 = 4$ then the 3-simplex T^3 satisfies the conditions. We proceed by induction. For $f_0 > 4$ let $P^* \subset R^3$ be a simplicial 3-polytope satisfying

$$f(P^*) = (f_0 - 1, 3(f_0 - 1) - 6, 2(f_0 - 1) - 4).$$

Let $V \in R^3$ be a point which is not on any 2-dimensional plane determined by vertices of P^* and which is beyond exactly one 2-face of P^*. Then, by theorem 5.2.1, the simplicial 3-polytope $P = \operatorname{conv}(P^* \cup \{V\})$ satisfies $f(P) = (f_0, 3f_0 - 6, 2f_0 - 4)$.

This completes the proof of theorem 1.

By duality it follows from theorem 1 that the set of f-vectors of simple 3-polytopes is $\{(2f_2 - 4, 3f_2 - 6, f_2) \mid f_2 \geq 4\}.$

By theorem 10.1.1, the maximal number of k-faces, for a given number of vertices, is attained for simplicial polytopes. Therefore equations (*)

yield, for every $P \in \mathscr{P}^3$, the relations

$$f_1(P) \leq 3f_0(P) - 6 \qquad f_2(P) \leq 2f_0(P) - 4;$$

by duality, we also have

$$f_1(P) \leq 3f_2(P) - 6 \qquad f_0(P) \leq 2f_2(P) - 4.$$

Note that in each pair of inequalities, either one follows from the other and Euler's relation.

We have established one half of the following theorem, due to Steinitz [1]:

2. $f(\mathscr{P}^3) = \{(f_0, f_0 + f_2 - 2, f_2) \mid 4 \leq f_0(P) \leq 2f_2(P) - 4 \text{ and } 4 \leq f_2(P) \leq 2f_0(P) - 4\}.$

We shall prove the remaining part of theorem 2 on hand of Fig. 10.3.1; it is obvious that the proof could easily be formalized.

In figure 10.3.1, which represents the (f_0, f_2)-plane, each 3-polytope P is marked by a sign at the appropriate point $(f_0, f_2) = (f_0(P), f_2(P))$. The

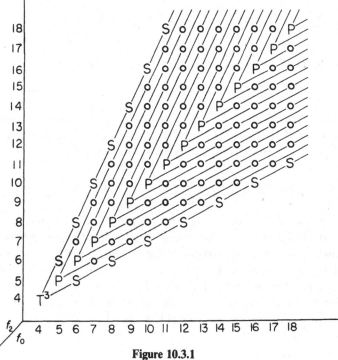

Figure 10.3.1

letter P corresponds to pyramids with an appropriate number of vertices, while S denotes a simplicial or a simple 3-polytope. Taking the convex hull of the union of a pyramid and a point which is beyond one of its triangular 2-faces only, we obtain a polytope with one vertex and two 2-faces more than the pyramid. This shows that there exist polytopes corresponding in the diagram to the lattice-points which are directly connected to the p's representing the pyramids and are in the upper half of the diagram. Since the resulting polytopes have, themselves, triangular 2-faces, the process may be continued indefinitely. By duality there follows the existence of polytopes corresponding to points in the lower half of the diagram. (They could be obtained also by 'cutting off' trivalent vertices.)

10.4   The Set $f(\mathscr{P}^4)$

It would be rather interesting to find a characterization of those lattice points in R^4 which are the f-vectors of 4-polytopes. This goal seems rather distant, however, in view of our inability to solve even such a small part of the problem as the lower bound conjecture for 4-polytopes.

In view of Euler's relation, the determination of $f(\mathscr{P}^4)$ is obviously equivalent to the determination of its projection onto any of the co-ordinate 3-spaces. Some information about $f(\mathscr{P}^4)$ is contained even in the projections of $f(\mathscr{P}^4)$ into the coordinate planes such as (f_0, f_1), etc. Some of those easier problems are completely solvable, and we shall now determine the sets $\prod_1 = \{(f_0, f_1) \mid (f_0, f_1, f_2, f_3) \in f(\mathscr{P}^4)\}$ and $\prod_3 = \{(f_0, f_3) \mid (f_0, f_1, f_2, f_3) \in f(\mathscr{P}^4)\}$.

The characterization of $\prod_3$ is quite easy.

1. *There exists a 4-polytope P with $f_0(P) = f_0$ and $f_3(P) = f_3$ if and only if the integers f_0 and f_3 satisfy $5 \leq f_0 \leq \frac{1}{2}f_3(f_3 - 3)$ and $5 \leq f_3 \leq \frac{1}{2}f_0(f_0 - 3)$.*

PROOF. The assertion that $f_0(P)$ and $f_3(P)$ satisfy the above relations whenever $P \in \mathscr{P}^4$ is quite obvious (see remark (2) on page 173). The existential part of the theorem shall be proved on hand of figure 10.4.1; as in section 10.3, the proof could easily be formalized. The symbols used in figure 10.4.1 have the following meaning:

T denotes the 4-simplex;

N denotes the cyclic polytope having the appropriate number of vertices;

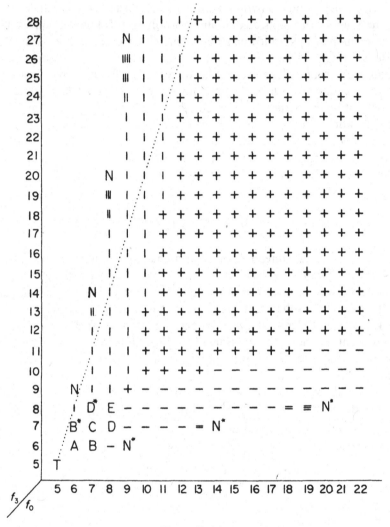

Figure 10.4.1

A, C, E denote the 2-fold 4-pyramids based on a quadrangle, pentagon, or hexagon;

B denotes the 4-pyramid based on the 3-prism with triangular base;

D denotes the 4-pyramid based on the 3-polytope a Schlegel diagram of which is shown in figure 10.4.2.

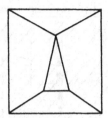

Figure 10.4.2

X^* denotes the dual of the polytope X.

| indicates a polytope P obtained from another polytope P' (which has one vertex and 3 facets less than P) as the convex hull of P' and a point which is beyond one and only one facet of P', the facet in question being a 3-simplex.

||, |||, ||||, $\cdots$ denotes a polytope P obtained from the cyclic polytope N with $f_0(N) = f_0(P) - 1$ as the convex hull of N and a point V, where V is beyond two, three, four, $\cdots$, facets of N taken from some set of $f_0(N) - 2$ facets of N having a common edge. (Compare exercise 4.8.22 and the construction used in the proof of theorem 7.2.4.)

$—, =, \equiv, \overline{\equiv}, \cdots$, indicate polytopes obtained in a fashion dual to |, ||, |||, ||||, $\cdots$.

$+$ indicates that the polytope is obtainable both by | and by $—$.

This completes the proof of theorem 1.

We turn now to the characterization of $\prod_1$.

2. *There exists a $P \in \mathscr{P}^4$ with $f_0(P) = f_0$ and $f_1(P) = f_1$ if and only if the integers f_0 and f_1 satisfy*

$$10 \le 2f_0 \le f_1 \le \tfrac{1}{2}f_0(f_0 - 1),$$

and (f_0, f_1) is not one of the pairs (6, 12), (7, 14), (8, 17), (10, 20).

PROOF The inequalities are obviously satisfied by $f_0(P)$ and $f_1(P)$ for every 4-polytope P. The existential part of the theorem follows from figure 10.4.3, by the same method as used in the proof of theorem 1; the notation is the same as in figure 10.4.1. In addition,

F denotes the 4-pyramid based on the 3-polytope a Schlegel diagram of which is shown in figure 10.4.4.

G denotes the dual of the simplicial polytope with $f_0 = 7, f_1 = 20$.

H denotes the polytope of exercise 4.8.6, the Schlegel diagram of which is shown in figure 10.4.5.

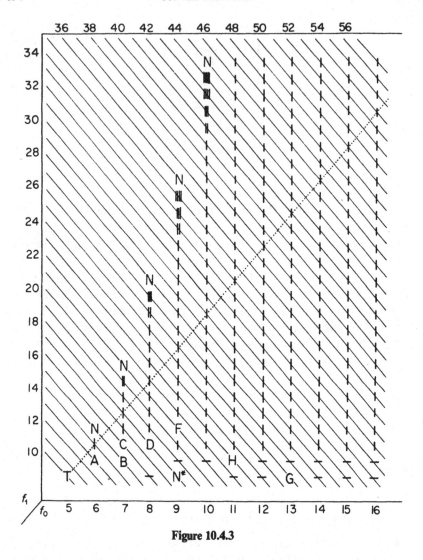

Figure 10.4.3

Note that, in figure 10.4.3, | indicates an increase of 4 edges and 1 vertex, while — indicates an increase of 3 vertices and 6 edges. In order to save space, an oblique system of coordinates (f_0, f_1) is used.

The most interesting feature of theorem 2 are the four exceptional pairs; we shall now prove that they are indeed exceptional. For the pairs (6, 12) and (7, 14) this follows from theorem 10.2.2. The other two cases are more complicated; we consider (10, 20) first.

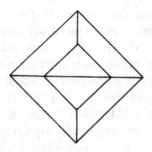

Figure 10.4.4

If there exists a 4-polytope P with $f_0(P) = 10$ and $f_1(P) = 20$, then

(i) $f_1 = 2f_0$ implies that P is simple;

(ii) by Euler's equation, $f(P) = (10, 20, n + 10, n)$ for some positive integer n.

Since $f_2 \geq 2f_3$, it follows that $n \leq 10$. But $n = 10$ is impossible, since it would imply that P is simplicial—contradicting the fact that only simplices are both simple and simplicial. Hence $n \leq 9$.

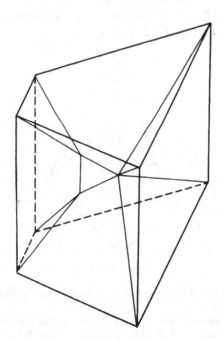

Figure 10.4.5

On the other hand, theorem 1 implies that $n \geq 7$†. However, $n = 7$ is impossible: by (i) the dual P^* of P is simplicial, and if $n = 7$ theorem 6.2.5 implies $f_0(P) = f_3(P^*) \geq 11$. Thus if P exists, we must have $n = 8$ or $n = 9$. Since P is simple, all its facets are simple and therefore have an even number of vertices. If some facet F would have 8 vertices, then $f(F) = (8, 12, 6)$; it follows that $n = 9$, P is a 4-pyramid based on F, and $f_0(P) = 9 \neq 10$. Hence all facets of P must be 3-simplices or 3-prisms based on triangles—but the only simple 4-polytopes with this property are the 4-simplex, and the 4-prism based on the 3-simplex, neither of which satisfies $(f_0, f_1) = (10, 20)$. Thus there exists no 4-polytope with $f_0 = 10$ and $f_1 = 20$.

Assuming next that P is a 4-polytope with $f_0(P) = 8$ and $f_1(P) = 17$, Euler's relation implies that $f(P) = (8, 17, n + 9, n)$ for some positive n. Since $f_2 \geq 2f_3$ it follows that $n \leq 9$. Also, P has either 6 vertices each of which is of valence 4 (i.e. incident with 4 edges) and two vertices of valence 5, or it has 7 vertices of valence 4 and one of valence 6.

If we had $n = 9$ then P would be simplicial. If AB is an edge of P such that both A and B have valence 4 (see the Schlegel diagram in figure 10.4.6) it follows that the omission of the vertices C, D, E disconnects the graph‡ of P between A,B and the remaining three vertices of P—in contradiction to theorem 11.3.2. Thus $n \leq 8$.

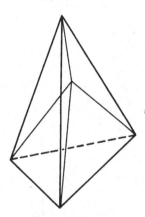

Figure 10.4.6

† It should be noted that if the lower bound conjecture were established, we could end the proof here by a reference to it.

‡ The reader is referred to section 11.3 for the terminology and facts used here and below. It is possible to avoid the use of theorem 11.3.2, but its application shortens the proof appreciably.

If we had $n = 8$, then $f(P) = (8, 17, 17, 8) = f(P^*)$, where P^* is the dual of P. Hence P^* also has 6 vertices of valence 4 and two of valence 5, or 7 of valence 4 and one of valence 6. In the first case, P has two facets F_1, F_2, each of which has five 2-faces; thus F_1, F_2 are either 3-pyramids with quadrangular bases, or 3-prisms with triangular bases; however, the latter type would require at least four such facets and thus is impossible. Hence F_1 and F_2 are 3-pyramids, and since the other facets of P are simplices, F_1 and F_2 must have the quadrangle as a common 2-face. But then the edges of P not contained in $F_1 \cup F_2$ are incident only to the apices of F_1 and F_2 (among the vertices of F_1 and F_2), again contradicting theorem 11.3.2.

Thus 7 facets of P are simplices, and by duality, seven of its vertices are of valence 4. Then the facet of P which is not a 3-simplex must be a 3-polytope F with six triangular 2-faces, and all but one vertex of F have valence 3. Since such F clearly does not exist, it follows that $n \leq 7$.

If we had $n = 7$, then $f(P) = (8, 17, 16, 7)$. Using Euler's relation for 3-polytopes and theorem 4.2.1, it is easily seen that neither P nor P^* are 4-pyramids. Thus facets of P^* have at most 5 vertices, hence each of them is either a 3-simplex, or a 3-bipyramid based on a triangle, or a 3-pyramid based on a quadrangle; a check of the valences of the vertices of P implies that P^* has either one bipyramidal facet, or two pyramidal ones, and it is easy to verify that neither alternative is realizable. Thus $n \leq 6$.

However, considering P^* (with $f(P^*) = (n, n + 9, 17, 8)$) we clearly have $f_1(P^*) = n + 9 \leq \frac{1}{2}n(n - 1)$, thus $n \geq 6$. Hence $n = 6$ and P^* is a neighborly polytope; therefore P^* is simplicial and $f_2(P^*) = 2f_3(P^*)$, contradicting $f_2(P^*) = 17$. Thus no 4-polytope has 8 vertices and 17 edges, and the proof of theorem 2 is complete.

10.5   Exercises

1. Show that $\varphi_k^*(v, d)$ is, for all d, v, and k, less than or equal to the bound given by the 'lower bound conjecture'. To this effect construct simplicial d-polytopes P_v^d with v vertices such that $f_k(P_v^d)$ equals to the value given by the 'lower bound conjecture' for all $v > d > k > 0$.

2. Construct d-polyhedra having n facets and $n - d + 1$ vertices.

3. Prove that the positions of the polytopes indicated in figures 10.4.1 and 10.4.3 by $=, \equiv, \cdots, \|, \|\|, \cdots$ are correct.

4. Prove that $\frac{1}{2}f_2(P) < f_1(P) < 2f_2(P)$ for every $P \in \mathscr{P}^4$, and that $c = 2$ is the smallest real number such that $f_1(P) \le cf_2(P)$ for every $P \in \mathscr{P}^4$.

5. It would be interesting (and probably not too hard) to determine the set $\prod_2 = \{(f_0(P), f_2(P)) \mid P \in \mathscr{P}^4\}$. The characterization of the set $\prod = \{(f_1(P), f_2(P)) \mid P \in \mathscr{P}^4\}$ seems to be more complicated; it may be conjectured that conv $\prod = \text{conv}\{(f_1(P), f_2(P)) \mid P$ or its dual are neighborly$\}$.

10.6   Additional notes and comments

For a recent survey on face numbers as well as on concepts extending f-vectors (such as "flag f-vectors" and the "cd-index") with a list of key open problems, we refer to Billera–Björner [a].

The upper bound theorem.
The "rather frustrating" upper bound problem discussed in section 10.1 was solved completely in 1970 by McMullen [c]; his proof made substantial use of h-vectors (see the notes in section 9.9) and of the shellability construction of Bruggesser–Mani [a] (see the notes in section 8.7). The upper bound theorem was later extended by Stanley [a] to general simplicial spheres, using algebraic techniques; here the key result is that the "face ring" (or "Stanley–Reisner ring") of a sphere is Cohen–Macaulay. For extensions to larger classes of simplicial complexes see Novik [a] and Hersh–Novik [a].

The lower bound theorem.
The lower bound problem discussed in section 10.2 ("one of the more challenging open problems") was solved by Barnette [b] [d], also in 1970, and he himself extended the solution to general simplicial manifolds [c]. For a connection to rigidity theory see Kalai [a]. See also Blind–Blind [c] and Tay [a].

The g-theorem.
In 1970, McMullen [d] formulated the "g-conjecture": a combinatorial characterization of the possible f-vectors of simplicial polytopes ("McMullen's conditions" on the f-vector). This was a rather daring conjecture, and its proof nine years later by Billera–Lee [a] and Stanley [b] [c] represents the most spectacular achievement of modern polytope theory.

A complete statement of the g-theorem, in Björner's [a] matrix formulation, is as follows: The f-vectors $f = (1, f_0, \ldots, f_{d-1})$ of simplicial d-polytopes are exactly the vectors of the form $g \cdot M_d$, where M_d is the non-negative matrix of size $([\frac{d}{2}] + 1) \times (d + 1)$ given by

$$M_d := \left(\binom{d+1-j}{d+1-k} - \binom{j}{d+1-k} \right)_{0 \le j \le d/2, \, 0 \le k \le d}$$

and $g = (g_0, \ldots, g_{[d/2]})$ is an *M-sequence*. This means that g is a non-negative integer vector that satisfies $g_0 = 1$ and $g_{k-1} \ge \partial^k(g_k)$ for $0 < k \le \frac{d}{2}$; here the *upper boundary operator* ∂^k is defined by

$$\partial^k(n) := \binom{a_k - 1}{k - 1} + \binom{a_{k-1} - 1}{k - 2} + \cdots + \binom{a_i - 1}{i - 1}$$

in terms of the unique binomial expansion of n as

$$n = \binom{a_k}{k} + \binom{a_{k-1}}{k-1} + \cdots + \binom{a_i}{i}, \qquad a_k > a_{k-1} > \cdots > a_i \geq i > 0.$$

For given P, the vector $g = g(P)$ is determined by the f- resp. h-vector as $g_k = h_k - h_{k-1}$ for $0 < k \leq \frac{d}{2}$, with $g_0 = 1$.

The g-theorem subsumes the upper bound theorem and the lower bound theorem. Another consequence, that $g_k \geq 0$ holds for all k, is known as the *generalized lower bound theorem*, as conjectured by McMullen–Walkup [a]. It is non-trivial; except for the case $k = 1$ (Kalai [a], Tay [a]) we have no elementary proof for this. The inequalities $g_k \geq 0$ are valid much more generally for every simplicial $(d-1)$-sphere that is a subcomplex of the boundary complex of a simplicial $(d+1)$-polytope: This was proved via algebraic shifting by Kalai [g], and via commutative algebra (Cohen–Macaulay rings) by Stanley [e]. See below for references about these topics.

It remains an open problem to characterize the simplicial d-polytopes with $g_k = 0$. McMullen–Walkup [a] conjecture that they are characterized by the property that they can be triangulated without introducing new faces of dimension less than $d - k$ (*k-stacked polytopes*). This part of McMullen–Walkup's "generalized lower bound conjecture" has been proved for some special cases (including simplicial d-polytopes with at most $d + 3$ vertices) by Lee [a], who established a very interesting interpretation of the g_k-coefficients in terms of winding numbers in the Gale-diagram. Additional connections that may turn out to be extremely useful were established by Welzl [b] and by McMullen [l].

See Björner [b] and Björner–Linusson [a] for further consequences of the g-theorem.

Beyond the g-theorem.

While the sufficiency proof for the g-theorem uses intricate combinatorics, the necessity proof by Stanley for the g-theorem relies on heavy algebraic geometry machinery, notably the hard Lefschetz theorem for a compact toric variety with only finite quotient singularities that may be associated with a rational simplicial polytope (see Ewald [a], Fulton [a]). There have been serious efforts to find a proof for the g-theorem that gets by without algebraic geometry, and to establish its validity beyond the realm of convex polytopes. While the first challenge has been met by McMullen [h] [i], the hope for the second has not yet materialized. In particular, it is not known whether the g-theorem generalizes to fans (that is, to star-shaped simplicial polytopes). McMullen's conditions might be true for all simplicial $(d-1)$-spheres (see McMullen [d, p. 569]); they do hold for $d \leq 5$ (Walkup [a]), and for simplicial spheres with at most $d + 3$ vertices (Mani [b]).

The connections between the combinatorics of rational polytopes (extremal properties of f-vectors), commutative algebra (via Stanley–Reisner rings), and algebraic geometry (the cohomology of the associated toric varieties) are surprising, but they are certainly not coincidental. The basic correspondences, as used to prove the upper bound theorem for spheres and the g-theorem, have been extended, and applied to a number of other questions involving f-vectors of polytopes and related objects; see, for instance, Stanley's [h] proof of the Bárány–Lovász and Björner lower bound conjectures on centrally-symmetric simple (or simplicial) polytopes. Such results are summarized in Stanley [g, Chap. III] and [i, Sect. 2]. A powerful related method is Kalai's theory of algebraic shifting; see Björner–Kalai [a] and Kalai [k].

Flag vectors.

For general d-polytopes, it has turned out to be very fruitful to study the flag f-vector, which counts for each $S \subseteq \{0, \ldots, d-1\}$ the number of chains $F_1 \subset \cdots \subset F_r$ of faces with $\{\dim F_1, \ldots, \dim F_r\} = S$. Similarly to the h-vectors of simplicial polytopes, *flag h-vectors* are defined from the flag f-vectors via a certain linear transformation. Bayer–Billera [a] stated the *generalized Dehn–Sommerville equations* and proved that they are necessary and sufficient to define the linear span of all flag h-vectors. (Kalai [c] gave an alternative sufficiency proof.) The dimension of the linear span of the flag h-vectors (flag f-vectors) is F_d, the d-th Fibonacci number ($F_0 = F_1 = 1$).

The *cd-index*—devised by J. Fine—is a linearly equivalent encoding of the flag f-vector by a vector of length F_d that is obtained exploiting the generalized Dehn–Sommerville equations. Both the flag h-vector of an arbitrary polytope and its cd-index are non-negative. See Bayer–Klapper [a] and Stanley [g, Sect. III.4]. For lower bounds on the components of the cd-index for odd-dimensional simplicial manifolds see Novik [b].

The *toric g-vector* is a generalization of the g-vector of a simplicial polytope to arbitrary polytopes along different lines. Stanley [d] showed that the toric g-vectors of polytopes that can be realized by rational coordinates are non-negative.

4-Polytopes.

The case of $d = 4$, as discussed in section 10.4, remains interesting, and may slowly get within reach. Here the set of f-vectors is 3-dimensional. The flag vectors form a 4-dimensional set, which is described in some detail in Bayer [a]; see also Höppner–Ziegler [a]. As suggested by exercise 10.5.5, the remaining 2-dimensional coordinate projections of the f-vectors to the coordinate planes were determined by Barnette–Reay [a] and Barnette [e].

However, the picture is still quite incomplete. In particular, we do not know whether the parameters of *fatness* $\phi(P) := (f_1 + f_2)/(f_0 + f_3)$ and *complexity* $\gamma(P) := f_{03}/(f_0 + f_3)$ are bounded for 4-polytopes. However, they satisfy $\phi(P) \le 2\gamma(P) - 2$ and $\gamma(P) \le 2\phi(P) - 2$. See Eppstein–Kuperberg–Ziegler [a] and Ziegler [d].

A complete answer is still not available even for special classes such as cubical 4-polytopes, or 2-simple 2-simplicial 4-polytopes. In both cases, the main obstacle seems to be a lack of versatile construction methods. The flag vectors for the second case are characterized by the "extremal" property that they satisfy the valid inequality $\phi(P) \le 2\gamma(P) - 2$ with equality.

Special lower bound problems.

Blind–Blind [a] proved that the d-cubes have componentwise minimal f-vectors among all d-polytopes without a triangle 2-face. Furthermore, if a triangle-free d-polytope for some k has the same number of k-faces as the d-cube, then it is combinatorially equivalent to the d-cube.

There is no such result for polytopes without triangles and quadrilaterals: according to Kalai [f], for $d \ge 5$ every d-polytope has a triangle or a quadrilateral 2-face—see the notes in section 11.6.

Among the 3-polytopes without triangles or quadrilaterals, the dodecahedron has componentwise minimal f-vector. The corresponding question for 4-polytopes is open.

CHAPTER 11

Properties of Boundary Complexes

In the present chapter we shall study in more detail the facial structure of polytopes by investigating some properties of the boundary complexes of polytopes, and some of their subcomplexes.

We recall from earlier sections some definitions, and we supplement them by additional ones.

A *complex* $\mathscr{C} = \{C\}$ is a finite collection of polytopes in R^n such that every face of a member of $\mathscr{C}$ is itself a member of $\mathscr{C}$, and the intersection of every two members of $\mathscr{C}$ is a face of both of them. A complex formed by all the (proper and improper) faces of a polytope P is denoted by $\mathscr{C}(P)$; the complex consisting of all the faces of P different from P is denoted by $\mathscr{B}(P)$ and called the *boundary complex* of P. For a complex $\mathscr{C}$ the *k-skeleton* $\mathrm{skel}_k \mathscr{C}$ is the complex consisting of all members of $\mathscr{C}$ which have dimension at most k. If $\mathscr{C} = \mathscr{C}(P)$ for some polytope P we shall write $\mathrm{skel}_k P$ for $\mathrm{skel}_k \mathscr{C}(P)$. Two complexes $\mathscr{C}_1$ and $\mathscr{C}_2$ are *combinatorially equivalent* (or *isomorphic*) provided there exists a biunique, incidence-preserving correspondence between the members of $\mathscr{C}_1$ and those of $\mathscr{C}_2$. For a complex $\mathscr{C}$ we shall denote by set $\mathscr{C}$ the set of all points of R^n belonging to at least one member of $\mathscr{C}$. A complex $\mathscr{C}$ is a *refinement* ('convex subdivision' in the terminology used by Lefschetz [1]) of a complex $\mathscr{K}$ if there exists a homeomorphism ψ of set $\mathscr{C}$ onto set $\mathscr{K}$ such that for every $K \in \mathscr{K}$ there exists a subcomplex $\mathscr{C}_K \subset \mathscr{C}$ satisfying $\psi^{-1}(K) = \mathrm{set}\ \mathscr{C}_K$. The homeomorphism ψ is called a *refinement map*.

For example, the complex $\mathscr{K}_1$ consisting of two triangles with a common edge is a refinement of the complex $\mathscr{K}_2$ consisting of one triangle. Note, however, that the 1-skeleton of $\mathscr{K}_1$ is not a refinement of the 1-skeleton of $\mathscr{K}_2$.

Clearly, if $\mathscr{K}_1$ is a refinement of $\mathscr{K}_2$, and if $\mathscr{K}_i'$ is a complex combinatorially equivalent to $\mathscr{K}_i, i = 1, 2$, then $\mathscr{K}_1'$ is a refinement of $\mathscr{K}_2'$.

If $\mathscr{P}$ is a family of polytopes we shall say that a *k-complex* $\mathscr{C}$ is $(\mathscr{P})$-*realizable* provided the *k-skeleton* $\mathrm{skel}_k P$ of some $P \in \mathscr{P}$ is combinatorially equivalent to $\mathscr{C}$.

The characterization of $(\mathscr{P}^2)$-realizable *k-complexes* is trivial. The only interesting case is $k = 1$, and $\mathscr{C}$ is $(\mathscr{P}^2)$-realizable if and only if $\mathscr{C}$ is a (simple) *circuit*, i.e. the members of $\mathscr{C}$ are $\varnothing$, the r distinct points V_1, $V_2, \cdots, V_r = V_0, r \geq 3$, and the r edges conv$\{V_{i-1}, V_i\}$, $i = 1, 2, \cdots, r$. The characterization of $(\mathscr{P}^3)$-realizable complexes may also be formulated in a rather simple manner; we shall consider it in chapter 13.

No characterization of $(\mathscr{P}^d)$-realizable complexes is known for $d \geq 4$. The present chapter is mainly devoted to an exposition of the known necessary conditions for $(\mathscr{P}^d)$-realizability of complexes. Those conditions may clearly be reformulated in such a manner as to state properties of polytopes, or of their boundary complexes.

11.1   Skeletons of Simplices Contained in $\mathscr{B}(P)$

We shall start with the following result of Grünbaum [15], which is a sharpening of theorem 3.2.4 (T^d denotes, as before, the d-simplex).

1. *For every d-polytope P the complex $\mathscr{B}(P)$ is a refinement of $\mathscr{B}(T^d)$.*

An equivalent formulation of theorem 1 is

1′. *For every d-polytope P the complex $\mathscr{C}(P)$ is a refinement of $\mathscr{C}(T^d)$.*

The proof of theorem 1 proceeds by induction on d; the case $d = 1$ being trivial, we shall assume $d \geq 2$. Let $V \in$ vert P and let H be a hyperplane which strictly separates V and conv(vert $P \sim \{V\}$). Then $P_0 = P \cap H$ is a $(d - 1)$-polytope, and by the inductive assumption $\mathscr{B}(P_0)$ is a refinement of $\mathscr{B}(T^{d-1})$, while $\mathscr{C}(P_0)$ is a refinement of $\mathscr{C}(T^{d-1})$. Mapping link$(V; \mathscr{B}(P))$ onto $\mathscr{B}(P_0)$ by rays issuing from V, it is clear that link$(V; \mathscr{B}(P))$ is a refinement of $\mathscr{B}(P_0)$ and therefore also of $\mathscr{B}(T^{d-1})$. The same radial mapping shows that st$(V; \mathscr{B}(P))$ is a refinement of st$(V^*; \mathscr{B}(T^d))$, where V^* is a vertex of T^d. The refinement map from set link$(V; \mathscr{B}(P))$ to set $\mathscr{B}(T^{d-1}) =$ rel bd T^{d-1} is obviously extendable to a homeomorphism from set ast$(V; \mathscr{B}(P))$ to set $\mathscr{C}(T^{d-1}) = T^{d-1}$. This shows that ast$(V; \mathscr{B}(P))$ is a refinement of $\mathscr{C}(T^{d-1})$ in such a fashion as to agree on the subcomplex link$(V; \mathscr{B}(P))$ of the complex ast$(V; \mathscr{B}(P))$ with the previously determined structure of link$(V; \mathscr{B}(P))$ as a refinement of the subcomplex $\mathscr{B}(T^{d-1})$ of $\mathscr{C}(T^{d-1})$. Since

$$\mathscr{B}(P) = \text{st}(V; \mathscr{B}(P)) \cup \text{ast}(V; \mathscr{B}(P)),$$

and

$$\mathscr{B}(T^d) = \text{st}(V^*, \mathscr{B}(T^d)) \cup \text{ast}(V^*; \mathscr{B}(T^d)),$$

and since ast$(V^*; \mathscr{B}(T^d)) = \mathscr{C}(T^{d-1})$, the above construction implies that $\mathscr{B}(P)$ is a refinement of $\mathscr{B}(T^d)$. This completes the proof of theorem 1.

As an immediate consequence of theorem 1 we have:

2. *For every d and k such that* $0 < k < d$, *each* $(\mathscr{P}^d)$-*realizable k-complex contains a refinement of the k-skeleton* skel$_k$ T^d *of* T^d.

In order to derive additional criteria for $(\mathscr{P}^d)$-realizability of k-complexes, we recall the following result of van Kampen [1] and Flores [1].

3. set skel$_k$ T^{2k+2} *is not homeomorphic with any subset of the 2k-dimensional space* R^{2k}.

The proof of theorem 3 is somewhat outside the domain of the present book. However, it seems that the published proofs of it are rather inaccessible. Therefore we shall present a proof of theorem 3, but for the sake of continuity we defer it to the next section.

Let now $a(d, k)$, for $1 \le k \le d$, denote the integer m such that set skel$_k$ T^d is homeomorphic to a subset of R^m but is not homeomorphic to any subset of R^{m-1}. Similarly, let $b(d, k)$ denote the m such that skel$_k$ T^d is combinatorially equivalent to a complex in R^m but is not combinatorially equivalent to any complex in R^{m-1}. Clearly $a(d, k) \le b(d, k)$ for all d and k. Using theorem 3, we shall obtain the following result of van Kampen [1] (see also Chrislock [1]).

4. *For all d and k*, $1 \le k \le d$, *we have*
  (i) $a(k, k) = b(k, k) = k$
 (ii) $a(k + 1, k) = b(k + 1, k) = k + 1$
(iii) $a(d, k) = b(d, k) = d - 1$ for $k + 2 \le d \le 2k + 2$
(iv) $a(d, k) = b(d, k) = 2k + 1$ for $d \ge 2k + 2$.

PROOF Clearly $a(k, k) = b(k, k) = k$ by the invariance of dimension; clearly, also $b(k + 1, k) \le k + 1$. On the other hand, set skel$_k$ T^{k+1} is homeomorphic to the k-sphere and therefore, by the Borsuk–Ulam theorem on antipodes (see Borsuk [1], Alexandroff–Hopf [1]), it is not homeomorphic to any subset of R^k. Thus $k + 1 \le a(k + 1, k)$, and (ii) is proved. For $d \ge k + 2$, the Schlegel diagram of T^d contains a subcomplex combinatorially equivalent to skel$_k$ T^d; therefore $b(d, k) \le d - 1$ for all $d \ge k + 2$. Applying theorem 3 we see that

$$2k + 1 \le a(2k + 2, k) \le b(2k + 2, k) \le 2k + 1;$$

this establishes the case $d = 2k + 2$ of (iii) and (iv). On the other hand, clearly $a(d, k) \le a(d + 1, k) \le a(d, k) + 1$ and, for $k + 2 \le d \le 2k + 2$, $2k + 1 = a(2k + 2, k) \le a(d, k) + (2k + 2 - d)$; hence

$$d - 1 \le a(d, k) \le b(d, k) \le d - 1,$$

and (iii) is proved. The Schlegel diagrams of cyclic $(2k + 2)$-polytopes with d vertices show that $b(d, k) \leq 2k + 1$ whenever $d \geq 2k + 2$. This completes the proof of (iv), and with it also the proof of theorem 4.

From theorem 4, using Schlegel diagrams, there follows

5. *If $P \in \mathscr{P}^d$ and if $\mathrm{skel}_k P$ contains a refinement of $\mathrm{skel}_k T^{d+1}$, then $k < [\frac{1}{2}d]$.*

This is a generalization of theorem 7.1.4. A useful reformulation of theorem 5 is

6. *If a k-complex $\mathscr{C}$ is $(\mathscr{P}^d)$-realizable, where $k \geq [\frac{1}{2}d]$, then $\mathscr{C}$ does not contain any refinement of $\mathrm{skel}_k T^{d+1}$.*

In analogy to the definitions of $a(d, k)$ and $b(d, k)$ we define, for each complex $\mathscr{C}$, two integers $a(\mathscr{C})$ and $b(\mathscr{C})$, where $a(\mathscr{C})$ is the least possible dimension of a Euclidean space which contains a subset homeomorphic with set $\mathscr{C}$, while $b(\mathscr{C})$ is the least possible dimension of a Euclidean space containing a complex combinatorially equivalent with $\mathscr{C}$.

Theorem 4 implies that $a(\mathscr{C}) = b(\mathscr{C})$ whenever $\mathscr{C}$ is a skeleton of some simplex. One could ask whether $a(\mathscr{C}) = b(\mathscr{C})$ for every complex $\mathscr{C}$, or at least for every simplicial complex $\mathscr{C}$. The answer is affirmative if $\mathscr{C}$ is a 1-complex. We shall obtain this result as a corollary of theorem 13.1.1.

However, an example of Cairns [1] may be modified to show:

7. *There exist simplicial 2-complexes $\mathscr{C}$ such that $a(\mathscr{C}) = 3$ while $b(\mathscr{C}) \geq 4$.*

Cairns' original example (a related example was given by van Kampen [2]) is a 3-complex which is homeomorphic to the 3-simplex T^3 (and is combinatorially equivalent to a simplicial subdivision of T^3 considered as a topological complex), but is not combinatorially equivalent to any complex in R^3. The variant of the example suitable for our purposes is as follows:

We start from the 2-complex $\mathscr{C}_0$ indicated in figure 11.1.1 by the heavy lines. $\mathscr{C}_0$ has 7 vertices, 15 edges, and 10 2-faces ($A_0 B_1 B_2$, $A_0 B_2 B_3$, $A_0 B_3 B_1$, $A_1 B_1 B_2$, $A_1 B_2 B_4$, $A_1 B_4 B_3$, $A_1 B_3 B_1$, $CB_2 B_3$, $CB_3 B_4$, $CB_4 B_2$), and it is obviously homeomorphic to the 2-sphere. We enlarge $\mathscr{C}_0$ to a 2-complex $\mathscr{C}_1$ by the addition of the edges CA_0, CA_1, and CB_1, and of the nine triangles determined by C and the nine edges of $\mathscr{C}_0$ incident to A_0, A_1, or B_1.

The second stage of the construction consists of the 2-complex $\mathscr{B}_0$ indicated in figure 11.1.2 by heavy lines. $\mathscr{B}_0$ is obviously combinatorially

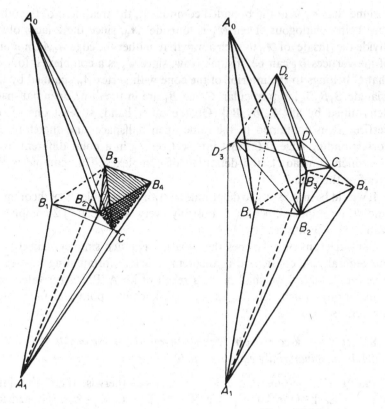

Figure 11.1.1   **Figure 11.1.2**

equivalent to $\mathscr{C}_0$, its vertices being denoted in the same way as the corresponding vertices of $\mathscr{C}_0$. We enlarge $\mathscr{B}_0$ to a 2-complex $\mathscr{B}_1$ by the addition, first, of the 7 edges A_0A_1, D_1D_2, D_2D_3, D_3D_1, CD_1, CD_2, CD_3, followed by the addition (not shown in the drawing) of the six triangles $D_1D_2B_3$, $D_2D_3B_1$, $D_3D_1B_2$, $B_1B_2D_3$, $B_2B_3D_1$, $B_3B_1D_2$, and the six edges B_iD_j, $1 \leq j, i \leq 3$, $i \neq j$. The idea behind this construction is to make the circuits $A_0A_1B_1$ and $D_1D_2D_3$ linked. Let $\mathscr{C}$ be the 2-complex obtained as the union of $\mathscr{B}_1$ and $\mathscr{C}_1$, with $\mathscr{B}_0$ and $\mathscr{C}_0$ identified. Clearly set $\mathscr{C}$ is homeomorphic to a subset of R^3. We shall show that $\mathscr{C}$ is not combinatorially equivalent to any complex $\mathscr{K}$ contained in R^3. Indeed, assuming that such a $\mathscr{K}$ exists, its subcomplex $\mathscr{K}_0$, which corresponds to $\mathscr{C}_0 = \mathscr{B}_0$, is homeomorphic to the 2-sphere. Therefore each of the subcomplexes $\mathscr{B}_1'$ and $\mathscr{C}_1'$ of $\mathscr{K}$, which correspond to $\mathscr{B}_1$ and $\mathscr{C}_1$, is contained in one of the connected components of the complement of set $\mathscr{K}_0$ in R^3. We shall assume·that $\mathscr{C}'_1$ is in the bounded component, the treatment of the other case being analogous. Then $\mathscr{B}'_1$ is 'outside' $\mathscr{X}_0$, since the 2-faces of $\mathscr{C}'_1$ divide the 'inside' of $\mathscr{X}_0$ in such a way that neither the edge A_0A_1, nor any of the vertices D_1 can be 'inside'. Now, since $\mathscr{C}'_1$ is a complex, it follows that C belongs to the interior of the cone with vertex A_0 spanned by the triangle $B_1B_2B_3$; in particular, C and B_1 are in the same open halfspace determined by aff$\{A_0, B_2, B_3\}$. On the other hand, at least one of the vertices D_1 is contained in the same open halfspace, and therefore the corresponding edge D_iC must intersect set $\mathscr{X}_0$ in a point different from C—which contradicts the definition of complexes. This completes the proof of theorem 7.

It would be interesting to determine $b(\mathscr{C})$ for the complex $\mathscr{C}$ of theorem 7; since $\mathscr{C}$ is simplicial, $b(\mathscr{C}) \leq 5$. Possibly every (simplicial ?) 2-complex $\mathscr{C}$ with $a(\mathscr{C}) = 3$ satisfies $b(\mathscr{C}) \leq 4$.

For skeletons of polytopes the situation is much simpler than it is in the general case considered in theorem 7. Before formulating this result (theorem 9 below) we shall prove a result of M. A. Perles (private communication), which generalizes the well-known topological imbedding theorem (see exercise 4.8.25).

8. *Let $\mathscr{C}$ be a k-complex in R^d; then there exists a k-complex $\mathscr{C}'$ in R^{2k+1} which is combinatorially equivalent to $\mathscr{C}$.*

PROOF   We assume that $d > 2k + 1$, since otherwise there is nothing to prove. Each of the flats $H_{ij} = $ aff$(C_i \cup C_j)$, for $C_i, C_j \in \mathscr{C}$, has dimension at most $2k + 1 \leq d - 1$, hence R^d contains a one-dimensional subspace L which is neither contained in nor parallel to any of the H_{ij}. Let H be a $(d - 1)$-dimensional subspace of R^d not containing L, and let π be the projection of R^d onto H parallel to L. Then the d-complex $\{\pi C \mid C \in \mathscr{C}\}$ is combinatorially equivalent to $\mathscr{C}$ and contained in R^{d-1}. An easy induction completes the proof. Incidentally, the proof shows that the combinatorial equivalence between $\mathscr{C}$ and $\mathscr{C}'$ is induced by a piecewise affine map.

Returning to skeletons of polytopes, we note that theorems 2 and 4 imply the relations $b(\text{skel}_k P) \geq a(\text{skel}_k P) \geq a(d, k) = b(d, k)$ for every d-polytope P. However, we have the stronger result

9. *If P is a d-polytope then*

$$b(\text{skel}_k P) = a(\text{skel}_k P) = a(d, k) = b(d, k).$$

PROOF   If $2k + 2 \leq d$, then the above remark implies $a(\text{skel}_k P) \geq a(d, k) = 2k + 1$ while theorem 8 implies $b(\text{skel}_k P) \leq 2k + 1$; hence the desired equation. If $k + 2 \leq d \leq 2k + 2$ then $\mathrm{skel}_k P$ contains a refinement of $\mathrm{skel}_k T^d$ and thus by theorem 4 $a(\mathrm{skel}_k P) \geq a(d, k) = d - 1$; but the Schlegel diagram of P shows that $b(\mathrm{skel}_k P) \leq d - 1$, and again the theorem follows. Finally, if $k = d - 1$, all the numbers considered are equal to d.

Exercises

1. Let P_1 and P_2 be d-polytopes in R^d, combinatorially equivalent under a mapping φ of $\mathscr{F}(P_1)$ onto $\mathscr{F}(P_2)$. Prove that there exists a piecewise affine homeomorphism A of R^d onto itself which induces φ (i.e. such that $A(F) = \varphi(F)$ for every $F \in \mathscr{F}(P_1)$).

2. Show that the maps in theorems 1 and 9 may be assumed to be piecewise affine.

3. Let P be a d-polytope and let $\varnothing \subset F^0 \subset F^1 \subset \cdots \subset F^{d-1} \subset F^d = P$ be a *tower* of faces of P (where F^k is a k-face; see exercise 3.1.15). Let T^d be a d-simplex with vertices $v_0, \cdots, v_d$, and let $T^k = \mathrm{conv}\{v_0, \cdots, v_k\}$. Then there exists a piecewise affine refinement map ψ from P to T^d such that

(i) $\psi(F^k) = T^k$ for each $k = 0, \cdots, d$;

(ii) $\psi^{-1}(\mathrm{conv}\{v_0, \cdots, v_{i-1}, v_i, v_j\})$ is a face of P for all i, j with $0 \leq i < j \leq d$.

4. Let P be a d-polytope, let $\varnothing \subset F^0 \subset F^1 \subset \cdots \subset F^{d-1} \subset F^d = P$ be a tower of faces of P, and let V be a vertex of P not contained in F^{d-1}. Prove the following result (the author is unable to do so for $d \geq 4$) which is a simultaneous generalization of theorem 1 and of theorem 11.3.2:

There exists a piecewise affine refinement map ψ of P onto the d-simplex T^d such that $\psi(V)$ and $\psi(F^i)$, $i = 0, 1, \cdots, d$, are faces of T^d.

5. Prove that if P is a centrally symmetric 3-polytope different from the octahedron, then $\mathscr{B}(P)$ is a refinement of the boundary complex of the 3-cube. Moreover, the refinement map may be chosen in such a way that symmetric points are mapped onto symmetric points, the map being piecewise affine.

It may be conjectured that for each d there exists a finite family $\{P_i \mid 1 \leq i \leq n(d)\}$ of centrally symmetric d-polytopes such that for each centrally symmetric d-polytope P the complex $\mathscr{B}(P)$ is a refinement of at least one of the complexes $\mathscr{B}(P_i)$, $1 \leq i \leq n(d)$. A similar result probably holds for polytopes invariant under any finite group of linear transformations.

It would also be interesting to investigate the conjecture that the boundary complex of every centrally symmetric simple d-polytope is a refinement of the boundary complex of the d-cube.

6. Let P be a d-polytope in R^d.

(i) If $d = 3$ and P is simple, prove that there exists $\varepsilon = \varepsilon(P) > 0$ with the following property: whenever $P' \subset R^3$ is at Hausdorff distance less than ε from P, then $\mathscr{B}(P')$ is a refinement of $\mathscr{B}(P)$.

(ii) Show that (i) does not hold if P is not simple, even if P' has only one vertex more than P.

It may be conjectured that (i) is true for all d.

7. Let $\mathscr{C}$ be a connected d-complex in R^n, $1 \leq d \leq n$. We call $\mathscr{C}$ *simple* if and only if each k-face, $0 \leq k \leq d$, of $\mathscr{C}$ is contained in precisely $d + 1 - k$ different d-faces of $\mathscr{C}$. Prove that a d-complex $\mathscr{C}$ is simple if and only if $\mathscr{C}$ is combinatorially equivalent to $\mathscr{C}(P)$ for some simple $(d + 1)$-polytope P. If $d \geq 1$, prove that each simple d-complex coincides with the boundary complex of a simple $(d + 1)$-polytope.

8. The following definition of an 'abstract (cell) complex', and the results mentioned below, are due to M. A. Perles (private communication). The definition is an attempt to achieve for complexes what the usual definition of abstract simplicial complexes does for simplicial complexes.

An *abstract d-complex* $\mathscr{F}$ is a finite lower semilattice (i.e. a partially ordered set, with greatest lower bounds for any set of elements) of height $d + 1$, with the property that for each $a \in \mathscr{F}$ there exists a polytope K such that the lattice $[0, a] = \{b \in \mathscr{F} \mid 0 \leq b \leq a\}$ is isomorphic to $\mathscr{F}(K)$.

A *realization* of an abstract d-complex $\mathscr{F}$ in R^n is a complex $\mathscr{K} = \{K_a \mid a \in \mathscr{F}\}$ in R^n, together with a function φ assigning to each $a \in \mathscr{F}$ a polytope $\varphi(a) = K_a$, such that

(i) $\varphi([0, a]) = \{K_b \mid b \in \mathscr{F}, 0 \leq b \leq a\} = \mathscr{F}(K_a)$ for each $a \in \mathscr{F}$.

(ii) $K_a \cap K_b = K_{a \wedge b}$ for all $a, b \in \mathscr{F}$.

Clearly, d-complexes, and abstract simplicial d-complexes, are abstract d-complexes. However, not every abstract d-complex has realizations by some d-complex in R^n. Show that the following are examples of abstract 2-complexes which are not realizable by 2-complexes in any Euclidean space. (Use exercise 7.)

(1) The boundary complex of the regular dodecahedron, in which diametrally opposite points are identified (i.e. the abstract 2-complex $\mathscr{C}_1$ obtained from figure 11.1.3 by identifying equally denoted points and segments). Similar examples of abstract d-complexes not realizable in any Euclidean space may be derived (by identifying diametrally opposite points) from the boundary complex of each centrally symmetric, simple $(d + 1)$-polytope P having the property that no facet of P meets two

opposite facets of P. (The abstract 2-complex $\mathscr{C}_1$ may clearly be realized by a '2-complex' in the projective plane.)

(2) The abstract 2-complex $\mathscr{C}_2$ obtained from figure 11.1.4 by identifying equally designated points and segments. $\mathscr{C}_2$ is, in an obvious meaning, homeomorphic to the torus.

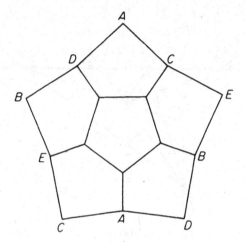

Figure 11.1.3

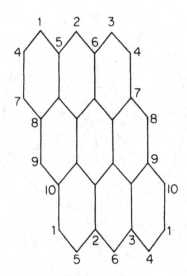

Figure 11.1.4

The next example uses the existence of projectively unique polytopes which have facets that are not projectively unique (see exercise 5.5.3). The method of construction, and the notation, are similar to those used in the proof of theorem 5.5.4.

Let K_1 be the 9-polytope with 13 vertices the Gale-diagram of which is the 13-tuple (see figure 11.1.5(a))

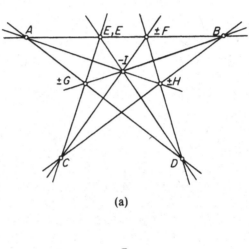

(a)

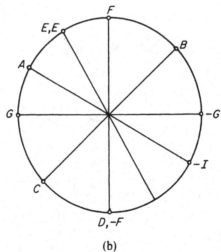

(b)

Figure 11.1.5

$$\left(\frac{A}{\|A\|}, \frac{B}{\|B\|}, \frac{C}{\|C\|}, \frac{D}{\|D\|}, \frac{\cdot E}{\|E\|}, \frac{E}{\|E\|}, \frac{F}{\|F\|}, \right.$$

$$\left. -\frac{F}{\|F\|}, \frac{G}{\|G\|}, -\frac{G}{\|G\|}, \frac{H}{\|H\|}, -\frac{H}{\|H\|}, -\frac{I}{\|I\|} \right)$$

and let F_1 be the facet of K_1 corresponding to the coface $(H, -H)$ (a Gale-diagram of F_1 is shown in figure 11.1.5(b)). Let K_2 be the 9-polytope with 13 vertices, the Gale-diagram of which is the 13-tuple (see figure 11.1.6(a))

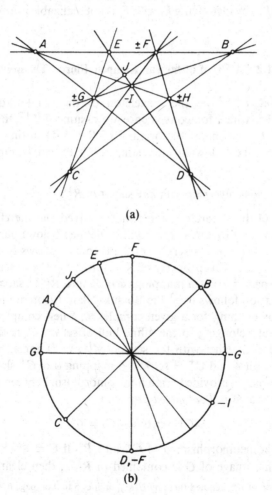

(a)

(b)

Figure 11.1.6

$$\left(\frac{A}{\|A\|}, \frac{B}{\|B\|}, \frac{C}{\|C\|}, \frac{D}{\|D\|}, \frac{E}{\|E\|}, \frac{F}{\|F\|}, -\frac{F}{\|F\|}, \right.$$

$$\left. \frac{G}{\|G\|}, -\frac{G}{\|G\|}, \frac{H}{\|H\|}, -\frac{H}{\|H\|}, -\frac{I}{\|I\|}, \frac{J}{\|J\|} \right)$$

and let F_2 be the facet of K_2 corresponding to the coface $(H, -H)$ (a Gale-diagram of F_2 is shown in figure 11.1.6(b)). Then both K_1 and K_2 are projectively unique, while F_1 and F_2 are combinatorially equivalent but not projectively equivalent. Thus the abstract 9-complex obtained from $\mathscr{F}(K_1) \cup \mathscr{F}(K_2)$ by identifying F_1 with F_2 is not realizable in any Euclidean space.

11.2   A Proof of the van Kampen–Flores Theorem

Let $\mathscr{K} = \mathrm{skel}_n T^{2n+2}$ be the n-skeleton of the $(2n + 2)$-dimensional simplex T^{2n+2}, where for convenience we assume T^{2n+2} to be regular and to have its centroid at the origin 0 of R^{2n+2}. Denoting $K = \mathrm{set}\, \mathscr{K}$ we shall prove the following interesting result of van Kampen [1] and Flores [1]:

1. *K is not homeomorphic with any subset of R^{2n}.*

The proof of this theorem is somewhat involved and therefore we first give an outline of it (the new terms will be defined below; our exposition follows Flores [1]*). The assertion of the theorem follows from $\mathscr{K}$ being $(2n + 1)$-entangled. This property of $\mathscr{K}$ results from the observation that for a metric space $\hat{K}$ certain mappings $\hat{\varphi}$ of $\hat{K}$ into R^{2n+1} satisfy $0 \in \hat{\varphi}(\hat{K})$. The last assertion follows from the Borsuk–Ulam theorem on antipodes.

We start by defining, for a given complex $\mathscr{C}$, a new complex $\mathscr{C}^+$. Let v be a point not belonging to the affine hull of set $\mathscr{C}$. $\mathscr{C}^+$ is the complex consisting of $\mathscr{C}$ together with the sets $\mathrm{conv}(\{v\} \cup F)$, for all members F of $\mathscr{C}$. Let $C = \mathrm{set}\, \mathscr{C}$ and $C^+ = \mathrm{set}\, \mathscr{C}^+$. A mapping φ of C^+ shall be called a *C-homeomorphism* provided φ is a homeomorphism between C and $\varphi(C)$. We shall say that $\mathscr{C}$ is *d-entangled* provided

$$\varphi(C) \cap \varphi(C^+ \sim C) \neq \varnothing$$

for every C-homeomorphism φ of C^+ into R^d. If $C \subset R^{d-1}$, or if some homeomorphic image of C is contained in R^{d-1}, then clearly $\mathscr{C}$ is not

---

* The definition of K^* (denoted there by $U(K_n)$) in lines 5 to 9 on page 6 of Flores [1] is not correct.

d-entangled. Therefore, in order to show that C is not homeomorphic to any subset of R^{d-1}, it is enough to prove that $\mathscr{C}$ is d-entangled.

We turn now to the definition of $\hat{K}$. The points of $\hat{K}$ are the (ordered) pairs $(a\,;b)$ such that

   (i) $a, b \in K^+$;

   (ii) at least one of a, b belongs to K;

   (iii) there exist disjoint members F_a, F_b of $\mathscr{K}$ such that $a \in \mathrm{conv}(\{v\} \cup F_a)$ and $b \in \mathrm{conv}(\{v\} \cup F_b)$. Defining the distance between $(a\,;b)$ and $(a'\,;b')$ as $\rho(a, a') + \rho(b, b')$, $\hat{K}$ becomes a compact metric space. Points $(a\,;b)$ and $(a'\,;b')$ of $\hat{K}$ are called *opposite* provided $a = b'$ and $b = a'$.

Let φ be a K-homeomorphism of K^+ into R^{2n+1}. For $(a\,;b) \in \hat{K}$ we define

$$\hat{\varphi}(a\,;b) = \varphi(b) - \varphi(a).$$

Clearly $\hat{\varphi}$ is a continuous mapping of $\hat{K}$ into R^{2n+1}, and $\hat{\varphi}(a\,;b) = -\hat{\varphi}(b\,;a)$. If we assume that $\mathscr{K}$ is not $(2n+1)$-entangled and that φ is a K-homeomorphism of K^+ into R^{2n+1} such that $\varphi(K) \cap \varphi(K^+ \sim K) = \varnothing$, it follows that $0 \notin \hat{\varphi}(\hat{K})$. Indeed, otherwise for some $(a\,;b) \in \hat{K}$ we would have $0 = \hat{\varphi}(a\,;b) = \varphi(b) - \varphi(a)$. Therefore $a \in K$ and $b \in K$, and since φ is a K-homeomorphism, $a = b$ contradicting condition (iii) in the definition of K. We shall prove that $0 \in \hat{\varphi}(\hat{K})$ for every φ; therefore $\mathscr{K}$ must be $(2n+1)$-entangled, and the theorem follows.

In order to achieve this, let $\mathscr{K}^*$ be the complex formed by all sets $\mathrm{conv}(F_i \cup (-F_j))$, where F_i and F_j are disjoint members of $\mathscr{K}$, and let $K^* = \mathrm{set}\ \mathscr{K}^*$. Then (see exercise 4.8.26) K^* is homeomorphic to the $(2n+1)$-sphere S^{2n+1} by the radial projection which carries opposite (i.e. symmetric with respect to 0) points of K^* onto antipodal points of S^{2n+1}. By the Borsuk–Ulam theorem on antipodes, every mapping of S^{2n+1} into R^{2n+1} carries some pair of antipodal points of S^{2n+1} onto the same point of R^{2n+1}. Therefore, some pair of opposite points of K^* are mapped onto the same point by every mapping of K^* into R^{2n+1}.

On the other hand, K^* is homeomorphic to $\hat{K}$ by a mapping π carrying opposite points onto opposite points. Indeed, each point of $\hat{K}$ is of the form $(\lambda a + (1-\lambda)v\,;\mu b + (1-\mu)v)$, where a and b are points of K contained in disjoint members of $\mathscr{K}$, $0 \le \lambda, \mu \le 1$, and $\max\{\lambda, \mu\} = 1$. Then it is easily checked that the following definition of π satisfies all the requirements:

$$\pi(\lambda a + (1-\lambda)v\,;\mu b + (1-\mu)v) = \begin{cases} (1 - \tfrac{1}{2}\mu)a - \tfrac{1}{2}\mu b & \text{for } \lambda = 1 \\ \tfrac{1}{2}\lambda a - (1 - \tfrac{1}{2}\lambda)b & \text{for } \mu = 1. \end{cases}$$

It follows that for every mapping ψ of K into R^{2n+1} there exists a pair of opposite points $(a_\psi; b_\psi)$ and $(b_\psi; a_\psi)$ of $\hat{K}$ such that $\psi(a_\psi; b_\psi) = \psi(b_\psi; a_\psi)$. In particular, this applies to the map $\hat{\varphi}$ for every K-homeomorphism φ of K^+ into R^{2n+1}. But $\hat{\varphi}$ satisfies $\hat{\varphi}(x; y) = -\hat{\varphi}(y; x)$ for every $(x; y) \in \hat{K}$. Therefore $\hat{\varphi}(a_\varphi; b_\varphi) = 0$ for every φ. This shows that $\mathcal{K}$ is $(2n + 1)$-entangled, and thus completes the proof of theorem 1.

11.3 d-Connectedness of the Graphs of d-Polytopes

We begin by recalling some definitions and facts from graph-theory.

By the usual definition, a graph $\mathcal{G}$ is a pair $(\mathcal{N}, \mathcal{E})$ consisting of a (finite) set $\mathcal{N} = \{N\}$ of *nodes* (or *vertices*) of $\mathcal{G}$, together with a subset $\mathcal{E}$ of the set $\{\{N_i, N_j\} \mid N_i, N_j \in \mathcal{N}\}$ of pairs of elements of $\mathcal{N}$. (According to the usual conventions of set-theory, this notation implies $N_i \neq N_j$.) The members $\{N_i, N_j\}$ of $\mathcal{E}$ are the *edges* of $\mathcal{G}$. The nodes N_i and N_j contained in an edge $E = \{N_i, N_j\}$ are called the *endpoints* of E, and are said to be *incident* with E, or *connected* by E, or *adjacent*. (Some authors find it more convenient to use 'oriented graphs', in which 'edges' are *ordered pairs* (N_i, N_j) of elements of $\mathcal{N}$, so that (N, N) is possible, and (N_i, N_j) is in general different from (N_j, N_i); others find it useful to allow 'multiple edges' having the same endpoints. We shall not use here any of these notions.)

Every 1-complex clearly determines a graph in the above sense. Conversely, for any given graph $\mathcal{G}$ there exists a 1-complex $\mathcal{C}$ which is combinatorially equivalent to $\mathcal{G}$ (i.e. there exists a biunique incidence-preserving correspondence between the members of $\mathcal{C}$ and those of $\mathcal{G}$). Indeed, such a 'realization' of $\mathcal{G}$ is given (in R^3) by an appropriate subcomplex of the Schlegel diagram of the cyclic polytope $C(v, 4)$, where $v = \text{card } \mathcal{N}$.

Thus 'graphs' and '1-complexes' are equivalent notions, and may be used interchangeably. We shall call the 1-skeleton of a complex $\mathcal{C}$ or $\mathcal{C}(P)$ the *graph* of $\mathcal{C}$ or of P; in the latter case we shall denote it by $\mathcal{G}(P)$.

A *path* $\mathcal{P}$ with endpoints A, B in a graph $\mathcal{G}$ is a subgraph of $\mathcal{G}$ having as nodes the nodes $V_0 = A, V_1, \cdots, V_{n-1}, V_n = B$ of $\mathcal{G}$ and as edges the edges $\{V_{i-1}, V_i\}$, $i = 1, \cdots, n$, of $\mathcal{G}$. n is called the *length* of $\mathcal{P}$. Two paths $\mathcal{P}_1$ and $\mathcal{P}_2$ with common endpoints A, B are called *disjoint* provided the intersection $\mathcal{P}_1 \cap \mathcal{P}_2$ consists of A and B only. A graph $\mathcal{G}$ is *connected* provided for each pair of its nodes there is a path in $\mathcal{G}$ having these nodes as endpoints.

A graph $\mathcal{G}$ is *k-connected* provided for every pair of nodes of $\mathcal{G}$ there exist k pairwise disjoint paths in $\mathcal{G}$ having these nodes as endpoints. The following theorem of Whitney is one of the fundamental results in graph theory; we only formulate it and refer the reader to the proofs in Whitney [1], Berge [1], Ore [1], Dirac [3]:

1. *A graph $\mathcal{G}$ with at least $k + 1$ nodes is k-connected if and only if every subgraph of $\mathcal{G}$, obtained by omitting from $\mathcal{G}$ any $k - 1$ or fewer nodes and the edges incident to them, is connected.*

A node V of a graph $\mathcal{G}$ is called *n-valent* provided V is incident to n edges of $\mathcal{G}$; $\mathcal{G}$ is called *n-valent* provided each of its nodes is *n*-valent. If $\mathcal{G}$ is a k-connected graph then the valence of each node of $\mathcal{G}$ is obviously at least k. The converse, however, does not hold.

Let $\mathcal{G}$ be the graph of a d-polytope P. Then clearly each node of $\mathcal{G}$ is at least d-valent. But in this particular case the following stronger result holds:

2. *If P is a d-polytope then $\mathcal{G}(P)$ is d-connected.*

This result is due to Balinski [1]; chronologically, it is the first (non-trivial) necessary condition for the $(\mathcal{P}^d)$-realizability of a graph.

The proof of theorem 2 uses Whitney's theorem in the following way. Let $V_1, \cdots, V_{d-1}$ be some $d - 1$ vertices of P and let $\mathcal{G}^*$ be the subgraph of $\mathcal{G}(P)$ obtained by the omission of $V_1, \cdots, V_{d-1}$ and of the edges incident with them. We shall prove theorem 2 by showing that $\mathcal{G}^*$ is connected. We have to distinguish a number of possibilities. Denoting $M = \text{aff} \{V_1, \cdots, V_{d-1}\}$, we have either (i) $M \cap \text{int } P = \emptyset$, or (ii) $M \cap \text{int } P \neq \emptyset$. In case (i) let $F = M \cap P$ be the face (of dimension at most $d - 1$) of P determined by $\{V_1, \cdots, V_{d-1}\}$, let H be a supporting $(d - 1)$-hyperplane of P such that $H \cap P = F$, and let H^+ be the other supporting hyperplane of P parallel to H. By exercise 3.1.4, for every vertex V of P either $V \in H^+$ or there exists a vertex V' of P, joined by an edge to V, such that V' is nearer to H^+ than V. It follows that each vertex of P which is not in $\{V_1, \cdots, V_{d-1}\}$, is connected by a path in $\mathcal{G}^*$ to some vertex of $H^+ \cap P$. Since $H^+ \cap P$ is a polytope, its graph is a connected subgraph of $\mathcal{G}^*$; hence $\mathcal{G}^*$ is connected in case (i).

In case (ii) let H be any $(d - 1)$-hyperplane containing M and at least one vertex V_d of P which is not in M. Denoting by H^+, H^- the two supporting hyperplanes of P parallel to H, the reasoning used in case (i) applies separately to the part of P contained in the slab determined by H and H^+, and to the part of P contained in the slab determined by H and H^-. Therefore, each of the corresponding subgraphs of $\mathcal{G}^*$ is connected, and since they have the common node V_d the graph $\mathcal{G}^*$ is connected. This completes the proof of theorem 2.

Considering only 1-complexes, the results on $(\mathscr{P}^d)$-realizability obtained so far may be formulated as follows.

If a graph $\mathscr{G}$ is $(\mathscr{P}^d)$-realizable then

(i) $\mathscr{G}$ contains a refinement of the complete graph with $d + 1$ nodes (i.e. of $\mathscr{G}(T^d)$); moreover, for each edge of $\mathscr{G}$ it is possible to find a refinement map carrying that edge onto an edge of $\mathscr{G}(T^d)$.

(ii) $\mathscr{G}$ is d-connected;

(iii) if $d = 3$ then $\mathscr{G}$ does not contain any refinement of the graph of T^4.

These conditions are not sufficient to guarantee the $(\mathscr{P}^d)$-realizability of a graph $\mathscr{G}$. Indeed, if $d = 3$ an example satisfying (i), (ii), (iii) but is not $(\mathscr{P}^3)$-realizable is provided by the graph $\mathscr{G}_0$ with 6 vertices $\{A_1, A_2, A_3, B_1, B_2, B_3\}$ and 9 edges $\{\{A_i, B_j\} \mid i, j = 1, 2, 3\}$. As we shall see in Chapter 13, for $d = 3$ conditions (ii), (iii) and (iv) are sufficient for the $(\mathscr{P}^3)$-realizability of $\mathscr{G}$ where (iv) is the condition:

(iv) $\mathscr{G}$ does not contain any refinement of $\mathscr{G}_0$.

It would be very interesting to find the higher-dimensional analogues of condition (iv).

For $d \geq 4$ the condition (iii) becomes void, and the remaining conditions (i) and (ii) are not sufficient for the $(\mathscr{P}^d)$-realizability of a graph $\mathscr{G}$. A simple example to that effect (see Grünbaum–Motzkin [2]) is that of a complete graph with 8 nodes from which seven edges forming a simple circuit are omitted. (See figure 11.3.1, where only the omitted edges are shown.) It is easily checked that this $\mathscr{G}$ satisfies (i) and (ii) with $d = 4$. Assuming $\mathscr{G}$ realized by a 4-polytope P, let F be any 3-face of P which does not contain V_8. Considering the graph $\mathscr{G}'$ with 7 nodes obtained from

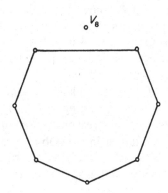

Figure 11.3.1

$\mathscr{G}$ by omitting V_8 and the edges incident to it, it is easily checked (by applying conditions (i), (ii), (iii) and (iv) for the case $d = 3$) that the graph of F is obtained from $\mathscr{G}'$ by omitting one of its nodes and the edges incident to it. Then F is necessarily of the combinatorial type represented by the Schlegel diagram in figure 11.3.2. Since F contains only 6 vertices, there must exist another 3-face F' of P which does not contain V_8. The face F' is necessarily of the same combinatorial type as F. But now, $F \cap F'$ contains five of the vertices $V_1, \cdots, V_7$—which is impossible since F has no pentagonal 2-face.

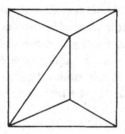

Figure 11.3.2

It should be mentioned that the conditions (i) and (ii) are independent. Indeed, the graphs of 3-dimensional antiprisms are 4-connected but contain no refinement of $\mathscr{G}(T^4)$; refinements of $\mathscr{G}(T^d)$ may obviously fail to be d-connected.

The following graph-theoretical problem arises in connection with conditions (i) and (ii):

What is the least $n = n(k)$ such that every n-connected graph contains as subgraph a refinement of $\mathscr{G}(T^k)$ (the complete graph with $k + 1$ nodes)?

It is easily seen that $n(k) = k$ for $k = 1, 2, 3$ (Dirac [1]). The following short proof of $n(3) = 3$ is due to J. Isbell. Let $\mathscr{G}$ be 3-connected and V a node of $\mathscr{G}$. Deleting V one obtains a 2-connected graph, which contains a circuit $V_0, V_1, \cdots, V_m = V_0$ (where $m \geq 3$). Let $\mathscr{G}^*$ be a graph obtained from $\mathscr{G}$ by adjoining a new node W and the edges $\{W, V_i\}$, $i = 1, \cdots, m$. Then $\mathscr{G}^*$ is clearly 3-connected, and therefore V may be connected to W by three pairwise disjoint paths. The circuit $V_0, \cdots, V_m$ and the parts of those paths between V and the circuit yield a refinement of $\mathscr{G}(T^3)$ contained in $\mathscr{G}$.

Thus, the first open problem is the determination of $n(4)$. The graph of the icosahedron shows that $n(4) > 5$. It is easy to show that $n(k) > k + 1$ for every $k \geq 4$ (Dirac [2]). A related result of Halin [1] is:

3. *The* 1-*skeleton of the d-octahedron (which is* $(2d - 2)$-*connected) contains a refinement of* $\mathcal{G}(T^k)$ *if and only if* $k < [\frac{3}{2}d]$.

It may be conjectured that the graph of every centrally symmetric simplicial *d*-polytope contains a refinement of the graph of the $([\frac{3}{2}d] - 1)$-simplex.

Halin's result clearly implies that $n(k) \geq [\frac{4}{3}k]$. It should be noted, however, that even the finiteness of $n(k)$, $k \geq 4$, has not been established so far.

Many generalizations of theorem 2 were obtained by T. Sallee [2]. Without proof we shall quote some of his results.

For a *d*-polytope *P*, and for $0 \leq r < s < d$, let $\mathcal{G}(P; r, s)$ be the (r, s)-*incidence graph* of *P*; that is, the nodes of $\mathcal{G}(P; r, s)$ correspond to the *r*-faces and the *s*-faces of *P*, two nodes determining an edge if and only if one of them corresponds to an *r*-face and the other to an *s*-face containing the *r*-face.

$\alpha(P; r, s)$ is the largest integer *k* such that $\mathcal{G}(P; r, s)$ has at least $k + 1$ *r*-nodes (nodes corresponding to *r*-faces), and the removal of any $k - 1$ nodes from $\mathcal{G}(P; r, s)$ does not disconnect $\mathcal{G}(P; r, s)$ between any two *r*-nodes. $\beta(P; r, s)$ is defined similarly, with '*s*-nodes' substituted for '*r*-nodes'. We also define

$$\alpha(d; r, s) = \min\{\alpha(P; r, s) \mid P \in \mathscr{P}^d\}$$

and

$$\beta(d; r, s) = \min\{\beta(P; r, s) \mid P \in \mathscr{P}^d\}.$$

The following results are among those proved in Sallee [2].

4. $\alpha(P; r, s) = \beta(P^*; d - 1 - s, d - 1 - r)$, *where* P^* *is the dual of* P^*.

5. $\beta(d; r, s) \leq \beta(d'; r, s)$ *for* $d \leq d'$.

6. $\max\left\{s + 1, \min\left\{d - r, \binom{s + 1}{r + 1}\right\}\right\} \leq \beta(d; r, s) \leq \min\left\{\binom{s + 1}{r + 1}, \binom{d}{s}\right\}$.

As a special case of theorem 6 we have $\beta(d; r, s) = s + 1$ if $r = 0$, or $s = d - 1$, or $s = r + 1$. Sallee [2] conjectured that $\beta(d; r, s) = \min\left\{\binom{s + 1}{r + 1}, \binom{d}{s}\right\}$ holds for all $0 \leq r < s < d$. This is confirmed, in part, by his result

7. $\beta(d; r, s) = \binom{s + 1}{r + 1}$ *if* $\binom{s}{r + 1} + r + 1 \leq d$.

11.4 Degree of Total Separability

The aim of the present section is to derive an additional necessary condition for the $(\mathscr{P}^d)$-realizability of graphs. It extends Balinski's theorem 11.3.2, and uses the notion of total separation of one set of nodes of a graph by another set of nodes. This notion, and theorem 1 below, are due to Klee [14].

Let $\mathscr{G}$ be a graph, M a set of nodes of $\mathscr{G}$, and A, B two nodes of $\mathscr{G}$ which do not belong to M. We shall say that M *separates* A from B (in $\mathscr{G}$) provided every path with endpoints A and B contains some node from M. A set N of nodes of $\mathscr{G}$ is said to be *totally separated* by a set M of nodes of $\mathscr{G}$ provided $M \cap N = \varnothing$ and every two members of N are separated by M. The nth *degree of total separability* of $\mathscr{G}$, denoted by $s_n(\mathscr{G})$, is defined as the largest cardinality of a set of nodes of $\mathscr{G}$ which are totally separated by some set of n nodes of $\mathscr{G}$.

For example, if $\mathscr{G}$ is a circuit then $s_n(\mathscr{G}) \leq n$, with equality holding provided $\mathscr{G}$ has at least $2n$ nodes.

The quantity interesting us is $s(n,d)$, defined by $s(n,d) = \max\{s_n(\text{skel}_1 P)| P$ is d-polytope$\}$.

We recall from chapter 10 that $\mu(n, d)$ denotes the maximal possible number of facets of a d-polytope with n vertices. We have seen in section 10.1 that $\mu(n, d) = f_{d-1}(C(n, d))$ for $n \leq d + 3$ and $n \geq [d/2]^2 - 1$; if the 'upper bound conjecture' is true, this relation holds for all n and d.

The interest in $s(n, d)$ stems from the fact that Klee succeeded with the help of this notion to disprove a conjecture of Grünbaum–Motzkin [3] (see section 12.2).

Klee's result [14] on total separability may be formulated as follows:

1. *For all n and d,*

$$s(n,d) = \begin{cases} 1 & \text{if } n \leq d - 1 \\ 2 & \text{if } n = d \\ \mu(n, d) & \text{if } n \geq d + 1 \end{cases}$$

The proof of theorem 1 for $n \leq d$ follows at once from theorem 11.3.2. Let us therefore assume that $n \geq d + 1$.

For any d-polytope P, let P^K denote the d-polytope obtained by adding sufficiently low pyramidal caps on all facets of P; such polytopes are known as Kleetopes. To be more precise, a Kleetope P^K is the convex hull of the union of P with $f_{d-1}(P)$ additional points, each of the 'new' points being beyond exactly one facet of P and sufficiently near to the centroid of the facet so that every segment determined by two of the 'new' points intersects into P. By theorem 5.2.1, each 'new' vertex is incident only to such edges of P^K which have as the other endpoint a vertex of P. Therefore, in the graph of P^K, the 'new' vertices are totally separated by the vertices of P. Choosing for P any d-polytope with n vertices and $\mu(n, d)$ facets, it follows that $s_n(\mathscr{G}(P^K)) \geq \mu(n, d)$; hence $s(n, d) \geq \mu(n, d)$.

In order to show that (for $n \geq d + 1$) the relation $s(n, d) \leq \mu(n, d)$ holds, we proceed as follows. Let P be a d-polytope such that $s_n(\mathscr{G}(P)) = s(n, d)$, and let V be a set of n vertices of P such that the set W of vertices of P not belonging to V contains an $s(n, d)$-membered subset which is totally separated by V in the graph of P. Let now P_0 be a d-polytope obtained from P by successively *pulling* each of the vertices in V (see Section 5.2), so as to form in the new positions a set V_1. By theorem 5.2.2 the edges of P_0 are of the two types:

(i) edges of P, having both endpoints in W;

(ii) edges having at least one endpoint in V_1.

Let $P_1 = \operatorname{conv} V_1$; since $n \geq d + 1$ it follows from the definition of pulling that P_1 is a d-polytope. For a facet F of P_1 let W_F be the set of all vertices of P_0 which are (in relation to P_1) beyond F. Then $W_F \subset W$, and using the method applied in the proof of theorem 11.3.2, it is immediate that no two points of W_F are separated (in the graph of P_0) by V_1. Since the construction of P_0 implies that two points of W are separated by V in the graph of P if and only if they are separated by V_1 in the graph of P_0, and since $W \subset \bigcup_F W_F$, it follows that $s(n, d) \leq s_n(\mathscr{G}(P_0)) \leq f_{d-1}(P_1) \leq \mu(n, d)$, as claimed.

This completes the proof of theorem 1.

A number of unsolved problems are related to theorem 1. For example if, in the definition of $s(n, d)$, the polytope P is assumed to belong to some subfamily $\mathscr{P}$ of $\mathscr{P}^d$, numbers $s(n, \mathscr{P})$ are obtained. From the proof of theorem 1 it follows at once that $s(n, \mathscr{P}^d_s) = s(n, d)$. However, if $\mathscr{P}$ is the family of all simple d-polytopes the value of $s(n, \mathscr{P})$ is not known; similarly, the problem is open in case $\mathscr{P}$ is the family of all centrally symmetric d-polytopes.

11.5 d-Diagrams

Sections 11.3 and 11.4 were concerned with the properties of 1-skeletons of d-polytopes. In the present section we shall consider the other extreme, namely the properties of $(d - 1)$-skeletons (i.e. boundary complexes) of d-polytopes. The most obvious property of the boundary complex

$\mathscr{B}(P)$ of every d-polytope P is that set $\mathscr{B}(P)$ is homeomorphic to the $(d-1)$-sphere S^{d-1}. As we have seen in theorem 11.1.1, this may be strengthened to the assertion that $\mathscr{B}(P)$ is a refinement of $\mathscr{B}(T^d)$. However, Cairns' example (see section 11.1) may easily be modified to show that there exists a (topological) simplicial subdivision of S^3 which is not combinatorially equivalent to the boundary complex of any 4-polytope. Therefore it would be desirable to find other properties of the boundary complexes, which do take into account the convexity of its elements.

One property of the boundary complex of each d-polytope which could be considered in this context is its representability by a Schlegel diagram in R^{d-1}. This brings us to the idea of trying to define, independently of Schlegel diagrams, objects which 'look like' Schlegel diagrams and to use them in order to determine additional properties of the boundary complexes. In section 3.3 we defined objects of this type which we called d-diagrams. As we shall see in section 13.2, 2-diagrams are indeed combinatorially equivalent to boundary complexes of 3-polytopes. However, in higher dimensions the situation is different and we have

1. *There exist simplicial 3-diagrams which are not combinatorially equivalent to Schlegel diagrams of 4-polytopes.*

Theorem 1 may be proved by various examples. We shall describe here the (chronologically first) example found in 1965 (Grünbaum [16]); though it contains an unnecessarily large number of simplices, it has the advantage over the smaller known example (see below) of being describable easily and without computations.

We begin by describing the construction of a 3-diagram $\mathscr{D}$, which will then be shown to have the properties required by theorem 1.

The construction starts with the complex shown in figure 11.5.1 which contains the following 3-simplices (denoted by their vertices): *ABGH, GPQV, GRSV, GPTV, GSTV, GHPT, GHST.* (In order to allow a clearer representation of the other simplices, *ABGH* is in figure 11.5.1 contracted to half its length in direction *AB*.) Adjoining the simplices *AGPQ, AGHP, BGRS, BGHS*, and their faces, a complex $\mathscr{C}_1$ is obtained (figure 11.5.2). Four copies of $\mathscr{C}_1$, joined along the common edge *AB*, form a new complex $\mathscr{C}_2$ (figure 11.5.3). Using three copies of $\mathscr{C}_2$ we form $\mathscr{C}_3$ (figure 11.5.4). Completing $\mathscr{C}_3$ by two copies of the simplicial decomposition of a 'small' cube (figure 11.5.5) we obtain a simplicial complex $\mathscr{C}_4$. The outward appearance of $\mathscr{C}_4$ is that of a cube with square pyramids attached on parts of each of its 2-faces; however, the structure of $\mathscr{C}_4$ is different since among its members are the edges *AB, CD,* and *EF.* Adding simplices having one

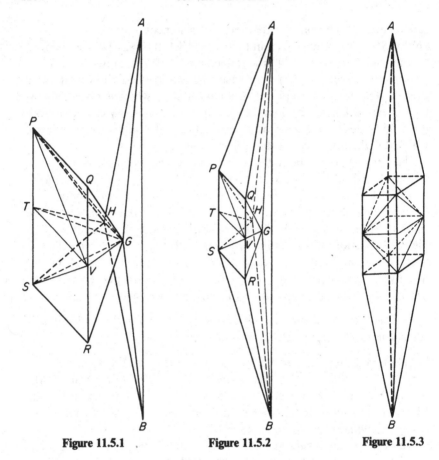

Figure 11.5.1   Figure 11.5.2   Figure 11.5.3

vertex in the set A, B, C, D, E, F, and base on the face of the 'big' cube near to that vertex, we obtain the complex $\mathscr{C}_5$ (figure 11.5.6). Adjoining to $\mathscr{C}_5$ the 24 simplices of the type AEV_1W_1, and the 8 simplices of the type $ACEV_1$, a complex $\mathscr{C}_6$ is obtained (figure 11.5.7). Applying a projective transformation to $\mathscr{C}_6$ it is transformed to the shape indicated in figure 11.5.8, in which form we may complete it to a 3-diagram $\mathscr{D}$, for which $\mathscr{C}$ is the union of $\mathscr{C}_6$ and the simplices containing 0. The detailed structure of $\mathscr{D}$ is of interest only to the extent that it shows that $\mathscr{D}$ contains: (i) the simplices $OACE, OAED, OADF, OAFC, OBCF, OBDE, OBEC, OBFD$, and their faces; (ii) the edges AB, CD, EF.

Let us now assume that there exists a 4-dimensional polytope P such that $\mathscr{D}$ is combinatorially equivalent to the Schlegel diagram of P; we assume the vertices of P labeled by the same letters as the corresponding vertices of $\mathscr{D}$. Then P contains as faces the simplices mentioned in (i) and (ii) above. Let P^* be the convex hull of the vertices 0, A, B, C, D, E, F of P. Then P^* has, obviously, 7 vertices and since it contains as faces all the 21 edges determined by its vertices, it is a neighborly polytope.

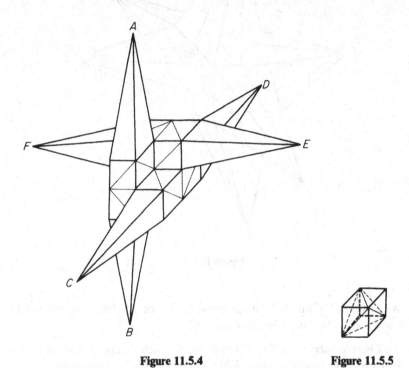

Figure 11.5.4   Figure 11.5.5

According to a theorem of Gale [4] (see theorem 7.2.3), the combinatorial structure of 4-dimensional neighborly polytopes with 7 vertices is completely determined and coincides with that of the cyclic polytope $C = C$ (7, 4). Using Gale's evenness condition (see section 4.7) it is easily seen that each vertex of C is incident to edges which are incident to three 2-faces, as well as to edges which are incident to four, and to five, 2-faces. (Compare exercise 4.8.22.) This completes the proof of theorem 1, since, in $\mathscr{D}$, the vertex 0 is contained only in faces the vertices of which belong to the set $\{0, A, B, C, D, E, F\}$ and are therefore faces of P and of P^*— but each edge of $\mathscr{D}$ incident with 0 is incident with four triangles of $\mathscr{D}$.

This completes the proof of theorem 1.

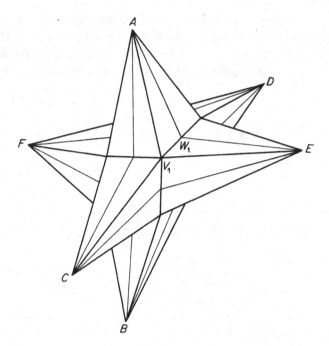

Figure 11.5.6

A considerably smaller 3-diagram which is not a Schlegel diagram was found recently (Grünbaum–Sreedharan [1]):

2. *There exists a simplicial 3-diagram $\mathscr{D}'$ with 8 vertices, which is not combinatorially equivalent to a Schlegel diagram of any 4-polytope.*

Denoting the vertices of $\mathscr{D}'$ by A, B, C, D, E, F, G, H, let the basis of $\mathscr{D}'$ be the 3-simplex $ABCD$, and let $\mathscr{D}'$ contain the following 19 additional simplices: $ABCF, ABDF, ACDG, ACFH, ACGH, ADEF, ADEG, AEFH, AEGH, BCDH, BCEF, BCEH, BDFG, BDGH, BEFG, BEGH, CDGH, CEFH, DEFG$. This diagram is indeed realizable in R^3, for example by choosing as vertices the points having the following cartesian coordinates:

$$A = (0, 215, 0) \qquad E = (82, 55, 114)$$
$$B = (0, 0, 0) \qquad F = (4, 79, 7)$$
$$C = (279, 0, 0) \qquad G = (94, 44, 142)$$
$$D = (0, 0, 333) \qquad H = (117, 47, 98).$$

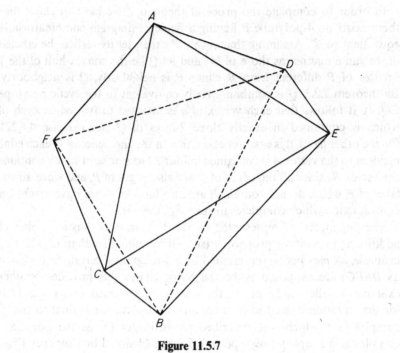

Figure 11.5.7

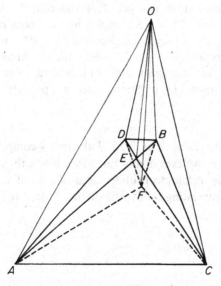

Figure 11.5.8

In order to complete the proof of theorem 2 we have to show that there exists no 4-polytope P having a Schlegel diagram combinatorially equivalent to $\mathscr{D}'$. Assuming that such a P exists, let its vertices be labeled in the same manner as those of $\mathscr{D}'$, and let Q be the convex hull of the 7 vertices of P different from B. Since P is neighborly, Q is neighborly. By theorem 7.2.3, Q is combinatorially equivalent to the cyclic polytope $C(7, 4)$. It follows that each vertex of q is incident to two edges each of which is contained in exactly three 2-faces of Q (see exercise 4.8.22). On the other hand, it is easily checked that in $\mathscr{D}'$, and hence in P, each edge incident to the vertex A is contained in four 2-faces none of which contains the vertex B; since all the edges of Q are also edges of P, and since all the faces of P which do not contain B are also faces of Q, we have reached a contradiction which completes the proof of theorem 2.

The 3-complex $\mathscr{M}$ represented by the 3-diagram $\mathscr{D}'$ has a number of additional interesting properties (see Grünbaum–Sreedharan [1]). For example, $\mathscr{M}$ *may not* be represented by a 3-diagram if certain facets (such as $DEFG$) are supposed to be the basis. Hence this provides another example, smaller and simpler than the one discussed in section 11.1, for the situation described in theorem 11.1.7. Also, $\mathscr{M}$ is dual to the 3-complex $\mathscr{M}^*$ which was described by Brückner [3] as the boundary complex of a simple 4-polytope with 8 facets (denoted by Brückner P^8_{39}); however, such a 4-polytope does not exist (since its dual would be the 4-polytope P discussed in the proof of theorem 2). Moreover, even Brückner's claim that $\mathscr{M}^*$ is a 3-complex has not been corroborated so far; as a matter of fact, it may be conjectured that $\mathscr{M}^*$ is not a 3-complex, or at least is not representable by a 3-diagram. (It should be noted that $\mathscr{M}^*$, as well as $\mathscr{M}$, are representable—with arbitrary facets as basis—by 'topological 3-diagrams', i.e. if 'curved faces' are permitted.)

Exercises

1. Using the 3-diagrams $\mathscr{D}$ or $\mathscr{D}'$, find abstract 4-complexes which may not be realized by complexes in any Euclidean space, though they may be realized by suitable topological subdivisions of a 4-cell in R^4.

2. It would be interesting to investigate the conjecture that $\mathscr{M}^*$ is not a 3-complex.

11.6 Additional notes and comments

The van Kampen–Flores theorem.
This is a fundamental topological result. It may elegantly be derived from the Borsuk–Ulam theorem via "deleted joins". We refer to Matoušek [a] for a modern treatment that also gives references to the extensive literature in connection with this topic as well as various interesting applications.

Embedding 2-complexes.
Examples of 2-dimensional simplicial complexes $\mathscr{C}$ that have an embedding into $\mathbb{R}^3$ (that is, with $a(\mathscr{C}) = 3$), but without a straight embedding (that is, with $b(\mathscr{C}) \geq 4$) may be found among the triangulated Möbius strips of Brehm [a]. The problem whether all triangulated 2-tori have straight, intersection-free embeddings into R^3 is still unsolved. Brehm–Schild [a] have shown that all tori have such embeddings into R^4. On the other hand, there is an orientable triangulated 2-manifold of genus 6 on 12 vertices that has no straight embedding into R^3, according to Bokowski–Guedes de Oliveira [a]. This provides a negative answer to the conjecture of exercise 13.2.3. See also the interesting necessary conditions for linear realizability by Novik [c].

Embedding higher-dimensional complexes.
Many embedding problems get easier in higher dimensions. So, while the problem of embeddability of d-dimensional simplicial complexes into R^{2d} is intricate for $d \leq 2$, for $d \geq 3$ there are necessary and sufficient conditions for embeddability of d-complexes in R^{2d} that were conjectured by van Kampen and Flores, and proved by Shapiro [a] and by Wu [a].

An extension of Balinski's theorem.
A "directed" version of Balinski's theorem was given by Holt–Klee [c]: If P is a d-polytope and φ is a linear function in general position, then there are d vertex-disjoint φ-monotone paths from the minimal vertex to the maximal vertex of P. (Using projective transformations, one can easily derive Balinski's theorem from this.)

Non-polytopal diagrams and spheres.
While the existence of non-polytopal diagrams was a substantial new fact for Grünbaum's book, we now have many examples, and we may assume that in some sense "most" diagrams are not polytopal. The simplest examples are the Brückner sphere and the Barnette sphere; they both appear with short proofs in Ewald [a, Sect. III.4]. Mihalisin–Williams [a] show that each of the two

non-polytopal simplicial 3-spheres with 8 vertices can be realized in 4-space (without further subdivision) as the boundary complex of a 4-cell with nice geometric properties. In one case, the 4-cell may even be star-shaped.

Whitney's/Menger's theorem.
Whitney's theorem (page 213) is commonly attributed to Menger [a]; see e. g. Diestel [a, Sect. 3.3]. This theorem and its variants (for edge-connectivity, directed versions, etc.) are fundamental results for graph theory and also for combinatorial optimization (network flows, multi-commodity flows, etc.).

High connectivity implies subdivisions of large complete graphs.
The parameter $n(k)$ is defined on page 215 as the smallest n such that every n-connected graph contains a subdivision of $K_{k+1} = \mathcal{G}(T^k)$. It is now known that this parameter is finite for all k: Indeed, every n-connected graph has minimal degree at least n, and it is a theorem of Bollobás–Thomason and of Komlós–Szemerédi [a] (see Diestel [a, Thm. 8.1.1]) that every graph of average degree at least ck^2 contains a subdivision of K_{k+1}, for some constant $c > 0$.
   In particular, we know that $n(5) = 6$. Indeed, any 6-connected graph on N vertices has at least $3N$ edges, and thus by a deep theorem of Mader [a] it contains a subdivision of K_5; see Diestel [a, Sect. 8.3].

Unavoidable small faces in the low-dimensional skeleton.
Kalai [f] showed that every d-polytope, $d \geq 5$, has a triangle or quadrilateral 2-face. Further, he conjectured that for each dimension k there exists a finite list $L(k)$ of k-polytopes and a dimension $d(k) > k$ such that each $d(k)$-polytope has a k-face combinatorially equivalent to some member of $L(k)$. He even suggests that it may suffice for $L(k)$ to consist of merely the k-simplex and the k-cube. Some results in this direction have been obtained by the use of FLAGTOOL, a computer program by G. Meisinger. For example, there is a finite list of 3-polytopes such that every *rational* 9-polytope has a 3-face in the list. (See Kalai–Kleinschmidt–Meisinger [a] and Meisinger–Kleinschmidt–Kalai [a].)

Minimal centrally symmetric polytopes.
Exercise 11.1.5 may be related to the conjecture by Kalai [e] that if a d-polytope is centrally symmetric, then it must have at least 3^d non-empty faces.
   Kalai conjectures that equality is achieved only by the *Hanner polytopes* that can be generated from an interval by taking products and dualization (which includes the cubes and the cross polytopes). One may speculate that this also provides the finite family needed for exercise 11.1.5.

CHAPTER 12

k-Equivalence of Polytopes

In chapter 11 we have investigated some properties of complexes realizable by skeletons of polytopes. The present chapter complements this by discussing the known results on the uniqueness of such realizations.

12.1 k-Equivalence and Ambiguity

From different points of view, one special case of $(\mathscr{P}^d)$-realizability of complexes is of particular interest. This is the question whether the k-skeleton of a polytope $P \in \mathscr{P}^d$ is combinatorially equivalent to the k-skeleton of a polytope P', where P' is not combinatorially equivalent to P. The simplest occurrence of this situation is provided by the fact that the cyclic d-polytopes $C(v, d)$ are $[\frac{1}{2}d]$-neighborly. This implies that for $2k + 2 \leq d$ the k-skeleton of $C(d + 1, 2k + 2)$ is combinatorially equivalent to the k-skeleton of T^d.

It is convenient to introduce the notion of k-equivalence of polytopes. Two polytopes P and P' are said to be k-*equivalent* provided $\text{skel}_k P$ is combinatorially equivalent to $\text{skel}_k P'$.

Obviously, d-equivalence, or $(d - 1)$-equivalence, of d-polytopes P and P' means the same as their combinatorial equivalence.

Generalizing the terminology introduced in Grünbaum–Motzkin [2] we shall say that a k-complex $\mathscr{C}$ is *dimensionally ambiguous* provided there exist polytopes P and P' with $\dim P \neq \dim P'$, such that $\mathscr{C}$, $\text{skel}_k P$ and $\text{skel}_k P'$ are combinatorially equivalent. As mentioned above, the complete k-complex with $d + 1$ vertices, $d > 2k + 2$, is dimensionally ambiguous since it is realizable by the k-skeletons of both T^d and $C(d + 1, 2k + 2)$. On the other hand, by theorem 7.1.4, the k-skeleton of T^d, $k \leq d \leq 2k + 2$, is not dimensionally ambiguous.

We shall discuss two additional notions.

A k-complex $\mathscr{C}$ is called *strongly d-ambiguous* provided there exist two d-polytopes P and P', not combinatorially equivalent, such that $\mathscr{C}$, $\text{skel}_k P$, and $\text{skel}_k P'$ are combinatorially equivalent.

For· example, by theorem 7.2.4, the complete graph with 8 nodes (i.e. $\text{skel}_1 T^7$) is strongly 4-ambiguous. On the other hand, theorem 7.2.3 implies that $\text{skel}_1 T^6$ is not strongly 4-ambiguous, although it is strongly 5-ambiguous (since it is combinatorially equivalent both to $\text{skel}_1 C(7, 5)$ and to the 1-skeleton of the 5-pyramid having as basis $C(6, 4)$).

A k-complex $\mathscr{C}$ is called *weakly d-ambiguous* provided there exists a d-polytope P and two combinatorial equivalences φ and ψ from $\mathscr{C}$ to $\text{skel}_k P$ such that the combinatorial equivalence $\varphi\psi^{-1}$ of $\text{skel}_k P$ to itself can not be extended to a combinatorial equivalence of $\mathscr{C}(P)$ to itself.

For example, if $\mathscr{C}$ is the complete graph with 6 nodes $V_1, \cdots, V_6$, $\mathscr{C}$ is combinatorially equivalent to $\text{skel}_1 C(6, 4)$ in essentially different ways. If φ maps the V_i's to six points on the moment curve in the natural order, and if ψ differs from φ by making V_2 correspond to the first and V_1 to the second of those six points on the moment curve, then the induced automorphism $\varphi\psi^{-1}$ of $\text{skel}_1 C(6, 4)$ is a 1-equivalence which is not extendable to an automorphism of $\mathscr{C}(C(6, 4))$. Indeed, the vertices of $C(6, 4)$ corresponding to V_2, V_3, V_4, V_5 under φ determine a facet of $C(6, 4)$, but the vertices $\psi(V_2), \psi(V_3), \psi(V_4), \psi(V_5)$ do not determine a facet of $C(6, 4)$.

The interest in k-equivalence and the different ambiguity concepts arose from the observation of dimensional ambiguity of the 1-skeletons of d-simplices, $d \geq 5$. Lack of examples to the contrary led Grünbaum–Motzkin [3] to the conjecture that for $d \geq 4$ every ($\mathscr{P}^d$)-realizable graph is ($\mathscr{P}^4$)-realizable. This conjecture has since been disproved (first by Klee [14] in 1963) but the following variant is still undecided.

Conjecture. If the k-complex $\mathscr{C}$ is ($\mathscr{P}^{d'}$)-realizable and ($\mathscr{P}^{d''}$)-realizable, where $d' \leq d''$, then $\mathscr{C}$ is ($\mathscr{P}^d$)-realizable for every d satisfying $d' \leq d \leq d''$.

The remaining sections of the present chapter are devoted to a discussion of the known partial results on ambiguity and k-equivalence.

12.2 Dimensional Ambiguity

One of the general results on dimensionally unambiguous complexes is the following immediate consequence of the results of section 11.1.

1. *If $k \geq [\frac{1}{2}d]$ and if P is a d-polytope, then $\text{skel}_k P$ is not dimensionally ambiguous.*

Indeed, the k-skeleton of each d-polytope P contains a refinement of $\text{skel}_k T^d$ and, by theorem 11.1.6, $\text{skel}_k P$ does not contain a refinement of $\text{skel}_k (T^{d+1})$ for $k \geq [\frac{1}{2}d]$. The first fact implies (again by theorem 11.1.6) that $\mathrm{skel}_k P$ is not ($\mathscr{P}^{d'}$)-realizable for $d' < d$, while the second fact implies (because of theorem 11.1.1) that $\mathrm{skel}_k P$ is not ($\mathscr{P}^{d''}$)-realizable for $d'' > d$.

The special cases $k = 1, 2$ of theorem 1 yield the corollary:

The graph of every 3-polytope is dimensionally unambiguous; the 2-skeleton of every d-polytope, $d \leq 5$, is dimensionally unambiguous.

The first of these assertions forms part of theorem 1 of Grünbaum–Motzkin [2]; the other part of the Grünbaum–Motzkin result will turn out to be a corollary to theorem 12.3.2 below.

The question about the dimensional ambiguity of $\mathrm{skel}_k P$, for $P \in \mathscr{P}^d$ and $k < [\frac{1}{2}d]$, is much more involved. It is probably related to the geometric structure of polytopes of highest dimension which are k-equivalent to P, but practically nothing is known in this connection. The one affirmative result is:

2. *For every k and d, $1 \leq k \leq d$, there exist d-polytopes P such that $\mathrm{skel}_k P$ is dimensionally unambiguous. Moreover, there exist d-polytopes with this property having arbitrarily many vertices.*

The main part of this theorem, the case $k = 1$, is due to Klee [14]; Klee's proof is based on the notion of total separation and uses theorem 11.4.1. We shall reproduce that proof below. A more direct proof, avoiding the notion of total separation, was given in Grünbaum [12].

In order to prove theorem 2, we first remark that it is sufficient to prove the theorem for $k = 1$, since any realization of the k-skeleton, $k \geq 1$, of a polytope P yields also a realization of the graph of P.

Let now P denote any simplicial d-polytope with v vertices such that P has the maximal possible number $\mu(v, d)$ of $(d - 1)$-faces. By theorem 10.1.3, if v is sufficiently large (for example, if $v \geq [\frac{1}{2}d]^2$) we may take $P = C(v, d)$. We claim that for all sufficiently large v the graph $\mathscr{G}$ of the Kleetope P^K over P (see section 11.4) is dimensionally unambiguous. Indeed, $\mathscr{G}$ is not ($\mathscr{P}^{d'}$)-realizable for $d' > d$ since each 'new' vertex of P^K is d-valent, while all vertices of d'-polytopes are at least d'-valent. On the other hand, it is clear from the construction of P^K that $s_v(\mathscr{G})$, the vth degree of total separability of $\mathscr{G}$, satisfies $s_v(\mathscr{G}) = s(v, d) = \mu(v, d)$ (see theorem 11.4.1). Therefore, $\mathscr{G}$ will not be ($\mathscr{P}^{d'}$)-realizable with $d' < d$ provided $s(v, d') < s(v, d)$. By theorem 11.4.1 this is equivalent with $\mu(v, d - 1) < \mu(v, d)$. Comparing theorems 10.1.3 and 9.6.1 we see that all these requirements are certainly met provided $v \geq \max\{2d; [\frac{1}{2}d]^2\}$. Thus for all such v, the graph of P^K, and hence all skeletons of P^K, are dimensionally unambiguous.

Using more careful estimates Klee [14] obtains a somewhat better bound· for v. However, even Klee's bound is not best possible since (see Grünbaum [12]) $(T^5)^K$ has a dimensionally unambiguous graph (with 12 nodes), while Klee's estimates guarantee only the existence of some such graph with 39 nodes. It may be shown that even the 11-node graph, obtained from $skel_1 (T^5)^K$ by omitting one of the 'new' vertices, is dimensionally unambiguous; it is not known, however, whether there are smaller dimensionally unambiguous graphs which are $(\mathscr{P}^5)$-realizable.

In Grünbaum [12] it was conjectured that the graph of P^K is dimensionally unambiguous for every d-polytope P. Though this conjecture can be shown to be false (see exercise 1), there persists the impression that $skel_k P^K$ is dimensionally unambiguous provided dim P is not too large in comparison to k. The first interesting open problem is the conjecture that the graph of P^K is dimensionally unambiguous for every $P \in \mathscr{P}^5$. 'Supporting evidence' for this conjecture may be found in exercise 2.

Exercises

1. Construct a $(\mathscr{P}_s^6)$-realization of the graph of $(T^7)^K$.
2. In the notation of section 10.2, show that if $\varphi_k^*(v, 5) = \phi_k^*(v, 5)$ (that is, if the 'lower bound conjecture' holds for simplicial 5-polytopes) then, for every $P \in \mathscr{P}_s^5$, the graph of P^K is not $(\mathscr{P}_s^4)$-realizable.

12.3   Strong and Weak Ambiguity

The results on strong and weak ambiguity of complexes are even more fragmentary than those on dimensional ambiguity. Thus, for example, no instances are known of k-complexes which are strongly d-ambiguous but not weakly d-ambiguous.

The only positive result of a general nature is:

1. *For $d \geq 3$, every $(\mathscr{P}^d)$-realizable $(d - 2)$-complex is both strongly and weakly d-unambiguous.*

In other words, the $(d - 2)$-skeleton of a d-polytope determines the combinatorial type of the polytope.

The proof of theorem 1 is based on the fact that every subset of the n-sphere which is a homeomorphic image of the $(n - 1)$-sphere, divides the n-sphere into two parts. Let $\mathscr{C}$ be a $(d - 2)$-complex which is $(\mathscr{P}^d)$-realizable. Without loss of generality we may assume that $\mathscr{C} = skel_{d-2} P$ for some $P \in \mathscr{P}^d$. Let $\{F_i \mid 1 \leq i \leq f_{d-1}(P)\}$ be the facets of P, i.e. the $(d - 1)$-dimensional elements of $\mathscr{B}(P)$. For each i, $\mathscr{B}(F_i)$ is a subcomplex of $\mathscr{C}$

such that set $\mathscr{B}(F_i)$ is homeomorphic to the $(d - 2)$-sphere. The complexes $\mathscr{B}(F_i)$ obviously have the following property:

(*) If F is a face of P such that all the vertices of F belong to $\mathscr{B}(F_i)$, then either $F = F_i$ or $F \in \mathscr{B}(F_i)$.

Let P' be any d-polytope such that $\mathscr{C}$ is combinatorially equivalent to $\text{skel}_{d-2} P'$. We wish to show that P and P' are combinatorially equivalent; this will be accomplished by showing that the $(d - 2)$-equivalence of P and P', induced by $\mathscr{C}$, may be extended to a $(d - 1)$-equivalence. The subcomplex $\mathscr{B}_i$ of $\text{skel}_{d-2} P'$ which corresponds to $\mathscr{B}(F_i)$ is homeomorphic to a $(d - 2)$-sphere imbedded in bd P'. Therefore the complement of set $\mathscr{B}_i$ in bd P' consists of two connected components. If each of these components would contain at least one vertex of P', a contradiction would result. In $\mathscr{C}$ the corresponding vertices may be joined by a path missing $\mathscr{B}(F_i)$, but in $\mathscr{B}(P')$ every path connecting them meets B_i. Therefore, at least one of the components of bd P' determined by $\mathscr{B}_i$ contains no vertex of P'. But because of property (*) of $\mathscr{C}$, this component may not meet the relative interior of any k-face of P', $1 \leq k \leq d - 2$. It follows that this component meets only one $(d - 1)$-face F'_i of P', and $\mathscr{B}_i = \mathscr{B}(F'_i)$. This completes the proof of theorem 1. (For a more elementary proof see the exercises at the end of this section.)

It is easily seen that theorem 1 is best possible in the sense that for $k \leq d - 3$ there exist k-complexes which are strongly d-ambiguous. This assertion is trivial for $d = 3$ and $k = 0$; for $d = 4$ and $k = 1$ an example is provided by the combinatorially non-equivalent neighborly 4-polytopes with 8 vertices (chapter 7). Note that in these examples the different realizations are all by simplicial polytopes; non-simplicial examples are still easier to find.

Combining theorem 1 with theorem 12.2.1 we have

2. *For $d \geq 3$, the $(d - 2)$-skeleton of every d-polytope is dimensionally, strongly, and weakly unambiguous.*

The case $d = 3$ of this result is theorem 1 of Grünbaum–Motzkin [2].

Concerning the situation for $k \leq d - 3$, it was pointed out by M. A. Perles in a private communication that the claim, made in Grünbaum [12], about the strong d-unambiguity of the graphs of the Kleetopes P^K, is unfounded. Indeed, for $d \geq 4$, the moving of one of the 'new' vertices to a position such that it becomes beyond one facet of P and contained in the affine hull of another, yields a polytope with the same graph—although the resulting polytope is combinatorially different from P^K.

However, for every d and $k \geq 1$, there exist k-complexes with arbitrarily many vertices, which are strongly and weakly d-unambiguous. The following construction of such graphs is due to Perles.

Let P^n be an n-polytope such that the graph of P^n is dimensionally, strongly, and weakly unambiguous, and that none of its subgraphs is $(\mathscr{P}^n)$-realizable. As examples we may take for $n = 2$ any polygon as P^2, while for $n = 3$ every simple 3-polytope may serve as P^3. The graph $\mathscr{G}$ of the $(d - n)$-fold d-pyramid P^d with basis P^n is strongly and weakly d-unambiguous. Indeed, if $\mathscr{G}$ is combinatorially equivalent to the graph of the d-polytope P then, by exercise 3.1.5, there exists an n-face F of P such that the vertices of F correspond to some of the vertices of P^n. By the assumptions made about P^n, it follows that F is combinatorially equivalent to P^n. Therefore P is a $(d - n)$-fold d-pyramid with basis F, hence combinatorially equivalent with P^d. This proves our assertion.

In general, these graphs are dimensionally ambiguous. For example, let $d = 5$, $n = 2$, and let P^2 be a k-gon, $P^2 = \operatorname{conv}\{V_1, \cdots, V_k\}$, the other vertices of P^5 being A, B, C. Then the graph $\mathscr{G}^*$ obtained from $\mathscr{G}$ by omitting the vertex V_k is $(\mathscr{P}^4)$-realizable. Indeed, $\mathscr{G}^*$ is the graph of the 4-pyramid K with the 3-polytope represented in figure 12.3.1 as basis. The 4-polytope K has the facets $F_1 = \operatorname{conv}\{V_1, A, B, C\}$ and $F_2 = \operatorname{conv}\{V_{k-1}, A, B, C\}$. If V_k is a point in the complement of K, sufficiently near the centroid of the 2-face $\operatorname{conv}\{A, B, C\}$ of K in order to be only beyond the facets F_1 and F_2 of K then, by theorem 5.2.1, the 1-skeleton of $\operatorname{conv}(\{V_k\} \cup K)$ will be combinatorially equivalent to $\mathscr{G}$. Therefore $\mathscr{G}$ is both $(\mathscr{P}^4)$- and $(\mathscr{P}^5)$-realizable.

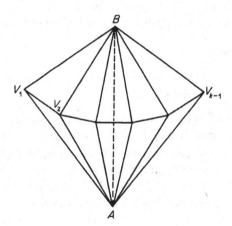

Figure 12.3.1.

Clearly, the same argument shows that, in general, the graphs constructed with $n = 2$, $d \geq 5$, are both $(\mathscr{P}^d)$- and $(\mathscr{P}^{d-1})$-realizable. (For $d = 4$, they are dimensionally unambiguous.) The dimensional ambiguity of the graphs constructed with $n \geq 3$ and $d \geq 5$ is still undecided.

Exercises

The following exercises lead to a proof of theorem 1 which is elementary in the sense that it uses no topological tools heavier than the Jordan curve theorem (and even that only in the special case of polygonal lines in the plane). The same approach leads also to interesting extensions of theorem 1. The exercises are based on a private communication from M. A. Perles. We start with definitions of the terms used in the exercises.

Let $\mathscr{F}$ be a finite partially ordered set (with order relation $\leq$). The *height* $h_{\mathscr{F}}(F)$ of an element $F \in \mathscr{F}$ is the largest h such that there exists a chain of the type $F_0 < F_1 < \cdots < F_{h+1} = F$, with all $F_i \in \mathscr{F}$. We define $\mathrm{skel}_k \mathscr{F} = \{F \in \mathscr{F} \mid h_{\mathscr{F}}(F) \leq k\}$; clearly, $\mathrm{skel}_k \mathscr{F}$ is a partially ordered set in the order relation inherited from $\mathscr{F}$. If $\mathscr{F}$ and $\mathscr{F}'$ are two partially ordered sets, an *isomorphism* φ of $\mathscr{F}$ onto $\mathscr{F}'$ is any one-to-one mapping of $\mathscr{F}$ onto $\mathscr{F}'$ such that both φ and φ^{-1} are order-preserving; a *k-equivalence* of $\mathscr{F}$ to $\mathscr{F}'$ is any isomorphism of $\mathrm{skel}_k \mathscr{F}$ to $\mathrm{skel}_k \mathscr{F}'$. A lattice $\mathscr{F}$ is *d-realizable* provided there exists a *d-polytope* P such that $\mathscr{F}$ is d-equivalent to $\mathscr{F}(P)$, the lattice of all faces of P. For a d-realizable lattice the height of each element obviously equals the dimension of the corresponding face of the polytope. Heights of elements of a lattice will be denoted by superscripts; thus F^1 or F_j^1 denote elements of height 1. The reformulation of theorem 1 for which a proof will be outlined in the following exercises is:

Every $(d-2)$-equivalence φ of a d-realizable lattice $\mathscr{F}$ to a d-realizable lattice $\mathscr{F}'$ may be extended to a d-equivalence (isomorphism) $\bar{\varphi}$ of $\mathscr{F}$ to $\mathscr{F}'$.

1. Show that the extension $\bar{\varphi}$ mentioned in the reformulation of theorem 1 is unique.

2. Show that for $d = 3$ the proof of theorem 1 as given in the text may be reformulated so as to require only the particular case of the Jordan theorem which deals with polygonal lines in the plane. Find an elementary proof for this case of Jordan's theorem.

3. Let $\mathscr{F} = \mathscr{F}(P)$ for a d-polytope, P and let $F^{d-4} \in \mathscr{F}$. Define $\mathscr{F}_{F^{d-4}} = \{F \in \mathscr{F} \mid F^{d-4} \leq F\}$; show that $\mathscr{F}_{F^{d-4}}$ is a 3-realizable lattice.

4. Let P, P' be d-polytopes, $d \geq 4$, and let φ be a $(d-2)$-equivalence between P and P' (and hence between $\mathscr{F} = \mathscr{F}(P)$ and $\mathscr{F}' = \mathscr{F}(P')$). If

$F'^{d-4} = \varphi(F^{d-4})$ for a $(d-4)$-face F^{d-4} of P, and if $\varphi_{F^{d-4}}$ is the restriction of φ to $\mathscr{F}_{F^{d-4}} \cap \text{skel}_{d-2}\mathscr{F}$, then $\varphi_{F^{d-4}}$ is a 1-equivalence between the 3-realizable lattices $\mathscr{F}_{F^{d-4}}$ and $\mathscr{F}_{F'^{d-4}}$. Using exercises 1 and 2, show that $\varphi_{F^{d-4}}$ has a unique extension $\tilde{\phi}_{F^{d-4}}$ which is a 3-equivalence between $\mathscr{F}_{F^{d-4}}$ and $\mathscr{F}'_{F'^{d-4}}$.

5. Let (*) denote the statement:

(*) If F^{d-1} and G^{d-1} are facets of the d-polytope P of exercise 4, and if $F^{d-4} \subset F^{d-1}$ and $G^{d-4} \subset G^{d-1}$, then $F^{d-1} = G^{d-1}$ if and only if $\tilde{\phi}_{F^{d-4}}(F^{d-1}) = \tilde{\phi}_{G^{d-4}}(G^{d-1})$.

Assuming the validity of (*), prove the reformulated theorem 1 by showing that the d-equivalence $\tilde{\phi}$ between $\mathscr{F}$ and $\mathscr{F}'$ may be defined as follows:

$$\tilde{\phi}(F^k) = \varphi(F^k) \text{ for } k \le d-2, \ F^k \in \mathscr{F};$$

$$\tilde{\phi}(F^{d-1}) = \tilde{\phi}_{F^{d-4}}(F^{d-1}), \text{ where } F^{d-4} \text{ is any } (d-4)\text{-face of } F^{d-1};$$

$$\tilde{\phi}(P) = P'.$$

6. Show that (*) is valid if and only if the following assertion (**) is valid:

(**) If F^{d-1} is a facet of P, and if F^{d-4} and G^{d-4} are $(d-4)$-faces of F^{d-1}, then $\tilde{\phi}_{F^{d-4}}(F^{d-1}) = \tilde{\phi}_{G^{d-4}}(F^{d-1})$.

7. Let F^{d-1}, F^{d-4} and G^{d-4} be as in (**). Then there exists a finite sequence of $(d-4)$-faces F_i^{d-4} and $(d-3)$-faces F_j^{d-3} having the incidence properties indicated in the scheme

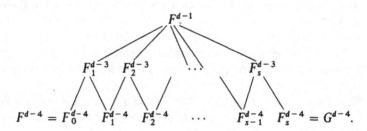

Hence (**) holds if and only if (***) is valid:

(***) If F^{d-1} is a facet of P, if F^{d-3} is a face of F^{d-1}, and if F^{d-4} and G^{d-4} are facets of F^{d-3}, then $\tilde{\phi}_{F^{d-4}}(F^{d-1}) = \tilde{\phi}_{G^{d-4}}(F^{d-1})$.

8. In order to prove (***) note that there exist $(d-2)$-faces of F^{d-2} and G^{d-2} of F^{d-1} such that $F^{d-2} \ne G^{d-2}$, having the incidence properties

indicated in the scheme

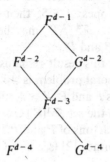

Considering the images of F^{d-1}, F^{d-2}, G^{d-2} under $\tilde{\varphi}_{F^{d-4}}$ and $\tilde{\varphi}_{G^{d-4}}$, note that images of $(d-2)$-faces of P under $\varphi_{F^{d-4}}$ and $\varphi_{G^{d-4}}$ coincide (if defined), and that two facets of a d-polytope P coincide if both contain two different $(d-2)$-faces of P. Deduce from this the validity of (***), thus completing the proof of the reformulation of theorem 1.

9. Let $\mathscr{F}^i(P)$ denote the set of all i-faces of the polytope P. Checking the various stages of the proof of the reformulated theorem 1, show the validity of the following generalization:

(i) If P and P' are polytopes of dimensions d and d', $(d, d' \geq 3)$, then any isomorphism between the partially ordered sets $\bigcup\limits_{i=d-4}^{d-2} \mathscr{F}^i(P)$ and $\bigcup\limits_{i=d'-4}^{d'-2} \mathscr{F}^i(P')$ is uniquely extendable to an isomorphism between the sets $\bigcup\limits_{i=d-4}^{d} \mathscr{F}^i(P)$ and $\bigcup\limits_{i=d'-4}^{d'} \mathscr{F}^i(P')$; moreover, $d' = d$.

Dually, this implies:

(ii) If P and P' are polytopes of dimension at least 3, any isomorphism between $\bigcup\limits_{i=1}^{3} \mathscr{F}^i(P)$ and $\bigcup\limits_{i=1}^{3} \mathscr{F}^i(P')$ is uniquely extendable to a 3-equivalence of $\mathscr{F}(P)$ and $\mathscr{F}(P')$.

10. (Compare exercise 3.2.3) Let P be a d-polytope, P' a d'-polytope, $0 \leq r < s < d$ and $0 \leq r' < s' < d'$, and let φ be an isomorphism between the partially ordered sets $\mathscr{F}^r(P) \cup F^s(P)$ and $\mathscr{F}^{r'}(P') \cup \mathscr{F}^{s'}(P')$. Then $s - r = s' - r'$ and φ is extendable to an isomorphism of $\bigcup\limits_{i=r}^{s} \mathscr{F}^i(P)$ to $\bigcup\limits_{i=r'}^{s'} \mathscr{F}^i(P')$.

11. Combining theorem 1 with exercises 9 and 10 prove:

Let P and P' be d-polytopes, $d \geq 5$; then every isomorphism φ of $\mathscr{F}^1(P) \cup \mathscr{F}^{d-2}(P)$ to $\mathscr{F}^i(P') \cup \mathscr{F}^{d-2}(P')$ may be extended to a combinatorial equivalence between P and P'.

It is not known whether this result remains valid for $d = 4$. Indeed, even the following more general problem is still unsolved:

Do there exist d-polytopes P and P', $d \geq 4$, such that an isomorphism φ of the set $\mathscr{F}^1(P) \cup \mathscr{F}^2(P)$ to the set $\mathscr{F}^1(P') \cup \mathscr{F}^2(P')$ is not extendable to an isomorphism of the 2-skeletons of P and P'?

12. Use a theorem of Whitney [2] (see also Ore [1], theorem 15.4.1) to prove the following result:

If P and P' are polytopes, and if φ is a one-to-one correspondence between $\mathscr{F}^1(P)$ and $\mathscr{F}^1(P')$ such that for all $E_1, E_2 \in \mathscr{F}^1(P)$

$$E_1 \cap E_2 \neq \varnothing \text{ if and only if } \varphi(E_1) \cap \varphi(E_2) \neq \varnothing,$$

then φ is extendable to an isomorphism of the 1-skeletons of P and P'.

A somewhat analogous result, which may be deduced from a theorem of P. Kelly [1], is:

If P and P' are k-neighborly polytopes, and if φ is a one-to-one correspondence between $\mathscr{F}^{k-1}(P)$ and $\mathscr{F}^{k-1}(P')$ such that for all F_1, $F_2 \in \mathscr{F}^{k-1}(P)$

$$\dim(F_1 \cap F_2) = k - 2 \text{ if and only if } \dim(\varphi(F_1) \cap \varphi(F_2)) = k - 2,$$

then φ is extendable to an isomorphism of the $(k-1)$-skeletons of P and P'.

It would be interesting to find common generalizations of these results.

12.4 Additional notes and comments

Algorithmic aspects.

For any class of d-polytopes whose r-skeleta are strongly d-unambiguous it is reasonable to ask how one can reconstruct the entire combinatorial structure (in terms of the vertex-facet incidences, say) of such a polytope from its r-skeleton. Furthermore, the question for the complexity of the corresponding reconstruction problem arises.

By theorem 12.3.1 the class of all d-polytopes is an example for such a class for $r = d - 2$. From Perles' proof presented in exercises 12.3.1–9 one can derive an algorithm for the reconstruction problem that runs in polynomial time. Its main ingredient is a subroutine for finding a planar embedding of a graph (see, e. g., Mohar–Thomassen [a, Sect. 2.7]).

Simplicial polytopes.

Perles proved that one can reconstruct the entire face lattice of a simplicial d-polytope from its $[d/2]$-skeleton (see Kalai [j, Thm. 17.4.19]; see also Dancis [a]). In particular, the $[d/2]$-skeleton of every simplicial d-polytope is strongly (and weakly) d-unambiguous; it is dimensionally unambiguous by theorem 12.2.1. Thus, the claim "non-simplicial examples are still easier to find" on page 229 needs amplification: Simplicial examples with $k = d - 3$ do not exist for $d \geq 5$.

Simple polytopes.

Blind–Mani [a] showed that the graphs of simple d-polytopes are strongly and weakly d-unambiguous. This gives an alternative proof of the existence of strongly and weakly d-unambiguous k-skeleta of d-polytopes with arbitrarily many vertices (compare Perles' construction on page 230).

Blind's and Mani's proof did not provide a reconstruction algorithm. Soon afterwards, Kalai [d] found a short, elegant, and constructive proof of the result, which, however, did not settle the computational complexity of the reconstruction problem. Joswig–Kaibel–Körner [a] showed that the problem can be formulated as a combinatorial optimization problem that has a strongly dual problem (in the sense of combinatorial optimization); in particular, this leads to polynomial size certificates for the correctness of the reconstruction.

Given the graph of a simple d-polytope, the subgraphs that correspond to facets are induced, non-separating, $(d-1)$-regular, and $(d-1)$-connected. For $d \leq 3$ these four conditions are sufficient to characterize the facet subgraphs (Whitney [2]), but not for $d \geq 4$ (as had been conjectured by Perles; see Haase–Ziegler [a]). However, it may be that for $d = 4$ it suffices to additionally require planarity. For computational experiments on the reconstruction problem for the graphs of simple polytopes see Achatz–Kleinschmidt [a].

Zonotopes.
Björner–Edelman–Ziegler [a] showed that the graph of a d-dimensional zonotope is strongly and weakly d-unambiguous. In particular, this proves the conjecture on page 397.

Cubical polytopes.
Joswig–Ziegler [a] showed that the $([d/2] - 1)$-skeleton of the d-cube is dimensionally ambiguous. Moreover, they constructed a cubical and a noncubical 4-polytope with the graph of the 5-cube. The cubical polytope is $\mathrm{conv}((T \times 2T) \cup (2T \times T)) \subset R^4$, for the square $T = [-1, 1]^2 \subset R^2$. Schlegel diagrams of both polytopes are shown below.

Thus, the graph of the 5-cube is dimensionally ambiguous and strongly 4-ambiguous. Moreover, the cubical polytope has a combinatorial automorphism group that is smaller than the automorphism group of the graph of the 5-cube. Hence, the graph of the 5-cube is also weakly 4-ambiguous.

The following question emerges from these results: Is the $[d/2]$-skeleton of the d-cube strongly and weakly d-unambiguous?

Extending Kalai's [d] methods for simple polytopes, Joswig [a] proved that the graphs of the duals of capped cubical d-polytopes are strongly and weakly d-unambiguous. In fact, he derived an algorithm for the corresponding reconstruction problem.

The analogous question for the duals of general cubical polytopes is open; the case of cubical zonotopes was solved by Babson–Finschi–Fukuda [a].

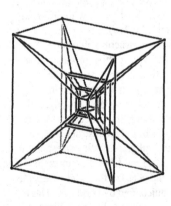 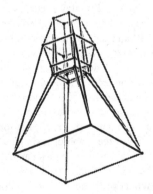

CHAPTER 13

3-Polytopes

Our knowledge about 3-polytopes far exceeds our knowledge of higher-dimensional polytopes. This may be explained in part by the 'experimental accessibility' of 3-polytopes and the fact that the research of their properties goes back to antiquity. A deeper reason is that, from many combinatorial points of view, 3-polytopes may be replaced by planar graphs. The various techniques of graph theory are therefore applicable to the study of 3-polytopes.

The main graph–theoretic tool we shall apply in this context are various types of 'reductions' of planar graphs. A 'reduction' consists of the deletion of certain edges and nodes of a given planar graph, and their replacement by suitable new edges and nodes, the details of the construction varying with the goal to be achieved. In proofs of nonexistence we shall show that a reduction is possible in every case—thus we shall be using Fermat's 'method of infinite descent'. In other proofs the reductions will be applicable to all but certain exceptional cases—thus the proof will essentially be by induction.

13.1 Steinitz's Theorem

The most important and deepest known result on 3-polytopes is the following theorem due to E. Steinitz (see Steinitz [6]. Steinitz–Rademacher [1]) and called by him 'the fundamental theorem of convex types'. The formulation of the result here is in terms different from those used by Steinitz. He works with a special notion of 2-complexes, but the graph-theoretic terminology we use permits clearer and shorter proofs.

1. *A graph $\mathscr{G}$ is $(\mathscr{P}^3)$-realizable if and only if $\mathscr{G}$ is planar and 3-connected.*

The planarity and 3-connectedness of the graph of any $P \in \mathscr{P}^3$ are obvious: any Schlegel diagram of P provides an imbedding of $\mathrm{skel}_1 P$ in the plane, while theorem 11.3.1 guarantees its 3-connectedness. In the proof that these properties are sufficient for the $(\mathscr{P}^3)$-realizability lies the depth of Steinitz's arguments.

We shall first give an outline of the proof, introducing at the same time an appropriate terminology.

The proof proceeds by induction on the number of edges $e = e(\mathscr{G}) = f_1(\mathscr{G})$ of $\mathscr{G}$. The assumption that $\mathscr{G}$ is 3-connected implies $e \geq 6$, with equality holding if and only if $\mathscr{G}$ is the complete graph with 4 nodes. Since in this case $\mathscr{G}$ is clearly $(\mathscr{P}^3)$-realizable by any 3-simplex, we may in the remaining part of the proof assume $e \geq 7$. The steps of the proof are as follows:

(i) Using a 'double counting of incidences' we show that each graph $\mathscr{G}$ considered has 3-*valent elements* (i.e. 3-valent vertices, or triangular faces).

(ii) For any given $\mathscr{G}$ and any 3-valent element of $\mathscr{G}$ we construct a new graph $\mathscr{G}^*$, also planar and 3-connected, such that from any realization of $\mathscr{G}^*$ by a 3-polytope P^* a 3-polytope P realizing $\mathscr{G}$ may be constructed. The procedures for obtaining $\mathscr{G}^*$ from $\mathscr{G}$ shall be called *elementary transformations*, or *reductions*.

(iii) If $\mathscr{G}$ contains a 3-valent vertex incident to a triangular face then an elementary transformation of $\mathscr{G}$ yields a $\mathscr{G}^*$ which has less edges than $\mathscr{G}$. Thus in this case induction takes over and the proof is completed. If $\mathscr{G}$ does not contain such an incidence, we shall show that there exists a finite sequence of elementary transformations such that the transformed graph contains a 3-valent vertex incident to a triangular face, i.e. is of a type to which the former argument applies.

(i) We first introduce a notation (different from the one used in previous chapters) for the number of faces of different kinds of a 3-polytope P. The number of vertices, edges, and 2-faces of P shall be denoted by $v = v(P)$, $e = e(P)$, and $p = p(P)$. The number of k-valent vertices of P shall be denoted by $v_k = v_k(P)$, while $p_k = p_k(P)$ is the number of k-gonal 2-faces of P.

Thus $v = \sum_{k \geq 3} v_k$ and $p = \sum_{k \geq 3} p_k$; Euler's equation becomes, in the new notation, $v - e + p = 2$.

Counting the number of incidences of edges and k-gonal faces we obtain $2e = \sum_{k \geq 3} k p_k$; similarly, counting incidences of edges and k-valent vertices there results $2e = \sum_{k \geq 3} k v_k$. Combining those expressions with the previous equations it follows that

$$\sum_{k \geq 3} k v_k + \sum_{k \geq 3} k p_k = 4e = 4v + 4p - 8 = 4 \sum_{k \geq 3} v_k + 4 \sum_{k \geq 3} p_k - 8.$$

Consequently

$$v_3 + p_3 = 8 + \sum_{k \geq 5} (k - 4)(v_k + p_k) \geq 8;$$

therefore every 3-polytope P has at least eight 3-valent elements.

Clearly, the above computations apply equally well to every connected planar graph (imbedded in the 2-sphere) which has no 2-valent vertices or digonal faces; in particular, it applies to 3-connected planar graphs.

(ii) In either of the following two cases we shall say that a graph $\mathscr{G}^*$ is obtained from a graph $\mathscr{G}$ (with $e(\mathscr{G}) > 6$) by an *elementary transformation* (or a *reduction*):

(1) A trivalent node of $\mathscr{G}$ and the edges incident to it are deleted, and the three nodes connected to it in $\mathscr{G}$ are pairwise connected by 'new' edges (unless some of them are already connected in $\mathscr{G}$, in which case there is no need for the corresponding 'new' edge). The four possible cases are represented in figure 13.1.1, where $\mathscr{G}$ is to the left of the $\mathscr{G}^*$ resulting from it by an elementary transformation ω_i, $i = 0, 1, 2, 3$, of this type.

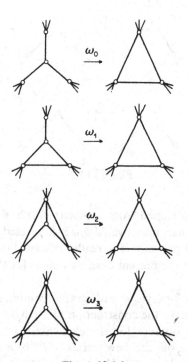

Figure 13.1.1

(2) ·The three edges of a trigonal face are deleted and the three nodes joined to a 'new' node. If any 2-valent node would result it is omitted and the two edges incident to it are replaced by a single edge. The four possible elementary transformations η_i, $i = 0, 1, 2, 3$, of this type are schematically indicated in figure 13.1.2.

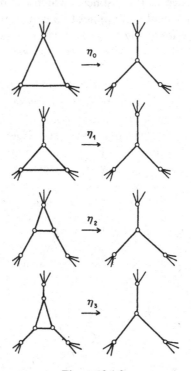

Figure 13.1.2

Every graph $\mathscr{G}^*$ obtained from a planar graph $\mathscr{G}$ by an elementary transformation is planar; moreover, if $\mathscr{G}$ is 3-connected, it is easily checked that $\mathscr{G}^*$ has the same property. The reader is invited to check the validity of these assertions in the present case, as well as for the reductions used in the next section.

Now, if P^* is any 3-polytope with graph (combinatorially equivalent to) $\mathscr{G}^*$, we shall describe the construction of a polytope P with graph $\mathscr{G}$. The construction is obvious for the elementary transformations η_i of figure 13.1.2: 'Cutting off' the 'new' vertex of P^* by an appropriate plane yields the required polytope P. The construction is also very simple in case of the reductions ω_1, ω_2, and ω_3: P is the convex hull of the union of P^* with an appropriate point V. The point V is beyond the facet of P^* corresponding to the 'new' triangle in $\mathscr{G}^*$, and it is beneath all the other facets of P^* in case of an ω_3 reduction, while in the case of ω_2 or ω_1 reductions V is beneath all but one (or two) of them, V being in the affine hull of each of those exceptional facets. If the elementary transformation is ω_0, the polytope P is the convex hull of the union of P^* with the point V determined as the intersection of the planes of the three facets of P^* adjacent to the 'new' triangle, provided V is beyond this triangular facet of P^*. A complication arises, however, if the three planes are parallel or intersect in a point which is *beneath* the triangular facet of P^*. In this case it is necessary to apply to P^* a projective transformation such that for the transform of P^* the former construction becomes applicable.

Therefore in all cases the $(\mathscr{P}^3)$-realizability of the elementary transform $\mathscr{G}^*$ of $\mathscr{G}$ implies the $(\mathscr{P}^3)$-realizability of $\mathscr{G}$. Since $\mathscr{G}^*$ contains i edges less than $\mathscr{G}$ if it is obtained from $\mathscr{G}$ by ω_i or η_i, it follows that the inductive proof of the theorem is completed for all $\mathscr{G}$ to which ω_i or η_i is applicable, $i = 1, 2, 3$. Thus the final part of the proof will consist in showing that given a graph $\mathscr{G}$ to which none of the reductions ω_i or η_i, $i = 1, 2, 3$, is applicable, it is possible to obtain a graph to which some of them are applicable by performing on $\mathscr{G}$ a finite sequence of reductions ω_0 and η_0. This is the most intricate part of Steinitz's proof, and for it we need certain results on 4-valent planar graphs. We interrupt here the proof of theorem 1 in order to establish the properties of 4-valent graphs we shall use. The connection with the proof of theorem 1 will become obvious below.

Throughout the remaining part of the present section, let $\mathscr{C}$ denote a 4-valent, 3-connected, planar graph.

An edge AB of $\mathscr{C}$ has a *direct extension* BC provided the edges AB and BC separate the other two edges incident to B.

A path $A_0 A_1 \cdots A_n$ in $\mathscr{C}$ (where the node A_i is joined to A_{i-1} by an edge, $i = 1, \cdots, n$) is called a *geodesic arc* provided $A_{i-1} A_i$ has $A_i A_{i+1}$ as direct extension, for $1 \le i < n$; for a *closed geodesic*, $A_0 = A_n$, and $A_{n-1} A_n$ has $A_0 A_1$ as direct extension.

A subgraph $\mathscr{L}$ of $\mathscr{C}$ is called a *lens* provided:

(i) $\mathscr{L}$ consists of a simple closed path $\mathscr{Q}$: $A_0 A_1 \cdots A_n B_0 B_1 \cdots B_m A_0$ (called the boundary of $\mathscr{L}$) and all the nodes and edges of $\mathscr{C}$ contained in one of the connected components of the complement of $\mathscr{Q}$ in the 2-sphere (called *inner* nodes and edges of $\mathscr{Q}$);

(ii) $\mathcal{Q}$ is formed by two geodesic arcs $A_0A_1 \cdots A_nB_0$ and $B_0B_1 \cdots B_mA_0$, such that no inner edge of $\mathcal{Q}$ is incident to the *poles* A_0 and B_0 of $\mathcal{Q}$. (See figure 13.1.3, where (a) and (b) are lenses, while (c) is not.)

A lens $\mathcal{L}$ in $\mathcal{C}$ is called *indecomposable* provided no lens of $\mathcal{C}$ is properly contained in $\mathcal{L}$.

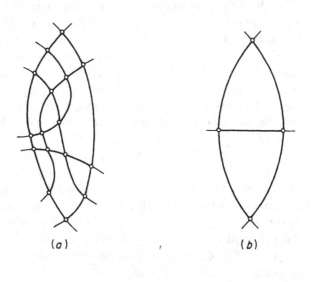

(a)   (b)

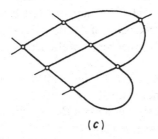

(c)

Figure 13.1.3

Every $\mathcal{C}$ contains at least one indecomposable lens. If $\mathcal{L}$ is an indecomposable lens, then $n = m$, and each node A_i, $1 \leq i \leq n$, is joined to a unique B_j, $1 \leq j \leq n$, by a geodesic arc $\mathcal{Q}_i$ contained in $\mathcal{L}$; we shall call $\mathcal{Q}_i$ the *cut* determined by A_i. The cuts $\mathcal{Q}_i$ and $\mathcal{Q}_k$, $i \neq k$, intersect in at most one inner node of $\mathcal{L}$. Each inner edge of $\mathcal{L}$ belongs to one cut $\mathcal{Q}_i$, each inner node to exactly two cuts.

All the above assertions follow at once by considering the class of all subgraphs of $\mathscr{G}$ which consist of a simple circuit composed of at most two geodesic arcs, and of all the nodes and edges of $\mathscr{G}$ contained in one of the connected components of the circuit's complement in the 2-sphere. The members of this class, minimal with respect to the number of nodes, are indecomposable lenses.

We need the following lemma:

2. *Every indecomposable lens $\mathscr{L}$ contains a triangular face incident to the boundary $\mathscr{Q}$ of $\mathscr{L}$.*

PROOF If $\mathscr{L}$ has no inner nodes, then the face of $\mathscr{L}$ incident to A_0 is a triangle. If $\mathscr{L}$ has inner nodes, let $D_1, \cdots, D_r$ be all the inner nodes such that D_i is a neighbor of some A_k. Let $h(D_i)$ denote the number of faces of $\mathscr{L}$ contained in the triangular region determined by the two cuts $\mathscr{Q}_j$ and $\mathscr{Q}_k$ intersecting at D_i, and by the arc of $\mathscr{Q}$ with endpoints A_jA_k. As easily seen, if $h(D_i) = \min\{h(D_1), \cdots, h(D_r)\}$, then D_i, A_j, and A_k determine a triangular face of $\mathscr{L}$. This completes the proof of the lemma.*

The number of faces in an indecomposable lens is obviously at least 2.

Let now $\mathscr{G}$ be a planar 3-connected graph; we define a new graph $I(\mathscr{G})$ in the following fashion:

The nodes of $I(\mathscr{G})$ are (interior) points of the edges of $\mathscr{G}$, one on each edge. Two nodes of $I(\mathscr{G})$ are joined by an edge if and only if the two edges of $\mathscr{G}$ corresponding to them have a common node and are incident to the same face (in $\mathscr{G}$). Clearly $I(\mathscr{G})$ is planar (and 3-connected), and every node of $I(\mathscr{G})$ has valence 4; the faces of $I(\mathscr{G})$ are in one-to-one correspondence with the union of the set of faces of $\mathscr{G}$ and the set of nodes of $\mathscr{G}$; i.e. $p(I(\mathscr{G})) = p(\mathscr{G}) + v(\mathscr{G})$. If, and only if, a node and a face of $\mathscr{G}$ are incident, the corresponding faces of $I(\mathscr{G})$ have a common edge. A k-gonal face of $I(\mathscr{G})$ corresponds to a k-gonal face of $\mathscr{G}$ or to a k-valent node of $\mathscr{G}$. We define $g(\mathscr{G})$ as the minimal number of faces in an indecomposable lens $\mathscr{L}$ in $I(\mathscr{G})$. (Though we shall not use this fact, we note that

$$2 \leq g(\mathscr{G}) \leq \tfrac{1}{2}p(I(\mathscr{G})) = \tfrac{1}{2}(p(\mathscr{G}) + v(\mathscr{G})) < e(\mathscr{G}).)$$

If $g(\mathscr{G}) = 2$, the corresponding indecomposable lens of $I(\mathscr{G})$ is necessarily that represented in figure 13.1.3(b), and therefore $\mathscr{G}$ contains a triangular face incident to a trivalent node; thus one of the elementary

* The following remark results at once from the above proof; we shall use it in section 13.2. For each of the geodesic arcs forming the boundary of an indecomposable lens with inner nodes, there exists a triangular face incident to it but not incident to the other arc.

transformations ω_i or η_i, $i = 1, 2, 3$, may be applied to $\mathcal{G}$, and the $(\mathcal{P}^3)$-realizability of $\mathcal{G}$ follows by induction.

In order to complete the proof of theorem 1 we shall show how to apply a reduction of type ω_0 or η_0 to a graph $\mathcal{G}$ for which $g(\mathcal{G}) > 2$ in order to obtain a graph $\mathcal{G}^*$ with $g(\mathcal{G}^*) < g(\mathcal{G})$.

This is accomplished by taking, in $I(\mathcal{G})$, an indecomposable lens $\mathcal{L}$ with $g(\mathcal{G})$ faces. Let T be a triangle in $\mathcal{L}$, incident to the boundary of $\mathcal{L}$; the existence of such T was established by lemma 2. Depending on whether T corresponds to a triangular face or to a trivalent node of $\mathcal{G}$, one of the two procedures η_0 or ω_0 is applicable.

If T is incident to only one of the two geodesic arcs forming the boundary of $\mathcal{L}$, then the transition from figure 13.1.4(a) to figure 13.1.4(b) illustrates the fact that $g(\mathcal{G}^*) < g(\mathcal{G})$ in case T corresponds to a triangular face of $\mathcal{G}$ (note that $\mathcal{L}$ is above the line denoted by L_1), while the opposite transition applies if T corresponds to a trivalent node of $\mathcal{G}$ (in this case $\mathcal{L}$ is beneath the line L_1). If T is incident to a pole of $\mathcal{L}$, the same relations persist, only $\mathcal{L}$ is in this case beneath the wedge formed by the lines L_2 and L_3, or above it, respectively.

This completes the proof of theorem 1.

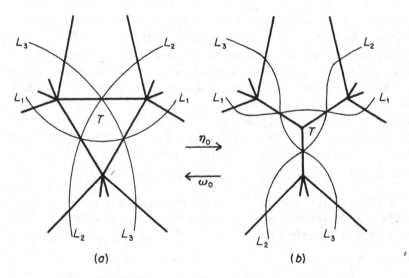

Figure 13.1.4. Heavy edges denote $\mathcal{G}$, light edges denote $I(\mathcal{G})$

The proof of theorem 1 could have been reformulated in such a fashion that either *only* η_i reductions, or *only* ω_i reductions, are used. This observation is based on the existence of polar polytopes (section 3.4) and on the following facts:

(i) Polar polytopes are dual to each other;

(ii) The graphs $\mathscr{G}_1$, $\mathscr{G}_2$ of dual polytopes P_1, P_2 are dual to each other in the sense of the theory of planar graphs;

(iii) For each i, the reductions η_i and ω_i are dual to each other;

(iv) Each member of a pair of polar polytopes is constructible from the other member of the pair;

(v) Dual graphs (or dual polytopes) have the same number of edges;

(vi) If $\mathscr{G}_1$ and $\mathscr{G}_2$ are dual graphs, then $I(\mathscr{G}_1) = I(\mathscr{G}_1)$; therefore indecomposable lenses are the same in $I(\mathscr{G}_1)$ and in $I(\mathscr{G}_2)$, and thus $g(\mathscr{G}_1) = g(\mathscr{G}_2)$. But nodes and faces are interchanged in $\mathscr{G}_1$ and $\mathscr{G}_2$, i.e. if a reduction η_i is applicable to one of them, ω_i may be applied to the other.

In other words, the induction may be carried out simultaneously for pairs of dual graphs.

Exercises

1. As noted by Steinitz–Rademacher [1], a very simple proof may be given for the special case of theorem 13.1.1 in which the graph $\mathscr{G}$ is assumed to be 3-valent. Only a single 'elementary transformation' ϑ is needed, which consists in deleting an edge AB and amalgamating the two pairs of edges incident to A and B to two edges (see figure 13.1.5). The possibility of applying ϑ to an edge AB of $\mathscr{G}$ clearly presupposes that $\mathscr{G}$ contains neither of the edges $A'A''$ and $B'B''$.

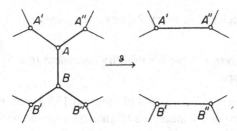

Figure 13.1.5

Prove theorem 13.1.1 for 3-valent $\mathscr{G}$ by showing:

(i) Every 3-valent, 3-connected planar G with more than 6 edges contains an edge AB to which ϑ is applicable; the resulting graph $\mathscr{G}^*$ is 3-valent and 3-connected.

(ii) From any realization of $\mathscr{G}^*$ by a 3-polytope P^* a realization of $\mathscr{G}$ by a 3-polytope P may be constructed.

2. Show that the above assertion (i) is valid for all 3-valent, 3-connected (not necessarily planar) graphs $\mathscr{G}$. (This result has been conjectured by V. Klee in a private communication.)

3. Dualize exercise 1, thus obtaining a characterization of graphs of simplicial 3-polytopes.

4. Show that every 3-connected graph is obtainable from the complete graph with 4 nodes by a finite number of 'enlargements'. (An 'enlargement' of a graph $\mathscr{G}$ is the addition of an edge, each endpoint of which is either a node of $\mathscr{G}$, or a new node introduced on an edge of $\mathscr{G}$.)

5. The following problem is open even in the first nontrivial case ($n = 5$): Let P be a 3-polytope, F an n-gonal facet of P, and let F' be a given n-gon. Does there exist a polytope P' which has F' as a facet, such that P' is combinatorially equivalent to P under a mapping which makes F' correspond to F? (It is conceivable that a suitable modification of the proof of theorem 13.1.1 could be used to solve the problem.)

13.2 Consequences and Analogues of Steinitz's Theorem

We mention first some corollaries of theorem 13.1.1.

The Schlegel diagram of any 3-polytope realizing a given graph $\mathscr{G}$ shows that

1. *Every 3-connected planar graph $\mathscr{G}$ is combinatorially equivalent to the 1-skeleton of some 2-diagram.*

Since the 1-skeleton of every 2-diagram is obviously 3-connected, we have:

2. *Every 2-diagram is combinatorially equivalent to a Schlegel diagram of some 3-polytope.*

Since all the steps in the proof of theorem 13.1.1 may equally well be carried out in *rational* 3-space, the affirmative solution to Klee's problem (see section 5.5) for 3-polytopes is given by

3. *Every combinatorial type of 3-polytopes may be realized in the rational 3-space.*

A slight modification of the proof in section 13.1 yields also the stronger result:

4. *For every 3-polytope $P \subset R^3$ and for each $\varepsilon > 0$ there exists a 3-polytope P^* combinatorially equivalent to P such that, in any pre-assigned cartesian system of coordinates in R^3, all vertices of P^* have rational coordinates, the distance between corresponding vertices of P and P^* being less than ε.*

The proof of Steinitz's theorem may be modified to yield characterizations of graphs of certain special families of 3-polytopes. For example, we have:

5. *A graph $\mathscr{G}$ is realizable by a centrally symmetric 3-polytope if and only if $\mathscr{G}$ is planar and 3-connected, and there exists an involutory mapping φ of $\mathscr{G}$ such that for each node N of $\mathscr{G}$, the nodes N and $\varphi(N)$ are separated by a circuit in $\mathscr{G}$.*

We shall only sketch a proof of theorem 5, leaving out the details. An equivalent formulation for the condition imposed on the involution φ is that a node N and its image $\varphi(N)$ are never incident to the same face of $\mathscr{G}$. The proof is essentially a repetition of the proof of theorem 13.1.1, with the following exceptions: The least possible value of $e(\mathscr{G})$ is 12; in case of equality $\mathscr{G}$ is either the graph of the octahedron, or the graph of the cube. The involution φ of $\mathscr{G}$ induces an involution (also denoted by φ) on $I(\mathscr{G})$; thus we may speak about a lens $\mathscr{L}$ in $I(\mathscr{G})$ and its image $\varphi(\mathscr{L})$, etc. The reductions are applied simultaneously to nodes D and $\varphi(D)$, or to triangular faces ABC and $\varphi(A)\varphi(B)\varphi(C)$. The polytope P is derived from P^* by performing the necessary changes in a fashion preserving central symmetry. If $\mathscr{G}^*$ results from $\mathscr{G}$ by a reduction of type η_1, the two new nodes are defined to be images of each other under φ. This is possible since those two nodes are not incident to the same face of $\mathscr{G}^*$ provided no node and its φ-image are incident to one face in $\mathscr{G}$. The necessity of applying a projective transformation to P^* in case of a reduction of type ω_0 does not arise if P^* is centrally symmetric.

The only complication in the induction is caused by the possibility that the chosen indecomposable lens $\mathscr{L}$ with a minimal number of faces is incident at its poles to $\varphi(\mathscr{L})$, while $\mathscr{L}$ has no inner nodes. (It can be shown that $\mathscr{L}$ and $\varphi(\mathscr{L})$ have at most the poles in common.) In this case (see figure 13.2.1) the ω_0 and η_0 reductions of $\mathscr{L}$ do not change the value of $g(\mathscr{G})$.

Now, if the complementary lenses $\mathscr{M}$ and $\varphi(\mathscr{M})$ are decomposable, we take an indecomposable lens $\mathscr{N}$ contained in $\mathscr{M}$, and perform the reductions ω_i or η_i indicated by $\mathscr{N}$. Since $\mathscr{N}$ is properly contained in $\mathscr{M}$,

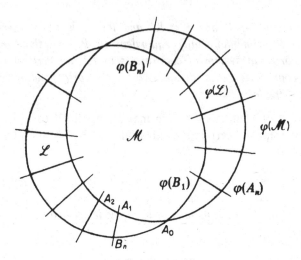

Figure 13.2.1

and therefore not incident to $\varphi(\mathcal{N})$, a finite number of reductions ω_0 and η_0 yields a graph to which some ω_i or η_i, $i \geq 1$, can be applied; a graph with fewer edges results, and induction takes over.

If $\mathcal{M}$ is indecomposable, we consider the cut of $\mathcal{M}$ determined by A_1 (see figure 13.2.1). Together with its extension $B_n A_1$, and the arc $B_n A_0 \varphi(B_1) \cdots \varphi(B_k)$ it determines a lens $\mathcal{L}^*$ which is indecomposable (since each proper sublens of it would be a proper sublens of $\mathcal{M}$). If $\mathcal{L}^*$ contains inner nodes, there are triangular faces of $\mathcal{L}^*$ incident to the cut determined by A_1 but not incident to $\mathcal{L} \cup \varphi(\mathcal{L})$. Performing a finite number of ω_0 and η_0 reductions (without increasing $g(\mathcal{G})$) all inner nodes may be eliminated from $\mathcal{L}^*$. In case the endpoint $\varphi(B_k)$ of the cut of $\mathcal{M}$ determined by A_1 is different from $\varphi(B_n)$, the value of g decreases, and induction takes over. Otherwise, we repeat the same argument for the cut of $\mathcal{M}$ determined by A_2. This leads to a graph $\mathcal{G}_0$, with $I(\mathcal{G}_0)$ containing the configuration of figure 13.2.2 and its φ-image. (Note that the nodes E and $\varphi(A_n)$ are necessarily different.) Therefore $\mathcal{G}_0$ contains the configuration of figure 13.2.3 or its dual, and thus is reducible to that of figure 13.2.4 or its dual, with fewer edges.

This completes the proof of theorem 5.

Another result of the same type is

6. *A graph $\mathcal{G}$ is realizable by a 3-polytope having a plane of symmetry if and only if $\mathcal{G}$ is planar and 3-connected, and there exists an involutory mapping φ of $\mathcal{G}$ which reverses the orientation of the faces.*

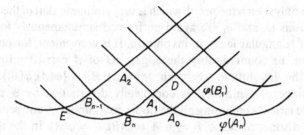

Figure 13.2.2

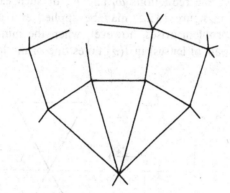

Figure 13.2.3

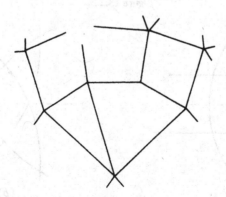

Figure 13.2.4

We again only sketch the proof, which is very similar to that of theorem 5. The reductions ω_i and η_i are again performed simultaneously for pairs of nodes, or triangular faces, in involution. It is convenient, for purposes of induction, to count doubly the edges AB of $\mathscr{G}$ carried into themselves by the involution, i.e. such that $\{A, B\} = \{\varphi(A), \varphi(B)\}$. Note that the 'plane of symmetry' is completely determined by $\mathscr{G}$ and φ; it is a simple circuit containing all the nodes fixed under φ, and separating all pairs of distinct nodes A, B with $A = \varphi(B), B = \varphi(A)$. In the illustrations the plane of symmetry shall be indicated by a dotted line.

Interference with the application of the reductions ω_i and η_i can occur only if the elements concerned are interrelated by φ. Even in those cases, either the reductions ω_i and η_i, or such easy variants as that represented in figure 13.2.5, may be applied and used to reduce $e(\mathscr{G})$ or $g(\mathscr{G})$. A problem arises, however, when the minimal indecomposable lens (or pair of lenses) in $I(\mathscr{G})$ takes one of the forms indicated in figure 13.2.6.

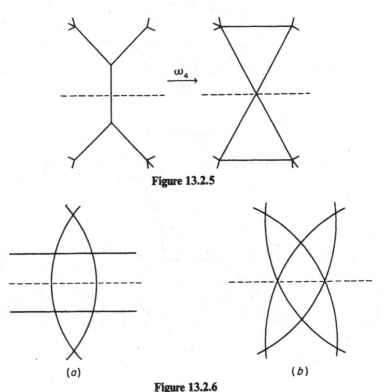

Figure 13.2.5

(a)   (b)

Figure 13.2.6

In the latter case, which corresponds in $\mathscr{G}$ to the configuration of figure 13.2.7(a) (or its dual), one may reduce it to the configuration of figure 13.2.7(b) (or its dual).

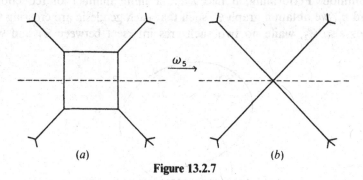

(a)   (b)

Figure 13.2.7

Therefore only the occurrence of a minimal lens $\mathscr{L}$ with three faces (figure 13.2.6(a)) has to be investigated. In this case $\mathscr{G}$ contains the configuration of figure 13.2.8, or its dual. A symmetric reduction of any of the types used so far would reduce the number of faces in $\mathscr{L}$ by at least two, i.e. would make $\mathscr{L}$ disappear as a lens.

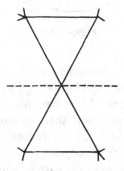

Figure 13.2.8

We apply practically the same idea we used in the proof of theorem 5, and consider the two geodesic lines passing through the cuts of $\mathscr{L}$. (We do not assert that they are distinct.) If they would intersect more than once, they would determine a lens, non-incident to $\mathscr{L}$, which would contain an indecomposable lens and reductions indicated by this lens (including ω_4 and ω_5) would after a finite number of steps decrease $e(\mathscr{G})$ or $g(\mathscr{G})$. Thus, the two geodesic lines are either different, each of them being simple, or (see figure 13.2.9) they form one closed geodesic with one selfintersection. In either case, we have two simple circuits $\mathscr{C}_1$ and $\mathscr{C}_2$, determined by the two cuts of $\mathscr{L}$, having at most one node in common. Performing, if necessary, a finite number of reductions ω_i and η_i, we obtain a graph $\mathscr{G}'$ such that each geodesic arc crossing $\mathscr{C}_1$ crosses also $\mathscr{C}_2$, while no two such arcs intersect between $\mathscr{C}_1$ and $\mathscr{C}_2$.

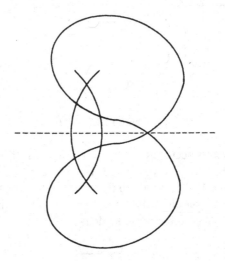

Figure 13.2.9

In other words, $\mathscr{C}_1$ and $\mathscr{C}_2$ determine an indecomposable 'singular lens'* with no inner nodes. Applying alternatingly ω_0 and η_0 reductions, one of the two triangular faces of this lens may be made adjacent to $\mathscr{L}$ (figure 13.2.10). The graph itself then contains the configuration of figure 13.2.11(a) (or its dual), and a reduction is possible (see figure 13.2.11(b)) to a graph with fewer edges.

Therefore there remains only the case in which $\mathscr{C}_1$ and $\mathscr{C}_2$ are closed geodesics, without common nodes (see figure 13.2.12). Let $\mathscr{L}_1$ and $\mathscr{L}_2$ be the extensions of the geodesic arcs bounding $\mathscr{L}$ into the region determined by $\mathscr{C}_1$. If the lens determined by $\mathscr{L}_i$ and $\mathscr{C}_1$ were decomposable, a reduction to a graph with fewer edges would be possible. Thus those lenses may be assumed to be indecomposable, and performing, if necessary, a

---

* The definition of 'singular lens' is obtained from that of 'lens' by permitting the poles to coincide. In other words, a singular lens has as its boundary a closed geodesic with exactly one self-intersection. It is easily checked that lemma 13.1.2 and its proof remain valid for singular lenses.

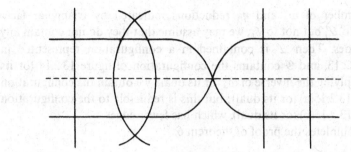

Figure 13.2.10

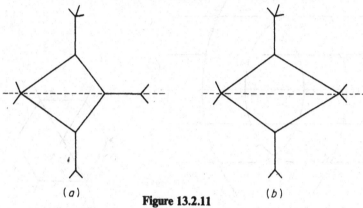

(a)　　Figure 13.2.11　　(b)

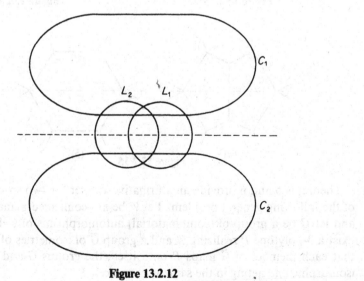

Figure 13.2.12

finite number of ω_0 and η_0 reductions indicated by triangular faces incident to $\mathscr{L}_i$ but not to $\mathscr{C}_1$, we may assume that they do not contain any inner nodes. Then $\mathscr{L}$ is contained in a configuration represented in figure 13.2.13, and $\mathscr{G}$ contains the configuration of figure 13.2.14 (or its dual). Applying the inverse of ω_4 (or its dual), we obtain the configuration of figure 13.2.15(a) (or its dual); but this is reducible to the configuration of figure 13.2.15(b) (or its dual), which has fewer edges.

This completes the proof of theorem 6.

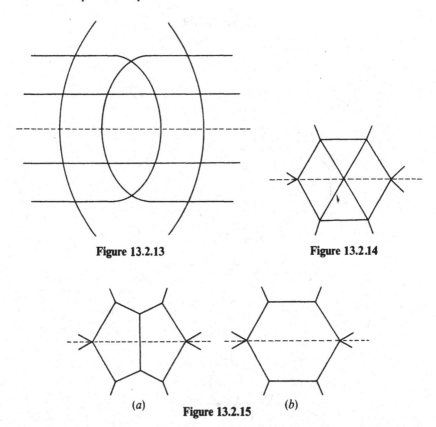

Figure 13.2.13   Figure 13.2.14

(a)   Figure 13.2.15   (b)

Theorems 5 and 6 provide an affirmative answer for two special cases of the following general problem: Let $\mathscr{G}$ be a 3-connected planar graph, and let G be a group of (combinatorial) automorphisms of $\mathscr{G}$. Do there exist a 3-polytope P realizing $\mathscr{G}$ and a group $\bar{G}$ of isometries of R^3 such that each member of $\bar{G}$ maps P onto itself, the groups G and $\bar{G}$ being isomorphic and acting in the same way on $\mathrm{skel}_1 P$.

Exercises

1. Show by a simple example that the following statement, which generalizes theorem 2, is false: Every 2-diagram is projectively equivalent to a Schlegel diagram of some 3-polytopes.

2. Derive from Steinitz's theorem the following result of Fáry [1]: Every planar graph with no 1- or 2-circuits has representations in the plane such that each edge is a segment.

3. Theorem 2 may be reformulated as follows: Each abstract 2-complex $\mathscr{C}$ homeomorphic to the 2-sphere is realizable by the boundary complex of a 3-polytope; in particular, $\mathscr{C}$ is realizable by a 2-complex in R^3.

It is well known (Dehn–Heegard [1], Seifert–Threlfall [1]) that every closed, orientable, topological 2-manifold $\mathscr{M}$ without boundary is imbeddable in R^3. In analogy to Steinitz's theorem, one may ask whether every abstract 2-complex $\mathscr{C}$ homeomorphic to such a 2-manifold $\mathscr{M}$ is realizable by a 2-complex in R^3, or in any Euclidean space. As mentioned above, such a realization is possible (in R^3) if $\mathscr{M}$ is of genus 0 (i.e. is a 2-sphere). If $\mathscr{C}$ is simple then (see exercise 11.1.7) it is not realizable by a 2-complex in any Euclidean space unless $\mathscr{M}$ is the 2-sphere. If $\mathscr{C}$ is simplicial, it is trivially realizable by a 2-complex in R^5; however, no example is known to contradict the conjecture that each simplicial $\mathscr{C}$ is realizable by a 2-complex in R^3. (For an interesting realization by a 2-complex in R^3 of the abstract 2-complex homeomorphic to the triangulation of the torus (genus 1) determined by 7 vertices and 21 edges, see Császár [1]. Another realization in R^3 of the same abstract 2-complex may be obtained by omitting suitable 2-faces from the 2-skeleton of the Schlegel diagram of the cyclic polytope $C(7, 4)$.)

13.3 Eberhard's Theorem

The problem of determining the possible f-vectors of d-polytopes, which we discussed in chapters 8, 9, and 10, may be refined to the question: what combinatorial types of $(d-1)$-polytopes, and how many of each type, may be combined to form the boundary complex of a d-polytope? This problem is not completely solved even for $d = 3$, but significant partial results are known; they form the subject of the present section and the following one.

We shall say that a sequence $(p_k) = (p_3, p_4, \cdots)$ of nonnegative integers is *3-realizable* provided there exists a *simple* (i.e. 3-valent) *3-polytope* P such that (using the notation introduced in section 13.1) we have

$p_k = p_k(P)$ for all $k \geq 3$. Since for simple 3-polytopes $v = v_3$, we have $3v = 2e = \sum_{k \geq 3} k p_k$. Together with $p = \sum_{k \geq 3} p_k$ and Euler's relation $v + p = e + 2$, this yields:

1. *A necessary condition for the 3-realizability of a sequence (p_k) is*

$$(*) \qquad\qquad 3p_3 + 2p_4 + p_5 = 12 + \sum_{k \geq 7} (k - 6)p_k.$$

Theorem 1 clearly implies that each simple 3-polytope contains at least four faces each of which has at most five edges. This fact has manifold extensions and applications in various combinatorial problems and in graph theory (see, for example, Lebesgue [1], Ringel [3], Grötzsch [1], Grünbaum [8]).

One of the interesting features of theorem 1 is that it contains no information about p_6. Nevertheless, it is well known (Eberhard [3], Brückner [2]; see also table 13.3.1) that of two sequences which differ only in p_6, one may be realizable while the other is not. (See section 13.4 for the known results on this question.) The only general result in the direction of determining to what extent is the condition of theorem 1 sufficient for the 3-realizability of a sequence (p_k) is the following theorem of Eberhard [3]:

2. *For every sequence $(p_k \mid 3 \leq k \neq 6)$ of nonnegative integers satisfying* $(*)$, *there exist values of p_6 such that the sequence $(p_k \mid k \geq 3)$ is 3-realizable.*

The proof of theorem 2 being somewhat involved, we shall first give a proof of an analogous but simpler result. It deals with 4-realizable sequences; a (finite) sequence $(p_k \mid k \geq 3)$ is said to be *4-realizable* provided there exists a 3-polytope P having only 4-valent vertices, such that $p_k = p_k(P)$ for all $k \geq 3$. As in the proof of theorem 1, it is easy to deduce from Euler's relation that all such P satisfy

$$(**) \qquad\qquad p_3 = 8 + \sum_{k \geq 5} (k - 4)p_k,$$

which is therefore a necessary condition for the 4-realizability of a sequence (p_k). We have

3. *For every solution $(p_k \mid 3 \leq k \neq 4)$ of equation $(**)$ a value p_4 exists such that the sequence $(p_k \mid k \geq 3)$ is 4-realizable.*

PROOF OF THEOREM 3   Let $(p_k \mid 3 \leq k \neq 4)$ be a solution of $(**)$. If a polytope is to be found 4-realizing (p_k), we may consider the p_3 triangles to be of two different types, each type fulfilling a specific purpose. Eight of the triangles serve to "close up" the polytope, their function being analogous to that of the eight triangular faces of the octahedron. The remaining $f_3 - 8$ triangles, teamed in groups of $k - 4$ with the different k-gons, $k \geq 5$, serve to 'compensate' for the 'excessive' number of edges of those k-gons. We shall prove the theorem by

(i) describing how to form certain standard 'building blocks,' each block consisting of a k-gon, $k \geq 5$, $k - 4$ triangles, and a suitable number of quadrangles;

(ii) Showing how those blocks may be combined to form a 'basis';

(iii) 'closing up' the basis by the remaining 8 triangles and a suitable number of quadrangles.

The final result will be a 3-connected 4-valent planar graph having p_k k-gonal 'countries' ($3 \leq k \neq 4$) and a certain number of quadrangles; by Steinitz's theorem 13.1.1 this is sufficient for the existence of a polytope 4-realizing the sequence (p_k) and thus proves theorem 3.

(i) To form a *k-block* we start from a square and mark on each of two of its consecutive edges $k - 4$ points, $A_1 A_2, \cdots, A_{k-4}$ respectively $B_1, \cdots, B_{k-4}$ (see figure 13.3.1, where $k = 7$). Connecting the corresponding points A_j, B_j by segments, we divide the square into one k-gon, $k - 4$ triangles and $\binom{k-4}{2}$ quadrangles. The square subdivided in this way is a *k-block*.

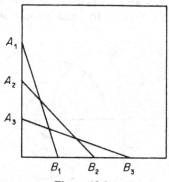

Figure 13.3.1

(ii) To form the 'basis' we take, for all $k \geq 5$, p_k k-blocks, arrange them diagonally (as in figure 13.3.2), and complete the basis (by the light lines in figure 13.3.2) to a quadrilateral shape. Note that the two members of one pair of consecutive 'edges' of this basis are subdivided into equally many parts, the same assertion being true for the other two 'edges'.

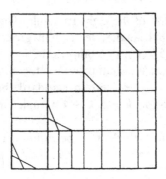

Figure 13.3.2

(iii) To 'close up' the basis we connect the corresponding points on the 'edges' of the quadrilateral by lines as indicated in figure 13.3.3. This clearly introduces, besides quadrangles, exactly 8 triangles (shaded in figure 13.3.3). Thus the proof of theorem 3 is completed.

PROOF OF THEOREM 2   Using a similar method, we shall now prove Eberhard's theorem. Equation (*) suggests the assignment of 'curvature units' such that each pentagon has one unit, each quadrangle two, and each triangle three units. Then our first aim will be to choose p_3' triangles, p_4' quadrangles, and p_5' pentagons from among the p_3, p_4, p_5 available ones, in such a way as to yield a total of 12 curvature units.

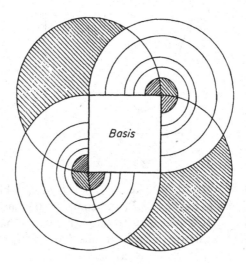

Figure 13.3.3

This is not always possible and we shall first dispose of the few cases in which p_3, p_4 and p_5 are such that this choice is impossible. Then necessarily either $p_3 = 3$, $p_4 = 2$, or $p_3 = 1$, $p_4 = 5$; in both cases $p_7 = 1$ and $p_k = 0$ for $k = 5$ and $k \geq 8$. Both solutions of (*) are realizable with $p_6 = 3$ (see figure 13.3.4).

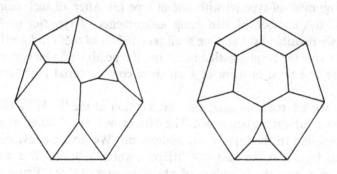

Figure 13.3.4

In the remaining part of the proof we may assume that we have chosen p_3' triangles, p_4' quadrangles, and p_5' pentagons which together have 12 curvature units; they will be used to "close up" the polytope. The remaining $p_3'' = p_3 - p_3'$, $p_4'' = p_4 - p_4'$, $p_5'' = p_5 - p_5'$ polygons of this type we shall endeavour to split into groups teamed with the k-gons, $k \geq 7$. As suggested by equation (*), it is convenient to group a k-gon ($k \geq 7$) with a_3 triangles, a_4 quadrangles, and a_5 pentagons, where a_3, a_4, a_5 are such that

(***) $$3a_3 + 2a_4 + a_5 = k - 6.$$

Our construction would be somewhat simpler if it were always possible to distribute the available $p_3'' + p_4'' + p_5''$ polygons with at most 5 sides among the $\sum_{k \geq 7} p_k$ k-gons in this manner. However, if p_5'' is relatively small, this is not always possible. Instead, we must take into consideration the possibility that for a given k-gon, either a_3 has a fractional part ($\frac{1}{3}$ or $\frac{2}{3}$) or that a_4 has a fractional part ($\frac{1}{2}$). It is easily seen that the case in which both a_3 and a_4 have non-zero fractional parts is easily transformed into a case in which at most one of them is not an integer. At the end of this 'planning' stage for our construction we are therefore left with assignments satisfying (***), among which we distinguish the following types:

(a) a_3, a_4, a_5 are integers;

(b) a_3 and a_5 are integers, a_4 has fractional part $\frac{1}{2}$;

(c) a_4 and a_5 are integers, a_3 has fractional part $\frac{1}{3}$;

(d) a_4 and a_5 are integers, a_3 has fractional part $\frac{2}{3}$.

Clearly, the number of assignments of type (b) is even, and we shall consider the corresponding polygons in pairs. Also, we shall pair off one assignment of type (c) with one of type (d). After all such pairs are formed, the number of remaining assignments of type (c) or (d) is obviously divisible by 3 and we shall take triples of them to form larger units. In case the surplus assignments are of type (d) we find it convenient to make one third of them of a different combinatorial type (denoted by (d^*)).

Now we are ready to start the construction of the 'building blocks', one block for each assignment. The blocks will be of different types, corresponding to the type of the assignment. We find it convenient to form all blocks in the shape of 'triarcs', where a 'triarc' is a simply-connected region the boundary of which consists of three 'Petrie arcs'† (that is, paths taking alternatingly the left or the right edge at each successive node) connected at their ends as in figure 13.3.5. (The dots denote endpoints of the Petrie arcs.)

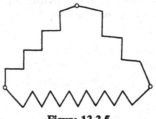

Figure 13.3.5

Each block, corresponding to an assignment of a k-gon, $k \geq 7$, and numbers a_3, a_4, a_5 satisfying (***), will consist of a suitable number of hexagons together with a k-gon and $3a_3 + 2a_4 + a_5 = k - 6$ pentagons.

For blocks of type (a), the $k - 6$ pentagons are in one row adjacent to the k-gon, and flanked by one hexagon on each side. (See figure 13.3.6, in which $k = 11$.)

For blocks of types (b) and (c), only $k - 7$ of the pentagons are in one row, the remaining pentagon being on one of the other arcs of the triarc. (See figure 13.3.7, where again $k = 11$.) In case $k = 7$ a block of type (b) or (c) may be taken to be the same as a block of type (a).)

†See Coxeter [1], pp. 24 and 223.

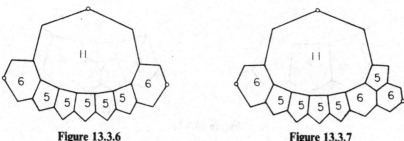

Figure 13.3.6   **Figure 13.3.7**

For blocks of type (d) $k - 8$ pentagons are in one row, the remaining two being on one of the other arcs of the triarc. (See figure 13.3.8 for an illustration of the case $k = 11$.) For blocks of type (d*) the two remaining pentagons are one on each of the remaining arcs of the triarc (figure 13.3.9). Again the lowest possible case, $k = 8$, is to be treated separately; for type (d) a block of type (a) may be used, while for type (d*) the block is represented in figure 13.3.10.

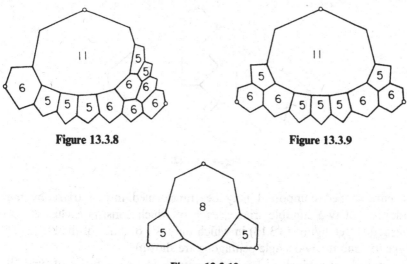

Figure 13.3.8   **Figure 13.3.9**

Figure 13.3.10

In the next stage the above blocks will be modified, preserving their shape as triarcs, by replacing $3[a_3]$ pentagons by $[a_3]$ triangles. This is achieved by inserting, instead of three successive pentagons, five hexagons and one triangle (see figure 13.3.11). From the remaining pentagons, we designate $2[a_4]$ as $[a_4]$ pairs such that the members of each pair have a common edge; at a later stage each such pair will yield a quadrangle.

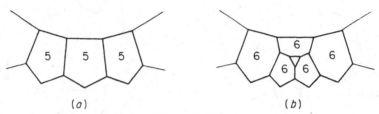

Figure 13.3.11

Now we shall take care of the peculiarities of blocks which are not of type (a). Two blocks of type (b) may be joined in such a way that their single pentagons have a common edge, which will enable us at a later stage to transform the pair into a quadrangle (see figure 13.3.12). A small complication arises however at this step (and in the analogous cases dealing with blocks of type (d)): the resulting compound of two blocks of type (b) is not a triarc. But this is easily remedied by noting that the

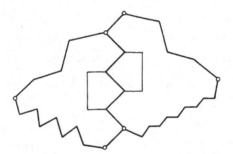

Figure 13.3.12

lozenge-shaped compound may be transformed into a triarc by the addition of two suitable triarcs, each of which consists exclusively of hexagons (see figure 13.3.13) in which only the outlines of the blocks of type (b), and the two single pentagons are shown).

Similarly, if a block of type (c) is combined with a block of type (d) we obtain three pentagons with a common vertex (see figure 13.3.14). 'Cutting off' this vertex (indicated by dashed lines in figure 13.3.14) yields one triangle and three hexagons. The addition of a row of hexagons (light lines in figure 13.3.14) yields a lozenge-shaped compound which may be transformed (as above) into a triarc. Completely analogous is the combination of two blocks of type (d) with one of type (d*) to yield two triangles (and six hexagons).

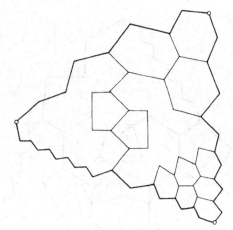

Figure 13.3.13

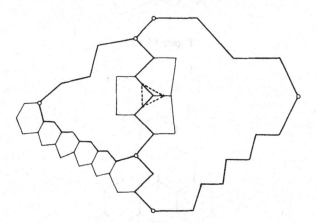

Figure 13.3.14

If three blocks of type (c) are to be combined, the three single pentagons yield a triangle (and hexagons) as illustrated in figure 13.3.15.

The following stage consists of including all the blocks of type (a) and all the compounds formed from blocks of the other types into one triarc. The possibility of this construction follows at once from the remark that two triarcs may be joined, together with a lozenge of hexagons, to form a new triarc (see figure 13.3.16 in which the outlines of the two triarcs are heavily drawn, while the lozenge-hexagons are lightly

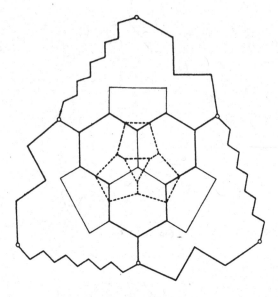

Figure 13.3.15

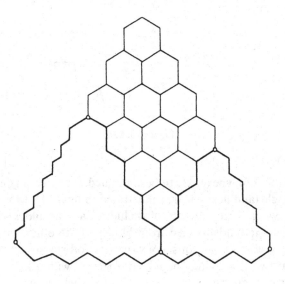

Figure 13.3.16

drawn). The triarc we obtain at the end of this stage contains a certain number of hexagons, p_k k-gons for all $k \geq 7$, p_3'' triangles, and $p_5'' + 2p_4''$ pentagons among which there are p_4'' pairs of pentagons with a common edge, destined to be transformed into p_4'' quadrangles.

Next we transform the triarc by inserting a hexagon instead of each edge (see figure 13.3.17, in which the result of this insertion is shown for the triarc of figure 13.3.6; the starting triarc is drawn in dashed lines).

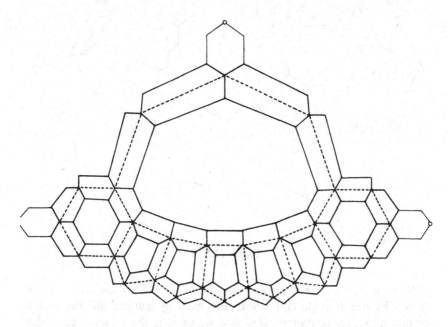

Figure 13.3.17

The addition of three hexagons (light lines in figure 13.3.17) yields a new triarc. The important aspect of this construction is that in the resulting triarc the number of edges of each arc is divisible by 4. Therefore, adding on one arc of the triarc a row of hexagons, we obtain the 'basis', which is again a triarc each arc of which has a number of edges which is of the form $4n + 2$. Placing at the center of each arc of the triarc three pentagons, and using three more pentagons and a suitable number of hexagons, the 'basis' may be 'closed up'. (See figure 13.3.18 which illustrates the case in which the 'basis' triarc has arcs consisting of 6, 10, and 14 edges, respectively.)

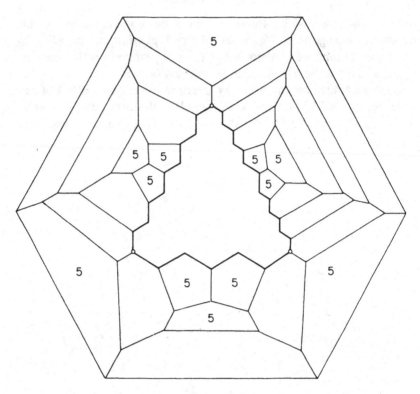

Figure 13.3.18

Now, if $p'_3 \leq 3$, we replace p'_3 triples of pentagons (adjacent to the 'base') by one triangle and 3 hexagons each by 'cutting off' the vertex common to three pentagons. If $p'_3 = 4$, we perform this on all three triples of pentagons and, in addition, we replace the outer hexagon and the three pentagons adjacent to it by the configuration of figure 13.3.19, which consists of one triangle and 6 hexagons. From the remaining $12 - 3p'_3 = 2p'_4 + p'_5$ pentagons, p'_4 pairs may be chosen which consist either of pentagons with a common edge, or of two pentagons joined by a chain of hexagons in which each member is joined to its neighbors along two opposite edges (figure 13.3.20). Note that the 'basis' contains p''_4 such pairs, in each of which the chain consists of a single hexagon.

The construction so far has yielded a polytope (or, rather, a 3-connected 3-valent planar graph) having the desired number p_k of k-gonal faces for $k = 3$ and $k \geq 7$, a certain number of hexagons, and containing $2p_4 + p_5$ pentagons, among which there are p_4 pairs of the type described above.

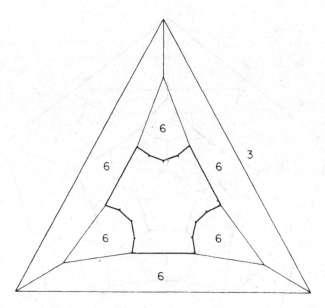

Figure 13.3.19

(Note that no hexagon is involved in more than one such pair.) The last stage of the construction will consist of a transformation of those p_4 pairs of pentagons into p_4 quadrangles (accompanied by an increase in the number of hexagons).

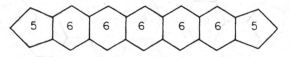

Figure 13.3.20

To achieve this, we first replace each vertex of the polytope by a hexagon (having three edges in common with other new hexagons, and three with faces which were incident to the vertex). Figure 13.3.21 illustrates this operation for the case of the pentagonal prism (the prism is indicated by dashed lines). Chains of the type of figure 13.3.20 are transformed into chains of the type represented in figure 13.3.22. A chain of the latter type is easily modified into another of the same type but having one *inter-mediate hexagon* less. (In figure 13.3.22 there are 3 intermediate hexagons.)

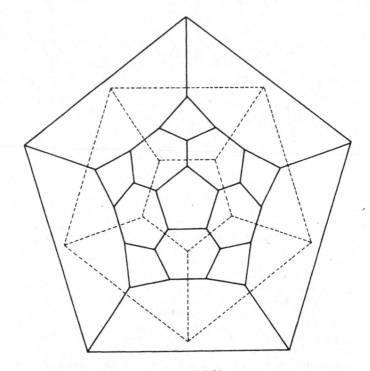

Figure 13.3.21

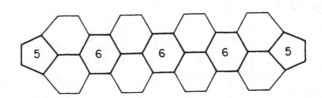

Figure 13.3.22

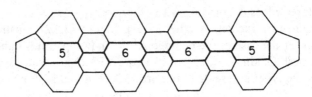

Figure 13.3.23

Figure 13.3.23 represents this modification as applied to the chain of figure 13.3.22. Repeating this process we arrive at chains with no intermediate hexagons (figure 13.3.24). However, these are at once transformable into a quadrangle (and two hexagons), as shown in figure 13.3.25.

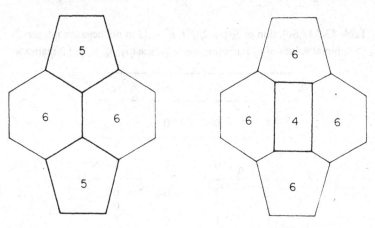

Figure 13.3.24   **Figure 13.3.25**

Modifying in this manner each of the p_4 pairs of pentagons in the polytope constructed above, we obtain a new polytope which has the desired number p_k of k-gonal faces for all $k \geq 3$, $k \neq 6$.

This completes the proof of theorem 2.

The proof of theorem 2 may be used to obtain upper bounds for the least value of p_6 rendering 3-realizable a given sequence (p_k)†. However, bounds obtained in this manner seem to be very poor.

CONJECTURE 1   There exists a constant c such that every sequence (p_k) satisfying (*) may be 3-realized for some value of p_6 satisfying

$$p_6 \leq c \sum_{3 \leq k \neq 6} p_k.$$

It is easy to find examples (see exercise 2) showing that if c exists, it must be greater than 1.

---

† It is well known that $p_6 = 0$ is not always possible. For example, there exist 19 sequences satisfying (*) such that $p_k = 0$ for all $k \geq 7$; the corresponding minimal values of p_6 have been determined experimentally (Eberhard [3], Brückner [2]) and found to vary between 0 and 3 (see table 13.3.1).

CONJECTURE 2 Every sequence (p_k) satisfying (*) may be 3-realized for some value of p_6 satisfying

$$p_6 < \max\{k \mid p_k \neq 0\}.$$

Similar conjectures may be formulated for 4-realizable sequences.

Table 13.3.1. Solution of $3p_3 + 2p_4 + p_5 = 12$ in nonnegative integers, and minimal values of p_6 rendering the sequence (p_3, p_4, p_5, p_6) 3-realizable

p_3	p_4	p_5	p_6
4	0	0	0
3	1	1	3
3	0	3	1
2	3	0	0
2	2	2	0
2	1	4	1
2	0	6	0
1	4	1	2
1	3	3	0
1	2	5	1
1	1	7	2
1	0	9	3
0	6	0	0
0	5	2	0
0	4	4	0
0	3	6	0
0	2	8	0
0	1	10	2
0	0	12	0

Exercises

1. Show that if a sequence (p_k) is 3-realizable for some value of p_6, then it is 3-realizable for infinitely many values of p_6. Formulate and prove the analogous statement concerning 4-realizability.

2. Show that $p_6 = 12$ is the least value of p_6 rendering 3-realizable the sequence (p_k), where $p_3 = 7$, $p_{15} = 1$, and $p_k = 0$ for $k \neq 3, 6, 15$.

3. If n is an integer ≥ 3, the sequence (p_k) determined by $p_{3n} = 1$, $p_3 = n + 2$, $p_k = 0$ for $k \neq 3, 6, 3n$, satisfies (*). Show that it is 3-realizable with $p_6 = 3n - 4 - (-1)^n$. If it were known that this is the least possible value for p_6, it would follow that the constant c in conjecture 1 is at least 3.

4. Show that the validity of conjecture 2 implies the validity of conjecture 1, with $c \leq 3$.

5. Let $(p_k \,|\, 3 \leq k \neq 6)$ be a sequence of nonnegative integers satisfying $\sum_k (6 - k)p_k = 12 + 2s$, where s is a nonnegative integer. Show that there exists a 3-polytope P, singular of degree s, such that $p_k(P) = p_k$ for all $k \neq 6$ (Eberhard [3]). (A 3-polytope P is *singular of degree s* provided $s = \sum_k (k - 3)v_k(P)$.)

6. The proofs of theorems 2 and 3 consist of quite disparate parts: (i) The construction of a 3-valent or 4-valent 3-connected graph imbedded in the plane, such that the resulting map has the correct number of faces of various kinds; (ii) An application of Steinitz's theorem to deduce the existence of a 3-polytope from the existence of the map. If one is interested in planar maps only, the problem may be generalized by allowing monogons and digons as faces. Equation (*) becomes

$$5p_1 + 4p_2 + 3p_3 + 2p_4 + p_5 = 12 + \sum_{k \geq 7} (k - 6)p_k\,;$$

there are 28 solutions with $p_1 + p_2 > 0$, $p_k = 0$ for $k \geq 7$, all of which are realizable by 3-valent planar maps. Equation (**) becomes

$$3p_1 + 2p_2 + p_3 = 8 + \sum_{k \geq 5} (k - 4)p_k\,;$$

there are 9 solutions with $p_1 + p_2 > 0$, $p_k = 0$ for $k \geq 5$, and they are all realizable by 4-valent planar maps. Probably it is possible to generalize the proofs of theorems 2 and 3 to situations in which monogons and digons are allowed; the main obstacle is probably not hard in principle, but only time-consuming: the number of 'exceptional cases' (similar to those we disposed of by figure 13.3.4 in case of theorem 2) is rather large. Most other steps of the proof should be easily modified.

7. Another problem worth considering is whether theorems 2 and 3 (or their generalizations mentioned in exercises 5 and 6) have valid analogues for maps on surfaces other than the plane (sphere).

8. If P is a centrally symmetric 3-polytope, then all $p_k(P)$ are even. Provided all p_k are even, one may probably assume the polytopes in theorems 2 and 3 to be centrally symmetric.

9. It is easily checked that if P is a simple 3-polytope such that $p_3(P) > 4$ then $v(P) \geq 3p_3(P)$. It would be interesting to find similar relations involving other $p_k(P)$, and also dealing with 4- or 5-valent polytopes, or with all 3-polytopes.

10. (Eberhard [3]; Brückner [2]) show that every simple 3-polytope with $p + 1 \geq 6$ facets can be obtained from a simple 3-polytope P with p facets by (at least) one of the three types of 'cutting off' parts of P, illustrated in figure 13.3.26: cutting off (i) a vertex; (ii) an edge; (iii) two edges with a common vertex.

11. (Klee) Show that if a 3-polytope has more faces than vertices, it has at least six triangular faces.

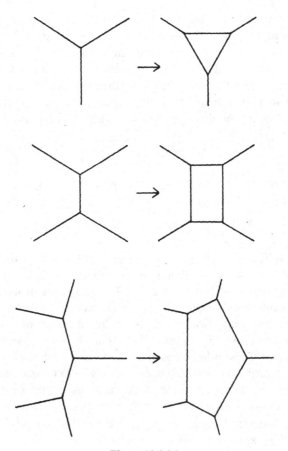

Figure 13.3.26

12. Let a sequence (p_k) be called 5-*realizable* provided there exists a 5-valent 3-polytope P such that $p_k = p_k(P)$ for all k.

(i) Show that a necessary condition for 5-realizability of (p_k) is

(****) $$p_3 = 20 + \sum_{k \geq 4} (3k - 10)p_k.$$

(ii) In contrast to the case of 3-realizable or 4-realizable sequences, (****) involves all the p_k's. However, (****) is not sufficient for 5-realizability. Prove this assertion with the example $p_3 = 22$, $p_4 = 1$, $p_k = 0$ for $k \geq 5$.

It would be interesting to find sufficient conditions for 5-realizability.

13. The following result of Barnette [3] (somewhat analogous to theorem 18.2.9) is very interesting and invites extensions in different directions:

If (p_k) is a 3-realizable sequence and if $p = \sum_{k \geq 3} p_3$, then for every positive integer m

$$\sum_{i \geq \frac{2p + 6m + 2}{m + 1}} p_i \leq m.$$

13.4   Additional Results on 3-Realizable Sequences

The present section deals with certain recent results concerning conditions for the 3-realizability of sequences $(p_k \mid k \geq 3)$ of nonnegative integers which satisfy the equation $\sum_{k \geq 3} (6 - k)p_k = 12$.

The first result determines the values of p_6 rendering realizable certain particular sequences† :

1. *Sequences* $(0, 6, 0, p_6)$ *and* $(0, 0, 12, p_6)$ *are realizable if and only if* $p_6 \neq 1$. *The sequence* $(4, 0, 0, p_6)$ *is 3-realizable if and only if* p_6 *is an even integer different from 2. The sequence* $(3, 1, 1, p_6)$ *is 3-realizable if and only if* p_6 *is an odd integer greater than* 1.

The existence assertions of theorem 1 may be established by easy examples (see exercise 1). The only non-trivial part is the assertion that p_6 is even if $(4, 0, 0, p_6)$ is 3-realizable, and odd if $(3, 1, 1, p_6)$ is 3-realizable. A proof of the former assertion, based on a complete determination of possible combinatorial types of realizations of $(4, 0, 0, p_6)$, was given

† When writing a sequence (p_k) explicitly, we shall write out only that part of its beginning which interests us, leaving out the infinite sequence of zeros following it.

in Grünbaum–Motzkin [4] (see exercise 2). Here we shall obtain those parts of theorem 1 as immediate consequences of a more general result on planar graphs (theorem 4 below).

Though we are mainly interested in simple 3-polytopes, or equivalently, in 3-valent 3-connected planar graphs, our results hold for all 3-valent connected planar graphs. As a matter of fact the proofs become simpler if we do not insist on 3-connectedness, and if we allow digons.

For easier formulation of the following results we need some definitions.

If k is an integer greater than 1 we shall say that a face of a graph is a *multi-k-gon* provided the number of its edges is a multiple of k.

Let $k \geq 2$ and $n \geq 0$ be integers. We shall denote by $\mathscr{G}(k, n)$ any 3-valent connected planar graph with the property that all but n of its faces are multi-k-gons, while the n exceptional faces are not multi-k-gons. We shall use $\mathscr{G}^*(k, 2)$ to denote any $\mathscr{G}(k, 2)$ having the additional property that the two exceptional faces have a common edge.

With this terminology we have the results:

2(k). *Let k be 2, 3, 4, or 5. Then there exists no graph of type $\mathscr{G}(k, 1)$ or $\mathscr{G}^*(k, 2)$.*

3(k). *Let k be 2, 3, 4, or 5. Then each graph $\mathscr{G}(k, 0)$ is 2-connected.*

4(k). *The number $p(\mathscr{G}(k, 0))$ of faces $\mathscr{G}(k, 0)$ satisfies:*

$$p(\mathscr{G}(3, 0)) \equiv p(\mathscr{G}(3, 2)) \equiv 0 \pmod 2;$$
$$p(\mathscr{G}(4, 0)) \equiv 2 \pmod 4;$$
$$p(\mathscr{G}(5, 0)) \equiv 2 \pmod{10}.$$

We start with the proof of theorem 2. The nonexistence of $\mathscr{G}(2, 1)$ and $\mathscr{G}(3, 1)$ follows readily by considering modulo 2 or modulo 3 the relation

(*) $$\sum_{k \geq 2} (6 - k)p_k = 12.$$

In order to prove the nonexistence of $\mathscr{G}^*(2, 2)$ we shall describe a number of *reductions*, i.e. changes to be performed on a given graph which yield a graph of the same type but having fewer edges. The nonexistence of $\mathscr{G}^*(2, 2)$ will be established by showing that at least one reduction may be applied to every $\mathscr{G}^*(2, 2)$.

The first reduction is applicable in the case there is in $\mathscr{G}^*(2, 2)$ a digon which does not have an edge in common with a triangle. The digon and the edges incident to it are deleted and replaced by a single edge (see figure 13.4.1). This reduces by 2 the number of edges of each of the two faces incident to the digon (or, if only one face is incident to it, the number of its edges is reduced by 4). Therefore the resulting graph is again of type $\mathscr{G}^*(2,2)$.†

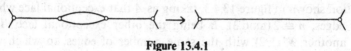

Figure 13.4.1

If the above reduction is not applicable, but there is a digon having an edge in common with a triangle, the reduction represented in figure 13.4.2 may be used.

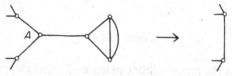

Figure 13.4.2

If the graph contains no digon, we choose a multi-2-face (denoted by A in figure 13.4.3) which has common edges with both exceptional faces (B and C in figure 13.4.3) and apply the transformation shown in figure 13.4.3. The resulting graph is again of type $\mathscr{G}^*(2,2)$ and has the same number of edges, but it contains a digon and thus one of the above reductions is applicable.

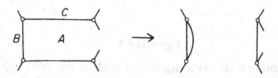

Figure 13.4.3

This completes the proof of nonexistence of graphs $\mathscr{G}^*(2,2)$.

We turn now to graphs $\mathscr{G}^*(3,2)$ and note that if the exceptional faces are an n-gon and an m-gon, then equation (*) implies that $n + m \equiv 0$ (mod 3).

If one of the exceptional faces is a digon, we use again the reductions shown in figures 13.4.1 and 13.4.2. A complication arises, however, with

† In most reductions to be discussed in the sequel, it will be left to the reader to check that the reduced graph is of the desired type.

the possibility that $\mathscr{G}^*(3,2)$ is not 2-connected and that the digon is adjacent to a single face along both its edges (then this face is the other exceptional face). In this case the reduction shown in figure 13.4.4 yields a $\mathscr{G}^*(3,2)$ with fewer edges. In case no digon is present we use the transformation shown in figure 13.4.3, taking as A that exceptional face which has n edges, $n \equiv 2 \pmod 3$, B being the other exceptional face. This yields another $\mathscr{G}^*(3,2)$ with the same number of edges, to which one of the above reductions may be applied.

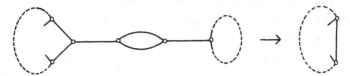

Figure 13.4.4

This completes the proof of nonexistence of graphs $\mathscr{G}^*(3,2)$.

Let now a graph $\mathscr{G}(4,1)$ be given, then the exceptional face must be even (i.e. have an even number of edges). If the exceptional face is a digon having a common edge with a quadrangle, the reduction shown in figure 13.4.5 yields a $\mathscr{G}(4,1)$ with fewer edges. (Note that each $\mathscr{G}(4,1)$ is also a $\mathscr{G}(2,0)$ and is therefore, by theorem 3(2), 2-connected.)

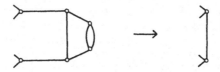

Figure 13.4.5

If the exceptional face is a digon, but each of the faces adjacent to it has at least 8 edges, we modify the graph as shown in figure 13.4.6 and obtain a $\mathscr{G}(4,1)$ with the same number of edges but having a digon adjacent to a quadrangle; thus the former reduction becomes applicable.

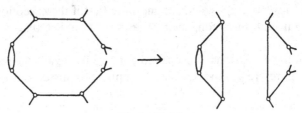

Figure 13.4.6

If the exceptional face is not a digon, we use the transformation shown in figure 13.4.3 (where A now denotes the exceptional face) in order to obtain a $\mathscr{G}(4, 1)$ having the same number of edges but containing a digon. Then the above reductions may be applied. This completes the proof of non-existence of graphs $\mathscr{G}(4, 1)$.

Considering graphs of type $\mathscr{G}^*(4, 2)$ we first note that by theorem 2(2) the two exceptional faces are even. Then the reduction shown in figure 13.4.7 (in which A and B are the exceptional faces) yields a $\mathscr{G}(4, 1)$.

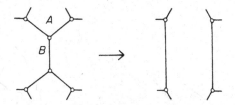

Figure 13.4.7

This completes the proof of theorem 2(4).

The proof of theorem 2(5) follows an analogous pattern, and we shall not discuss the details. The main idea is to use transformations similar to those in figures 13.4.3 and 13.4.6, but preserving multi-5-gons, in order to cut down to at most 6 edges the size of the exceptional face(s). Additional applications of such transformations lead to graphs $\mathscr{G}(5, 1)$ or $\mathscr{G}^*(5, 2)$ in which the exceptional face (or faces), and some pentagons, are connected by only one or two edges with the remaining part of the graph. The reduction to be applied at this stage consists of deleting those edges and the part of the graph containing the exceptional faces; it yields again a $\mathscr{G}(5, 1)$ or a $\mathscr{G}^*(5, 2)$. Figure 13.4.8 shows a typical example of this procedure.

This completes the proof of theorem 2.

The proof of theorem 3 is now easy. If $\mathscr{G}(k, 0)$ were not 2-connected it would contain a cut-edge, the deletion of which would disconnect the graph. After replacing the two edges incident to each endpoint of the cut-edge by a single edge, at least one of the connected components would be either a $\mathscr{G}(k, 1)$ or a $\mathscr{G}^*(k, 2)$. Since this is impossible by theorem 2(k), the proof of theorem 3 is completed.

We note that theorems 2 and 3 are best possible in the sense that there exist $\mathscr{G}(k, 2)$ which are not 2-connected. For $k = 2$ or 4 an example is

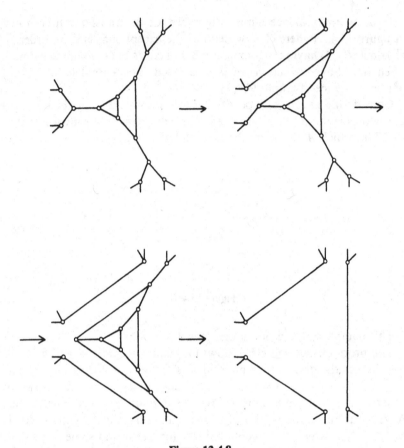

Figure 13.4.8

given in figure 13.4.9 while figures 13.4.10 and 13.4.11 show examples
for $k = 3$ and for $k = 5$.

Turning to the proof of theorem 4, we consider first a graph of type
$\mathcal{G}(3, 0)$. If the graph contains two triangles with a common edge, then the
graph is either the complete graph with 4 nodes (and 4 faces), or the
reduction shown in figure 13.4.12 may be applied.

If no edge of a triangle belongs to another triangle, the reduction to
be applied is shown in figure 13.4.13. If the reduced graph is connected
it has 2 multi-3-gons less; if it is not connected each of the connected
components is a $\mathcal{G}(3, 0)$ and the desired result again follows. Thus each
$\mathcal{G}(3, 0)$ has an even number of faces.

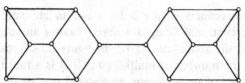

Figure 13.4.9

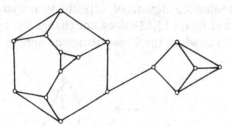

Figure 13.4.10

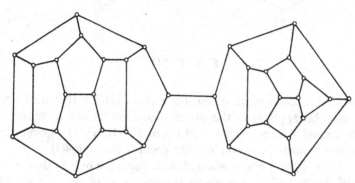

Figure 13.4.11

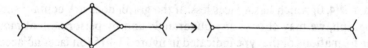

Figure 13.4.12

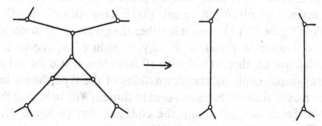

Figure 13.4.13

The proof of evenness of $p(\mathscr{G}(3,2))$ is not much more complicated. If one of the exceptional faces is a digon having an edge in common with a triangle, the reduction shown in figure 13.4.2 may be used. It decreases by 2 the number of multi-3-gons if A is a multi-3-gon, and it leaves the number of multi-3-gons unchanged but yields a $\mathscr{G}(3,0)$ if A is the other exceptional face.

If the graph contains a digon not adjacent to a triangle, the transformation shown in figure 13.4.14 does not change the number of faces and yields a $\mathscr{G}(3,2)$ to which the former reduction applies.

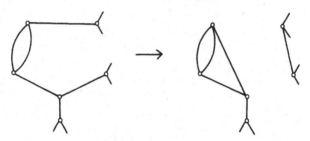

Figure 13.4.14

If the graph contains no digon, the transformation shown in figure 13.4.3 may be applied to the exceptional face having a number of edges $\equiv 2 \pmod 3$ (denoted by A in figure 13.4.3); a $\mathscr{G}(3,2)$ containing a digon is obtained. This completes the proof of theorem 4(3).

We now consider a graph $\mathscr{G}(4,0)$. If it contains the heavily drawn subgraph of figure 13.4.15, the graph is either that of the 3-cube (with $6 \equiv 2 \pmod 4$ faces), or it may be reduced as in figure 13.4.15, yielding another $\mathscr{G}(4,0)$ which has 4 faces less. If the graph does not contain such a subgraph, we may choose any quadrangle and by performing at most 3 transformations of the type indicated in figure 13.4.16 (on faces adjacent to the quadrangle) we reach the above configuration and the graph becomes reducible. This completes the proof of theorem 4(4).

Considering a graph $\mathscr{G}(5,0)$ we note that if it contains the heavily drawn subgraph of figure 13.4.17, then it is either the graph of the dodecahedron (with $12 \equiv 2 \pmod{10}$ faces), or it may be reduced as shown in figure 13.4.17, yielding another $\mathscr{G}(5,0)$ with 10 faces less. If it does not contain such a subgraph, applying transformations of the type shown in figure 13.4.18 at most 9 times to faces adjacent to those, a $\mathscr{G}(5,0)$ having the same number of faces but containing the configuration of figure 13.4.17 is reached. This completes the proof of theorem 4.

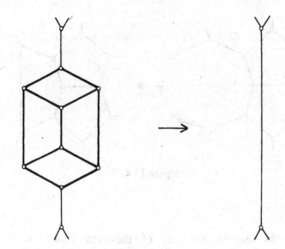

Figure 13.4.15

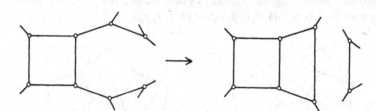

Figure 13.4.16

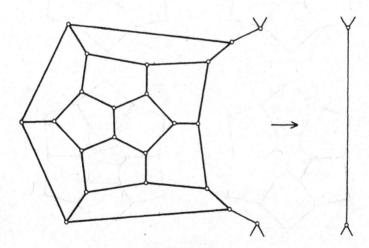

Figure 13.4.17

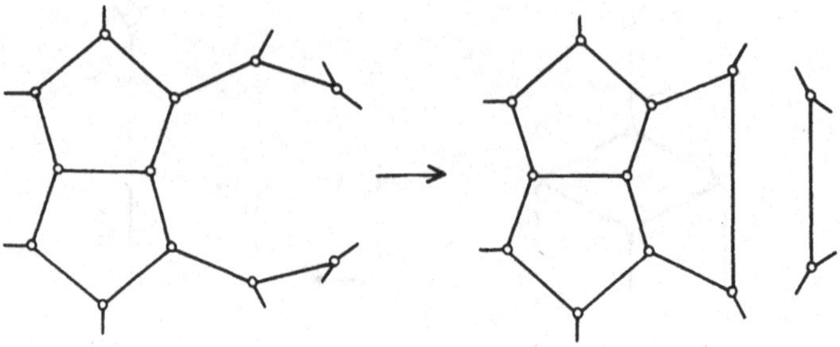

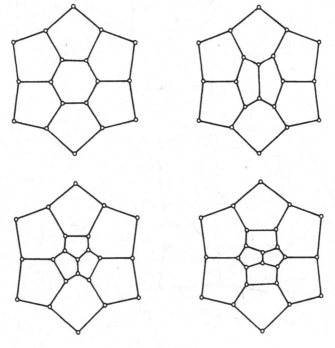

Figure 13.4.18

Exercises

1. Prove the existential part of theorem 13.4.1. For sequences $(0, 0, 12, p_6)$ use the 'pieces' represented in figure 13.4.19 and, if needed, additional 'belts' (figure 13.4.20). Find analogous constructions for sequences $(0, 6, 0, p_6)$, $(4, 0, 0, p_6)$, and $(3, 1, 1, p_6)$.

Figure 13.4.19

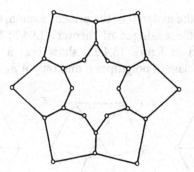

Figure 13.4.20

2. Show that all 3-realizations of $(4, 0, 0, p_6)$, $(p_6 \geq 0)$, are obtained in the following way (Grünbaum–Motzkin [4]): Take any two numbers $k \geq 1$ and $w \geq 0$ such that $p_6 = 2(k + w + kw)$, and consider a 'chain' (see figure 13.4.21) containing k hexagons. Enclose it by w 'belts', each of which consists of $2(k + 1)$ hexagons (see figure 13.4.22, where one such ring is represented for $k = 3$, together with the starting 'chain'); close the polytope by adding another 'chain' with k hexagons. (Note that the last chain may be added at different positions, yielding in general polytopes of different combinatorial types.)

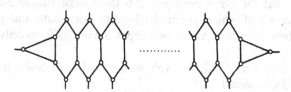

Figure 13.4.21

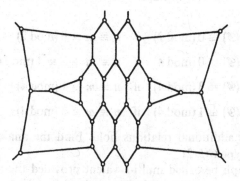

Figure 13.4.22

3. In analogy to the material of the present section, investigate 4-valent 3-polytopes. Find the analogue of theorem 13.4.1; by considering the 'pieces' represented in figure 13.4.23, show that a sequence $(8, p_4)$ is 4-realizable in this class of polytopes if and only if $p_4 \neq 1$.

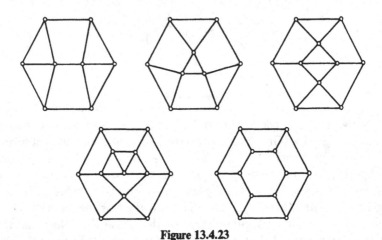

Figure 13.4.23

4. Show that for every even $p \neq 2, 6$ there exist simple 3-polytopes with p faces all of which are multi-3-gons. Formulate and prove the corresponding assertions about polytopes having as faces only multi-4-gons, or only multi-5-gons.

5. Show that every $\mathscr{G}(3, 0)$ graph without an edge belonging to two triangles is 3-connected.

6. Using the methods employed in the proofs of theorems 2 and 4 show that if $\mathscr{G}$ is a $\mathscr{G}(4, 2)$ graph with an n-gon and an m-gon as exceptional faces, then

$$p(\mathscr{G}) \equiv 0 \,(\mathrm{mod}\, 2) \quad \text{if} \quad n \equiv m \equiv 0 \,(\mathrm{mod}\, 2)$$

$$p(\mathscr{G}) \equiv 0 \,(\mathrm{mod}\, 4) \quad \text{if} \quad n \equiv m + 2 \equiv 1 \,(\mathrm{mod}\, 4)$$

$$p(\mathscr{G}) \equiv 3 \,(\mathrm{mod}\, 4) \quad \text{if} \quad n \equiv m \equiv 1 \,(\mathrm{mod}\, 4)$$

$$p(\mathscr{G}) \equiv 1 \,(\mathrm{mod}\, 4) \quad \text{if} \quad n \equiv m \equiv 3 \,(\mathrm{mod}\, 4).$$

Show that no additional relations hold. Find the analogous relations for graphs of type $\mathscr{G}(5, 2)$.

7. Let a graph be called multi-k-valent provided the valence of each of its nodes is a multiple of k. Prove the following generalization of

theorem 4 (due to Gallagher for 3-connected graphs; see Motzkin [8]): If $\mathscr{G}$ is a multi-3-valent connected planar graph such that all its faces are multi-k-gons, then the number of edges $e(\mathscr{G})$ satisfies

$$e(\mathscr{G}) \equiv 0\,(\mathrm{mod}\ 6) \quad \text{if} \quad k = 3$$

$$e(\mathscr{G}) \equiv 0\,(\mathrm{mod}\ 12) \quad \text{if} \quad k = 4$$

$$e(\mathscr{G}) \equiv 0\,(\mathrm{mod}\ 30) \quad \text{if} \quad k = 5.$$

(Hint: Reduce the general problem to the case of trivalent graphs, by using the 'polishing off' transformations shown in figures 13.4.24, 13.4.25, and 13.4.26.)

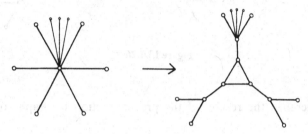

Figure 13.4.24

8. Denoting by $p_k(\mathscr{G})$ the number of k-gonal faces of the planar graph G and by $v_k(\mathscr{G})$ the number of k-valent nodes of G, prove the following generalization of a result of Motzkin [5]: If G is a connected graph with an odd number of edges, then

$$\sum_{k \equiv 0(\mathrm{mod}\ 3)} (p_k(G) + v_k(G)) \geq 3.$$

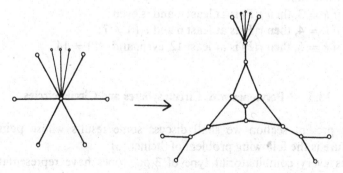

Figure 13.4.25

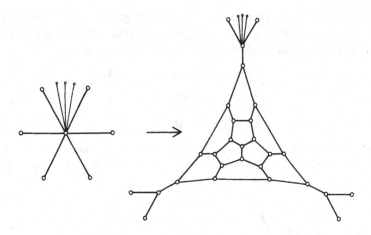

Figure 13.4.26

9. Generalize the results of the present section to graphs of genus $k > 0$.

10. There is an interesting connection between the well-known 'four-color problem' (see Ringel [3]) and simple 3-polytopes having only multi-3-gonal faces. Hadwiger [4] observed that the solution of the four-color problem is affirmative if and only if every simple 3-polytope P can be transformed into a simple 3-polytope all facets of which are multi-3-gons by 'cutting off' certain suitable vertices of P. ('Cutting off' a vertex means replacing it by a triangle, as in the first part of figure 13.3.26.)

11. (See Hawkins–Hill–Reeve–Tyrrell [1]) Let P be a k-valent 3-polytope; prove that
   (i) if $k = 3$, then $v(P)$ is at least 4 and is even;
   (ii) if $k = 4$, then $v(P)$ is at least 6 and $v(P) \neq 7$;
   (iii) if $k = 5$, then $v(P)$ is at least 12, even, and $v(P) \neq 14$.

13.5 3-Polytopes with Circumspheres and Circumcircles

In the present section we shall discuss some results whose point of departure is the following problem of Steiner [3]:

Does every combinatorial type of 3-polytopes have representatives all vertices of which belong to a sphere?

Combinatorial types which do have such representatives shall be called *inscribable*. *Circumscribable* combinatorial types are defined analogously.

The negative solution to Steiner's problem was given by Steinitz [8], in a beautiful paper which contains many additional important results. Theorems 1, 2, and 3 below are from Steinitz's paper [8]; the last of them establishes that a large family of combinatorial types is not circumscribable.

1. *A combinatorial type is inscribable if and only if the dual type is circumscribable.*

The proof is easy. If a 3-polytope is circumscribed to the unit sphere S^2, its polar is a 3-polytope inscribed into S^2. Also, if a 3-polytope P is inscribed into S^2 and if the center 0 of S^2 belongs to int P, then the polar of P is a 3-polytope circumscribed to S^2. However, if a 3-polytope P is inscribed into S^2 but $0 \notin$ int P, we first apply a projective transformation† which carries S^2 onto itself and maps some interior point of P onto 0. The resulting image P' is of the same type as P, but to P' the previous construction is applicable. This completes the proof of theorem 1.

2. *Let P be a 3-polytope circumscribed about a sphere. If $\mathscr{F}$ is a family of p^* facets of P such that no two members of $\mathscr{F}$ have an edge in common, then $p^* \leq \frac{1}{2}p(P)$. Moreover, $p^* = \frac{1}{2}p(P)$ if and only if each edge of P belongs to some member of $\mathscr{F}$.*

The proof is based on some simple facts from elementary geometry. Let F_1 and F_2 be facets of P with a common edge E. Denoting by T_1, T_2 the respective points of contact of F_1 and F_2 with the sphere and by 0 the center of the sphere, the plane aff$\{0, T_1, T_2\}$ is clearly orthogonal to the line aff E. Therefore the angle $\alpha(F_1, E)$ spanned at T_1 by E equals to the correspondingly defined angle $\alpha(F_2, E)$. Putting $\alpha(F, E) = 0$ if the facet F of P does not contain the edge E, we have therefore

$$\sum_E \sum_{F \in \mathscr{F}} \alpha(F, E) = \sum_E \sum_{F' \notin \mathscr{F}} \alpha(F', E),$$

the summation on E being over all E which are edges of members of $\mathscr{F}$.

---

† For example, the projective transformation T determined (for $|\alpha| < 1$ and $\beta = \sqrt{1 - \alpha^2}$) in cartesian coordinates by

$$T(x, y, z) = \left(\frac{x - \alpha}{1 - \alpha x}, \frac{\beta y}{1 - \alpha x}, \frac{\beta z}{1 - \alpha x} \right),$$

maps the unit sphere $\{(x, y, z) \mid x^2 + y^2 + z^2 = 1\}$ onto itself, while $T(\alpha, 0, 0) = (0, 0, 0)$.

If the range of E is extended to all the edges of P, there follows

(*) $$\sum_E \sum_{F \in \mathscr{F}} \alpha(F, E) \leq \sum_E \sum_{F' \notin \mathscr{F}} \alpha(F', E).$$

Since for each facet F^* of P we have $\sum_E \alpha(F^*, E) = 2\pi$, it follows that $2\pi p^* \leq 2\pi(p(P) - p^*)$. This establishes the inequality $p^* \leq \frac{1}{2}p(P)$; equality is possible only if equality holds in (*). In other words, $p^* = \frac{1}{2}p(P)$ is equivalent to the assertion that each edge of P belongs to some member of $\mathscr{F}$.

Now we are ready for

3. *Let P be any 3-polytope such that $v(P) \geq p(P)$. Let P' be the polytope obtained from P by 'cutting off' all vertices of P by planes which have no common points in P. Then the combinatorial type of P' is not circumscribable.*

Indeed, the polytope P' has $v(P) + p(P)$ facets; the $v(P)$-membered family $\mathscr{F}$ of facets of P' which correspond to the vertices of P consists of mutually disjoint polygons. Since the construction of P' guarantees that not all edges of P' belong to members of $\mathscr{F}$, lemma 2 implies the assertion of theorem 3.

It should be noted that the construction in theorem 3 yields *simple* 3-polytopes which are of noncircumscribable types. The dual types are, by theorem 1, simplicial polytopes of noninscribable types. This is interesting since a hasty consideration (see Brückner [2]) may leave the impression that every simplicial type is inscribable.

Generalizing Steiner's problem, the following question was (informally) posed by T. S. Motzkin:

Does every combinatorial type of 3-polytopes have representatives possessing circumcircles?

Here a polytope P is said to *possess circumcircles* provided for each facet F of P all the vertices of F are concyclic.

A negative answer to Motzkin's problem was given in Grünbaum [7]; it is based on the following result about 2-diagrams:

4. *If $\mathscr{D}^*$ is a 2-diagram obtained from the 2-diagram $\mathscr{D}$ on replacing some 3-valent inner vertex of $\mathscr{D}$ by the configuration of figure 13.5.1, then no 2-diagram combinatorially equivalent to $\mathscr{D}^*$ has circumcircles.*

(A 2-diagram is said to have circumcircles provided for each face F (including the basis) of the 2-diagram, all vertices of F are concyclic.)

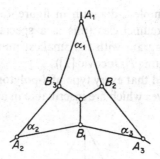

Figure 13.5.1

The proof of theorem 4 is very simple. Since in any quadrilateral inscribed into a circle the sum of the opposite angles is π, the sum $\alpha_1 + \alpha_2 + \alpha_3$ (see figure 13.5.1) equals π. If $\mathcal{D}^*$ had circumcircles, the points B_1, B_2, B_3 would be in the interior of $\mathrm{conv}\{A_1, A_2, A_3\}$, and therefore $\alpha_1 + \alpha_2 + \alpha_3$ would be less than π.

Since by theorem 13.2.2 every 2-diagram is the Schlegel diagram of some 3-polytope, and since a simple 3-polytope with circumcircles is inscribable, (see exercise 2) it follows that for every simple 2-diagram $\mathcal{D}$ the polytope corresponding to the 2-diagram $\mathcal{D}^*$ of theorem 4 is of a combinatorial type which not only fails to be inscribable, but even does not possess circumcircles.

Many problems related to the above results are still open. Thus, defining in an appropriate manner 3-polytopes (or 2-diagrams) with *incircles*, it is probably true that types without incircles exist. Similarly unsolved are the problems concerning *inellipses* or *circumellipses*.

The characterization of all inscribable types, or of all types with circumcircles, seems to be quite difficult. Probably easier is the following

CONJECTURE   A 3-polytope has a realization with circumcircles if and only if its Schlegel diagrams have circumcircles.

Exercises

1. Show that theorem 13.5.4 remains valid even if the 2-diagram $\mathcal{D}^*$ is obtained from $\mathcal{D}$ on replacing *any* 3-valent vertex of $\mathcal{D}$ by the configuration of figure 13.5.1.

2. Show that a simple 3-polytope has circumcircles if and only if it is inscribable. (Use the stereographic projection.) Generalize to d-polytopes.

3. Show that the simple 2-diagram in figure 13.5.2 does not have a representative with circumcircles (this is a special case of exercise 1) and that it is the 2-diagram with the smallest number of facets having this property (Grünbaum [7], Jucovič [2]).

It may be conjectured that every type of d-polytope with at most $d + 3$ facets has representatives which are inscribable in a $(d - 1)$-sphere.

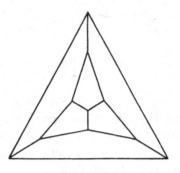

Figure 13.5.2

13.6 Remarks

3-polytopes and their constructions, properties and relationships have been investigated since the beginning of geometry. A particularly flourishing period was the second half of the nineteenth century. Mathematicians like Cayley and Kirkman made many contributions, but failed in their main aim—the determination of the numbers $c(p, 3)$, or $c_s(p, 3)$, of different combinatorial types of 3-polytopes, or of simple 3-polytopes, with p facets. Detailed historical surveys of these endeavors were given by Brückner [2] and Steinitz [6]. The determination of $c_s(p, 3)$ was empirically performed by Brückner [2] for $p \leq 10$; Hermes [1] tried to extend Brückner's work to $p = 11, 12$. Brückner [4] spent almost 8 years on the determination of $c_s(p, 3)$ for $p = 11, 12, 13$, and found that Hermes' enumeration was incomplete. However, even the results of Brückner [4] can not be considered final. A recent investigation of Grace [1] (see below) has uncovered a duplication already among Brückner's polytopes with 11 facets.

The values of $c(p, 3)$, for $p \leq 8$, have been determined by Hermes [1] (see table 2 on p. 424); unfortunately, these numbers seem not to have been checked independently. Hermes [1] also enumerated all the types of self-dual polytopes with at most 9 facets (see exercise 3.4.3), as well as certain other special families. Bouwkamp–Duijvestyn–Medema [1], utilizing a computer, determined all the combinatorial types of 3-polytopes with at most 19 edges.

A new attempt to determine $c_s(11, 3)$ was made by Grace [1], using a computer. While Grace's value $c_s(11, 3) = 1249$ is probably correct, it is still open to some doubts (see exercise 1 below).

Clearly, the experimental methods of determining $c_s(p, 3)$ for certain values of p, which were used in the papers mentioned above, cannot lead to a satisfactory solution of the general problem. Indeed, even fifty years ago the general question was considered to be quite hopeless. However, developments of Pólya's [1] enumeration technique during recent years have brought us rather close to this goal; it may be expected that a complete solution will be found in the not too distant future. For some partial results related to the determination of $c_s(p, 3)$, the reader should consult Tutte [7–10], W. G. Brown [1], Brown–Tutte [1], Rademacher [1], Mullin [1]. We shall mention here only one of the results of Tutte [7]. Let us consider *rooted, oriented* 3-polytopes; by this we mean polytopes on which an arbitrary vertex and one of its edges are distinguished as a *root*, and one of the two possible orientations about the root-vertex is chosen. Two rooted, oriented 3-polytopes are considered to be of the same type provided there exists a combinatorial equivalence between the polytopes under which the roots are mapped onto each other in an orientation-preserving manner. For the number $c_s^{(r)}(p, 3)$ of different types of rooted, oriented, simple 3-polytopes with p facets, Tutte [7] obtained

$$c_s^{(r)}(p, 3) = 2\frac{(4p - 11)!}{(p - 2)!(3p - 7)!}.$$

The relation between $c_s^{(r)}(p, 3)$, $c_s(p, 3)$, and the number $c_s^*(p, 3)$ of combinatorial types of simple 3-polytopes with p facets which have a trivial group of combinatorial automorphisms, is easily seen to be

$$c_s(p, 3) \geq \frac{c_s^{(r)}(p, 3)}{12(p - 2)} \geq c_s^*(p, 3).$$

It may be conjectured that

$$\lim_{p \to \infty} \frac{c_s^*(p, 3)}{c_s(p, 3)} = 1;$$

if this conjecture is true, then $c_s(p, 3)$ is asymptotically given by

$$c_s(p, 3) \sim \frac{1}{16} \sqrt{\frac{3}{2\pi}} p^{-\frac{5}{2}} \left(\frac{256}{27}\right)^{p-2}$$

A similar assumption regarding all 3-polytopes leads (see Tutte [9]) to the conjecture that the number of different combinatorial types of 3-polytopes with n edges is asymptotically given by

$$\frac{2}{243\sqrt{\pi}} n^{-\frac{7}{2}} 4^n.$$

It should be noted that the determination of $c_s(p, d)$ or of $c(p, d)$ for $d \geq 4$ and $p \geq d + 4$ is a problem of an entirely different order of magnitude than the determination of $c_s(p, 3)$ or $c(p, 3)$. The difficulty of the case $d \geq 4$ is intimately connected with, and may be appreciated on hand of, the results of sections 5.5, 11.1, and 11.5. The differences between complexes and topological complexes, as well as the realizability in the rational d-space or the lack of it, are non-essential for the enumeration of 3-polytopes, as shown by the results of sections 13.1 and 13.2. However, in higher dimensions each of those distinctions requires a separate enumeration, and no significant results are known about any of them.

More successful than the attempts at enumeration were different approaches to the construction of 3-polytopes, or of certain types of 3-polytopes, by standardized methods or transformations. The two most important sets of such transformations are those used by Eberhard [3] in the proof of the existence theorem 13.3.2, and the transformations ω_i and η_i used by Steinitz in the proof of his theorem 13.1.1. Different additional constructions of this kind are mentioned in various proofs and exercises of this chapter. Many other constructions may be found in Brückner [2]; see also Tutte [6].

It is remarkable how relatively unknown an important result may be even if it is the main topic of a monograph published in one of the best-known series. Steinitz's characterization of the boundary complexes of 3-polytopes (section 13.2), announced with outlines of proofs in Steinitz [6], and published in full in Steinitz–Rademacher [1], did not become a well-known proposition until a few years ago. As a matter of fact, except for the reproduction of one of Steinitz's proofs in Lyusternik [1] in 1956, there seems to be no mention or use of Steinitz's theorem in the literature prior to Grünbaum-Motzkin [1]. (The Steinitz-Rademacher book [1] is quoted by some authors, but only as a historical summary.)

Parts of Steinitz's theorem were rediscovered by different authors. For example, the special case mentioned in exercise 13.1.3 appears in T. A. Brown [2]. An unsuccessful attempt at proving theorem 13.2.1 was made by Stein [1]. This result was independently proved by Tutte [5, 11]; in particular, the second paper is very ingenious and contains additional interesting facts.

Steinitz gave three different proofs for his theorem (see Steinitz-Rademacher [1]). One of them is reproduced in Lyusternik [1]. Klee [18], in calling Steinitz's theorem the 'second landmark' of the theory of convex polytopes (Euler's theorem being the first landmark), discusses several variants of the theorem. The proof given in section 13.1 above is a modification of Steinitz's third proof. The fact (see section 11.5) that Steinitz's theorem does not generalize to higher dimensions emphasizes the difficulties encountered in any detailed analysis of polytopes of dimensions exceeding 3.

Theorem 13.3.2 is one of the oldest nontrivial results in the theory of 3-polytopes. It is the climax of Eberhard's book [3] which appeared in 1891. The theorem was practically forgotten for a long time, the only two references to it the present author was able to find being Brückner [2] in 1900 and Steinitz-Rademacher [1] (footnote on p. 8) in 1934. In section 13.3 we gave a relatively short proof of Eberhard's theorem; the only other proof seems to be Eberhard's [3], which relies heavily on the earlier parts of his book and is rather hard to follow.

The history of the results discussed in section 13.4 goes also back to Eberhard [3]. He wondered (on p. 84) whether there exist simple 3-polytopes with an odd number of faces each of which is a multi-3-gon. The problem was first solved by Motzkin [5], who proved theorem 13.4.4(3) for a simple 3-polytopes (i.e. for 3-connected planar graphs) using a group-theoretic method. In a similar manner, the corresponding special cases of theorems 13.4.4(4) and 13.4.4(5) were obtained by Gallagher (see Motzkin [8]). Additional proofs of results contained in theorem 13.4.4(3) were given by Kotzig [2] (using coloring arguments) and by Grünbaum [13]. For some related results see Coxeter [2]. Theorem 13.4.2(2) was conjectured by Minty and proved by Moon (see Minty [1]).

The material of section 13.5, though different in character from most other parts of the book by the decisively metric conditions, is a good example for various aspects of work on polytopes. Though Steinitz's [8] solution (theorem 13.5.3) of Steiner's problem came more than half a century after the problem was posed, the solution is completely elementary and characterized only by a high degree of ingenuity. It would have been easily comprehensible to Steiner himself. On the other hand, Steinitz obtained his result while working on a seemingly rather remote problem, the question of existence of an isoperimetrically best 3-polytope within each combinatorial type (see Steinitz [8], Fejes Tóth [1]). More than thirty years later, Steinitz's arguments formed the inspiration for a result of Klee [14] (theorems 11.4.1 and 12.2.2).

There are other results and open problems related in spirit to the contents of section 13.5. As an example we mention the following theorem of Ungar [1]:

> If $\mathcal{G}$ is a 3-valent, 3-connected planar graph such that no three faces of $\mathcal{G}$ have a multiply connected union, there exists an imbedding of $\mathcal{G}$ in the plane in which each bounded face is a rectangle, as is the complement of the unbounded face.

Other related subjects were treated by Cairns [2] and by Supnick [1].

Exercises

1. Grace [1] defines two simple 3-polytopes to be *equisurrounded* provided their facets have the same pattern of neighbors. More precisely, this means the following: With each k-gonal facet F we associate a sequence $(i_1, \cdots, i_k)$, where the integers $i_1, \cdots, i_k$ indicate that the facets neighboring to F are, in cyclic order, an i_1-gon, $\cdots$, an i_k-gon. Two sequences $(i_1, \cdots, i_k)$ and $(j_1, \cdots, j_k)$ are considered equivalent if one of them is a cyclic permutation of the other. (From Grace's description it is not clear whether reversal of orientation is permitted or not.) Two simple 3-polytopes are equisurrounded provided each associated sequence of one polytope occurs in the other as well, with the same multiplicity in both. Clearly, combinatorially equivalent polytopes are equisurrounded, but the converse does not hold. Grace [1] gives an example of two equisurrounded simple 3-polytopes with 18 facets which are not of the same combinatorial type. However, noting that equisurrounded polytopes with few vertices are of the same combinatorial type, Grace [1] uses equisurroundedness as a criterion for equivalence in his computer-assisted

determination of $c_s(11, 3)$. His result $c_s(11, 3) = 1249$ is still under some doubt since it is not known whether equisurroundedness implies equivalence for polytopes with $p = 11$ facets.

Show that the two polytopes with $p = 15$ represented in figure 13.6.1, as well as those with $p = 16$ represented in figure 13.6.2, are of different combinatorial types though they are equisurrounded—the first pair under the definition permitting reversal of the cyclic order, the second even if such a reversal is not permitted.

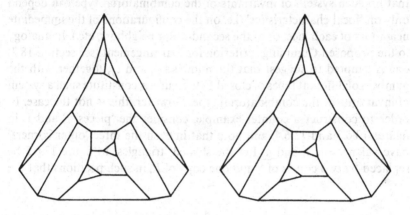

Figure 13.6.1

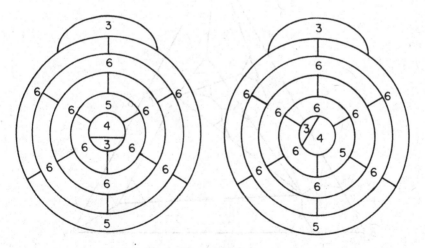

Figure 13.6.2

It may be conjectured that the above examples are minimal for the two variants of the definition of equisurroundedness.

2. Since Euler's times, the question was frequently posed whether there exist easily computable numerical characteristics (similar to the p_i's and v_j's) of 3-polytopes, such that the equality of the characteristics implies combinatorial equivalence. No such characteristics (short of a complete schematic description of the polytope) seem to be known. From the result of exercise 13.4.2 (compare also exercise 1 above) it is easy to deduce that no such system of invariants of the combinatorial type can depend only on 'local characteristics' (i.e. on the configurations of the immediate neighbors of each face, or of the second-order neighbors, etc.). In analogy to the proposed Cummings criterion for 2-arrangement (see section 18.2) one is tempted to suggest that the numbers p_i and v_j, together with the numbers of different types of closed Petrie-curves, constitute such a system of invariants of the combinatorial type. However, this is not the case. In order to construct a counter-example, consider the 'pieces' S and L in figures 13.6.3 and 13.6.4, and note that in each the three 'outer corners' have valences 2, 3, and 4. Let the shaded 'triangles' in figure 13.6.5 be replaced by two copies of S and one copy of L, in such positions that the

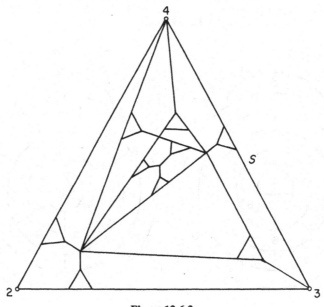

Figure 13.6.3

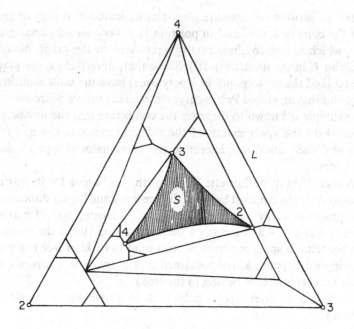

Figure 13.6.4

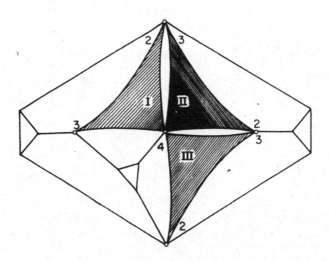

Figure 13.6.5

'corners' of various valences are placed as indicated. It is easy to check that if the copy of L is placed in position II a 3-connected planar graph results which is not combinatorially equivalent to the graph obtained by placing L in position I (or III). Show that, nevertheless, the graphs obtained (and the corresponding 3-polytopes) have the same numbers p_i and v_j, and that all closed Petrie-curves in each are simple 8-circuits.

No example is known to disprove the conjecture that the numbers p_i, together with the specification of the different types of closed Petrie-curves and their numbers, determine the combinatorial type of *simple* 3-polytopes.

3. A recent result of Barnette [1] is worth mentioning for its intrinsic interest and for the contrast between dimension 3 and higher dimensions:

The graph of every 3-polytope contains a *spanning tree* of maximal valence 3. For $d \geq 4$ and for any k there exist d-polytopes the graphs of which contain no spanning trees of valence at most k. (A *tree* is a graph containing no circuits; a tree contained in a graph $\mathscr{G}$ is a *spanning tree* provided all vertices of $\mathscr{G}$ belong to the tree.)

(Use Kleetopes over cyclic polytopes to prove the non-existence assertion.)

13.7 Additional notes and comments

Proofs of Steinitz's theorem.

There are three main, independent classes of proofs for the fundamental result of theorem 13.1.1 (for additional references see Ziegler [a, Lect. 4]):

1. Proofs "of Steinitz type" construct a polytope via simple operations, starting with a simplex. A particularly elegant proof of this type, by Truemper [a], uses ΔY-transformations. Based on exercise 13.1.4, Barnette–Grünbaum [a] constructed all 3-polytopes by repeatedly "bending" facets.

2. Tutte's *rubber band* method first produces a planar drawing by fixing the vertices of a triangular face in the plane, and then computing a stressed equilibrium embedding, where the graph is modeled as a network of rubber bands. (A triangle exists in the graph or in its dual.) Such a drawing can always be lifted to a polytope in R^3. The underlying rigidity theory may be traced back to work of Maxwell and Cremona in the 19th century; a modern version is Richter-Gebert [b, Sect. 13.1].

3. A third type of proofs proceeds via the primal-dual circle packing theorem due to Koebe, Andreev, and Thurston:

   Every 3-connected planar graph G has a representation by circles on the sphere with disjoint interiors, one circle for each node, such that adjacent nodes correspond to touching circles. The representation is unique up to Möbius transformations if one demands a simultaneous representation of the same type for the dual graph, with the same touching points, and with orthogonal intersections of primal and dual circles in these points.

   The circles of the representation determine facet planes for the 3-polytope and for its dual. For a proof of this result see Brightwell–Scheinerman [a]. A new derivation from a variational principle is Bobenko–Springborn [a]. See also Lovász [b].

Extensions and analogues.

A strong extension of Steinitz's theorem (obtained via a proof of type 3) is due to Schramm [b]: *Every combinatorial type of 3-polytope can be realized with all edges tangent to a 2-sphere.* (Compare this to theorem 13.5.3 and the remarks below it. Schramm [b] has a much stronger theorem of this type, for an arbitrary smooth convex body.) *Additionally, one can require that the origin is the barycenter of the tangency points; then the realization is unique up to orthogonal transformations.* In particular, every automorphism of $G(P)$ is then induced by a symmetry of P. (Mani [a] was the first to give this answer to the question on page 252; he used a type 1 proof.)

A type 2 proof yields that every combinatorial type of 3-polytope with n vertices can be realized with integer coordinates with a polynomial number (in n) of digits; this strengthens theorem 13.2.3. For example, all 3-polytopes that have a triangle facet may be represented with vertices in $\{0, 1, \ldots, 37^n\}^3$, by Richter-Gebert [b] and Stein [a]. (It is not clear whether integer coordinates of polynomial absolute value are sufficient.)

Another result by Steinitz with a type 2 proof is (see Richter-Gebert [b, Sect. 13.3]):
For every labeled 3-polytope P with e edges, the realization space *(i. e., the topological space of all polytopes in R^3 that are combinatorially equivalent to P, modulo affine motions) is an open $(e-6)$-dimensional ball.*

Barnette–Grünbaum [b] showed with a type 1 proof that one may prescribe the shape of one facet of a 3-polytope (see exercise 13.1.5). Similarly, any given cycle in the graph of a 3-polytope can be forced to form the "shadow boundary" for a projection to the plane (Barnette [a]). See Martini [a] for a survey on shadow boundaries of polytopes, and for connections to the notion of antipodality discussed in section 19.3.

Mihalisin–Klee [a] proved a directed Steinitz theorem:
Let G be a planar 3-connected graph and let $\mathcal{O}$ be an acyclic orientation of G with a unique sink that also induces unique sinks on all 2-faces. If there are three disjoint monotone paths from the (then) unique source to the sink, then there is a 3-polytope P in R^3 whose graph is isomorphic to G, and a linear function on R^3 that induces the orientation $\mathcal{O}$ on G.

For general d, a characterization of the orientations of the graphs of d-polytopes that are induced by linear functions is out of sight. However, using Gale-transforms, Mihalisin [a] gave such a characterization for d-polytopes with at most $d + 3$ vertices.

Eberhard's theorem.
Conjecture 1 on page 267 was proved by Fisher [a], with $c = 3$. With respect to Conjecture 2 on page 268, Grünbaum [a] proved that if a sequence (p_k) satisfies (*) and has $p_3 = p_4 = 0$, then it can be realized for every $p_6 \geq 8$.

The simple 3-polytopes with $p_i = 0$ for $i \notin \{5, 6\}$ resemble the structure of certain molecules (*Fullerenes*). We refer to Brinkmann–Dress [a].

Passing from simple to general 3-polytopes P, one considers pairs of face vectors $(p_3(P), p_4(P), \ldots)$ and vertex vectors $(v_3(P), v_4(P), \ldots)$. Any pair of sequences (p_k) and (v_k) obtained in this way from a 3-polytope satisfies the equations (a) $\sum_{k \geq 3}(4-k)p_k + \sum_{k \geq 3}(4-k)v_k = 8$ and (b) $\sum_{k \geq 3}(6-k)p_k + \sum_{k \geq 3}(6-2k)v_k = 12$ (see pages 236–237).

In extending Eberhard's theorem from simple to general 3-polytopes, it is natural to seek conditions which, when added to (a) or (b), will guarantee that the pair of sequences can actually be obtained from some 3-polytope.

Grünbaum [b] proved that if (a) holds and the sums $\sum_{k\geq 3} kp_k$ and $\sum_{k\geq 3} kv_k$ are both even, then there exists a 3-polytope P with $p_k(P) = p_k$ and $v_k(P) = v_k$ for all $k \neq 4$. Jucovič [a] showed that under these conditions, all sufficiently large integers can play the role of p_4.

The situation with respect to (b) is more complicated. Jendrol'–Jucovič [a] proved that if (b) holds then the following conditions are equivalent: (i) there exists a 3-polytope P with $p_k(P) = p_k$ for all $k \neq 6$ and $v_k(P) = v_k$ for all $k \neq 3$; (ii) $\sum_{2\nmid k} p_k > 0$ or $\sum_{3\nmid k} v_k \neq 1$.

For sequences (p_k) and (v_k) satisfying (b) and (ii), Jendrol' [b] provided detailed information concerning the set of all integers t such that setting $p_6 = t$ and

$$v_3 = \tfrac{1}{3}\Big(\sum_{k\geq 3} kp_k - \sum_{k\geq 4} kv_k\Big)$$

results in a pair (p_k), (v_k) that is realized by some 3-polytope.

Roudneff [a] showed that if there is a 3-polytope that has (p_k) as its face vector and (v_k) as its vertex vector, and if $\sum_{k\geq 6} p_k \geq 3$, then one has (c) $\sum_{k\geq 7}(6-k)p_k \geq 12 - \sum_{k\geq 3} v_k$. For $\sum_{k\geq 6} p_k \geq 4$, equality in (c) is attained precisely for 3-polytopes that can be obtained from simple 3-polytopes by a sequence of truncations of a specified sort.

Enumeration of combinatorial types.

Based on programs by Brinkmann–McKay [a] for enumerating planar graphs, Royle [a] provides $c_s(p,3)$ for $p \leq 21$ and $c(p,3)$ for $p \leq 13$, together with databases and generation algorithms. (Compare Dillencourt [a].)

Asymptotically, we also have precise knowledge about the numbers of 3-polytopes (answering the conjectures on pages 289–290): We refer to Bender–Wormald [a], and to the exposition in Klee–Kleinschmidt [b, Sect. 5].

Polytopes inscribed in a sphere.

For $n > d$, let $I(d,n)$ denote the collection of all d-polytopes that are formed as the convex hull of n points on the unit sphere $S^{d-1} \subset R^d$. A remarkable, complete characterization of the combinatorial types that are realized for some n by members of $I(3,n)$, based on a transfer to the Klein model of hyperbolic 3-space, was given by Hodgson–Rivin–Smith [a].

There is also an extensive literature about quantitative extremal problems that involve members of $I(d,n)$. It was motivated in part by the fact that, when $d = 2$ and n is fixed, the extremal configurations for the measures mentioned below (and for many others) are precisely the vertex-sets of regular n-gons inscribed in the unit circle. Our attention here is confined to the case $d = 3$ and to the problems of maximizing the following two measures over all members P of $I(3,n)$: (a) the volume of P; (b) the length of P's shortest edge. Like these

two measures, the many other measures that have been considered all lead
to partial results and to many open problems both for $d = 3$ and for higher
dimensions. See Saff–Kuijlaars [a] for relevant references.

When n is 4, 6, or 12, problem (a) is solved by placing the points at the
vertices of an inscribed Platonic solid (see Fejes Tóth [1,4]). In view of the
2-dimensional situation, this is not surprising. However, for $n = 8$ the solution
is not provided by the 3-cube, but instead by an 8-vertex 3-polytope with 3 different edge-lengths that was first discovered by a computer search (Grace [a])
and later proved by Berman–Hanes [a] to provide, up to rotation, the unique
global maximum for the volume. The volume problem has been solved only
for $n \leq 8$ and for $n = 12$. Conjectures for 9 and 10 points were given by
Berman–Hanes [b].

Problem (b) is usually phrased as the problem of arranging n points on the
unit sphere so as to maximize the minimum distance between two points of
the arrangement (the *misanthrope problem*). It's not hard to show that this is
equivalent to finding an n-vertex inscribed polytope that maximizes the length
of the shortest edge. The problem has been completely solved only for $n \leq 12$
and for $n = 24$. The cases of $n = 24$ and of $n = 10, 11$ were settled by Robinson
[a] and Danzer [a], respectively. Danzer has a comprehensive collection of
references to earlier results and to conjectures for many values of n. Among
the conjectures, a particularly attractive one was proposed by Robinson [b], to
the effect that when n is 24, 48, 60, or 120, the extreme for $n - 1$ points is the
same as for n points ("$n - 1$ misanthropes are not better off than n"). That is
true for $n = 6$ and $n = 12$, but for $n = 24$ it was disproved by Tarnai–Gáspár [a].
Tarnai [a] provides some excellent illustrations of spherical point-distributions
arising in nature. See also Melnyk–Knop–Smith [a].

Order dimension.

The *order dimension* of a partially ordered set Π is the least number of linear
orderings whose intersection is Π. Brightwell–Trotter [a] proved that the order
dimension of the face lattice $\mathscr{F}(P)$ of a 3-polytope P equals four, and the order
dimension of $\mathscr{F}(P) \setminus \{F_0\}$ is three, whenever F_0 is a vertex or a facet of P. (See
also Felsner [a] and Miller [a].)

Higher dimensions.

Analogs of Steinitz's theorem in higher dimensions probably do not exist: It
is NP-hard to decide whether a given lattice is the face lattice of a 4-polytope;
one cannot prescribe the shape of a facet, or even of a 2-face; some rational
4-polytopes cannot be realized with rational coordinates of polynomial coding
lengths: All of this follows from Richter-Gebert's [b] universality theorem.

CHAPTER 14

Angle-sums Relations; the Steiner Point

The subject matter of the present chapter differs from the foregoing parts of the book in being dependent on the structure of R^d as a Euclidean space. In spirit, however, the topic is closely related to the material of chapters 8 and 9.

Using appropriate definitions we shall show (in sections 14.1 and 14.2) that the 'angle-sums' α_k of d-polytopes satisfy relations analogous to the Euler and Dehn–Sommerville equations for the numbers of k-faces, f_k.

The subject is very old, but also relatively new. The simplest case, $d = 2$, of theorem 14.1.1 was known to Euclid: it is the formula for the sum of angles in a planar n-gon. On the other hand, the case $d = 3$ appears only in 1874 in a paper of Gram [1]—only to disappear until rather recent times.

In section 14.3 the Steiner point of a polytope is defined, and analogues of the Euler and Dehn–Sommerville equations are proved for this vector-valued function. Though the Steiner point of smooth convex curves was defined and investigated by Steiner already in 1840, the type of properties discussed here was discovered only very recently by G. C. Shephard (see section 14.4).

14.1 Gram's Relation for Angle-sums

Let P be a d-dimensional, convex polytope in Euclidean d-space R^d, and let F be a k-face of P; we consider P as a d-dimensional face of itself. We denote by $C(F, P)$ the cone with vertex at the centroid G_F (or any other relatively interior point) of F, spanned by P. The angle of P at F, denoted by $\varphi(F, P)$ or simply by $\varphi(F)$, is the fraction of E^d taken up by $C(F, P)$; more precisely, $\varphi(F)$ is the ratio of the $(d - 1)$-content of $S^{d-1} \cap C(F, P)$ to the $(d - 1)$-content of S^{d-1}, where S^{d-1} is a $(d - 1)$-sphere centered at G_F. For $k = 0, 1, \cdots d$, we define the kth angle-sum $\alpha_k(P)$ of P by $\alpha_k(P) = \sum \varphi(F; P)$, the summation being extended over all the k-faces F of P. (In particular, $\alpha_d(P) = 1$ and $\alpha_{d-1}(P) = \frac{1}{2} f_{d-1}(P)$.)

We shall establish an angle-sums analogue of Euler's equation; the case $d = 3$ is due to J. P. Gram [1].

1. *For every d-dimensional convex polytope P the angle-sums satisfy* $\sum_{i=0}^{d-1} (-1)^i \alpha_i(P) = (-1)^{d-1}$.

The proof of theorem 1 is divided into three parts. In the first part we establish the theorem for simplices. The second part is rather straight-forward: If P is a d-polytope and 0 an interior point of P, let $P_1, \cdots, P_f$, where $f = f_{d-1}(P)$, be the f d-pyramids having 0 as common apex, and a $(d-1)$-face of P as base. We show that if each P_i, $1 \le i \le f$, satisfies the theorem then P satisfies the theorem. The first two parts of the proof are thus sufficient in order to establish the theorem for simplicial poly-topes. In order to complete the proof of the theorem we have, therefore, to establish its validity for all d-pyramids (over arbitrary $(d-1)$-dimen-sional bases). This is accomplished in the third part of the proof.

PROOF OF THE THEOREM—PART I   Let P be a d-simplex, F^k a k-face of P, and let $C(F^k, P)$ be the cone associated with F^k. The angle $\varphi(F^k) = \varphi(F^k, P)$ of P at F^k can obviously be defined in the following way, equiva-lent to the definition used above: Let S^{d-1} be the unit $(d-1)$-sphere and let $V(F^k)$ be the subset of S^{d-1} consisting of all the unit vectors v such that $C(F^k, P)$ contains a ray parallel to v (in the same sense); in the notation used in section 2.4, $V(F^k) = S^{d-1} \cap \mathrm{cc}\, C(F^k, P)$. Then $\varphi(F^k)$ is the ratio of the $(d-1)$-content $\mu(V(F^k))$ of $V(F^k)$ to the $(d-1)$-content $\mu(S^{d-1})$ of S^{d-1}, μ denoting the $(d-1)$-dimensional Lebesgue measure.

Let $C_0, \cdots, C_d$ be the half-spaces of R^d determined by the $(d-1)$-faces $F_0^{d-1}, \cdots, F_d^{d-1}$ of P and containing P. Then clearly $C_i = C(F_i^{d-1}, P)$ and, more generally,

$$C(F^k, P) = \bigcap_{F_i^{d-1} \supset F^k} C_i;$$

also, for each m, $1 \le m \le d$, and each m-tuple $(i_1, \cdots, i_m)$ there exists a unique F^{d-m} such that

$$\bigcap_{j=1}^{m} C_{i_j} = C(F^{d-m}, P),$$

with different F^{d-m} corresponding to different m-tuples. Obviously

$$\bigcap_{i=0}^{d} C_i = P.$$

From the definition of V it is immediate that for every m-tuple $(i_1, \cdots, i_m)$, $1 \le m \le d$, we have

$$\bigcap_{j=1}^{m} V(C_{i_j}) = V\left(\bigcap_{j=1}^{m} C_{i_j}\right),$$

while

$$\bigcap_{i=0}^{d} V(C_i) = \varnothing.$$

Since hemispheres of S^{d-1} are measurable, we have

$$\mu(S^{d-1}) = \mu\left(\bigcup_{i=0}^{d} V(C_i)\right)$$

$$= \sum_{i=0}^{d} \mu(V(C_i)) - \sum_{\substack{i,j=0 \\ i \ne j}}^{d} \mu(V(C_i) \cap V(C_j)) + \cdots$$

$$\qquad + (-1)^d \mu(V(C_0) \cap V(C_1) \cap \cdots \cap V(C_d))$$

$$= \sum_{i=0}^{d} \mu(V(C_i)) - \sum_{\substack{i,j=0 \\ i \ne j}}^{d} \mu(V(C_i \cap C_j)) + \cdots$$

$$\qquad + (-1)^d \mu(V(C_0 \cap C_1 \cap \cdots \cap C_d))$$

$$= \sum_{i=0}^{d} \mu(V(F_i^{d-1})) - \sum_{\vartheta} \mu(V(F_\vartheta^{d-2})) + \sum_{\gamma} \mu(V(F_\gamma^{d-3})) + \cdots$$

$$\qquad + (-1)^d \mu(\varnothing)$$

$$= \mu(S^{d-1}) \cdot \{\alpha_{d-1}(P) - \alpha_{d-2}(P) + \alpha_{d-3}(P) + \cdots + (-1)^{d-1} \alpha_0(P)\}.$$

Thus $\sum_{i=0}^{d-1} (-1)^i \alpha_i(P) = (-1)^{d-1}$ as claimed, and the theorem is proved for all simplices P.

PROOF OF THE THEOREM—PART II We intend now to extend the validity of the theorem from simplexes to a wider class of polytopes. Let P be a d-polytope, 0 a point of int P, and let P_j, for $j \in J = \{1, \cdots, f_{d-1}(P)\}$, be the d-pyramids with common apex 0, spanned by the $(d-1)$-faces of P. We shall show that P satisfies the theorem provided the theorem is valid for each P_j.

We note, first, that

$$\sum_{j \in J} \alpha_0(P_j) = \alpha_0(P) + 1.$$

since the angles at 0 of the polytopes P_j add up to the full angle, while at any vertex of P the angles of the P_j incident with it add up to the angle of P at the vertex. Similarly, for each k with $1 \leq k \leq d - 1$, we have

$$\sum_{j \in J} \alpha_k(P_j) = \alpha_k(P) + f_{k-1}(P).$$

Therefore, denoting $f_{-1}(P) = 1$, we have

$$\sum_{k=0}^{d-1} (-1)^k \alpha_k(P) = \sum_{k=-1}^{d-2} (-1)^k f_k(P) + \sum_{k=0}^{d-1} (-1)^k \sum_{j \in J} \alpha_k(P_j)$$

$$= \sum_{k=-1}^{d-2} (-1)^k f_k(P) + \sum_{j \in J} \sum_{k=0}^{d-1} (-1)^k \alpha_k(P_j)$$

$$= \sum_{k=-1}^{d-2} (-1)^k f_k(P) + \sum_{j \in J} (-1)^{d-1}$$

$$= \sum_{k=-1}^{d-1} (-1)^k f_k(P) = (-1)^{d-1},$$

the last equation using Euler's formula.

Thus, provided we establish the theorem for all d-dimensional pyramids, its general validity is proved. At any rate, parts I and II of the proof already establish the theorem for all simplicial polytopes.

PROOF OF THE THEOREM—PART III   In this final part of the proof our aim is to establish the validity of the theorem for d-dimensional pyramids. The idea is to use decompositions, similar to those used in part II, but such that the final products of the decomposition are d-simplices, for which the theorem holds by part I of the proof. As it turns out, the relationships between a d-dimensional pyramid and any of its decomposition into d-simplices may be quite complicated, and it is in general not feasible to give the connections between the angle-sums of the pyramid and those of the simplices. Therefore our method shall be somewhat roundabout.

We shall show that the validity of the theorem for m-fold d-pyramids implies its validity for $(m - 1)$-fold d-pyramids, where $2 \leq m \leq d - 1$. Since part I of the proof established the theorem for $(d - 1)$-fold d-pyramids, this will complete the proof of the theorem for all d-pyramids and thus, by part II, for all d-polytopes.

In the proof we shall use the following notation. Given a d-polytope P and an m-face $F = F^m$ of P, we shall denote by F_i^k, $i \in I(k)$, all those k-faces of P which contain F. Thus $I(k) = \emptyset$ for $k < m$; $I(m)$ contains just one element and $\{F_i^m \mid i \in I(m)\} = \{F^m\}$; similarly for $m = d$,

$$\{F_i^d \mid i \in I(d)\} = \{P\}.$$

Let $\mathscr{C}^k$ be the complex consisting of all F_i^k, $i \in I(k)$, and their faces. For $k \geq m$, $\mathscr{C}^k$ is a k-complex; for $k < m$ obviously $\mathscr{C}^k = \{\varnothing\}$. In the notation of section 8.4, $\mathscr{C}^k = \text{st}_k(F; P)$. It is easily checked that for each $k, m < k \leq d$, and for each $i \in I(k)$, we have $\mathscr{C}(F_i^k) \cap \mathscr{C}^{k-1} = \text{st}(F; F_i^k)$, the star of F in F_i^k. Denoting by $A \sim B$ the set-theoretic difference, it follows that for each j, $0 \leq j \leq k - 1$,

$$f_j(\mathscr{C}(F_i^k) \sim \mathscr{C}^{k-1}) = f_j(F_i^k) - f_j(\text{st}(F; F_i^k)).$$

Using theorem 8.4.2 we have

$$\sum_{j=1}^{k} (-1)^j f_{j-1}(\mathscr{C}(F_i^k) \sim \mathscr{C}^{k-1}) = \sum_{j=0}^{k-1} (-1)^j f_j(\text{st}(F; F_i^k)) - \sum_{j=0}^{k-1} (-1)^j f_j(F_i^k)$$

$$= 1 - (1 - (-1)^k) = (-1)^k.$$

We shall use this relation in the sequel.

Let now 0 be a relatively interior point (e.g. the centroid) of the m-face F of the d-polytope P, $m = 2$. Let P_s, $s \in S$, be the pyramids with apex 0 spanned by the $(d - 1)$-faces of K which do not contain F. Obviously, card $S = f_{d-1}(P) - f_{d-1}(\text{st}(F; P))$. The crucial point to note here is that if P is a $(d - m)$-fold d-pyramid over F then each $P_s, s \in S$, is a $(d - m + 1)$-fold d-pyramid. Thus our proof shall be completed by showing that P satisfies the theorem provided each P_s satisfies it.

Let F^* be any k-face of one (or more) of the P_s. Then, if $0 \notin F^*$, it follows that F^* is a face of P and that $\varphi(F^*, P) = \sum \varphi(F^*, P_s)$, the summation being extended over all P_s which contain F^*. If, on the other hand, $0 \in F^*$ then F^* is not a face of P; in this case F^* is the convex hull of 0 and some $(k - 1)$-face F^{**} of P. Let $H = H(F^*)$ be the (uniquely determined) face of P of smallest possible dimension which contains F^* (and therefore, since $0 \in F^*$ and 0 is in the relative interior of F, $H \supset F$). Since $P = \bigcup_{s \in S} P_s$ we clearly have $\bigcup_{P_s \supset F^*} C(F^*, P_s) = C(H(F^*), P)$ and therefore

$$\sum_{P_s \supset F^*} \varphi(F^*, P_s) = \varphi(H(F^*), P).$$

It follows that the sum $\sum_{s \in S} \alpha_k(P_s)$ can be evaluated as follows.

(i) For $k > 0$, the k-faces F^* which do not contain 0 contribute $\beta_k^{(1)} = \sum_{F^*} \varphi(F^*, P)$; the k-faces F^* which contain 0 contribute

$\beta_k^{(2)} = \sum_{F*} \varphi(H(F^*), P)$. Each k-face F^* of the latter type is determined by some $(k-1)$-face F^{**} of P; in order to obtain all the k-faces F^* for which $H(F^*)$ is a certain fixed F_i^n, $i \in I(n)$, we have to consider all the $(k-1)$-faces of this F_i^n which do not belong to any F_j^{n-1}, $j \in I(n-1)$. In other words, breaking $\beta_k^{(2)}$ up into partial sums in each of which $H(F^*) = F_i^n$ is fixed, we have

$$\beta_k^{(2)} = \sum_{i, n} \sum_{\{F^* | H(F^*) = F_i^n\}} \varphi(H(F^*), P)$$

$$= \sum_{n = \max\{m, k\}} \sum_{i \in I(n)} \varphi(F_i^n, P) \cdot f_{k-1}(\mathscr{C}(F_i^k) \sim \mathscr{C}^{k-1}).$$

Regarding $\beta_k^{(1)}$ we note that exactly those k-faces of P which contain F fail to be faces of some P_s. Therefore we have $\beta_k^{(1)} = \alpha_k(P) - \sum_{i \in I(k)} \varphi(F_i^k, P)$.

Combining those results we find

$$\sum_{s \in S} \alpha_k(P_s) = \beta_k^{(1)} + \beta_k^{(2)}$$

$$= \alpha_k(P) + \sum_{n = \max\{m, k\}}^{d} \sum_{i \in I(n)} \varphi(F_i^n, P) \cdot [f_{k-1}(\mathscr{C}(F_i^n) \sim \mathscr{C}^{n-1}) - \delta_{n, k}],$$

where $\delta_{n, k}$ is, as usual, Kronecker's delta.

(ii) For $k = 0$ the above verbal reasoning applies as well. However, instead of making ad hoc notational conventions, we state the obvious result separately in the form

$$\sum_{s \in S} \alpha_0(P_s) = \alpha_0(P) + \varphi(F, P).$$

Applying now the assumption that the theorem holds for all the polytopes P_s, and using the above formulae and Euler's relations for polytopes and for stars, we have

$$(-1)^{d-1}[f_{d-1}(P) - f_{d-1}(\text{st}(F; P))] = (-1)^{d-1} \text{card } S$$

$$= \sum_{s \in S} \sum_{k=0}^{d-1} (-1)^k \alpha_k(P_s) = \sum_{k=0}^{d-1} (-1)^k \sum_{s \in S} \alpha_k(P_s)$$

$$= \alpha_0(P) + \varphi(F, P) + \sum_{k=1}^{d-1} (-1)^k \{\alpha_k(P) + \sum_{n = \max\{m, k\}}^{d} \sum_{i \in I(n)} \varphi(F_i^n, P)$$

$$\times [f_{k-1}(\mathscr{C}(F_i^n) \sim \mathscr{C}^{n-1}) - \delta_{n, k}]\}$$

$$= \sum_{k=0}^{d-1} (-1)^k \alpha_k(P) + \varphi(F, P) + \sum_{k=1}^{d-1} \sum_{i \in I(n)} \varphi(F_i^n, P)$$

$$\times \{[\sum_{k=1}^{n} (-1)^k f_{k-1}(\mathscr{C}(F_i^n) \sim \mathscr{C}^{n-1})] - (-1)^n\}$$

$$+ \varphi(P,P) \sum_{k=1}^{d-1} (-1)^k f_{k-1}(\mathscr{C}(P) \sim \mathscr{C}^{d-1})$$

$$= \sum_{k=0}^{d-1} (-1)^k \alpha_k(P) + \varphi(F,P) + \varphi(F^m,P)[-(1-(-1)^m)-(-1)^m]$$

$$+ \sum_{n=m+1}^{d-1} \sum_{i \in I(n)} \varphi(F_i^n,P)[(-1)^n - (-1)^n]$$

$$+ \varphi(P,P) \sum_{k=1}^{d-1} (-1)^k [f_{k-1}(P) - f_{k-1}(\text{st}(F,P))].$$

Thus, ignoring zeros, taking into account $\varphi(P,P) = 1$, and transposing, we have

$$\sum_{k=0}^{d-1} (-1)^k \alpha_k(P) = \sum_{k=0}^{d-1} (-1)^k [f_k(P) - f_k(\text{st}(F;P))]$$

$$= 1 - (-1)^d - 1 = (-1)^{d-1}.$$

This completes the proof of theorem 1.

As a complement to theorem 1 we have the following result, first proved by Höhn [1]:

2. *If an equation*

$$\sum_{i=0}^{d} (-1)^i \beta_i \alpha_i(P) = 0$$

holds for all d-polytopes P then $\beta_0 = \beta_1 = \cdots = \beta_d$.

PROOF Let P be a d-polytope, 0 a point of int P, and let P_j, for $j \in J = \{1, \cdots, f_{d-1}(P)\}$ be the d-pyramids with common apex 0, spanned by the $(d-1)$-faces of P. As noted in part II of the proof of theorem 1, we have

$$\sum_{j \in J} \alpha_i(P_j) = \alpha_i(P) + f_{i-1}(P) \quad \text{for} \quad i = 0, 1, \cdots, d-1,$$

and obviously also

$$\sum_{j \in J} \alpha_d(P_j) = f_{d-1}(P).$$

Under the hypothesis of theorem 2, we have therefore

$$0 = \sum_{j \in J} \sum_{i=0}^{d} (-1)^i \beta_i \alpha_i(P_j)$$

$$= (-1)^d \beta_d f_{d-1}(P) + \sum_{i=0}^{d-1} (-1)^i \beta_i [\alpha_i(P) + f_{i-1}(P)]$$

$$= (-1)^d \beta_d f_{d-1}(P) - \sum_{i=-1}^{d-2} (-1)^i \beta_{i+1} f_i(P) - (-1)^d \beta_d \alpha_d(P)$$

$$= -\{(-1)^d \beta_d + \sum_{i=-1}^{d-1} (-1)^i \beta_{i+1} f_i(P)\} = 0.$$

The last equation holds, by assumption, for every d-polytope P. But by theorem 8.1.1 (see section 8.2) Euler's relation is (up to a constant factor) the only linear relation holding for the f-vectors of all d-polytopes. Therefore all β_i's are equal, and the proof of theorem 2 is completed.

14.2   Angle-sums Relations for Simplicial Polytopes

The analogy between the numbers $f_i(P)$ and $\alpha_i(P)$ for a d-polytope P extends further than the similarity of form of the equations of Euler and Gram. In the present section we shall discuss the angle-sums analogues of the equations of Dehn–Sommerville.

The first step in the following theorem of Poincaré [3]:

1. *For every d-simplex T, the angle-sums satisfy the mutually equivalent systems of equations*

$$\sum_{j=0}^{d-k} (-1)^{d+j} \binom{d-j}{k} \alpha_j(T) = \sum_{j=0}^{k-1} (-1)^{j+1} \binom{d-j}{d+1-k} \alpha_j(T), 0 \le k \le d+1,$$

and

$$\sum_{j=m}^{d} (-1)^j \binom{j+1}{m+1} \alpha_j(T) = (-1)^d \alpha_m(T), \qquad -1 \le m \le d,$$

where $\alpha_{-1}(T) = 0$.

We begin the proof by considering, in R^d, d independent $(d-1)$-hyperplanes through the origin 0. Let one of the halfspaces determined by each of the hyperplanes be distinguished as the positive halfspace of that halfplane. The d given hyperplanes define in R^d regions of different types. We shall be interested in two of the types.

(i) Each of the 2^d simplicial cones determined by the d hyperplanes is an *orthant*. Every orthant is determined by a sequence of d plus or minus signs, the ith sign indicating whether the orthant is in the positive halfspace of the ith hyperplane, or not. We distinguish $d + 1$ classes of orthants. For $0 \le k \le d$, a k-*orthant* is an orthant which is determined by k plus signs and $(d - k)$ minus signs. Therefore there are $\binom{d}{k}$ k-orthants.

Using the notation of section 14.1, let $\omega_k = \sum_\Omega \varphi(\Omega)$ denote the sum of the contents of all k-orthants Ω. (Thus ω_d is the content of the single 'positive' orthant, ω_{d-1} is the sum of the contents of the d $(d-1)$-orthants, etc.). Since an orthant Ω and its reflection in the origin $-\Omega$ are congruent, they have the same content. It follows that

$$\omega_k = \omega_{d-k} \quad \text{for} \quad k = 0, 1, \cdots, d \qquad (1)$$

(ii) The second type of regions determined by the d hyperplane through 0 which interests us here are the *wedges*. For $0 \le k \le d$, a k-wedge is the intersection of some k of the positive halfspaces. Thus the single d-orthant is also the only d-wedge. A 1-wedge is a halfspace, the only 0-wedge is the whole R^d.

Every k-wedge is the union of k-orthants, $(k + 1)$-orthants, $\cdots$, the d-orthant. More precisely, a given k-wedge is the union of all those i-orthants, $k \le i \le d$, the i plus signs of which include the k plus signs determining the wedge. Consequently, any given i-orthant is contained in $\binom{i}{k}$ different k-wedges. Let σ_k denote the sum of the contents of all the k-wedges; then, for example, $\sigma_0 = 1$ and $\sigma_1 = \frac{1}{2}d$. As an immediate consequence of the above we have

$$\sigma_k = \sum_{i=k}^{d} \binom{i}{k} \omega_i \quad \text{for} \quad k = 0, \cdots, d. \qquad (2)$$

Inverting (2) we obtain

$$\omega_i = \sum_{j=i}^{d} (-1)^{j+1} \binom{j}{i} \sigma_j \quad \text{for} \quad i = 0, \cdots, d. \qquad (3)$$

Taking (1) into account there results

$$\sum_{i=k}^{d} (-1)^i \binom{i}{k}\sigma_i = \sum_{i=d-k}^{d} (-1)^{i+d} \binom{i}{d-k}\sigma_i$$

$$\text{for} \quad k = 0, 1, \cdots, [\tfrac{1}{2}(d-1)]; \quad (4)$$

the $[\frac{1}{2}(d+1)]$ equations are obviously independent.

Also, from (1) and (2) it follows that

$$\sigma_k = \sum_{i=k}^{d} \binom{i}{k}\omega_{d-i} = \sum_{i=0}^{d}\sum_{j=0}^{d} (-1)^{i+j+d} \binom{i}{k}\binom{j}{d-i}\sigma_j,$$

and an easy computation yields

$$\sigma_k = \sum_{j=0}^{k} (-1)^j \binom{d-j}{d-k}\sigma_j \qquad \text{for} \quad k = 0, \cdots, d. \qquad (5)$$

Let now T be a spherical d-simplex contained in the d-sphere $S^d \subset R^{d+1}$. The $d+1$ $(d-1)$-faces of T determine in R^{d+1} a family $\mathscr{H}$ of $d+1$ d-hyperplanes; for each of these hyperplanes we designate the halfspace containing T as positive. By a slight modification of the above notation, let $\alpha_k = \alpha_k(T)$ denote the sum of the d-contents $\varphi(F)$ at all k-faces F of T. Clearly $\alpha_k = \sigma_{d-k}$ for $0 \le k \le d$, the wedges being those determined by $\mathscr{H}$ in R^{d+1}. Also, σ_{d+1} is the d-content of T (as fraction of the d-content of S^d); we find it convenient to define $\alpha_{-1} = \alpha_{-1}(T) = \sigma_{d+1}$. (Note that $\alpha_d = 1, \alpha_{d-1} = \frac{1}{2}(d+1)$.)

Taking equations (4) and (5) for $d+1$ instead of d, and substituting $\sigma_i = \alpha_{d-i}$ for $-1 \le i \le d$, we obtain

$$\sum_{j=-1}^{d-k} (-1)^{d+j} \binom{d-j}{k}\alpha_j = \sum_{j=-1}^{k-1} (-1)^{j+1} \binom{d-j}{d+1-k}\alpha_j$$

$$\text{for} \quad 0 \le k \le d+1$$

and

$$(-1)^d \alpha_m = \sum_{j=m}^{d} (-1)^j \binom{j+1}{m+1}\alpha_j \qquad \text{for} \quad -1 \le m \le d. \qquad (6)$$

Keeping the vertices of T fixed and letting the radius of S^d increase to infinity we obtain the Euclidean case, in which $\alpha_{-1}(T) = 0$, and the first of the above equations simplifies to

$$\sum_{j=0}^{d-k} (-1)^{d+j}\binom{d-j}{k}\alpha_j = \sum_{j=0}^{k-1} (-1)^{j+1}\binom{d-j}{d+1-k}\alpha_j,$$

$$0 \leq k \leq d+1. \tag{7}$$

· This completes the proof of theorem 1.

For $m = -1$ equation (6) reduces to Gram's equation (theorem 14.1.1) for the d-simplex T. Thus the above is a different version for the first part of the proof of theorem 14.1.1.

Noting that $\alpha_d = 1$ and $f_m(T) = \binom{d+1}{m+1}$ it follows that equation (6) may be written as

$$\sum_{j=m}^{d-1} (-1)^j\binom{j+1}{m+1}\alpha_j(T) = (-1)^d(\alpha_m(T) - f_m(T)).$$

In this formulation the equations are valid not only for every d-simplex T, but for all simplicial d-polytopes. Indeed, the following result due to M. A. Perles* holds:

2. *Every simplicial d-polytope $P \in \mathscr{P}_s^d$ satisfies the equations*

$$\sum_{i=m}^{d-1} (-1)^i\binom{i+1}{m+1}\alpha_i(P) = (-1)^d(\alpha_m(P) - f_m(P))$$

for $m = -1, 0, 1, \cdots, d-1$.

We omit the proof of theorem 2 which is completely analogous to the second part of the proof of theorem 14.1.1. The difference between the proofs is that all the pyramids P_j appearing now are d-simplices, so that theorem 1 is applicable, and that the Dehn–Sommerville equations (theorem 9.2.1) are used instead of Euler's relation.

14.3   The Steiner Point of a Polytope†

Associated with every closed bounded convex set in E^n is a point known as its Steiner point or curvature-centroid. This point has many interesting properties, some of which will be proved here. In the case of a polytope,

* Private communication.
† This section and the last part of section 14.4 have been written by G. C. Shephard.

the Steiner point is conveniently defined as a sum involving the external angles of the polytope at its vertices; hence its inclusion in this chapter.

The main results are theorems 2 and 3, in which a number of linear relations are established between the Steiner points of a polytope and of its faces. These (like theorems 14.1.1 and 14.2.1) are closely analogous to the Euler and Dehn–Sommerville relations. However they are of a different nature being vector, as opposed to scalar, identities.

Let P be any d-polytope in R^n, and let S^{n-1} denote the unit $(n-1)$-sphere centered on the origin 0. For any vertex F^0 of P, denote by $C^*(F^0, P)$ the subset of R^n consisting of all those vectors $y \neq 0$ for which the normal supporting hyperplane $L(P, y)$ of P intersects P in the vertex F^0 (see exercise 2.2.8). Then $C^*(F^0, P)$ is a polyhedral cone, its facets being perpendicular to the edges of P that meet at F^0. Hence

$$V(F^0, P) = C^*(F^0, P) \cap S^{n-1}$$

is a convex spherical polytope in S^{n-1}. The ratio of the $(n-1)$-content of $V(F^0, P)$ to the $(n-1)$-content of S^{n-1} is denoted by $\psi(F^0, P)$ and is, for obvious geometrical reasons called the *external angle* of P at F^0. The *Steiner point* of P is defined by

$$s(P) = \sum_{j=1}^{f_0} v_j \psi(F_j^0, P) \tag{1}$$

where v_j is the position vector of the vertex F_j^0 of P ($j = 1, \cdots, f_0$). Since $\{V(F_j^0, P)\}_{j=1,\cdots,f_0}$ covers S^{n-1} except for a set of measure zero, we see

$$\sum_j \psi(F_j^0, P) = 1$$

and hence $s(P)$ is independent of the position of the origin 0, in other words

$$s(TP) = Ts(P)$$

for all congruence transformations T. It is also easily verified that the value of $\psi(F_j^0, P)$ is independent of the dimension of the space in which P lies. We shall make use of this fact in the proof of theorem 2.

1. *Let P, Q be any two convex polytopes in E^n and λ, μ be any real numbers. Then*

$$s(\lambda P + \mu Q) = \lambda s(P) + \mu s(Q),$$

addition on the left being vector addition of the polytopes.

PROOF   Clearly $s(\lambda P) = \lambda s(P)$ so it will suffice to prove

$$s(P + Q) = s(P) + s(Q). \tag{2}$$

Let P have vertices F_i^0 with position vectors v_i, and Q have vertices G_j^0 with position vectors w_j. Writing $V_{ij} = V(F_i^0, P) \cap V(G_j^0, Q)$ we see that

$$\{\text{cl } V(F_i^0, P)\} \qquad \{\text{cl } V(G_j^0, Q)\} \, \{\text{cl } V_{ij}\}$$

are three coverings of S^{n-1} by closed spherical polytopes, the third being a common refinement of the first two. Let τ_{n-1} be the $(n-1)$-content of S^{n-1}, and $\tau_{n-1}\psi_i^P$, $\tau_{n-1}\psi_j^Q$, $\tau_{n-1}\psi_{ij}$ be the $(n-1)$-contents of $V(F_i^0, P)$, $V(G_j^0, Q)$ and V_{ij} respectively. Then from the definition,

$$s(P) + s(Q) = \sum_i v_i\psi_i^P + \sum_j w_j\psi_j^Q = \sum_{i,j} (v_i + w_j)\psi_{ij}.$$

If $u \in V_{ij}$, then $L(P, u) \cap P = F_i^0$ and $L(Q, u) \cap Q = G_j^0$. We deduce that if $V_{ij} \neq \varnothing$ then $v_i + w_j$ is the position vector of a vertex of $P + Q$ and ψ_{ij} is the external angle of $P + Q$ at this vertex. Hence

$$\sum_{i,j} (v_i + w_j)\psi_{ij} = s(P + Q)$$

and (2) is proved. This completes the proof of the theorem.

2. *Let P be a convex d-polytope and, for $0 \leq j \leq d - 1$ let $F_i^j (i = 1, \cdots, f_j)$ be its j-faces. Then,*

$$(1 + (-1)^{d-1})s(P) = \sum_{i=1}^{f_0} s(F_i^0) - \sum_{i=1}^{f_1} s(F_i^1) + \cdots$$

$$+ (-1)^{d-1} \sum_{i=1}^{f_{d-1}} s(F_i^{d-1}) \tag{3}$$

PROOF   We may assume that P lies in E^d. For any vertex F_i^0 of P consider the region $V(F_i^0, P)$. This is an open spherical convex polytope in S^{d-1} and its $(d-1)$-content may be computed from the well-known formula of Sommerville [2, p.157]. In this way we obtain for $\psi(F_i^0, P)$ the expression

$$(1 + (-1)^{d-1})\psi(F_i^0, P) = \sum_{j=0}^{d-1} (-1)^{d-j-1}\alpha_i^j, \tag{4}$$

where α_i^j denotes the sum of the $(d - j - 1)$-dimensional solid angles subtended by the polytope $V(F_i^0, P)$ at its j-faces. (In this formula $\alpha_i^{d-2} = \frac{1}{2}m$, where m is the number of $(d - 2)$-faces of $V(F_i^0, P)$ and $\alpha_i^{d-1} = 1$ for all i.) Now the $(d - 2)$-faces of $V(F_i^0, P)$ lie in hyperplanes through the center of S^{d-1} which are perpendicular to the edges of P meeting at F_i^0. Hence the solid angle at a j-face of $V(F_i^0, P)$ is bounded by the hyperplanes perpendicular to the edges of a $(d - j - 1)$-face of P meeting at F_i^0, and so is equal to the external angle at F_i^0 of that face. Thus

$$\alpha_i^j = \sum_{k=1}^{f_{d-j-1}} \psi(F_i^0, F_k^{d-j-1})$$

where $\psi(F_i^0, F_k^{d-j-1})$ is put equal to zero if F_i^0 is not a vertex of F_k^{d-j-1}. Substitute these values of α_i^j in (4), multiply by v_i, and sum for i from 1 to f_0. Using (1) we obtain

$$(1 + (-1)^{d-1})s(P) = \sum_{i=1}^{f_0} v_i \left(\sum_{j=0}^{d-1} (-1)^{d-j-1} \left(\sum_{k=1}^{f_{d-j-1}} \psi(F_i^0, F_k^{d-j-1}) \right) \right)$$

$$= \sum_{j=0}^{d-1} (-1)^{d-j-1} \left(\sum_{k=1}^{f_{d-j-1}} \left(\sum_{i=1}^{f_0} v_i \psi(F_i^0, F_k^{d-j-1}) \right) \right)$$

$$= \sum_{j=0}^{d-1} (-1)^{d-j-1} \left(\sum_{k=1}^{f_{d-j-1}} s(F_k^{d-j-1}) \right).$$

This is (3) and concludes the proof of the theorem.

Theorem 1 is analogous to the Euler identity. The next theorem gives the analogues of the Dehn–Sommerville relations for simple polytopes. To obtain these, we apply (3) to an r-face F^r of a simple d-polytope P, giving

$$(1 + (-1)^{r-1})s(F^r) = \sum' s(F_i^0) - \sum' s(F_i^1) + \cdots + (-1)^{r-1} \sum' s(F_i^{r-1}) \quad (5)$$

where $\sum'$ means summation over those suffixes i for which the face $F_i^j (j < r)$ is incident with F^r. Now sum relation (5) over all the r-faces of P. Since each j-face is incident with $\binom{d - j}{r - j}$ r-faces, each term $s(F_i^j)$

occurs $\begin{pmatrix} d - j \\ r - j \end{pmatrix}$ times in the sum, and we obtain

$$(1 + (-1)^{r-1}) \sum s(F_i^r) = \begin{pmatrix} d \\ r \end{pmatrix} \sum s(F_i^0) - \begin{pmatrix} d - 1 \\ r - 1 \end{pmatrix} \sum s(F_i^1) + \cdots$$

$$+ (-1)^{r-1} \begin{pmatrix} d - r + 1 \\ 1 \end{pmatrix} \sum s(F_i^{r-1}), \quad (6)$$

where $\sum$ means summation over all the faces of P of the dimension indicated by the superscript. Putting $r = 1, \cdots, d$ we obtain d relations of type (6). These are not linearly independent:

3. *For the simple d-polytope P, there are exactly $[\frac{1}{2}(d + 1)]$ linearly independent relations of type (6), namely those corresponding to the values*

$$r = 1, 3, 5, \cdots, m$$

where m is the largest odd integer not exceeding d.

PROOF   Rewrite equations (6) in the form

$$0 = (-1)^r \begin{pmatrix} d \\ d - r \end{pmatrix} \sum s(F_i^0) + (-1)^{r-1} \begin{pmatrix} d - 1 \\ d - r \end{pmatrix} \sum s(F_i^1) + \cdots$$

$$+ (-1)^1 \begin{pmatrix} d - r + 1 \\ d - r \end{pmatrix} \sum s(F_i^{r-1}) + (1 + (-1)^{r-1}) \sum s(F_i^r),$$

and denote the right side of this equation by (S_r^d) $(r = 1, \cdots, d)$. If d is even, then it is simple to verify that

$$2(S_d^d) + \sum_{r=1}^{d-1} (S_r^d) = 0. \quad (7)$$

If r is even and $r < d$, then

$$\sum_{i=0}^{r-1} \begin{pmatrix} d - r + i \\ 1 + i \end{pmatrix} (S_{r-i}^d) = 0. \quad (8)$$

To see this, we notice that for $0 \leq k \leq r - 1$, the coefficient of $\sum s(F_i^k)$ in the left side of (8) is, after slight simplification, equal to

$$\sum_{i=-1}^{r-k} (-1)^i \begin{pmatrix} d - r + i \\ 1 + i \end{pmatrix} \begin{pmatrix} d - k \\ r - k - i \end{pmatrix}.$$

But this is the coefficient of x^{r-k+1} in the formal product of

$$(1 + x)^{d-k} = \binom{d-k}{0} + \binom{d-k}{1}x + \cdots + \binom{d-k}{d-k}x^{d-k}$$

by

$$-(1 + x)^{-d+r} = -\binom{d-r-1}{0} + \binom{d-r}{1}x - \binom{d-r+1}{2}x^2 + \cdots$$

and so is zero. Hence (8) is proved.

Relations (7) and (8) show that (S_r^d) is, for even r, linearly dependent on (S_j^d) $(j = 1, \cdots, r - 1)$, and so the equations $(S_r^d) = 0$ (r even) are redundant. The remaining equations are linearly independent since the matrix of coefficients is of triangular form, (S_r^d) containing no term in $\sum s(F_i^j)$ for $j > r$. This proves theorem 3.

14.4 Remarks

Many authors have considered relations between angle-sums, and various related notions—mostly unaware of much of the previous work. The first nontrivial result seems to be de Gua's [1] observation that theorem 14.1.1 holds for every 3-simplex. (This result was rediscovered many times; see, for example, Gaddum [1, 2], where additional references may be found.) The next achievement was Gram's [1] proof of theorem 14.1.1 for all 3-polytopes; this paper was, however, completely forgotten till very recently. More than thirty years later Dehn [1] and Poincaré [3] made, independently of each other, contributions to the subject.

Poincaré [3] obtained the second system of equations of theorem 14.2.1, while Dehn [1] proved (for $d \leq 5$) parts of theorem 14.2.2. Dehn's work was extended by Sommerville [1, 2], whose proof of 14.2.1 was followed here. Sommerville also obtained a set of equations equivalent to that of theorem 14.2.2, but more complicated in form. He also gave a proof of theorem 14.1.1; however, his version of the third part of the above proof is invalid. The subdivisions he uses lead, in dimensions greater than 3, to sets of convex polytopes of different dimensions which not only fail to be polytopes of types for which the theorem has already been established, but even fail to be complexes. It is interesting to note that although Sommerville [1] discusses the analogy between the angle-sums relations and the Dehn–Sommerville equations (for simplicial polytopes), and although he mentions both forms of the latter system (see section 9.2), he fails to obtain the second set of equations of theorem 14.2.1 and consequently misses the simple equations of theorem 14.2.2.

Höhn [1] gives a new derivation for the second set of equations of theorem 14.2.1. He also proves that the system contains $[\frac{1}{2}d] + 1$ independent equations, and shows that every other linear relation among the angle-sums of all d-simplices is a consequence of these. (Using a different notation, Höhn disregards the angle-sums $\alpha_d = 1$ and $\alpha_{d-1} = \frac{1}{2}(d + 1)$: therefore his system contains only $[\frac{1}{2}d]$ independent equations.) Although seemingly not aware of Sommerville [1], Höhn mentions as well-known theorem 14.1.1, and proves the uniqueness theorem 14.1.2.

Still another set of equations equivalent to those of theorem 14.2.1, and involving the Bernoulli numbers, was given by Peschl [1]. A simple algebraic proof of the equivalence of the two sets of equations of theorem 14.2.1 was given by Sprott [1]; Guinand [1] gave another short proof, including the equivalence of Peschl's system with the others.

As shown by the above proof of theorem 14.2.1, the results of the present chapter may be extended to the spherical geometry, as well as to the elliptic and hyperbolic geometries. For some steps in these directions, as well as for relations to other geometric problems see Dehn [1], Sommerville [1, 2], Höhn [1], Peschl [1], Coxeter [4], and Böhm [1].

Perles–Shepherd [1] recently obtained simpler proofs of the results of sections 14.1 and 14.2, along with many new results concerning angle-sums. Among their results is the following analogue of theorem 14.2.2, which deals with cubical d-polytopes (compare sections 4.6 and 9.4):

1. *Every cubical d-polytope P satisfies the equations*

$$\sum_{i=m}^{d-1} (-1)^i 2^{i-m} \binom{i}{m} \alpha_i(P) = (-1)^d(\alpha_m(P) - f_m(P))$$

for $m = 0, 1, \cdots, d - 1$.

It may be noted that the assumption that the 'content' φ is determined by the Lebesgue measure on the $(d - 1)$-sphere is unnecessarily special. The measure μ used in the definition of φ could be an arbitrary finitely additive measure invariant under reflection in the origin and such that $\mu(S^{d-1}) \neq 0$. It seems that no interesting applications of this fact have been made so far.

Steiner [3] introduced the *Krümmungsschwerpunkt* (or 'Steiner point' as we have called it) in 1840 in connection with an extremal problem for plane convex regions. Assuming that the boundary bd K of K was of class C^2, he defined $s(K)$ as the centroid of bd K, each point of which carries a weight equal to its curvature. In $d > 2$ dimensions there is a similar definition for smooth convex bodies: $s(K)$ is the centroid of bd K, each point of which carries a weight equal to its Gauss curvature.

In 1918 Kubota [1] showed that the Steiner point of a plane convex region is characterized by the fact that if $s(K)$ is taken as origin, the coefficients of $\cos \theta$ and $\sin \theta$ in the Fourier expansion of the supporting function $H(u(\theta), K)$ are both zero. The three dimensional analogue was established by Gericke [1] in 1940. These properties may be used to define the Steiner point, a procedure which has the advantage that no smoothness conditions on bd K need be assumed. Further, the additivity property

$$s(\lambda_1 K_1 + \lambda_2 K_2) = \lambda_1 s(K_1) + \lambda_2 s(K_2)$$

is an easy deduction, though this was not stated explicitly until 1963, see Grünbaum [9].

The results of Kubota and Gericke also lead to the definition of $s(K)$ for arbitrary closed bounded convex sets in R^d given by Shephard [6]:

$$s(K) = \frac{1}{\sigma_n} \int_{S^{d-1}} uH(u, K) \, d\omega, \qquad \sigma_n = \int_{S^{d-1}} \langle u, a \rangle^2 \, d\omega,$$

where u is a variable unit vector, a is any fixed unit vector, $H(u, K)$ is the supporting function of K and $d\omega$ is an element of surface area of the unit sphere S^{d-1} centered at 0. From this definition the additivity of $s(K)$ is obvious, and it also shows that $s(K)$ is a uniformly continuous function of K in the Hausdorff metric. It is because of this uniformity that $s(K)$ can be defined in a continuous manner on the set of all compact convex sets in R^d, and not only on those of maximum dimension. (By contrast, the centroid of K is not uniformly continuous and so does not have this property; see Shephard–Webster [1].)

The equivalence of the above integral definition with that given in section 14.3 for polytopes is proved in Shephard [8]. Here also the proofs of theorems 14.3.2 and 14.3.3 first appear.

It is not difficult to show that the linear relation of theorem 14.3.2 is the only one that holds for the Steiner points of an arbitrary polytope. It is an open question whether those of theorem 14.3.3 are the only ones that hold for all simple polytopes.

On the other hand, G. C. Shephard (private communication) recently established the following remarkable result:

2. *The Steiner point is the only point-valued function defined for all convex bodies which is additive, uniformly continuous, and commutes with similarity transformations.*

There exist other quantities associated with a convex polytope and its faces, which satisfy relations of the type of the Euler equation. One such quantity is the *mean width* (Shephard [11]; see Bonnesen–Fenchel [1] for the definition). Since the mean width of compact convex sets in R^d is clearly an additive, uniformly continuous and similarity–covariant function, it follows that it is possible to associate d-dimensional convex bodies to all compact convex sets in R^d, in such a manner that the function has all the properties just mentioned, and satisfies for polytopes an Euler-type equation as well; for a compact, convex set $C \subset R^d$ such an associated convex body is the d-ball centered at the Steiner point of C and having radius equal to the mean width of C (or a constant multiple of it). It may be conjectured that no other set-valued function defined for all compact convex subsets of R^d has all those properties.

Another interesting result on Steiner points is the 'valuation property' (Sallee [1]):

3. *If $K_1, K_2, K_1 \cup K_2$ are compact convex sets then*

$$s(K_1 \cup K_2) + s(K_1 \cap K_2) = s(K_1) + s(K_2).$$

This property can be used as a starting point for extending the definition of the Steiner point to non-convex sets (Sallee [1], Shephard [8]).

14.5   Additional notes and comments

Gram's equation.
A beautiful short proof of theorem 14.1.1, due to Shephard [a] and Welzl [a],
reduces the claim to Euler's equation by a probabilistic argument:

Let z be chosen uniformly at random from the unit sphere $S^{d-1} \subset R^d$, and
denote by π^z the orthogonal projection of R^d along the line $R \cdot z$. Suppose
that for every proper face F of P we have $\dim(\pi^z(F)) = \dim F$ (this happens
with probability one). Under this assumption, the projection $\pi^z(F)$ of a proper
face F of P is a face of the projected polytope $\pi^z(P)$ if and only if $(F + R \cdot z) \cap
P = F$, which in turn is equivalent to $z, -z \notin C(F,P)$. Hence, the probability
that $\pi^z(F)$ is a face of $\pi^z(P)$ equals $1 - 2\varphi(F,P)$. Therefore, the expected value
of the number $f_i(\pi^z(P))$ of i-faces of $\pi^z(P)$ is $\sum(1 - 2\varphi(F,P))$, where the sum
ranges over all i-faces of P, i.e., that expected value equals $f_i(P) - 2\alpha_i(P)$.

By linearity of expectation, theorem 14.1.1 thus follows from Euler's equations for $\pi^z(P)$ and P.

Steiner point.
The Steiner point of a polytope may be interpreted as the expected value of the
random variable that maps a random direction $z \in S^{d-1}$ to the highest point of P
in direction z (which is unique with probability one). It has interesting invariance properties that are not shared, say, by the *centroid* (the barycenter of the
vertex set), or by the center of gravity: One example is the fact (theorem 14.3.1)
that computing the Steiner point commutes with forming Minkowski sums,
which may be derived from linearity of expectation.

Calculating the Steiner point of a polytope (even if it is specified by both a
$\mathscr{V}$-and an $\mathscr{H}$-description) seems to be hard, because it involves volume computations (see the notes in section 15.5).

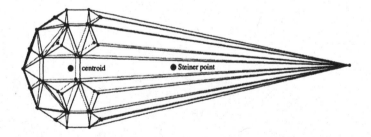

Figure 14.1: Steiner point vs. centroid

The Steiner point has also proved to be useful for approximation algorithms for shape matching (Aichholzer–Alt–Rote [a]).

Sallee [a] answered a question of Grünbaum [9] by producing, on the collection of all compact convex subsets of R^d, a function that has all the standard properties of the Steiner point *except* that it is not continuous with respect to the Hausdorff metric.

In addition to Steiner points and other notions of "centers" of convex bodies surveyed in Grünbaum [9], several other sorts of centers have proved to be particularly relevant to questions of optimization and computational complexity. For a survey of these, see Kaiser–Morin–Trafalis [a].

Valuations.

The topic of valuations and their characterization—as exemplified by Shephard's theorem 14.4.2—is of great importance, for example for convex geometry, for geometric measure theory and geometric probability (see for example Klain–Rota [a]), for the decomposition theory of polyhedra as discussed in Chapter 15, but also for the question of dissections (related to Hilbert's third problem; see Boltianskii [a]). We refer to McMullen [j] for a comprehensive survey.

CHAPTER 15

Addition and Decomposition of Polytopes*

In this chapter we discuss two methods of 'adding together' two polytopes to form a third. The first of these, vector addition or Minkowski addition as it is sometimes called, has already been used in the preceding chapters. The second, called Blaschke addition, will form the topic of section 15.3. One of the main problems is to find criteria for deciding whether a given polytope is *decomposable*, that is to say, can be expressed as a 'sum' (in either sense) of other polytopes. In section 15.4 the extension of these results to general convex sets will be discussed, and references given to the relevant literature.

15.1 Vector Addition

Let Q and R be given polytopes in E^d. Then there are three equivalent ways in which the *vector sum* $Q + R$ can be defined (see exercise 1):

(a) $Q + R = \{x + y \mid x \in Q \text{ and } y \in R\}$.

(b) Let $H(Q, u)$ and $H(R, u)$ be the supporting functions of Q and R (see section 2.2). Then $Q + R$ is defined to be the convex set whose supporting function is given by the equation

$$H(Q + R, u) = H(Q, u) + H(R, u)$$

(see exercise 2.2.8).

(c) Let q_i $(i = 1, \cdots, n)$ be the (position vectors of the) vertices of Q, and r_j $(j = 1, \cdots, m)$ be the vertices of R. Then we define

$$Q + R = \operatorname{conv}\{q_i + r_j \mid i = 1, \cdots, n \quad \text{and} \quad j = 1, \cdots, m\}.$$

If $P = Q + R$, then Q and R are called *summands* of P. Since a polytope is characterized by the fact that its supporting function is piecewise linear (see exercise 3.1.19), it follows that every summand of a polytope is a polytope.

---

* Chapter 15 was written by G. C. Shephard.

316

As in section 2.2, for each $v \neq 0$, write $L(Q, v)$ for the supporting hyperplane of Q with outer normal v. Then $L(Q, v) \cap Q$ is a face of Q, which will be denoted by $F(Q, v)$. The first theorem gives an expression for the faces of a polytope in terms of the faces of its summands: it is an immediate consequence of exercise 2.2.8.

1. *If $P = Q + R$, then for each $v \neq 0$,*

$$F(P, v) = F(Q, v) + F(R, v).$$

It is clear that the definition of a vector sum depends upon the position of the origin, so that, if T is any nonzero translation, then $T(Q + R) \neq T(Q) + T(R)$. This is inconvenient since we are usually interested only in the 'shapes' of the polytope and its summands, and not upon their position relative to an arbitrarily chosen origin. Consequently it seems more natural, in some ways, to consider vector addition as operating on translation classes $[P]_T$ of polytopes, instead of on the polytopes themselves, defining $[Q]_T + [R]_T$ to be $[Q + R]_T$. There is a simple manner in which this apparent complication can be avoided. We shall make the convention that, unless otherwise stated, every polytope considered in this section will be translated so that its Steiner point (see section 14.3) lies at the origin. Then the additivity of the Steiner point has the consequence that the conditions $[P]_T = [Q]_T + [R]_T$ and $P = Q + R$ are equivalent.

The scalar multiple λP of a polytope P by a real number λ can be defined in three equivalent ways analogous to the three definitions of the vector sum given above. Thus,

(a) $\lambda P = \{\lambda x \mid x \in P\}$

(b) $H(\lambda P, u) = \lambda H(P, u)$ for all u.

(c) $\lambda P = \text{conv}\{\lambda p_i \mid p_i \text{ is a vertex of } P\}$.

The following properties of vector addition and scalar multiplication are easily verified:

$$P_1 + P_2 = P_2 + P_1, \qquad P_1 + (P_2 + P_3) = (P_1 + P_2) + P_3,$$

$$\lambda(P_1 + P_2) = \lambda P_1 + \lambda P_2, \qquad (\lambda\mu)P = \lambda(\mu P),$$

$$(\lambda + \mu)P = \lambda P + \mu P \qquad \text{for} \quad \lambda\mu \geq 0.$$

Notice that the last relation does not hold if $\lambda\mu < 0$, so that we do not obtain a vector space. (For a discussion of methods by which the set of all closed bounded convex sets can be converted into a vector space, see Ewald–Shephard [1].)

If $\lambda > 0$ then any translate of λP is said to be *positively homothetic* to P. If $0 \leq \lambda \leq 1$, then λP is trivially a summand of P for

$$P = \lambda P + (1 - \lambda)P.$$

A polytope is said to be *decomposable* if it possesses a summand which is not positively homothetic to the polytope. Thus a decomposable polytope is one that can be expressed as a vector sum in a nontrivial manner.

We now consider the problem of characterizing decomposable and indecomposable polytopes. In $d = 2$ dimensions this is simple: every convex polygon P is decomposable unless it is a triangle, which is indecomposable. To see this, notice that every polygon Q is a summand of P if its edges are parallel to the edges of P and do not exceed them in length. The next theorem generalizes this result to $d > 2$ dimensions.

Let P and Q be two given polytopes with the property that

$$\dim F(P, v) = \dim F(Q, v) \qquad (1)$$

for all $v \neq 0$. Then there will be a one-to-one correspondence between the r-faces of P and the r-faces of Q $(r = 0, \cdots, d - 1)$ in which $F(P, v)$ corresponds to $F(Q, v)$. Let $p_1, \cdots, p_m$ be the vertices of P and $q_1, \cdots, q_m$ be the corresponding vertices of Q. If an edge of P joins p_i to p_j, then a parallel edge of Q will join q_i to q_j. Hence

$$\lambda(p_i - p_j) = q_i - q_j \qquad (2)$$

where $\lambda > 0$ is a real number whose value depends on i and j. If, instead of (1) we impose the weaker condition

$$\dim F(P, v) \geq \dim F(Q, v) \qquad (3)$$

for all $v \neq 0$, then similar considerations will apply, except that the correspondence is no longer one-to-one, and several vertices of P may correspond to the same vertex of Q. It will be convenient to continue using q_i for the vertex of Q that corresponds to p_i, with the understanding that $q_1, \cdots, q_m$ may not all be distinct. Relations (2) also hold in this case if we put $\lambda = 0$ when $q_i = q_j$. If, in addition to (3), all the values of λ defined by (2) satisfy $0 \leq \lambda \leq 1$, then we shall write $P \geq Q$.

2. *The polytope Q is a summand of P if and only if $P \geq Q$.*

PROOF   The necessity of the condition is clear. For if Q is a summand of P, then by theorem 1, for each v, $F(Q, v)$ is a summand of $F(P, v)$ and so (3) holds. If v is chosen so that $F(P, v)$ is the edge joining p_i to p_j, then $F(Q, v)$ will be either the edge joining q_i to q_j, or q_i and q_j will coincide.

In either case theorem 1 implies that $|q_i - q_j| \leq |p_i - p_j|$ and so $0 \leq \lambda \leq 1$ in equation (2). Hence $P \geq Q$.

To prove the condition is sufficient, we shall show that if $P \geq Q$, it is possible to construct a polytope R such that $P = Q + R$. Let P have k edges and write $e_1, \cdots, e_{2k}$ for vectors, one in each direction, along these edges, then

$$\lambda_1 e_1, \cdots, \lambda_{2k} e_{2k} \qquad (0 \leq \lambda_i \leq 1) \qquad (4)$$

will be vectors along the corresponding edges of Q. Now translate P and Q so that the origin coincides with one of the vertices of P and also with the corresponding vertex of Q. Then the position vector of any vertex p_i of P can be written as a sum

$$p_i = e_{i_1} + \cdots + e_{i_s} \qquad (5)$$

of at most k edge vectors, corresponding to an edge path joining the origin to the vertex p_i. There will be many such expressions for the position vector of each vertex, so we shall suppose that one such representation is chosen for each vertex. Now let

$$q_i = \lambda_{i_1} e_{i_1} + \cdots + \lambda_{i_s} e_{i_s}, \qquad (6)$$

$$r_i = (1 - \lambda_{i_1}) e_{i_1} + \cdots + (1 - \lambda_{i_s}) e_{i_s}. \qquad (7)$$

These expressions will be said to be *analogous* to (5) if they have the same set of suffixes. q_i is clearly the vertex of Q corresponding to p_i. Also, since $r_i = p_i - q_i$, the point r_i depends only on p_i and not on the particular representation (5) that is chosen. Define R to be the convex hull of the m points r_i given by all the expressions (7) analogous to the chosen representations (5) of the vertices p_i. We shall complete the proof by showing that for any $u \neq 0$,

$$H(P, u) = H(Q, u) + H(R, u) \qquad (8)$$

and so $P = Q + R$ by definition (b).

Let p_u be any vertex of $F(P, u)$. Then the corresponding vertex q_u will be a vertex of $F(Q, u)$ and

$$H(P, u) = \langle u, p_u \rangle, \qquad H(Q, u) = \langle u, q_u \rangle.$$

Write $r_u = p_u - q_u$. Then (8) is equivalent to

$$H(R, u) = \langle u, r_u \rangle.$$

Since the expression for r_u is analogous to that for p_u, it follows that $r_u \in R$, and so $H(R, u) \geq \langle u, r_u \rangle$. Hence it will suffice to show that strict inequality leads to a contradiction. Assume therefore that

$$H(R, u) = \langle u, r_* \rangle > \langle u, r_u \rangle = H(P, u) - H(Q, u) \qquad (9)$$

for some vertex r_* of R. We can express r_* in the form (7) and write p_* and q_* for the vertices of P and Q given by the analogous expressions (5) and (6). Define an edge path $e_{j_1}, \cdots, e_{j_t}$ on P in the following manner. Let e_{j_1} be any edge of which p_* is an end-point and $\langle u, e_{j_1} \rangle > 0$ (so that e_{j_1} joins p_* to a vertex p_{**} of P nearer to $L(P, u)$). This is clearly possible by the convexity of P, and

$$p_{**} = p_* + e_{j_1}.$$

Select e_{j_2} starting from p_{**} in a similar manner. Continuing thus, after a finite number of steps we obtain an edge-path joining p_* to some vertex p_0 of P lying in $L(P, u)$, and so

$$p_0 = p_* + e_{j_1} + \cdots + e_{j_t},$$

and

$$\langle u, e_{j_k} \rangle > 0 \qquad (k = 1, \cdots, t).$$

If the vertex q_0 corresponds to p_0 then

$$q_0 = q_* + \lambda_{j_1} e_{j_1} + \cdots + \lambda_{j_t} e_{j_t},$$

and

$$\begin{aligned} H(P, u) - \langle u, p_* \rangle &= \langle u, (p_0 - p_*) \rangle \\ &= \langle u, (e_{j_1} + \cdots + e_{j_t}) \rangle \\ &\geq \langle u, (\lambda_{j_1} e_{j_1} + \cdots + \lambda_{j_t} e_{j_t}) \rangle \\ &= \langle u, (q_0 - q_*) \rangle \\ &= H(Q, u) - \langle u, q_* \rangle. \end{aligned}$$

Thus $H(P, u) - H(Q, u) \geq \langle u, (p_* - q_*) \rangle = \langle u, r_* \rangle = H(R, u)$ which is a contradiction. Hence (9) is false, (8) is true and the theorem is proved.

The next two theorems, which are consequences of theorem 2, will enable us to deduce that every simplicial polytope is indecomposable, whereas every simple polytope, with the exception of a simplex, is decomposable. Further applications of these theorems will be given in the exercises.

3. *If all the 2-faces of a d-polytope P are triangles, then P is indecomposable.*

PROOF Let v be chosen so that $F(P, v)$ is a triangular 2-face of P. If Q is a summand of P, then $F(Q, v)$ is, by theorem 1, a summand of $F(P, v)$ and so is a triangle homothetic to $F(P, v)$. Consequently the constants λ_i associated with all the edges of $F(P, v)$ in (2) are all equal. This reasoning applies to every triangular 2-face of P, and since any two edges of P can be joined by a 'chain' of triangles, each consecutive pair having an edge in common, we deduce that the constants λ_i associated with all the edges of P have the same value, which we may denote by λ_*.

It follows from (5) and (6) that for each i,

$$q_i = \lambda_* p_i,$$

and so Q is homothetic to P, the ratio of similarity being λ_*. Thus every summand of P is homothetic to P, and so P is indecomposable. This completes the proof of theorem 3.

4. *Except for the d-simplex, every simple d-polytope P is decomposable.*

PROOF   It is easy to show that, with the exception of the simplex, every simple polytope P has a facet $F(P, u_0)$ with the property that at least one edge e_* of P is disjoint from $F(P, u_0)$. Let $u_1, \cdots, u_f$ be the outward unit normals to the other facets of P, so that P can be defined as the set of points x satisfying the inequalities

$$\langle x, u_i \rangle \leq H(P, u_i) \qquad (i = 0, 1, \cdots, r).$$

Let Q' be the set of points

$$\langle x, u_0 \rangle \leq H(P, u_0) - \varepsilon$$

$$\langle x, u_i \rangle \leq H(P, u_i) \qquad (i = 1, \cdots, r).$$

Then for ε sufficiently small, it is clear that Q' will have the property

$$\dim F(P, v) = \dim F(Q', v)$$

for all v. (In fact ε must be chosen less than the smallest positive distance of any vertex of P from $L(P, u_0)$). If $Q = \delta Q'$, where δ is chosen so small that each edge of Q is shorter than the corresponding edge of P, then $Q \leq P$ and so Q will be a summand of P. We complete the proof by showing that Q is not homothetic to P.

At least one edge of Q' (for example one that has exactly one vertex in common with $F(Q, u_0)$) will be shorter than the corresponding edge of P, whereas at least one edge (for example e_*) will have length equal to the corresponding edge of P. Hence Q' is not homothetic to P, and therefore neither is Q. By theorem 2, P is decomposable and theorem 4 is proved.

Another property of polytopes closely related to decomposability is that of reducibility.

Let K be any convex set, and write $-K$ for $(-1)K$, the reflection of K in the origin. Then

$$K + (-K) = \{x_1 - x_2 \mid x_1, x_2 \in K\}$$

is called the *difference* set of K. A convex set H is said to be *reducible* if it is the difference set of a convex set K which is not homothetic to H. Since

$$-H = (-K) + K = K + (-K) = H,$$

central symmetry in the origin is a necessary condition for reducibility. Obviously decomposability is also a necessary condition. A sufficient condition is given in the next theorem.

5. *A centrally symmetric polytope P is reducible if and only if it possesses a summand Q which is not centrally symmetric.*

PROOF   Suppose first that P is the difference set of a convex set R. Then R must be a polytope since it is a summand of P. If every summand of P were centrally symmetric, then R would be centrally symmetric and

$$P = R + (-R) = 2R.$$

Thus P and R would be homothetic. Since this applies to every R, P would be irreducible and the necessity of the given condition is established.

If, on the other hand P possesses a noncentrally symmetric summand Q, then $P = Q + R$ for some polytope R and $P = (-P) = (-Q) + (-R)$. Thus

$$P = \tfrac{1}{2}(Q + R) + \tfrac{1}{2}((-Q) + (-R))$$
$$= \tfrac{1}{2}((-Q) + R) + \tfrac{1}{2}(Q + (-R))$$

and so P is the difference set of $\tfrac{1}{2}((-Q) + R)$. If $(-Q) + R$ were homothetic to $Q + R$, then, since the widths of these two sets in any direction are equal, we could deduce that $(-Q) + R = Q + R$. But this is impossible since $-Q \neq Q$ by assumption. Thus $\tfrac{1}{2}((-Q) + R)$ is not homothetic to $Q + R = P$, so P is reducible and theorem 5 is proved.

This theorem, along with theorem 2, enables us to decide whether a given polytope is reducible or not. For example in E^2 every centrally symmetric polygon is reducible unless it is a parallelogram. This follows from the fact that every summand of a parallelogram is either a parallelogram or a line segment, and so is centrally symmetric. On the other hand it is easy to construct a noncentrally symmetric summand of any centrally symmetric $2n$-gon if $n > 2$. We now generalize these statements to $d > 2$ dimensions.

A *zonotope* in R^d is defined to be the vector sum of a finite number of line segments. It is centrally symmetric and, if no d of the line segments are parallel to a hyperplane, it is a cubical polytope (see section 4.6).

6. *A zonotope P is irreducible if and only if it is the vector sum of line segments no three of which are coplanar (i.e. are parallel to a two-dimensional plane).*

PROOF If no three of the line segments are coplanar, then all the 2-faces of P will be parallelograms. If Q is a summand of P, then by theorem 1 and the above remarks, every 2-face of Q will be a parallelogram. It is simple to show that Q must be a parallelotope and so is centrally symmetric. We deduce that P is irreducible.

On the other hand, if n of the line segments are parallel to a 2-plane, then P will have a 2-face F (parallel to the given 2-plane) which is a $2n$-gon. But F is reducible and so has a summand Q which is not centrally symmetric. Q is also a summand of P, and so, by theorem 5, P is reducible.

Exercises

1. Show that if P and Q are polytopes in R^d, then the three definitions of the vector sum $P + Q$ are equivalent.

2. Prove that every polygon can be expressed as a vector sum of line segments and triangles. Is the expression unique?

3. Show that the relation

$$T(\lambda P + \mu Q) = \lambda T(P) + \mu T(Q)$$

holds for all translations T if and only if $\lambda + \mu = 1$.

4. Prove that k-fold d-pyramids (section 4.2), k-fold d-bipyramids (section 4.3) and d-pyramidoids (exercise 4.8.1) are indecomposable. Show also that a d-prismoid (section 4.4) is indecomposable if no edge of the base P_1 is parallel to an edge of the base P_2.

5. Find a necessary and sufficient condition for a simple polytope to be reducible.

6. If $v_1, \cdots, v_n$ are outward unit vectors normal to the facets $F_1, \cdots, F_n$ of a d-polytope P, then the relation

$$V_d(P) = \frac{1}{d} \sum_{i=1}^{n} H(P, v_i) V_{d-1}(F_i) \tag{10}$$

may be used to define the d-content or volume $V_d(P)$ inductively. (If the origin is an interior point of P, this formula corresponds to the process of dissecting P into pyramids as in section 14.1 and then using the well-known expression for the volume of each of these pyramids.) Use (10) and theorem 1 to prove that $V_d(\lambda P + \mu Q)$ is a homogeneous polynomial of degree d in λ and μ. More generally, prove that for d d-polytopes $P_1, \cdots, P_d$,

$$V_d(\lambda_1 P_1 + \cdots + \lambda_d P_d) = \sum_{\substack{i_1 + \cdots + i_d = d \\ 0 \leq i_j \leq d}} \frac{d!}{i_1! \cdots i_d!} V_{i_1 \cdots i_d} \lambda_1^{i_1} \cdots \lambda_d^{i_d}.$$

The coefficients $V_{i_1 \cdots i_d}$ are called the *mixed volumes* of the given convex polytopes.

15.2 Approximation of Polytopes by Vector Sums

In section 15.1 we examined the conditions under which a polytope was decomposable with respect to vector addition. Here we consider the related problem of deciding whether a given polytope can be approximated arbitrarily closely in the Hausdorff metric (see section 1.2) by vector sums of polytopes of some prescribed type. For example, it is easily deduced from the results of section 15.1 that *every* bounded convex region in the plane can be approximated arbitrarily closely by vector sums of triangles and line segments, so it is a natural question to ask whether every convex body in R^3 can be approximated by vector sums of tetrahedra, triangles and line segments. In the section we shall prove that this is not the case, and, for example, an octahedron cannot be approximated in this way. Although many of the results we prove will hold for general convex bodies, the main theorem will be stated for polytopes only since this is the most interesting and significant case.

We begin by introducing some notation. As in section 1.2 $\rho(K_1, K_2)$ will be used for the Hausdorff distance between two closed bounded convex sets K_1, K_2 in R^d. Thus ρ is a metric on $\mathscr{P}$, the set of all polytopes in R^d. By a *class* of convex polytopes, we mean any subset $\mathscr{K}$ of $\mathscr{P}$, and, since $\mathscr{P}$ is a metric space, it is meaningful to speak about a class $\mathscr{K}$ as being closed, bounded, compact, etc. (Thus $\mathscr{K}$ is bounded if there exists a $\lambda > 0$ such that $K_i \subset \lambda B$ for all $K_i \in \mathscr{K}$). If $\mathscr{K}_1$ and $\mathscr{K}_2$ are closed bounded subsets of $\mathscr{P}$, then we may define a metric $\rho^*(\mathscr{K}_1, \mathscr{K}_2)$ in an analogous way to that in which ρ was defined:

$$\rho^*(\mathscr{K}_1, \mathscr{K}_2) = \max(\sup_{K_1 \in \mathscr{K}_1} \inf_{K_2 \in \mathscr{K}_2} \rho(K_1, K_2), \sup_{K_2 \in \mathscr{K}_2} \inf_{K_1 \in \mathscr{K}_1} \rho(K_2, K_1)).$$

In the following, whenever we refer to such concepts as limits, continuity, etc., it will be understood that these are defined relative to the euclidean distance in R^d, relative to the metric ρ, or relative to the metric ρ^*, as the case may be.

The properties we shall be discussing relate to homothety classes of polytopes, rather than to polytopes themselves, and for this reason it will be convenient to select one definite polytope from each homothety class. We do this as follows. Define $\mathscr{C} \subset \mathscr{P}$ to be the class of all polytopes P in R^d whose diameter diam $P = 1$, and whose Steiner point (see section 14.3) coincides with the origin. Then $\mathscr{C}$ contains precisely one polytope from each homothety class except for the class of 0-polytopes. Since the Steiner point of a polytope is a relative interior point of P, (this follows immediately from the definition in section 14.3), we deduce that if $K \in \mathscr{C}$ then $K \subset B$, and so $\mathscr{C}$ is bounded. Further, if $\{K_i\}$ is any infinite sequence of polytopes of $\mathscr{C}$, then by Blaschke's theorem (see section 2.1) there exists a subsequence converging to a polytope K. By continuity diam $K = 1$, and $s(K) = 0$, so $K \in \mathscr{C}$ and we deduce that $\mathscr{C}$ is a compact subset of $\mathscr{P}$.

Now let $\mathscr{K}$ be any closed subset of $\mathscr{C}$. We notice that $\mathscr{K}$ is compact, and each $K \in \mathscr{K}$ is of dimension at least 1. Write $\sum \mathscr{K}$ for the set of all polytopes which can be written as finite vector sums

$$\lambda_1 K_1 + \cdots + \lambda_r K_r,$$

where $K_i \in \mathscr{K}$, and $\lambda_i \geq 0$. From theorem 14.3.1, the Steiner points of all these polytopes lie at the origin. Hence $(\sum \mathscr{K}) \cap \mathscr{C}$ consists of all those vector sums which have diameter one, and we write

$$\sigma(\mathscr{K}) = \mathrm{cl}((\sum \mathscr{K}) \cap \mathscr{C})$$

for its closure.

If $P \in \mathscr{C}$ is a given polytope, then we shall say that P is *approximable* by the class $\mathscr{K}$ if there exist members of $\sum \mathscr{K}$ arbitrary close to P, or, equivalently,

$$P \in \sigma(\mathscr{K}).$$

We now state the main theorem.

1. *Let $\mathscr{K}$ be a given class of polytopes which is a closed subset of $\mathscr{C}$. If P is an indecomposable polytope, and is approximable by the class $\mathscr{K}$, then $P \in \mathscr{K}$.*

The condition that $\mathscr{K}$ is closed is clearly essential. For example, a regular tetrahedron T in R_3 can be approximated by the class $\mathscr{T}$ of all tetrahedra which are not regular, and so if $T \in \mathscr{C}$ then $T \in \sigma(\mathscr{T})$, but $T \notin \mathscr{T}$.

The proof of the theorem depends essentially upon the fact stated in exercise 3.1.19, that the supporting function $H(P, u)$ of P is piecewise linear. It falls into two parts:

(i) First we shall show that if $P \in \sigma(\mathscr{K})$, we can delete from $\mathscr{K}$ all those polytopes whose supporting functions are not linear in the same regions as $H(P, u)$ to produce a class $\mathscr{K}_{D(P)} \subset \mathscr{K}$ with the property $P \in \sigma(\mathscr{K}_{D(P)})$.

(ii) Secondly we shall show that if P is indecomposable then $\mathscr{K}_{D(P)}$ contains exactly one set, namely P itself.

The proof of (1) is the more difficult; it depends on the following lemmas 2–6, which shall be established first.

2. *σ is a continuous function (with respect to ρ^*) on the classes $\mathscr{K} \subset \mathscr{C}$.*

PROOF It is easy to establish that if $x = (x_1, \cdots, x_d)$ is a unit vector in R^d, then $|x_1| + \cdots + |x_d| \geq 1$. Now each polytope $K_i \in \mathscr{C}$ is of diameter one, and so contains a unit line segment. We deduce that the sums of the lengths of the projections of K_i on the coordinate axes is at least one, and therefore the sum of the lengths of the projections of

$$\lambda_1 K_1 + \cdots + \lambda_p K_p$$

on the coordinate axes is at least $\sum_{i=1}^{p} \lambda_i$. One of these projections is therefore at least $\sum_{i=1}^{p} \lambda_i / d$ in length, and we deduce

$$\operatorname{diam}(\lambda_1 K_1 + \cdots + \lambda_p K_p) \geq \sum_{i=1}^{p} \frac{\lambda_i}{d}.$$

Thus if $\lambda_1 K_1 + \cdots + \lambda_p K_p \in \mathscr{C}$, it follows that $\sum_{i=1}^{p} \lambda_i \leq d$.

Let $\mathcal{K}, \mathcal{K}'$ be any two subsets of $\mathscr{C}$ with $\rho^*(\mathcal{K}, \mathcal{K}') < \varepsilon$. This means that for each $K_i' \in \mathcal{K}'$ there exists a $K_i \in \mathcal{K}$ such that

$$K_i' \subset K_i + \varepsilon B, \qquad K_i \subset K_i' + \varepsilon B,$$

and for each $K_i \in \mathcal{K}$ there exists a $K_i' \in \mathcal{K}'$ satisfying the same conditions. Hence for any vector sum $\sum_{i=1}^{p} \lambda_i K_i' \in \mathscr{C}$,

$$\sum_{i=1}^{p} \lambda_i K_i' \subset \sum_{i=1}^{p} \lambda_i K_i + \varepsilon \left(\sum_{i=1}^{p} \lambda_i \right) B \subset \sum_{i=1}^{p} \lambda_i K_i + d\varepsilon B.$$

Similarly,

$$\sum_{i=1}^{p} \lambda_i K_i \subset \sum_{i=1}^{p} \lambda_i K_i' + d\varepsilon B$$

and so

$$\rho\left(\sum_{i=1}^{p} \lambda_i K_i, \sum_{i=1}^{p} \lambda_i K_i' \right) \leq d\varepsilon.$$

Since this is true for any vector sums belonging to $(\sum \mathcal{K}) \cap \mathscr{C}$ and $(\sum \mathcal{K}') \cap \mathscr{C}$, we deduce

$$\rho^*((\sum \mathcal{K}) \cap \mathscr{C}, (\sum \mathcal{K}') \cap \mathscr{C}) \leq d\varepsilon$$

and so

$$\rho^*(\sigma(\mathcal{K}), \sigma(\mathcal{K}')) \leq d\varepsilon.$$

This shows that σ is continuous and lemma 2 is proved.

Let C be a pointed polyhedral convex cone with at least d facets whose apex lies at the origin 0. Then C is the intersection of at least d closed half-spaces, and it has at least d edges incident with 0. Let $a_1, \cdots, a_r$ be unit vectors from 0 along these edges, so that $\sum_{i=1}^{r} a_i = a$ is an interior point of C. For each such C and each $K \in \mathscr{C}$ we define a function ϕ_C by

$$\phi_C(K) = \left(\sum_{i=1}^{r} H(K, a_i) - H(K, a) \right) \Big/ \left(\sum_{i=1}^{r} H(K, a_i) \right).$$

We notice that $K \in \mathscr{C}$ implies that the origin is a relative interior point of K, so the denominator is strictly positive and ϕ_C is defined. Also, by the convexity of the supporting function $H(K, u)$, the numerator is non-negative and we deduce that $\phi_C(K) \geq 0$. The properties of ϕ_C are summarized in the next lemma; ϕ_C may be regarded as a measure of how far the supporting function of K departs from linearity in the cone C.

3. *If* ϕ_C *is defined as above, and* $K, K_1, K_2 \in \mathcal{K} \subset \mathcal{C}$, *then the following properties hold:*

(i) $\phi_C(K) = \phi_C(\lambda K)$ *for all* $\lambda > 0$,

(ii) $\min\{\phi_C(K_1), \phi_C(K_2)\} \leq \phi_C(K_1 + K_2) \leq \max\{\phi_C(K_1), \phi_C(K_2)\}$, *and similarly for any finite number of sets of* $\mathcal{K}$,

(iii) $\phi_C(K)$ *is a continuous function of* K,

(iv) $\phi_C(K) = 0$ *if and only if* $H(K, u)$ *is a linear function of* u *in the region* C.

The proofs of these statements are straightforward. (i) and (iii) follow immediately from the definition and the fact that the supporting function is continuous ((iii) would not be true if $\mathcal{K}$ contained a sequence of sets which converged to a single point.) (ii) arises from the fact that

$$H(K_1 + K_2, u) = H(K_1, u) + H(K_2, u),$$

so

$$\phi_C(K_1 + K_2) = \frac{\sum H(K_1 + K_2, a_i) - H(K_1 + K_2, a)}{\sum H(K_1 + K_2, a_i)}$$

$$= \frac{\{\sum H(K_1, a_i) - H(K_1, a)\} + \{\sum H(K_2, a_i) - H(K_2, a)\}}{\{\sum H(K_1, a_i)\} + \{\sum H(K_2, a_i)\}}.$$

Since the terms in braces are all positive, this lies between the numbers

$$\frac{\sum H(K_1, a_i) - H(K_1, a)}{\sum H(K_1, a_i)} \quad \text{and} \quad \frac{\sum H(K_2, a_i) - H(K_2, a)}{\sum H(K_2, a_i)}.$$

that is, between $\phi_C(K_1)$ and $\phi_C(K_2)$.

For (iv) assume first that $H(K, u)$ is linear in C. Then

$$H(K, \sum a_i) = \sum H(K, a_i)$$

and so $\phi_C(K) = 0$. If $H(K, u)$ is not linear then

$$H(K, \sum \lambda_i a_i) < \sum \lambda_i H(K, a_i)$$

for some $\lambda_1, \cdots, \lambda_r \geq 0$ not all zero. Suppose without loss of generality that $\lambda_1 = \max_i \lambda_i$. Then

$$\lambda_1(a_1 + \cdots + a_r) = (\lambda_1 a_1 + \cdots + \lambda_r a_r) + \sum_{j=2}^{r} (\lambda_1 - \lambda_j)a_j.$$

The coefficients on the right are nonnegative so, by convexity,

$$\lambda_1 H(K, a) \le H(K, \sum \lambda_i a_i) + \sum_{j=2}^{r} (\lambda_1 - \lambda_j) H(K, a_j)$$

$$< \sum_{j=1}^{r} \lambda_j H(K, a_j) + \sum_{j=2}^{r} (\lambda_1 - \lambda_j) H(K, a_j)$$

$$= \lambda_1 \sum H(K, a_i)$$

and so $\phi_C > 0$. This completes the proof of lemma 3.

For given C and $\varepsilon \ge 0$, we define $\mathcal{K}_C^{(\varepsilon)}$ to consist of the subset of $\mathcal{K}$ for which $\phi_C(K) \le \varepsilon$. Since $\mathcal{K}$ is closed and $\phi_C(K)$ is continuous on $\mathcal{C}$ by 3(iii), $\mathcal{K}_C^{(\varepsilon)}$ is a closed set or is empty. By 3(iv) $\mathcal{K}_C^{(0)}$ consists of those sets whose supporting functions are linear on C.

4. *If $K \in \sigma(\mathcal{K})$, $\phi_C(K) = 0$, and K is indecomposable, then, for all $\varepsilon > 0$, $K \in \sigma(\mathcal{K}_C^{(\varepsilon)})$.*

PROOF If $K \in \sigma(\mathcal{K})$ there exists a sequence of sets $\{K_i\}$ with $K_i \in \sum \mathcal{K}$ such that $K_i \to K$ as $i \to \infty$. Each K_i is a vector sum of sets from $\mathcal{K}$ so write

$$K_i = \sum_{j=1}^{r(i)} \lambda_j K'_{ij} + \sum_{j=1}^{s(i)} \mu_j K''_{ij} \qquad (\lambda_j, \mu_j > 0)$$

where $\phi_C(K'_{ij}) > \varepsilon$, $\phi_C(K''_{ij}) \le \varepsilon$ (all i, j). Thus if

$$K'_i = \sum_{j=1}^{r(i)} \lambda_j K'_{ij}, \qquad K''_i = \sum_{j=1}^{s(i)} \mu_j K''_{ij},$$

using 3(ii) we see that $\phi_C(K'_i) \ge \varepsilon$, $\phi_C(K''_i) \le \varepsilon$, and

$$K_i = K'_i + K''_i.$$

As $i \to \infty$, the sequences $\{K'_i\}$ and $\{K''_i\}$ are clearly bounded, so by Blaschke's theorem we may select a subsequence such that $K'_i \to K^*$ and $K''_i \to K^{**}$ as $i \to \infty$ through this subsequence. Hence, in the limit, $K = K^* + K^{**}$.

There are now two possibilities. Either K^* consists of a single point, or it does not. In the former case $K = K^{**}$ and $K^{**} \in \sigma(\mathcal{K}_C^{(\varepsilon)})$ which proves the result. In the latter case, $\phi_C(K^*)$ is defined, and by 3(i) and 3(ii) $\phi_C(K^*) \ge \varepsilon$. But then K is expressed as the vector sum of two sets of which K^* is not homothetic to K since $\phi_C(K^*) \ge \varepsilon$, but $\phi_C(K) = 0$. Thus K is decomposable. This contradicts the hypothesis, and this second case cannot arise. This completes the proof of lemma 4.

In particular, lemma 4 shows that under the given hypotheses, however small ε may be, $\mathscr{K}_C^{(\varepsilon)}$ cannot be empty.

5. *If* $K \in \sigma(\mathscr{K})$, $\phi_C(K) = 0$ *and* K *is indecomposable, then* $K \in \sigma(\mathscr{K}_C^{(0)})$.

PROOF   Let $\varepsilon \to 0$ through some sequence of values. Then because ϕ_C is continuous, $\mathscr{K}_C^{(\varepsilon)} \to \mathscr{K}_C^{(0)}$, and since, by lemma 2, σ is continuous $\sigma(\mathscr{K}_C^{(\varepsilon)}) \to \sigma(\mathscr{K}_C^{(0)})$. (Notice that $\mathscr{K}_C^{(0)} \neq \varnothing$ being the intersection of a decreasing sequence of compact sets.) However, by lemma 4, $K \in \sigma(\mathscr{K}_C^{(\varepsilon)})$ for all $\varepsilon > 0$, so $K \in \sigma(\mathscr{K}_C^{(0)})$. This proves lemma 5.

Now let P be any d-polytope in R^d. Then the Steiner point $s(P)$, which we take as origin, is an interior point of P and

$$H(P, y) = \sup_i \langle y, v_i \rangle$$

where v_i $(i = 1, \cdots, k)$ are the k vertices of P. This shows that $H(P, y)$ is a piecewise linear function of y and the number of regions of linearity is equal to k. Let the region where $H(P, y) = \langle y, v_i \rangle$ be denoted by C_i $(i = 1, \cdots, k)$. Then it is easily seen that C_i is bounded by a number of hyperplanes which pass through the origin and are perpendicular to the edges of P that meet at v_i. The cones $\{C_1, \cdots, C_k\}$ form a dissection of R^d, which will be denoted by $D(P)$. We notice that each C_i is a cone of the type denoted in the above discussion by C, and that $\phi_{C_i}(P) = 0$ for $i = 1, \cdots, k$.

By application of lemma 5 to each of the cones we deduce:

6. *If* $P \in \sigma(\mathscr{K})$ *and* P *is indecomposable, then*

$$P \in \sigma(\mathscr{K}_{C_1}^{(0)} \cap \cdots \cap \mathscr{K}_{C_k}^{(0)}).$$

Let us denote the right side by $\sigma(K_{D(P)})$, where $\mathscr{K}_{D(P)}$ consists of those sets of $\mathscr{K}$ whose supporting functions are linear on all of the regions $C_1, \cdots, C_k$. Thus if $P \in \sigma(\mathscr{K})$ and P is indecomposable, then $P \in \sigma(\mathscr{K}_{D(P)})$. This completes the first stage (i) in the proof of theorem 1.

It should be mentioned that it is not necessary for P to be indecomposable for lemmas 4, 5, and 6 to be true. This additional assumption leads to some simplification and is justified by the fact that these results will be needed only for indecomposable polytopes.

We remark that if $K \in \mathscr{K}_{D(P)}$ then $D(P)$ is a refinement of $D(K)$ and so each such K has k or fewer vertices. The proof of theorem 1 will be completed by showing that if P is indecomposable and $K \in \mathscr{K}_{D(P)}$ then K must be homothetic to P.

7. *If P is any polytope and $K \in \mathcal{K}_{D(P)}$ then K is homothetic to a summand of P.*

Consider first the case $D(K) = D(P)$. Then associated with each cone $C_i \in D(P)$ there is a vertex of K which will be denoted by w_i. From the geometrical description of $D(P)$ given above we see that two vertices v_i, v_j of P are joined by an edge if and only if the corresponding cones C_i and C_j have a facet in common, and then this edge is perpendicular to this facet. For the same reason, w_i and w_j will be joined by an edge parallel to that joining v_i to v_j. We deduce that a one-to-one correspondence exists between the r-faces of P and the r-faces of K ($r = 0, \cdots, d - 1$) and that corresponding faces are parallel. Hence if λ is sufficiently small $\lambda K \leq P$ and so by theorem 15.1.2, λK is a summand of P.

If $D(P)$ is a proper refinement of $D(K)$ similar considerations apply except that we must regard two vertices w_i and w_j of K as coinciding if $C_i \cup C_j$ is contained in one of the convex cones of $D(K)$. Then the above statements hold except that if an edge of P joins v_i to v_j, then K either has a parallel edge joining w_i to w_j, or else w_i and w_j coincide. As before $\lambda K \leq P$ and λK is a summand of P. This completes the proof of lemma 7.

From lemma 7 we deduce that if P is indecomposable, and $K \in \mathcal{K}_{D(P)}$, then K must be homothetic to P. If K and P belong to $\mathscr{C}$, then $K = P$ and statement (ii) is proved, completing the proof of theorem 1.

If $\mathcal{K} \subset \mathscr{C}$ and $\sigma(\mathcal{K}) = \mathscr{C}$, then $\mathcal{K}$ may be called a *universal approximating class*. We have already remarked that the set of all triangles and line segments form a universal approximating class in E^2. Theorem 1 has the important consequence:

8. *There exist no nontrivial closed universal approximating classes $\mathcal{K} \subset \mathscr{C}$ in $d \geq 3$ dimensions.*

Here 'nontrivial' means $\mathcal{K} \neq \mathscr{C}$. To establish theorem 8 we need only note that the simplicial polytopes (which are indecomposable) are dense in $\mathscr{C}$ if $d \geq 3$. Theorem 1 then implies that $\mathcal{K}$ is dense in $\mathscr{C}$, and so, if $\mathcal{K}$ is closed, $\mathcal{K} = \mathscr{C}$.

15.3 Blaschke Addition*

Let P be a k-polytope in R^d, and let R^k be the k-dimensional subspace parallel to aff P. Denote by $f(P) \geq k + 1$ the number of facets of P, and with each facet F_i ($1 \leq i \leq f(P)$) associate a vector $u_i \in E^k$ as follows:

* In the present section, the use of the letter u will not be restricted to unit vectors.

(i) If $k = 1$, i.e. P is a line segment with end points F_1 and F_2, we put $u_1 = F_1 - F_2$ and $u_2 = F_2 - F_1$. (Here we are using F_1 and F_2 to denote the position vectors of the end-points of P.)

(ii) If $k \geq 2$, then for $i = 1, \cdots, f(P)$, the direction of u_i is that of the outward normal to F_i, and its length $\|u_i\|$ is equal to the $(k - 1)$-content of F_i.

This definition associates a system $\mathscr{U}(P) = \{u_i \mid 1 \leq i \leq f(P)\}$ of vectors with every polytope P. If P_1 is a translation of P_2, then clearly $\mathscr{U}(P_1) = \mathscr{U}(P_2)$, and so $\mathscr{U}(P)$ may be regarded as being associated with the translation class of polytopes containing P, rather than with P itself.

A system $\mathscr{V} = \{v_i \mid 1 \leq i \leq n\}$ of non-zero vectors in R^k is called *equilibrated* if $\sum_{i=1}^{n} v_i = 0$, and no two of the vectors of V are positively proportional. $\mathscr{V}$ is called *fully equilibrated* in R^k provided it is equilibrated and spans R^k.

The following result of Minkowski regarding equilibrated systems of vectors is fundamental. We shall give a brief sketch of the proof in section 15.4. For further details the reader is referred to the original paper of Minkowski [1].

MINKOWSKI'S THEOREM   (i) *If P is a polytope in R^d, then $\mathscr{U}(P)$ is equilibrated. If P is a k-polytope, then $\mathscr{U}(P)$ is fully equilibrated in the subspace R^k parallel to aff P.*

(ii) *If $\mathscr{V}$ is a fully equilibrated system of vectors in R^k ($k \geq 2$), there exists a polytope P, unique within a translation, such that $\mathscr{V} = \mathscr{U}(P)$.*

We are now able to define the Blaschke sum $P \# Q$ of two polytopes P and Q. Since the definition is in terms of the associated systems of vectors $\mathscr{U}(P)$ and $\mathscr{U}(Q)$, the sum $P \# Q$ is only determined within a translation. For definiteness, therefore, it will be convenient to pick one polytope out of each translation class, for example that one whose centroid coincides with the origin. Throughout this section we shall restrict attention to polytopes that satisfy this condition.

Let P and Q be polytopes of dimension p and m respectively in R^d, and let their associated systems of vectors

$$\mathscr{U}(Q) = \{u_i \mid 1 \leq i \leq f(P)\}$$

$$\mathscr{U}(Q) = \{w_i \mid 1 \leq i \leq f(Q)\}$$

be numbered in such a way that u_i and w_i are positively proportional for i satisfying $1 \leq i \leq n$, while no other pair of vectors from $\mathscr{U}(P) \cup \mathscr{U}(Q)$

are positively proportional. Then the system

$$\mathscr{V} = \{v_i \mid 1 \le i \le f(P) + f(Q) - n\}$$

defined by

$$v_i = \begin{cases} u_i + w_i & \text{for } 1 \le i \le n, \\ u_i & \text{for } n < i \le f(P), \\ w_{i-f(P)+n} & \text{for } f(P) < i \le f(P) + f(Q) - n, \end{cases}$$

is equilibrated since each of $\mathscr{U}(P)$ and $\mathscr{U}(Q)$ is equilibrated. Moreover $\mathscr{V}$ spans a linear space of dimension $k \ge \max(p, m)$ and so is fully equilibrated in some R^k. By Minkowski's theorem there exists a unique polytope P' in this R^k with $\mathscr{U}(P') = \mathscr{V}$, and centroid at the origin. We write

$$P' = P \,\#\, Q$$

and say that P' is the *Blaschke sum* of P and Q.

It is convenient to define an associated multiplication by a scalar factor λ. If $\lambda = 0$, define $\lambda \times P$ to be a point; otherwise $\lambda \times P$ is that polytope for which $\mathscr{U}(\lambda \times P) = \{\lambda u_i \mid u_i \in \mathscr{U}(P)\}$. Again, by Minkowski's theorem, the existence and uniqueness of $\lambda \times P$ is assured. Clearly $(-1) \times P = -P$, and if P is k-dimensional for $k \ge 2$, then

$$\lambda \times P = \pm |\lambda|^{1/(k-1)} P$$

where the last is the usual scalar multiplication associated with vector addition (see section 15.1), and the indeterminate sign is that of λ. For $k = 1, \lambda \times P = |\lambda| P$.

The properties of #-addition and its associated $\times$-multiplication which are listed below are easily verified

$$P_1 \,\#\, P_2 = P_2 \,\#\, P_1, \qquad P_1 \,\#\, (P_2 \,\#\, P_3) = (P_1 \,\#\, P_2) \,\#\, P_3.$$

$$\lambda \times (P_1 \,\#\, P_2) = \lambda \times P_1 \,\#\, \lambda \times P_2, \qquad (\lambda\mu) \times P = \lambda \times (\mu \times P),$$

$$(\lambda + \mu) \times P = \lambda \times P \,\#\, \mu \times P \qquad \text{when } \lambda\mu \ge 0.$$

We shall also use the notation

$$\mathop{\#}_{i=1}^{n} P_i = P_1 \,\#\, P_2 \,\#\, \cdots \,\#\, P_n.$$

A polytope P is said to be *decomposable* with respect to Blaschke addition if it can be expressed in the form $P' \,\#\, P''$ where P' and P'' are not homothetic to P. The next theorem shows that, with the exception of a simplex, every polytope is decomposable in this sense, and can be expressed as a Blaschke sum of simplexes. The contrast with the theorems of section 15.1 is striking, for these show that the indecomposable polytopes with respect to vector addition are dense in the set of all polytopes. Theorem 1 is essentially a geometrical formulation of the fact that an equilibrated system of vectors is a superposition of minimal equilibrated systems.

1. *Every polytope P is expressible in the form*

$$P = \#_{i=1}^{m} P_i \tag{1}$$

where each P_i is a simplex. Further, if P is d-dimensional and $f(P) = n \geq d + 1$, there is a representation (1) with $m \leq n - d$.

PROOF   We use induction on n. The assertion is obvious for $d = 1$ and also for $d > 1$ when $n = d + 1$. Thus we may assume that $d > 1$ and $n > d + 1$, and that the proposition is true for every polytope P_1 with $f(P_1) < n$.

Let C be the convex hull of the points with position vectors $\mathscr{U}(P)$ and let u_{i_0} be any vector of $\mathscr{U}(P)$. Then for a suitable $\alpha_0 > 0$, $-\alpha_0 u_{i_0}$ lies on the boundary of C, and therefore, by exercise 2.3.8, is an interior point of some $(d_0 - 1)$-simplex whose vertices are vertices of C. Thus, for certain $u_{i_j} \in \mathscr{U}(P)$ and $\alpha_j > 0$,

$$-\alpha_0 u_{i_0} = \sum_{j=1}^{d_0} \alpha_j u_{i_j}. \tag{2}$$

If $\alpha = \max \alpha_j$, and $\beta_j = \alpha_j/\alpha$ for $0 \leq j \leq d_0$, then $0 < \beta_j \leq \max \beta_j = 1$ and (2) is equivalent to

$$\sum_{j=0}^{d_0} \beta_j u_{i_j} = 0.$$

Hence the system $\mathscr{U}_0 = \{\beta_j u_{i_j} \mid 0 \leq j \leq d_0\}$ is equilibrated and there exists a polytope P_0 with $\mathscr{U}(P_0) = \mathscr{U}_0$. Since $u_{i_0}, \cdots, u_{i_{d_0}}$ are linearly independent, the dimension of P_0 is d_0, and since the number of its $(d_0 - 1)$-faces is $d_0 + 1$, we deduce that P_0 is a simplex.

Let $\mathscr{U}_1$ be the system of vectors obtained from

$$\{(1 - \beta_j)u_{i_j} \mid 0 \leq j \leq d_0\} \cup \{u_i \mid i \notin \{i_j \mid 0 \leq j \leq d_0\}\}$$

by omitting the zero vectors. Then $\mathcal{U}_1$ is also equilibrated and there exists a polytope P_1 with $\mathcal{U}(P_1) = \mathcal{U}_1$. Let n_1 be the number of nonzero vectors in $\mathcal{U}_1$, so that if P_1 has dimension d_1, then n_1 is the number of its $(d_1 - 1)$-faces. If q is the number of $\beta_j < 1$, then

$$1 + n_1 = n - d_0 + q. \tag{3}$$

On the other hand, the q nonzero vectors of $\{(1 - \beta_j)u_i \mid 0 \le j \le d_0\}$ are linearly independent, so the intersection of the spaces R^{d_0} and R^{d_1} spanned by $\mathcal{U}_0$ and $\mathcal{U}_1$ has dimension at least q and,

$$d \le d_0 + d_1 - q. \tag{4}$$

By the inductive assumption, P_1 is expressible as the Blaschke sum of at most $n_1 - d_1$ simplexes, and therefore, by (3) and (4), P is decomposable into at most

$$1 + n_1 - d_1 = (n - d_0 + q) - d_1 \le n - d$$

simplexes. Thus the theorem is proved.

We omit the simple proof of the following result which has no analogue in the case of vector addition for $d \ge 3$:

2. *Every centrally symmetric polytope P is a Blaschke sum of parallelotopes. If P is k-dimensional and has $2m$ facets ($m \ge k$), then P is representable as a sum of $-\lfloor -m/p \rfloor$ p-dimensional parallelotopes where $1 \le p \le k$.*

In a certain sense, Blaschke addition seems more natural if the summands and the sum all have the same dimension; it is only under these conditions that the Blaschke sum of arbitrary convex sets can be defined (see section 15.4). Following this idea one is led to the question whether every polytope in R^k may be represented as a Blaschke sum of k-simplexes in R^k, or other 'standard' polytopes of dimension k. Without loss of generality we may take $k = d$. The example of the cube shows that simplexes alone will not suffice for this purpose. Indeed for every representation of the d-cube P as $P = P_1 \# P_2$ with P_1 and P_2 d-dimensional, we have

$$f(P_1) = f(P_2) = f(P).$$

Thus the bound $2d$ in the next theorem is the best possible.

3. *Every d-polytope P is representable in the form*

$$P = \overset{m}{\underset{i=1}{\#}} P_i$$

where each P_i is a d-polytope with at most $2d$ facets.

PROOF The proof is by induction both on the dimension d and on the number of facets $f(P)$ of P. The assertion is trivially true for the cases $d = 1$, and $d > 1, f(P) \leq 2d$. Thus we may assume that $d > 1, f(P) > 2d$.

The vectors $\mathscr{U}(P)$ span R^d, so the origin 0 is an interior point of the convex hull of the points with position vectors $\{u_i \mid 1 \leq i \leq f(P)\}$. By exercise 2.3.5, there exists a subset I of $\{1, 2, \cdots, f(P)\}$ which contains at most $2d$ integers and is such that 0 is the interior point of the convex hull of $\{u_i \mid i \in I\}$. Therefore, for suitable $\alpha_i > 0$, the system $\mathscr{U}_1 = \{\alpha_i u_i \mid i \in I\}$ is fully equilibrated in R^d. Obviously we may assume that the α_i are chosen so that $\max\{\alpha_i\} = 1$. Let $\mathscr{U}_2 = \{u_j \mid j \in J\}$ be the system obtained from

$$\{(1 - \alpha_i)u_i \mid i \in I\} \cup \{u_i \mid i \notin I\}$$

by the omission of zero vectors. Since $\mathscr{U}(P)$ and $\mathscr{U}_1$ are fully equilibrated, $\mathscr{U}_2$ is equilibrated, and there exist polytopes P_1 and P_2 such that $\mathscr{U}(P_1) = \mathscr{U}_1$ and $\mathscr{U}(P_2) = \mathscr{U}_2$.

If $\mathscr{U}_2$ is fully equilibrated, that is, if P_2 is d-dimensional, then the proof by induction is completed since $f(P_2) < f(P)$.

Suppose, however that $\mathscr{U}_2$ is not fully equilibrated; let R^k where $1 \leq k \leq d - 1$ be the space spanned by $\mathscr{U}_2$. Since P_2 is k-dimensional, by the inductive assumption, it may be represented in the form

$$P_2 = \#_{s=1}^{q} R_s$$

where each R_s is k-dimensional and $f(R_s) \leq 2k$. But

$$P = P_1 \# P_2 = \#_{s=1}^{q} \left[R_s \# \left(\frac{1}{q} \times P_1 \right) \right];$$

thus the theorem will be proved if we can establish it in the case $f(P_2) \leq 2k$, i.e. provided J has at most $2k$ elements.

Let π be that projection of R^d onto R^{d-k} which carries R^k onto 0. The projection of a fully equilibrated system is fully equilibrated. Thus $\{\pi(u_i) \mid i \in I\}$ is fully equilibrated in R^{d-k} (possibly some of the $\pi(u_i)$ are zero vectors and have to be omitted). As before, there exists a set $I_0 \subset I$ which has at most $2(d - k)$ integers, as well as positive numbers $\beta_i < \frac{1}{2}\alpha_i$ such that

$$\sum_{i \in I_0} \beta_i \pi(u_i) = 0,$$

where $\{\pi(u_i) \mid i \in I_0\}$ is fully equilibrated in R^{d-k}. Now the vector

$$u = \sum_{i \in I_0} \beta_i u_i$$

is in R^k. Since $\mathscr{U}_2$ is fully equilibrated in R^k, there is a β, $0 < \beta \leq 1$ for which

$$-\beta u = \sum_{j \in J} \gamma_j u_j$$

where $\gamma_j \geq 0$ and $0 < \max\{\gamma_j\} = \gamma < 1$. Consequently the two systems obtained by deletion of any zero vectors from the systems

$$\mathscr{U}_1^* = \{\beta \beta_i u_i \mid i \in I_0\} \cup \{(1 - \gamma + \gamma_j) u_j \mid j \in J\}$$

and

$$\mathscr{U}_2^* = \{(\alpha_i - \beta \beta_i) u_i \mid i \in I_0\} \cup \{\alpha_i u_i \mid i \in I \sim I_0\} \cup \{(\gamma - \gamma_j) u_j \mid j \in J\}$$

are both fully equilibrated in R^d. But the first system contains at most $2(d - k) + 2k$ vectors, and the second less than $f(P)$ since at least one of the coefficients $\gamma - \gamma_j$ is zero. Hence the inductive assumption may be applied to the d-polytopes P_1^*, P_2^* that correspond to the systems $\mathscr{U}_1^*$ and $\mathscr{U}_2^*$. Clearly $P = P_1^* \# P_2^*$ and so theorem 3 is proved.

In conclusion we remark that by a slight change in definitions, Blaschke addition in the plane can be made to coincide with vector addition of polygons and line segments. Only the definition of $\mathscr{U}(P)$ in the case of a segment P needs to be modified to read: if P is a segment, $\mathscr{U}(P)$ is a pair of opposite vectors, each of length equal to P and perpendicular to the carrier line of P. The definition of $\mathscr{U}(P)$ for proper polygons and the definition of $\#$-addition in terms of equilibrated systems remains unchanged. Then it is easily seen that, within a translation,

$$P_1 + P_2 = P_1 \# P_2$$

for all proper polygons or line segments P_1 and P_2. Theorem 1 then reduces to the property of polygons given in exercise 15.1.2 and theorem 3 gives a representation of a polygon as a sum of triangles and quadrilaterals.

15.4 Remarks

The properties of vector addition of convex sets, linear systems of convex sets (see below) and mixed volumes (see exercise 15.1.6) were studied extensively at the end of the nineteenth, and during the early part of the twentieth century. For an account of the results obtained up to 1933, the reader should consult Bonnesen–Fenchel [1]. One of the more important theorems discovered at this time was the *Brunn–Minkowski theorem*, which may be stated as follows. Let K_0, K_1 be any two closed, bounded convex sets in R^d, and for $0 \le \lambda \le 1$, define

$$K_\lambda = (1 - \lambda)K_0 + \lambda K_1.$$

The system of convex sets $\{K_\lambda\}_{0 \le \lambda \le 1}$ is called a *linear (vector) system*, and λ is the *system parameter*. Writing V_d for the d-dimensional volume or content, the Brunn–Minkowski theorem states that $[V_d(K_\lambda)]^{1/d}$ is a concave function of λ, or, equivalently,

$$[V_d(K_\lambda)]^{1/d} \ge (1 - \lambda)[V_d(K_0)]^{1/d} + \lambda[V_d(K_1)]^{1/d} \tag{1}$$

for $0 \le \lambda \le 1$. It further asserts that equality holds for all $0 \le \lambda \le 1$ (that is, $[V_d(K_\lambda)]^{1/d}$ is a linear function of λ) if and only if either (a) K_0 and K_1 lie in parallel hyperplanes (and so $V_d(K_\lambda) = 0$ for $0 \le \lambda \le 1$), or (b) K_0 and K_1 are homothetic.

At least four different proofs of this theorem are known: Brunn's proof by integration, Hilbert's proof using curvature functions, Blaschke's proof by symmetrization and Hadwiger–Ohmann's [1] proof which depends upon approximating K_λ by the union of a finite number of rectangular blocks. The first three of these proofs are to be found in Bonnesen–Fenchel [1], pp. 89, 102, and 72, and the last two in Eggleston [3], pp. 98 and 97. Several generalizations of the Brunn–Minkowski theorem are known: (a) to volumes of intersections of a convex set by pencils of half-hyperplanes, see Busemann [1], Barthel [1], Barthel and Franz [1], Ewald [1], (b) to arbitrary measurable sets, see Henstock–Macbeath [1], and (c) to mixed volumes, see Alexandrov [1], Fenchel [1, 2], and Busemann [2]. For applications of the Brunn–Minkowski theorem to the properties of mixed volumes, see Bonnesen–Fenchel [1], pp. 88–114 and Shephard [1].

Gale [1] conjectured many of the results on decomposition of polytopes given in section 15.1, but proofs were never published. Theorems 15.1.2, 15.1.3, and 15.1.4 first appear in Shephard [2]. For the results on reducibility of polytopes, see Grünbaum [5] and Shephard [5].

The theorems on approximations given in section 15.2 first appear in Shephard [4], although at an earlier date, Asplund had conjectured that an octahedron was not approximable by vector sums of simplexes (see Grünbaum [9]). It should be mentioned that an exact analogue of theorem 15.2.1 holds for Blaschke addition, but this is almost a vacuous assertion due to the powerful decomposition theorem 15.3.1.

The process we have called Blaschke addition was first described by Blaschke [1] for smooth convex sets, although, even earlier, the corresponding addition of polytopes occurs implicitly in the work of Minkowski [1]. Blaschke's method was generalized by Fenchel–Jessen [1] in the following manner. Associated with each d-dimensional closed bounded convex set K in R^d is an *area function* $S_K(\omega)$, which is a nonnegative totally additive set function on the Borel sets ω of the unit sphere S^{d-1} in R^d, defined as follows: $S_K(\omega)$ is the $(d-1)$-content of the set of boundary points of K, each of which has a supporting hyperplane with outer unit normal in ω. Then the Blaschke sum $K_1 \# K_2$ of two such convex sets is given by

$$S_{K_1 \# K_2}(\omega) = S_{K_1}(\omega) + S_{K_2}(\omega)$$

for all Borel sets ω. The existence of the convex set $K_1 \# K_2$ satisfying this condition is assured by a generalization of Minkowski's theorem. It will be seen readily that in the case of polytopes, the definition in terms of area functions coincides with that in terms of equilibrated systems of vectors given above. The treatment of section 15.3 follows closely that of Firey–Grünbaum [1], in which paper the decomposition theorems 15.3.1, 15.3.2, and 15.3.3 first occur.

We shall now indicate briefly how Minkowski's fundamental theorem quoted in section 15.3 may be proved. Let $e_1, \cdots, e_n$ be unit vectors parallel to the outward normals of the facets $F_1, \cdots, F_n$ of the polytope P, so that

$$\mathscr{U}(P) = \{e_i V_{d-1}(F_i) \mid 1 \le i \le n\}$$

where $V_{d-1}(F_i)$ is the $(d-1)$-content of F_i. Let the origin 0 be taken as an interior point of P, and let p_i be the perpendicular distance from 0 to F_i. Then $p_i = H(P, u_i)$ and the volume $V_d(P)$ is given by the equation

$$dV_d(P) = p_1 V_{d-1}(F_1) + \cdots + p_n V_{d-1}(F_n) \tag{1}$$

(see exercise 15.1.6). Any other interior point x of P is at distance $p_i - \langle x, e_i \rangle$ from F_i, so also

$$dV_d(P) = (p_1 - \langle x, e_1 \rangle)V_{d-1}(F_1) + \cdots + (p_n - \langle x, e_n \rangle)V_{d-1}(F_n). \tag{2}$$

Since this is true for all $x \in \operatorname{int} P$, we deduce from (1) and (2) that $\sum_{i=1}^{n} e_i V_{d-1}(F_i) = 0$, that is to say $\mathscr{U}(P)$ is equilibrated. The second assertion of (i) follows easily.

Part (ii) of the theorem is more difficult to prove. Let

$$\mathscr{V} = \{v_i \,|\, 1 \leq i \leq n\}$$

be a given fully equilibrated system in R^d and e_i be a unit vector in direction v_i, so that $\sum \|v_i\| e_i = 0$. Let a d-polytope $P(\lambda_1, \cdots, \lambda_n)$ be defined, for all $\lambda_i \geq 0$ by the inequalities $\langle x, e_i \rangle \leq \lambda_i \, (i = 1, \cdots, n)$ and let

$$\omega(\lambda_1, \cdots, \lambda_n) = |V_d(P(\lambda_1, \cdots, \lambda_n))|^{1/d}.$$

Then the inequalities

$$\lambda_i \geq 0 \qquad 0 \leq y \leq \omega(\lambda_1, \cdots, \lambda_n)$$

define a closed cone W in the $(n + 1)$-dimensional space with coordinates $(\lambda_1, \cdots, \lambda_n, y)$ and, by the Brunn–Minkowski theorem, it is easy to see that W is convex. Let W' be the closed bounded convex region which is the intersection of W with the hyperplane $y = 1$. Then the linear functional

$$\|v_1\| \lambda_1 + \cdots + \|v_n\| \lambda_n$$

will attain its minimum value at some point $(\lambda_1^*, \cdots, \lambda_n^*, 1)$ of W'. Then $P(\lambda_1^*, \cdots, \lambda_n^*)$ has facets whose $(d - 1)$-contents are proportional to the $\|v_i\|$ and so, for a suitable value of $v > 0$, $\mathscr{U}(v(P(\lambda_1^*, \cdots, \lambda_n^*))) = \mathscr{V}$. Thus $vP(\lambda_1^*, \cdots, \lambda_n^*)$ is the required polytope. For further details, and a proof of the uniqueness, see Minkowski [1].

The Brunn–Minkowski theorem has an analogue for Blaschke addition, namely, for any two convex sets K_0 and K_1 which are closed, bounded and d-dimensional,

$$[V_d((1 - \lambda) \times K_0 \,\#\, \lambda \times K_1)]^{(d-1)/d}$$

is a concave function of λ, and is linear if and only if K_0 and K_1 are homothetic (see Bonnesen–Fenchel [1], p. 124). The proof of this statement depends upon the theory of mixed volumes.

In conclusion we mention that besides vector addition and Blaschke addition, other methods of 'adding together' convex bodies have been described in the literature. For these the reader should consult the publications of Firey [1, 2, 3, 4].

15.5   Additional notes and comments

Minkowski sums.

See Schneider [b, Chap. 3] for a comprehensive treatment of Minkowski sums.

The Minkowski sum of two polytopes induces the multiplication in the polytope algebra introduced by McMullen [g] (see also Morelli [a] and Brion [a]). A fascinating application of that concept was an elementary proof by McMullen [h] for the necessity part of the g-theorem. (See also McMullen [i] [k] and the notes in section 10.6.) The decompositions of a simple polytope P play an important role in McMullen's proof; the crucial object of study here is the subalgebra of the polytope algebra that is generated by the Minkowski summands of P.

Decompositions.

Smilansky [b] extends much of the theory of polytope decomposability to unbounded polyhedral sets that are line-free and of full dimension. In this setting, Theorem 15.1.3 is extended as follows: If a polyhedral set P has a strongly connected set of triangular faces that touches all facets of P, then P is indecomposable.

Another theorem in that paper concerns the set $P(n, p)$ of all 3-polytopes that have n vertices and p facets. It asserts that if $p < n$ then every member of $P(n, p)$ is decomposable; if $n \leq p \leq 2n - 7$ then $P(n, p)$ has both decomposable and indecomposable members; and if $2n - 7 < p$ then every member of $P(n, p)$ is indecomposable.

Smilansky [a] showed that, while an indecomposable 3-polytope must have at least four triangular (hence indecomposable) facets, there are indecomposable 4-polytopes in which each facet is decomposable.

Theorem 15.1.4 ("except for simplices, all simple polytopes are decomposable") also yields an alternative proof for a result of Kaibel–Wolff [a]: Every simple 0/1-polytope is a product of 0/1-simplices.

Kallay [a] proved that if P is a polytope in R^d and T is a projective transformation of R^d that is admissible for P, then decomposability of P is equivalent to that of TP. However, he gave an example in which the set of summands of P is not combinatorially equivalent to the set of summands of TP.

Sallee [b] showed that, if K is a d-dimensional closed convex set whose boundary contains a neighborhood U such that U is "ε-smooth" for some $\varepsilon > 0$ (in particular, U is twice continuously differentiable) and U does not contain any line-segment, then K is decomposable. Of course the latter condition excludes polytopes, but the pair of conditions does apply to a very wide class of convex bodies.

Difference sets of simplices.
The difference set $T^d + (-T^d)$ of a d-simplex is an interesting polytope. It is a hexagon for $d = 2$ (regular if T^2 is an equilateral triangle), and it is known as the *cuboctahedron* for a (regular) tetrahedron T^3. See Rogers–Shephard [a] for a detailed analysis of the facial structure of $T^d + (-T^d)$ for arbitrary d, and Doehlert–Klee [a] for additional information and an application in the design of experiments.

Zonoids.
The polytopes that are limits (with respect to the Hausdorff metric) of zonotopes are called *zonoids* (see also Theorem 5.2.5). They can be characterized, e. g., as the ranges of vector-valued non-atomic measures. Properties of zonoids have been surveyed by Bolker [a] and Schneider–Weil [a], and sharpened results on the limiting process have been obtained by Bourgain–Lindenstrauss–Milman [a].

Algorithmic aspects.
Much attention has been paid to algorithmic questions concerning volume (see, e. g., Gritzmann–Klee [d]). Dyer–Frieze [a] proved that the problem to decide if a polytope (specified by a rational $\mathcal{V}$- or $\mathcal{H}$-description) has volume at most $\alpha \in Q$ (where α is part of the input) is #P-hard. Lawrence [a] showed that, in general, the coding size of the volume of a polytope specified by linear inequalities with rational coefficients is not bounded polynomially in the coding size of the inequality description.

It is easy to see that for fixed dimension the volume of a polytope can be computed in polynomial time.

In a seminal paper, Dyer–Frieze–Kannan [a] described a randomized algorithm that for a polytope P (given by $\mathcal{V}$- or $\mathcal{H}$-description) and for two positive rationals ε and β computes a number V with

$$\text{Prob}\left\{ (1 - \varepsilon)\,\text{vol}(P) \leq V \leq (1 + \varepsilon)\,\text{vol}(P) \right\} \geq 1 - \beta$$

in time polynomially bounded by the coding size of the description of P, $1/\varepsilon$, and $\log(1/\beta)$. Their algorithm extends to the much more general setting of convex bodies specified by membership oracles. For the concept of such oracles see Grötschel–Lovász–Schrijver [a]. Its crucial ingredient is a certain random walk, whose associated Markov chain is proved to be "rapidly mixing". Since then, algorithms of this kind have been applied to algorithmic counting problems quite successfully (see, e. g., Jerrum–Sinclair [a]).

Of course, the hardness results on the theoretical complexity of computing the volume carry over to the problem of computing (mixed) volumes; in this context Dyer–Gritzmann–Hufnagel [a] derive further hardness results. In particular, they prove that computing the volume of a zonotope (given by its segments) is #P-hard. See Girard–Valentin [a] for a problem in mixture management whose solution indeed requires computing the volumes of zonotopes.

On the other hand, Dyer–Gritzmann–Hufnagel [a] exploit Dyer, Frieze, and Kannan's randomized approximate volume algorithm for computing certain mixed volumes.

It was shown by Gritzmann–Hufnagel [a] that the algorithmic problem corresponding to part (ii) of Minkowski's theorem (page 332) is #P-hard, while it can be solved in polynomial time in any fixed dimension.

Blaschke sum and oriented matroids.
As Grünbaum remarks before theorem 15.3.1, the fact that every polytope is a Blaschke sum of some simplices is "essentially a geometric formulation of the fact that an equilibrated system of vectors is a superposition of minimal equilibrated systems". Nowadays this can be identified as a special instance of the fact that every covector of an oriented matroid arises from a conformal composition of cocircuits (see Björner et al. [a, Prop. 3.7.2]).

Brunn–Minkowski theory.
The interplay of Minkowski sum and volume, creating the concept of mixed volumes, is a backbone of the Brunn–Minkowski theory. We refer to Schneider's book [b] for an extensive treatment of this rich theory and to his article [c] for a special treatment directed towards polytopes. See also Gardner [a].

CHAPTER 16

Diameters of Polytopes*

This and the next chapter are concerned with paths on polytopes. References are given, unsolved problems are stated, and some outlines of proofs are included. Most of the results have previously appeared in the literature but a few of them are new.

A *path* on a polyhedron† P is a sequence $(x_0, x_1, \ldots, x_k)$ of successively adjacent vertices of P. The integer k is the *length* of the path, which is said to *join* x_0 and x_k. When x and y are two vertices of P the *distance* $\delta_P(x, y)$ is defined as the length of the shortest path joining x and y on P. This number always exists, for the graph formed by the vertices and bounded edges of P is always connected and is in fact d-connected when P is a d-polytope (theorem 11.3.2, due to Balinski [1]). For each vertex x of P the x-*radius* of P is defined as

$$\rho_x(P) = \max\{\delta_P(x, y) \mid y \quad \text{a vertex of } P\},$$

and the *radius* and *diameter* of P are defined respectively as

$$\rho(P) = \min\{\rho_x(P) \mid x \quad \text{a vertex of } P\}$$

and

$$\delta(P) = \max\{\rho_x(P) \mid x \quad \text{a vertex of } P\}.$$

Thus $\delta(P)$ is the smallest integer k such that any two vertices of P can be joined by a path of length $\leq k$. Note that

$$\rho(P) \leq \delta(P) \leq 2\rho(P),$$

so that one who is interested only in the order of magnitude of $\rho(P)$ or $\delta(P)$ may work with whichever function proves to be more tractable.

We shall be concerned primarily with the minimum and the maximum of $\delta(P)$ as P ranges over certain important classes of polytopes. The general pattern of investigation of the function δ would seem to be appropriate also for many other functions of polytopes, and the definitions to follow can be applied to any such function (for example, to ρ in place of δ). For $n > d > 1$ let $m_v(\delta, d, n)$ and $M_v(\delta, d, n)$ denote respectively the

---

* This chapter was written by Victor Klee.

† In the present chapter, *polyhedron* means polyhedral set, i.e. intersection of finitely many halfspaces.

341

minimum and the maximum of $\delta(P)$ as P ranges over all d-polytopes with n vertices. Similarly, $m_f(\delta, d, n)$ and $M_f(\delta, d, n)$ are respectively the minimum and the maximum of $\delta(P)$ as P ranges over all d-polytopes with n facets. The subscript v or f tells whether n indicates the number of vertices or the number of facets. The numbers $m_v^v(\delta, d, n)$, $M_v^v(\delta, d, n)$, $m_f^v(\delta, d, n)$ and $M_f^v(\delta, d, n)$ are defined in the same way for simple d-polytopes, and the numbers $m_v^f(\delta, d, n)$, $M_v^f(\delta, d, n)$, $m_f^f(\delta, d, n)$ and $M_f^f(\delta, d, n)$ for simplicial d-polytopes. The superscript v or f restricts attention to simple polytopes (those whose *vertex* figures are simplices) or to simplicial polytopes (those whose *facets* are simplices). For example, $M_f^v(\delta, d, n)$ is the maximum diameter of simple d-polytopes with n facets. Here and in other cases involving superscripts, it may happen that there are no polytopes of the sort in question. (For example, no simple 3-polytope has an odd number of vertices.) We shall neglect this possibility in the discussion below, restricting our attention implicitly to the case in which polytopes of the sort in question do exist.

16.1 Extremal Diameters of d-Polytopes

Because of its connection with linear programming (to be described later) the function $M_f^v(\delta, \cdot, \cdot)$ is especially deserving of study. Unfortunately, it is also especially intractable. Most of the known results concerning minima and maxima of diameters of polytopes are summarized below, where M_v stands for $M_v(\delta, d, n)$ and so forth. ($[\tau]$ and $]\tau[$ are respectively the greatest integer $\leq \tau$ and the least integer $\geq \tau$.) A few of the statements below must be modified slightly when $d = 2$ or $n = d + 1$, but these cases are trivial and will be ignored.

1. $M_v = M_v^f = \left[\dfrac{n-2}{d}\right] + 1.$

2. $m_v = m_v^f = 2$ when $d = 3, m_v = m_v^f = 1$ when $d \geq 4.$

3. $M_v^v = \left[\dfrac{n-2}{d}\right] + 1$ when $d \leq 3,$

$\geq (d-1)\left[\dfrac{n-2}{2^d-2}\right] + 1$ when $n \geq 2^d.$

4. m_v^v *is between* $m_v^v(\rho, d, n)$ *and twice this integer;*

$$m_v^v(\rho, d, n) \geq]\log_{d-1}((d-2)n + 2)/d[,$$

with equality when $n \equiv 2 \bmod(d-1).$

5. $M_f = M_f^v = \left[\dfrac{(d-1)n}{d}\right] - d + 2$ when $d \leq 3$ or $n \leq d + 5$,

except that $M_f(\delta, 4, 9) = 5$.

$(d-1)\left[\dfrac{n}{d}\right] - d + 2 \leq M_f \leq 3^{d-5}(6n - 10d + 19)$ *for* $d \geq 4$.

6. $m_f \leq m_f^v$;   *both are* ≤ 2, *are* $= 2$ *when* $d = 3$, *but when* $d \geq 4$ *are* $= 1$ *for infinitely many values of* n.

7. $M_f^v \geq \left[\dfrac{n - 2d}{2^d - 2}\right] + 2$,   *with equality when* $d \leq 3$;

$M_f^v \leq n - d$,   *and* $\leq \dfrac{n + 2d(d-2)}{d(d-1)}$ *if the lower bound conjecture is*
true.

8. m_f^v *is between* $m_f^v(\rho, d, n)$ *and twice this integer;*

$]\log_{d-1}((d-2)n - d^2 + 3d)/2[\leq m_f^v(\rho, d, n)$

$\leq]\log_{d-1}((d-1)(d-2)v - d^3 + 3d^2 - 2)/d[,$

with equality on the right if the lower bound conjecture is true and in par-
ticular when $d = 3$.

A role in several of these results is played by the d-polytopes $P(d, j)$,
where $P(d, j)$ is generated by $j + 1$ $(d - 1)$-simplices in R^d, situated in
parallel hyperplanes so that successive simplices are antihomothetic and
the relative boundary of each simplex is in the boundary of $P(d, j)$. (Figure
16.1.1 is a Schlegel diagram of $P(3, 2)$). The simplicial d-polytope $P(d, j)$
has $d(j + 1)$ vertices, $(2^d - 2)j + 2$ facets, diameter j (when $j \geq 2$) and
facet-diameter $(d - 1)j + 1$. By the *facet-diameter* of a polytope P is
meant the diameter of its polar, which of course is the smallest integer k
such that any two facets F and G of P can be joined by a sequence of
facets $F_0 = F, F_1, \ldots, F_k = G$ in which the intersection of any two
successive facets is a face of dimension $d - 2$.
   For the equality 1 note that

$$\left[\frac{n-2}{d}\right] + 1 \leq M_v^f(\delta, d, n) \leq M_v(\delta, d, n) \leq \left[\frac{n-2}{d}\right] + 1.$$

The second inequality is obvious and the third (first noted by Grünbaum–
Motzkin [1]) follows from the fact that since the graph of a d-polytope

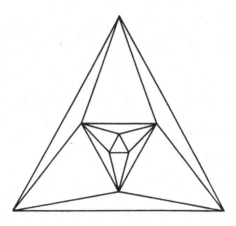

Figure 16.1.1

is d-connected, any two of its vertices can be joined by d independent paths. The first inequality follows from a consideration of the polytopes formed from $P(d, j)$ by adding pyramidal caps over certain facets.

By pulling at vertices any d-polytope can be deformed into a simplicial d-polytope whose facet-diameter is at least that of the original polytope. By polarity it follows that $M_f(\delta, d, n) = M_f^v(\delta, d, n)$. The lower bounds in statements 3, 5, and 7 follow from properties of the polytopes $P(d, j)$ or their polars, or of other polytopes closely related to these. For equality when $d = 3$ use these polytopes, Euler's theorem, and the fact that $f = 2v - 4$ for simplicial 3-polytopes while $v = 2f - 4$ for simple 3-polytopes. (Figure 16.1.2 is the graph of a simple 3-polytope of 16

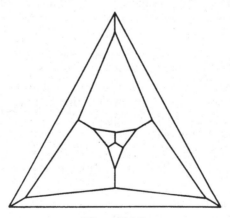

Figure 16.1.2

vertices and diameter 5, obtained from the polar of $P(3, 2)$ by truncating the polar at one of its vertices.) The first upper bound in statement 7 follows from the fact that on a simplicial polytope, a shortest path joining two vertices does not revisit any facet. (See the discussion of W_v paths below.) The second upper bound in statement 7 is based on 1 and the lower bound conjecture. This disposes of 3 and 7, but the discussion of 5 will be continued later. Most of the above results have been taken from Klee [12].

For statements 2 and 6 when $d \geq 4$, consider the neighborly polytopes. For the 3-dimensional case, consider pyramids and bipyramids and recall that the 3-simplices are the only 2-neighborly 3-polytopes.

To establish statements 4 and 8 let us consider an arbitrary simple d-polytope P and vertex x of P. Let $r = \rho_x(P)$ and for $0 \leq i \leq r$ let $v(i)$ denote the number of vertices y of P such that $\delta_P(x, y) = i$. Let $f(i)$ denote the number of facets F of P such that $\min\{\delta_P(x, y) \,|\, y \in F\} = i$. It is easily verified that

$$v(0) = 1, v(1) = d, \cdots, v(i) \leq (d-1)v(i-1), \cdots, \qquad (*)$$

$$v(r) \leq (d-1)v(r-1)$$

$$f(0) = d, f(1) \leq d, \cdots, f(i) \leq v(i), \cdots, f(r) \leq v(r)/d, \qquad (**)$$

where the last inequality follows from the fact that the facets counted by $f(r)$ are pairwise disjoint and each has at least d vertices. Suppose P has v vertices and f facets in all. Then

$$v = \sum_{i=0}^{r} v(i) \leq 1 + d \sum_{j=0}^{r-1} (d-1)^j = 1 + d\frac{(d-1)^r - 1}{d-2}$$

and consequently

$$r \geq \log_{d-1}((d-2)v + 2)/d. \qquad (***)$$

If the lower bound conjecture is true then $v \geq (d-1)(f-d) + 2$ and from (***) it follows that

$$r \geq \log_{d-1}((d-1)(d-2)f - d^3 + 3d^2 - 2)/d.$$

Without assuming the lower bound conjecture we see from (*) and (**) that

$$f = \sum_{i=0}^{r} f(i) \leq d + d \sum_{j=0}^{r-2} (d-1)^j + (d-1)^{r-1} = \frac{d^2 - 3d + 2(d-1)^r}{d-2},$$

whence $r \geq \log_{d-1}((d-2)f - d^2 + 3d)/2$.

To complete the proof of statements 4 and 8 we start with a d-simplex $Q(d, d + 1)$ and a distinguished vertex x of this simplex. The remaining d vertices are designated as being of level 1. For $d + 1 < j \leq 2d + 1$, $Q(d, j)$ is obtained from $Q(d, j - 1)$ by truncating the latter at one of its level 1 vertices and replacing this vertex by a facet which has one vertex at distance 1 from x and $d - 1$ vertices at distance 2 from x. The latter vertices are designated as level 2 vertices of $Q(d, j)$, so that $Q(d, 2d + 1)$ has $d(d - 1)$ level 2 vertices. For $2d + 1 < j \leq 2d + 1 + d(d - 1)$,

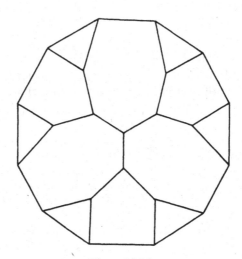

Figure 16.1.3

$Q(d, j)$ is obtained from $Q(d, j - 1)$ by truncating the latter at one of its level 2 vertices and thus introducing $d - 1$ level 3 vertices. (Figure 16.1.3 is a Schlegel diagram of $Q(3, 13)$.) Continuing in this manner, we see that for each $f > d$ the d-polytope $Q(d, f)$ has f facets and its x-radius is equal to r, where r is the largest integer for which

$$f - (d + d \sum_{j=0}^{r-2}(d - 1)^j) = k > 0.$$

In the notation of statements (*) and (**) we have
$$v(0) = 1, \quad f(0) = d, \quad v(1) = f(1) = d, \cdots, v(i) = f(i) = d(d - 1)^{i-1}, \cdots,$$
ending when $k = 1$ with

$$v(r - 1) = d(d - 1)^{r-2}, \quad v(r) = d(d - 1)^{r-1}$$
$$f(r - 1) = d(d - 1)^{r-2}, \quad f(r) = 1,$$

and when $k > 1$ with

$$v(r - 1) = d(d - 1)^{r-2}, \qquad v(r) = k(d - 1)$$

$$f(r - 1) = k.$$

It can be seen that the radius of $Q(d,f)$ is actually equal to its x-radius, and then the remaining assertions of statements 4 and 8 are derived by straightforward computation.

Since the proof of statement 1 was so easy, and since radii seem to be more tractable than diameters in connection with 4 and 8, it is interesting to recall the observation of Jucovič–Moon [1] that $M_v(\rho, 3, n) \geq \lceil n/4 \rceil + 1$ and their conjecture that equality holds for $n \geq 6$. It seems plausible also that $M_f(\rho, 3, n) = \lceil n/2 \rceil$, but both of these conjectures are open. Note that a k-gonal prism is a simple 3-polytope with $2k$ vertices, $k + 2$ facets, and radius $\lceil k/2 \rceil + 1$.

16.2 The Functions Δ and Δ_b

Among the various functions considered above the one most intensively studied (in terms of effort though not of success!) has been $M_f(\delta, \cdot, \cdot)$. The corresponding function for (not necessarily bounded) polyhedra has also been studied. In order to describe the results obtained, and in particular to complete our account of statement 16.1.5, we adopt a simpler notation. A polyhedron P is said to be *of class* (d, n) provided that $n > d > 1$ and P is a d-polyhedron which is pointed (has a vertex) and has exactly n facets. Then $\Delta(d, n)$ and $\Delta_b(d, n)$ are defined as the maxima of $\delta(P)$ as P ranges respectively over all polyhedra of class (d, n) and all polytopes of class (d, n). (Hence $M_f(\delta, d, n) = \Delta_b(d, n)$.) Our account of the functions Δ and Δ_b will be taken mainly from Klee–Walkup [1]. Some of the exposition is borrowed from Klee [20].

The special interest in the functions Δ and Δ_b stems in part from the connection of these functions with linear programming. A linear programming problem is that of maximizing or minimizing a linear function φ, the *objective function*, subject to a finite number of linear constraints. The polyhedron defined by the constraints is called the *feasible region* of the problem. It may be difficult to determine the exact class of this region from the constraints, but the form of the constraints does impose some immediate limitations on the class, since (for example) a region defined by n linear inequality constraints in d real variables is a polyhedron of dimension at most d and has at most n facets. Thus for the study of polyhedra in connection with linear programming it seems reasonable to group the polyhedra. according to class and to study the behavior, with respect to feasible regions of a given class, of the notions and procedures of linear programming.

If a linear function φ is bounded above on a pointed polyhedron P, the maximum of φ on P is attained at some vertex of P; if $\sup \varphi P = \infty$ then some vertex of P is incident to an unbounded edge E such that $\sup \varphi E = \infty$. The subject of linear programming is concerned with practical methods for finding such a vertex of P. Since P is given not in terms of its vertices but rather as the intersection of a finite family of halfspaces (corresponding to the linear constraints), it is generally not practical to examine all vertices of P. The most common procedures for the solution of linear programming problems are based upon various rules for the construction of paths on polyhedra. Having found a vertex of the feasible region, one applies the rule to produce a path leading from that vertex to a maximizing vertex. But if P is any polyhedron and x and y are vertices of P such that $\delta_P(x, y) = \delta(P)$, it is easy to construct a linear function φ whose maximum on P is attained only at y; if x is chosen as the initial vertex in solving the problem of maximizing φ on P, the resulting path will be of length at least $\delta(P)$ regardless of the rule by which it is formed. Thus $\Delta(d, n)$ represents, in a sense, the number of iterations required to solve the 'worst' linear program of n inequalities in d variables using the 'best' edge-following algorithm. (All the algorithms in current use produce paths along which the value of φ is increasing. For a discussion of the lengths of such paths, see the next chapter.)

The numbers $\Delta_b(d, 2d)$ are also of interest in connection with an exchange procedure for positive bases. Suppose X and Y are disjoint minimal positive bases for R^{d-1} (that is, sets of cardinality d positively spanning R^{d-1}), and suppose each d-pointed subset of $X \cup Y$ is linearly independent. Let C denote the set of all convex relations on $X \cup Y$, so that the members of C are those nonnegative functions γ on $X \cup Y$ such that

$$\sum_{p \in X \cup Y} \gamma(p) = 1 \quad \text{and} \quad \sum_{p \in X \cup Y} \gamma(p)p = 0.$$

Then C is a polytope of class $(d, 2d)$, there is a natural correspondence between the vertices of C and the minimal positive bases contained in $X \cup Y$ (Davis [3]), and vertices corresponding to bases A and B are adjacent if and only if the symmetric difference of A and B consists of two points. Hence there is a sequence of minimal positive bases,

$$X = X_0, X_1, \cdots, X_k = Y,$$

of length $k \leq \Delta_b(d, 2d)$, in which each X_i is obtained from its predecessor by the exchange of a single element. Further, $\Delta_b(d, 2d)$ is the smallest integer such that this is true for all X and Y as described. See Klee [21] for more details.

Clearly $\Delta(d, d + 1) = \Delta_b(d, d + 1) = 1$. The other known values of Δ and Δ_b are tabulated below, where asterisks indicate that each column is constant from the main diagonal downward (Klee–Walkup [1]) and thus provide another reason for emphasis on the numbers $\Delta(d, 2d)$ and $\Delta_b(d, 2d)$.

Δ

d \ $n-d$	2	3	4	5	
2	2	3	4	5	$\cdots$ $\Delta(2, n) = n - 2$
3	*	3	4	5	$\cdots$ $\Delta(3, n) = n - 3$
4		*	5	?	
5			*	?	

Δ_b

d \ $n-d$	2	3	4	5	6	
2	2	2	3	3	4	$\cdots$ $\Delta_b(2, n) = [n/2]$
3	*	3	3	4	5	$\cdots$ $\Delta_b(3, n) = [2n/3] - 1$
4		*	4	5	?	
5			*	5	?	
6				*	?	

A conjecture of W. Hirsch, reported by Dantzig [1], pp. 160 and 168, asserts that $\Delta_b(d, n) \leq n - d$. It will be called here the *bounded Hirsch conjecture*, and its special case $\Delta_b(d, 2d) = d$ will be called the *bounded d-step conjecture*. (As the table shows, the corresponding assertions for Δ are correct when $d \leq 3$ but false for $d = 4$.)

The work of Klee–Walkup [1] contains various 'reduction' theorems in addition to the one asserting that $\Delta(d, n) = \Delta(n - d, 2n - 2d)$ when $n \geq 2d$. For example, it is sufficient to consider simple polyhedra and simple polytopes when determining $\Delta(d, n)$ and $\Delta_b(d, n)$, and when $n \geq 2d$ it suffices to consider $\delta_P(x, y)$ for vertices x and y not on any common facet of P. In determining $\Delta(d, n)$ it suffices to consider vertices x and y which are incident to unbounded edges of P. We shall not discuss the proofs of these reduction theorems, but we do want to outline the proofs that $\Delta_b(4, 8) = 4$ while the numbers $\Delta(4, 8)$, $\Delta_b(4, 9)$ and $\Delta_b(5, 10)$ are all equal to 5.

A $(d_0, d_1, \cdots, d_k)$-*path* on a polyhedron P is a sequence $(F_0, F_1, \cdots, F_k)$ of faces of P such that F_i is of dimension d_i and F_i intersects F_{i-1}. Two vertices x and y are said to be *joined* by such a path provided $x \in F_0$ and $y \in F_k$. A face of P is called an x-*face* provided it is incident to x. A d-dimensional *Dantzig figure* is defined as an ordered triple (P, x, y) where P is a d-polyhedron with exactly d x-facets, exactly d y-facets, and $2d$ facets in all (see Dantzig [2]). The *edge-facet diagram* (or *ef-diagram*) of (P, x, y) is a directed bipartite graph having $2d$ nodes in all, each identified with a certain facet of P. The arcs of the *ef*-diagram represent the $(1, d - 1)$-paths joining x to y and y to x. For example, the diagram includes an arc from an x-facet F to a y-facet G if and only if the x-edge not in F terminates on G. Figure 16.2.1 depicts the Schlegel diagrams of two 3-dimensional Dantzig figures and also their *ef*-diagrams. (In the second Schlegel diagram the polyhedron is unbounded and the arrows represent unbounded edges.

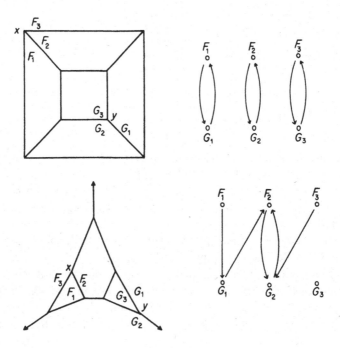

Figure 16.2.1

The *ef*-diagram is defined in terms of the $(1, d - 1)$-paths joining the distinguished vertices x and y of a Dantzig figure, but it also contains information about the $(1, d - 2, 1)$-paths joining x and y. Klee–Walkup show that on a simple Dantzig figure, such a $(1, d - 2, 1)$-path exists except when the *ef*-diagram has exactly two arcs (F_1, G_1) and (F_2, G_2) and two arcs (G_1, F_2) and (G_2, F_1) with $F_1 \neq F_2$ and $G_1 \neq G_2$ (as in Figure 16.2.2). The exceptional case requires in particular that only two

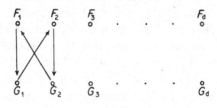

Figure 16.2.2

x-edges and only two y-edges are bounded. Thus a *bounded* simple Dantzig figure (P, x, y) always admits a $(1, d - 2, 1)$-path (f, Q, g) from x to y (figure 16.2.3). The $(d - 2)$-polytope Q has at most $2d - 2$ facets, and each of them is the intersection of two x-facets of P with a y-facet of P or of two y-facets with an x-facet. Let these two different sorts of facets of Q be assigned to classes $\mathscr{X}$ and $\mathscr{Y}$ respectively, and let X and Y consist of all vertices of Q which are entirely surrounded by $\mathscr{X}$-facets and $\mathscr{Y}$-facets respectively. It can be verified that card $\mathscr{X} \leq d - 1$ and X consists of all vertices of Q which are adjacent to x; similarly for $\mathscr{Y}$, Y and y. The existence of (f, Q, g) shows that both X and Y are nonempty. When $d = 4$ then Q is 2-dimensional and there is obviously a path of length 2 from a point of X to a point of Y. Hence it is possible to go from x to y in four 'steps' and the bounded 4-step conjecture is established. Similarly, when $d = 5$ it is proved by Klee–Walkup that Q admits a path of length 3

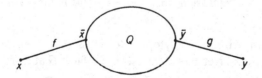

Q is a $(d - 2)$ − polytope with at most $2d - 2$ facets

Figure 16.2.3

joining a point of X to a point of Y, and this proves the bounded 5-step conjecture. On the other hand, they give an example of a system $(Q, \mathcal{X}, \mathcal{Y}, X, Y)$ in which Q is 4-dimensional and the above conditions are all satisfied even though Q does not admit any path of length 4 from a point of X to a point of Y. Thus their method does not apply to the bounded 6-step conjecture, which remains open. It is known only that $6 \leq \Delta_b(6, 12) \leq 9$.

For the bounded 4-step conjecture there is a much simpler proof than the one just indicated. Merely note that each x-edge terminates on a y-facet, where the latter is a 3-polytope having at most 7 facets and hence of diameter ≤ 3. Thus when (P, x, y) is a d-dimensional Dantzig figure with $d \leq 4$, every x-edge is the start of a path of length $\leq d$ from x to y. An example of Klee–Walkup shows this is not the case for $d \geq 5$.

Klee–Walkup construct a 4-dimensional simple Dantzig figure (P, x, y), necessarily unbounded, whose ef-diagram is as in figure 16.2.3 and which therefore admits no $(1, 2, 1)$-path from x to y. Since the figure is simple it follows easily that there is no $(1, 1, 1, 1)$-path from x to y and hence $\Delta(4, 8) \geq 5$. By considering products of P with itself it is seen that

$$\Delta(d, 2d) \geq d + [d/4].$$

The intersection of P with a suitable halfspace is a polytope of class $(4, 9)$ and diameter 5. From reasoning similar to the short proof of the bounded 4-step conjecture given above, using the fact that $\Delta(3, 7) = 4 = \Delta_b(3, 8)$, it then follows that

$$\Delta(4, 8) = 5 = \Delta_b(4, 9).$$

To complete our account of statement 16.1.5 we still must establish the upper bound on $\Delta_b(d, n)$. The argument for this is based on an idea of Barnette [2], who proved a similar theorem. Let us say that a pair of positive numbers (a, b) is d-admissible provided that $\Delta_b(d, n) \leq an - b$ for all $n > d$, and (d, m)-admissible provided that $\Delta_b(d, n) \leq an - b$ for all $n \geq \max(m, d + 1)$. We shall prove the following.

1. *If (a, b) is d-admissible then $(3a, 3a + 3b)$ is $(d + 1)$-admissible and $(3a, 5a + 3b)$ is $(d + 1, 3 + b/a)$-admissible.*

Consider an arbitrary polytope P of class $(d + 1, n)$, and vertices x and y of P. If an x-facet F of P intersects a y-facet G at a vertex q then

$$\delta_P(x, y) \leq \delta_F(x, q) + \delta_G(q, y) \leq 2\Delta_b(d, n - 1) \leq 2an - (2a + 2b),$$

for F and G are d-polytopes and each has at most $n - 1$ facets. Suppose, on the other hand, that no x-facet intersects a y-facet, and let $F_1, F_2, \cdots, F_k$

be a shortest sequence of facets such that $x \in F_1$, $y \in F_k$, and F_i intersects F_{i+1} for $1 \leq i < k$. Then of course $k \geq 3$. From the minimality of k it follows that none of the other facets of P can intersect more than three F_i's and that two F_i's cannot intersect unless they are neighbors in the sequence. Since each facet of F_i is the intersection of F_i with a facet of P, it follows that

$$\sum_{i=1}^{k} (\text{number of facets of } F_i) \leq 3(n - k) + 2(k - 2) + 2 = 3n - k - 2$$

and consequently

$$\delta_P(x, y) \leq (3n - k - 2)a - kb \leq 3an - (5a + 3b).$$

The maximum of $2an - (2a + 2b)$ and $3an - (5a + 3b)$ is at most $3an - (3a + 3b)$, and if $n \geq 3 + b/a$ the maximum is $3an - (5a + 3b)$. This completes the proof of theorem 1.

Now since $(2/3, 1)$ is 3-admissible it follows from theorem 1 that $(2, 19/3)$ is $(4, 9/2)$-admissible and hence $(2, 7)$ is 4-admissible. To establish the upper bound in statement 16.1.5 we show that $(2 \cdot 3^{d-4}, (10d - 19)3^{d-5})$ is d-admissible for all $d \geq 4$. This has been done for $d = 4$ and we proceed by induction. If the statement is known for d then theorem 1 implies that the pair

$$(3 \cdot 2 \cdot 3^{d-4}, 5 \cdot 2 \cdot 3^{d-4} + 3(10d - 19)3^{d-5})$$

$$= (2 \cdot 3^{(d+1)-4}, (10(d + 1) - 19)3^{(d+1)-5})$$

is $(d + 1, m)$-admissible for

$$m = 3 + \frac{(10d - 19)3^{d-5}}{2 \cdot 3^{d-4}} = \frac{10d - 1}{6}.$$

But then the pair in question is in fact $(d + 1)$-admissible, for when $d < n < m$ we have $n < 2d$ and hence

$$\Delta_b(d, n) \leq 1 + \Delta_b(d - 1, n - 1).$$

Note that theorem 1 applies also to admissibility as defined for Δ rather than Δ_b. From reasoning similar to that just employed, using the fact that $(1, 3)$ is 3-admissible for Δ, it follows that

$$\Delta(d, n) \leq 3^{d-4}(3n - 5d + 6)$$

for $n > d > 3$.

16.3 W_v **Paths**

An easy construction shows that always $\Delta(d, n) \geq n - d$, and it follows from results of Klee–Walkup [1] that the inequality is strict whenever $d \geq 4$ and $n - d \geq 4$. It remains to show $\Delta(3, n) = n - 3$, a result which follows from the existence of W_v paths on 3-polyhedra.

A path $(x_0, x_1, \cdots, x_k)$ on a polyhedron P is called a W_v *path* provided that it does not revisit any facet F—that is, provided $x_j \in F$ whenever $i < j < m$ and $x_i, x_m \in F$. The notion is related to the Hirsch conjecture, for on a polyhedron of class (d, n) any W_v path $(x_0, x_1, \cdots, x_k)$ is of length $\leq n - d$. To see this note that for $1 \leq i \leq k$ there is a facet F_i such that $x_{i-1} \in F_i$ but $x_i \notin F_i$. Further, the vertex x_k is incident to at least d facets $F_{k+1}, \cdots, F_{k+d}$. The W_v condition implies the listed facets are all distinct and hence $k + d \leq n$. Wolfe and Klee have conjectured (Klee [19]) that any two vertices of a polytope can be joined by a W_v path. Klee–Walkup [1] have proved that for simple polytopes this conjecture, the bounded Hirsch conjecture, and the bounded d-step conjecture are equivalent, though not necessarily on a dimension-for-dimension basis.

Let us show that any two vertices x and y of a 3-polyhedron P in R^3 can be joined by a W_v path, whence $\Delta(3, n) = n - 3$. If P is bounded let q be a vertex of P other than x or y, let H be a plane intersecting P only at q, and let π be a projective transformation of R^3 carrying H onto the plane at infinity. Then πP is an unbounded polyhedron and every W_v path from πx to πy on πP corresponds to a W_v path from x to y on P. Thus it suffices to consider the case in which P itself is unbounded. In this case there is a ray J in P and there are parallel planes H' and H in R^3 such that J intersects H' at a single point c, H is disjoint from P, and all vertices of P lie in the open strip S between H' and H. Let τ denote the transformation which carries each point s of S onto the intersection of H with the ray from c through s. Although τ is not defined on all of P, the full combinatorial structure of P is represented in $P \cap S$ and the boundary complex of $P \cap S$ is carried by τ onto an isomorphic cell-complex $\mathscr{K}$ in the plane H. Every W_v path from x to y in $\mathscr{K}$ corresponds to a W_v path from x to y on P. Let Π be a path from τx to τy in $\mathscr{K}$ such that the *Euclidean* length of Π is a minimum. If Π should revisit any cell K of $\mathscr{K}$, then (since K is convex) it would be possible to replace a portion of Π with a shorter path in the boundary of K. This implies that Π is a W_v path and completes the proof.

For additional information on W_v paths and related notions see Barnette [2] and Klee [17, 19]. Barnette shows that if two vertices of a

3-polytope do not share a facet they can be joined by three independent W_v paths. Klee [17] shows that if a linear form φ is bounded above on a 3-polyhedron P, then every vertex of P can be joined to a φ-maximizing vertex of P by a W_v path along which φ is steadily increasing.

16.4   Additional notes and comments

The Hirsch conjecture, the d-step conjecture, and the W_v-conjecture (which is also known as the *non-revisiting path conjecture*), presented in sections 16.2 and 16.3, are equivalent (page 354), though this is still not known to be true if the dimension is fixed. They have been of central interest and thus have been actively studied from various points of view, following the appearance of this book and the Klee–Walkup [a] paper. Nevertheless, the trio of conjectures remains as one of the key open problems of polytope theory.

An extensive survey, representing the state-of-the-art in 1986, is Klee–Kleinschmidt [a]. For later developments see Klee–Kleinschmidt [b] and Ziegler [a, Sect. 3.3].

Updated tables.
It seems that only two entries have been added to the tables on page 349 since the writing of the book: Goodey [a] proved $\Delta_b(4,10) = 5$ and $\Delta_b(5,11) = 6$.

General bounds.
The best currently available general upper bounds for $\Delta(d,n)$ and $\Delta_b(d,n)$ are

$$\Delta(d,n) \leq 2 \cdot n^{\log(d)+1}$$

due to Kalai [h] and Kalai–Kleitman [a] (see also Ziegler [a, Thm. 3.10]), and

$$\Delta_b(d,n) \leq \tfrac{1}{3} 2^{d-2}(n-d+\tfrac{5}{2})$$

by Barnette [f], improving on Larman [a]. The latter shows that in any fixed dimension the diameter is indeed bounded linearly in the number of facets.

With respect to lower bounds we know that the Hirsch conjecture is best possible for large dimensions (and bounded polyhedra), since $\Delta_b(d,n) \geq n - d$ holds for $n > d \geq 8$ (Holt–Klee [b], Fritzsche–Holt [a]).

Special cases.
An interesting class of polytopes for which the Hirsch conjecture is known to be true is given by the 0/1-polytopes (convex hulls of subsets of $\{0,1\}^d$), see the notes in section 4.9. Naddef [a] proved that a d-dimensional 0/1-polytope has diameter at most d with equality if and only if the polytope is the d-cube; an extension by Kleinschmidt–Onn [a] is that polytopes with vertices in $\{0,1,\ldots,k\}^d$ have diameter at most kd. Deza–Onn [a] showed that any d-polytope $P \subset R^d$ for which $P \cap Z^d$ is the set of vertices of P has diameter at most const $\cdot d^3$.

Moreover, the Hirsch conjecture and some of its relatives have been proved for various classes of polytopes that arise from combinatorial optimization problems. For relevant references, see the survey articles of Klee–Kleinschmidt [a] and Rispoli [a] and also the later papers Rispoli [b] and Rispoli–Cosares [a].

Klee [b] established the Hirsch conjecture for the duals of cyclic polytopes. Kalai [g] proved an upper bound of $d^2(n-d)^d \log n$ for the duals of d-dimensional neighborly polytopes with n vertices.

Monotone versions.

For a polyhedron $P \subset R^d$ and a linear objective function φ in general position, let $\delta(P, \varphi, v)$ be the length of a shortest increasing (with respect to φ) path joining v to the φ-maximum vertex of P. Let $\Delta^{\rightarrow}(d,n)$ be the maximal value that $\delta(P, \varphi, v)$ attains for all d-polyhedra P with at most n facets. Define $\Delta_b^{\rightarrow}(d,n)$ similarly for bounded polyhedra. Clearly, $\Delta(d,n) \leq \Delta^{\rightarrow}(d,n)$ and $\Delta_b(d,n) \leq \Delta_b^{\rightarrow}(d,n)$.

Todd [a] showed $\Delta_b^{\rightarrow}(d,n) \geq n - d + \min\{[d/4], [(n-d)/4]\}$, thus disproving the *monotone* Hirsch conjecture (which claimed $\Delta_b^{\rightarrow}(d,n) \leq n-d$). On the other hand, Kalai [h] derived the upper bound for the diameter mentioned above even for the directed setting:

$$\Delta^{\rightarrow}(d,n) \leq 2 \cdot n^{\log(d)+1}.$$

The *strong monotone* Hirsch conjecture claims that $\delta(P, \varphi, v_{\min}) \leq n - d$ holds for every (simple) d-polytope $P \subset R^d$ with at most n facets, every linear function φ in general position, and the φ-minimum vertex $v_{\min}$. This conjecture of Ziegler [a, Conj. 3.9] is still open.

Generalizations.

Mani–Walkup [a] found a simplicial 3-sphere whose dual cell-complex violates the W_v-conjecture; from this, they derived an 11-dimensional counterexample to the Hirsch conjecture for spheres. Altshuler [b], however, showed that these spheres are not polytopal.

Barnette [i] describes two-dimensional polyhedral manifolds (of genus eight) that do not allow W_v paths between certain pairs of vertices.

CHAPTER 17

Long Paths and Circuits on Polytopes*

A path $(x_0, x_1, \cdots, x_k)$ is called a *simple path* provided there is no repetition among the x_i's, and a *simple circuit* provided $x_k = x_0$ but there is otherwise no repetition. Any shortest path between two given vertices is simple. The present chapter deals with *longest* simple paths and circuits on polytopes and with some closely related notions.

A *Hamiltonian path* or *Hamiltonian circuit* for a polytope P is a simple path or simple circuit which involves all vertices of P. The study of Hamiltonian circuits on the regular dodecahedron was initiated by Kirkman and later popularized by Hamilton (see Tait [3], Ball [1]) as a 'game'. Since then there have appeared four lines of serious investigation of long paths and circuits on polytopes. One was stimulated by the conjecture of Tait [1,2,3] (1880, 1884) that every simple 3-polytope admits a Hamiltonian circuit and his proof of the four-color 'theorem' from this. The conjecture was supposedly proved by Chuard [1] in 1932 but a counterexample was finally given by Tutte [1] in 1946. Any such example can be used to construct a simple 3-polytope not admitting any Hamiltonian path. The minimal number of vertices for such examples is unknown and is of interest in connection with a classification scheme for organic compounds. A modification of Tait's conjecture is still of interest in connection with the four-color problem. These and related matters are discussed in the first section below.

When a polytope does not admit a Hamiltonian path, there are various ways of measuring how close it comes to admitting one. For example, one may consider the maximum number of vertices involved in a simple path on the polytope or the minimum number of disjoint simple paths covering all vertices. Such measurements are discussed in the second section below. They are of interest in connection with path-following search procedures which have been suggested for finding all vertices of a polytope (Balinski [2]).

For maximizing a linear form φ on a polytope P, the usual linear programming algorithms produce a path $(x_0, x_1, \cdots, x_k)$ such that

* This chapter was written by Victor Klee.

356

$\varphi(x_0) < \varphi(x_1) < \cdots < \varphi(x_k) = \max \varphi P$. As a guide to the amount of computation time required, it is desirable to estimate the lengths of such paths produced by various rules for progressing from one vertex to the next. Estimates of this sort are given in the third section below.

In most of the research concerning paths on polytopes attention has been directed at some large class of polytopes and theorems have been sought which would apply to all members of the class. When individual polytopes have appeared in studies of this sort it has been because of their special relationship to the class rather than because of any intrinsic importance which they might have. However, two lines of investigation have concerned paths on particular polytopes. One of these, mainly of historical interest, is Hamilton's 'game' mentioned earlier. The other, concerned with paths on cubes and motivated by certain coding problems, is discussed in the fourth section below.

17.1 Hamiltonian Paths and Circuits

For any graph G, $\mu(G)$ will denote the minimal number of pairwise disjoint simple paths (including those consisting of a single vertex) in G covering all vertices of G. Now suppose A and B are complementary nonempty sets of vertices of G, G_A is the subgraph formed by A together with all edges having both endpoints in A, and G_B is similarly defined with respect to B. Any simple path in G can be decomposed in a natural way into alternating sequences of simple paths in G_A and simple paths in G_B. Thus G admits no Hamiltonian circuit if $\mu(G_A) > \operatorname{card} B$ and admits no Hamiltonian path if $\mu(G_A) > 1 + \operatorname{card} B$. To construct a simplicial d-polytope P which admits no Hamiltonian circuit, start with a $(d-2)$-simplicial d-polytope Q whose number f_{d-1} of facets exceeds its number f_0 of vertices; then form the Kleetope $P = Q^K$ (see section 11.4) by adding a pyramidal cap over each facet of Q. If G denotes the graph of P, B the vertices of Q, and A the 'new' vertices of P, then $\mu(G_A) = f_{d-1} > f_0 = \operatorname{card} B$ and hence P admits no Hamiltonian circuit; if $f_{d-1} > f_0 + 1$ then P admits no Hamiltonian path. This construction, which is possible for all $d \geq 3$, is similar to those described by T. A. Brown [3] and Moon–Moser [1]. Taking for Q a triangular bipyramid and then a quadrangular bipyramid, the resulting Ps are simplicial 3-polytopes having f-vectors $(11, 27, 18)$ and $(14, 36, 24)$ respectively, the first admitting no Hamiltonian circuit and the second no Hamiltonian path. Coxeter [1], p. 8 (see also Coxeter–Rosenthal [1]), has similar examples whose f-vectors are $(11, 18, 9)$

and (14, 24, 12) (the rhombic dodecahedron), but his polytopes are not simplicial. It would be interesting to determine the minimum number of vertices, edges and facets for d-polytopes or simplicial d-polytopes not admitting Hamiltonian circuits or paths. In this direction, D. Barnette has proved that any 3-polytope with less than 11 vertices admits a Hamiltonian circuit.

Let us now describe the example of Tutte [1] showing that a simple 3-polytope need not have a Hamiltonian circuit. The first step is to notice that if Q is a pentagonal prism and A, B, C, D, E is a cyclic list of the five edges joining one pentagon to the other then Q does not admit a Hamiltonian circuit using both A and C. From this it follows that the graph Q' depicted below (figure 17.1.1) admits no Hamiltonian circuit using both

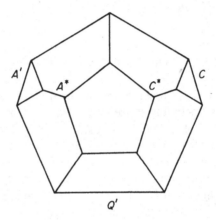

Figure 17.1.1

A^* and C^* and hence none using both A' and C'; this in turn implies that every Hamiltonian circuit of Q'' (in figure 17.1.2) uses the edge G. With w, x, y and z as in figure 17.1.2, let T denote the graph resulting from Q'' by removing w and the edges incident to it. Within combinatorial equivalence, T can be represented as in figure 17.1.3, and since every Hamiltonian circuit of Q'' uses G it is clear that T admits no Hamiltonian path from y to z.

The original example of Tutte [1] consists of three copies of T assembled as in figure 17.1.4. Any Hamiltonian circuit for this graph must (for $i = 1, 2, 3$) intersect T_i in a Hamiltonian path for T_i which has x_i as one of its endpoints. But then each of the edges ux_1, ux_2, and ux_3 is used by the

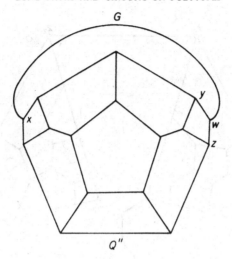

Figure 17.1.2

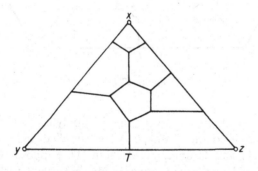

Figure 17.1.3

circuit and this is impossible. Tutte's example has 46 vertices. A modification M with only 38 vertices has been discovered independently by Lederberg [2], Bosak [1] and D. Barnette. It consists of two copies of T joined as in figure 17.1.5. If T_1 and T_2 are shrunk to points in M the resulting graph is combinatorially equivalent to a pentagonal prism Q, with A and C playing the same roles they played earlier. Any Hamiltonian circuit for M would use the edges u_1x_1 and u_2x_2, hence would give rise to a Hamiltonian circuit for Q using A and C. Thus M admits no Hamiltonian circuit.

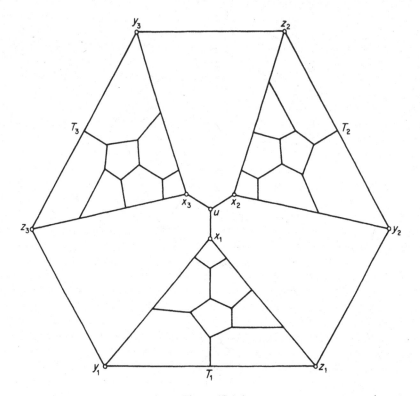

Figure 17.1.4

Balinski [1] asked whether every simple 3-polytope admits a Hamiltonian path, and counterexamples were supplied independently by T. A. Brown [1] and Grünbaum–Motzkin [1]. Suppose G is a graph which admits no Hamiltonian circuit, and the 3-valent vertex v of G is incident to edges E_1, E_2 and E_3. Let G_v denote the graph obtained from G by splitting v into three 1-valent vertices v_i incident only to E_i. Suppose G_v is a subgraph of a graph W and is separated from the rest of W by removal of the v_i's. Any Hamiltonian path for W which originates outside G_v or at one of the v_i's must use one or three of the edges E_i, for otherwise it generates a Hamiltonian circuit for G. Thus if G is the graph of figure 17.1.5 and three copies of G_v are joined after the pattern of figure 17.1.4, the resulting graph of 112 vertices admits no Hamiltonian path. (As it did for the earlier examples, Steinitz's theorem 13.1.1 guarantees that this graph really corresponds to a 3-polytope.) T. A. Brown has found (private communication) the example of figure 17.1.6, which has only 90 vertices.

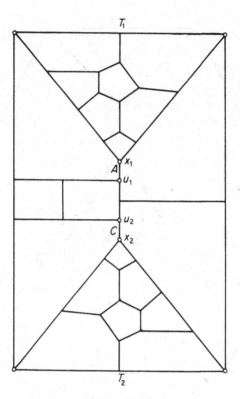

Figure 17.1.5

Each of the 'triangles' T_i has twelve vertices in addition to the three which are shown, and there is an isomorphism of T_i onto the graph of figure 17.1.3 carrying x_i, y_i and z_i onto x, y, and z respectively. Let B denote the entire graph and G [respectively H] the graph obtained from it by collapsing all the vertices x_i, y_i and z_i for $4 \leq i \leq 6$ [respectively $1 \leq i \leq 3$] and all edges joining these vertices into a single vertex u [respectively v]. Then G admits no Hamiltonian circuit, for it is isomorphic with Tutte's example (figure 17.1.4), and it is also seen that H admits no Hamiltonian circuit. Thus any Hamiltonian path for B uses one or three of the edges $[x_1, y_4]$, $[x_2, y_5]$ and $[x_3, y_6]$; indeed, it uses all three, for otherwise it would give rise to a Hamiltonian path for G starting at u and it can be seen no such path exists. We may assume the path starts in $T_1 \cup T_2 \cup T_3$ and first enters $T_4 \cup T_5 \cup T_6$ by means of the edge $[x_1, y_4]$. Suppose it leaves T_4 at z_4. Then it must later return to (and end in) T_4, for T_4 admits no Hamiltonian path from y_4 to z_4. If the path leaves

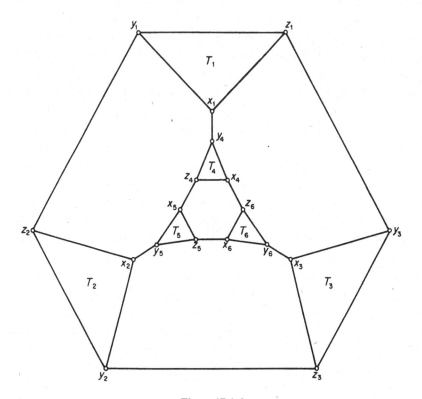

Figure 17.1.6

T_5 at z_5 it fails to use the edge $[x_2, y_5]$. If it leaves T_5 at y_5 it enters T_6 at y_6 and (in order to return to T_4) leaves T_6 at z_6, thus omitting at least one vertex of T_6 (since T_6 admits no Hamiltonian path from y_6 to z_6). A similar analysis disposes of the case in which the path leaves T_4 at x_4.

Lederberg [1,2] came to the problem of Hamiltonian circuits in seeking a systematic way of describing organic molecules which could facilitate the application in organic chemistry of modern methods of information retrieval. With most of the ring compounds there is associated (after some intermediate steps) a 3-valent ($\mathscr{P}^3$)-realizable graph or a combination of such graphs. It is important to have some sort of canonical representation for the graphs as well as an efficient algorithm for determining when two of them are isomorphic, and Hamiltonian circuits are useful in these connections. For a 3-valent graph, whose $2k$ vertices $x_1, x_2, \cdots, x_{2k}$ appear in this order on a Hamiltonian circuit, order the k chords of the circuit according to their first vertices and then list the sequence of spans of the successive chords. If the ith chord joins x_a to x_b (with $a + 1 < b$) the ith term of the sequence is $b - a - 1$. The graph is fully determined by this sequence of spans, which may thus serve as a canonical representation of the graph. It would be of interest to determine which sequences can be obtained in this way. (Other canonical representations may arise from other Hamiltonian circuits. Figure 17.1.7 gives the canonical representations associated with certain Hamiltonian circuits of the cube

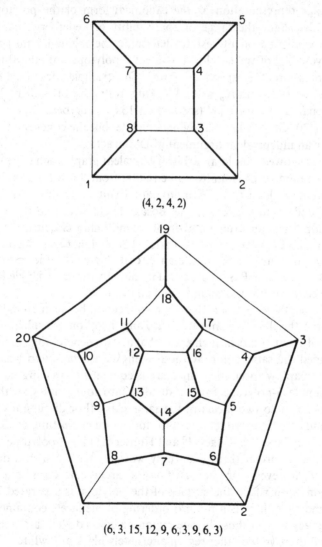

(4, 2, 4, 2)

(6, 3, 15, 12, 9, 6, 3, 9, 6, 3)

Figure 17.1.7

and the dodecahedron.) For another type of canonical representation and a discussion of isomorphism, see Lederberg [2]. (And for a different connection between polytopes and organic chemistry, see Schultz [1].)

The purposes for which Lederberg employs Hamiltonian circuits could perhaps be served by Hamiltonian paths, though the greater multiplicity of Hamiltonian paths for a given polytope leads to an increased number of canonical representations and hence to increased complexity in choosing a particular representation as *the* canonical form of the polytope. In order to determine the range of applicability of Lederberg's canonical forms, it would be of interest to determine the minimum number N_c [respectively N_p] of vertices for a 3-valent 3-polytope not admitting any Hamiltonian circuit [respectively path]. The examples described earlier show that $N_c \leq 38$ and $N_p \leq 90$. Working from a list of 3-valent 3-polytopes prepared by Grace [1] (see section 13.6), Lederberg [2] concludes that $N_c \geq 20$. Grace's list may be incomplete, but the conclusion is supported by an independent argument of D. Barnette.

Tutte [1] showed that in an arbitrary 3-valent graph, each edge is used by an even number of Hamiltonian circuits; from this it follows that the graph admits at least three Hamiltonian circuits if it admits any at all. For a related result see Kotzig [1]. Bosak [1] characterized the 3-valent graphs admitting an even number of Hamiltonian circuits. Extending an earlier theorem of Whitney [1], Tutte [3] proved that every 4-connected planar graph admits a Hamiltonian circuit. Note that the graph of a simplicial 3-polytope P is 4-connected if and only if every triangle formed from the edges of P is the boundary of a facet of P.

As was mentioned earlier, there is a connection between Hamiltonian circuits and the famous conjecture that any 'map' on a 2-sphere can be colored with four colors in such a way that no two neighboring 'countries' are assigned the same color. Without going into detail as to what constitutes a 'map', we note the conjecture is equivalent to saying the facets of any simple 3-polytope can be divided into four classes so that no edge is incident to two facets in the same class. (For discussions of the history and of various reductions of the four-color conjecture, see Ball [1], Franklin [1], Hasse [1], Ringel [1] and Hunter [1].) If a 3-polytope admits a Hamiltonian circuit there is an easy way of effecting such a division of its facets, for every edge not used by the circuit cuts across one of the two regions into which the surface of the polytope is separated by the circuit, and thus there is a natural ordering of the facets contained in a particular region; let those in one region be colored alternately red and green, and those in the other region alternately black and white.

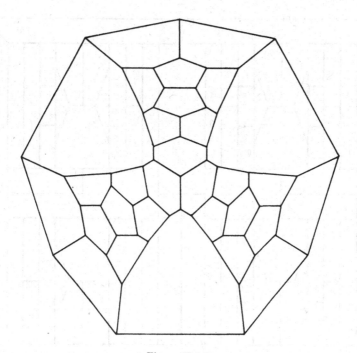

Figure 17.1.8

A simple 3-polytope is called *cyclically k-connected* if its graph cannot be broken into two separate parts, each containing a circuit, by the removal of fewer than k edges. The polytopes of figures 17.1.4 and 17.1.5 are cyclically 3-connected but not cyclically 4-connected. The four-color problem can be reduced to the case of cyclically 4-connected polytopes but Tutte [4] and Hunter [1] have produced such polytopes admitting no Hamiltonian circuits. Hunter found the cyclically 4-connected simple 3-polytope shown in figure 17.1.8 above, which has 58 vertices, no triangular or quadrangular facets, and which admits no Hamiltonian circuit. (It does admit a four-coloring, as shown by Hunter.) G. D. Birkhoff [1] reduced the four-color problem to cyclically 5-connected 3-polytopes, and Hunter [1] conjectured that these all admit Hamiltonian circuits, but a counter-example was recently found by Walther [1] (see figure 17.1.9).

In concluding this section, we should like to say something about Hamiltonian paths and circuits on simple polytopes of dimension $d > 3$. However, very little is known, and in particular it is unknown whether such polytopes always admit Hamiltonian paths or circuits. Two special

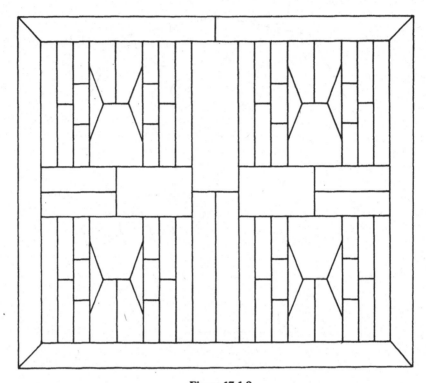

Figure 17.1.9

cases have been studied and will be referred to later—the polars of cyclic polytopes in section 17.2 and the cubes in section 17.4.

17.2   Extremal Path-Lengths of Polytopes

This section is devoted primarily to the *path-length* $\lambda(P)$, the maximum number of vertices in a simple path on P. (The name is due to Brown [3]. In terms of our earlier use of the word *length*, $\lambda(P)$ is one more than the length of the longest simple path on P.) Let us summarize what is known about the function λ, following the pattern of investigation applied to the diameter function δ in the preceding chapter. *The statements 1 to 8 below refer to λ, so that M_v means $M_v(\lambda, d, n)$, etc.*

1. $M_v = M_v^f = n$.

2. $M_v^v \leq n$, *with equality if $d \leq 3$ or $n \equiv 2 \bmod(d - 1)$ or every simple d-polytope admits a Hamiltonian path.*

3. $m_v \leq m_v^f \leq \begin{cases} 2(d + 2)n^{\log_d 2} \\ a(d)n^{1/\lceil d/2 \rceil} \ (\text{for some constant } a(d) \text{ depending on } d \text{ but} \end{cases}$

not on n).

$$2 \log_2 n - 5 \leq m_v \leq m_v^f \leq \begin{cases} \dfrac{2n + 11 + i}{3} & \text{for} \quad n - 2 \equiv i \bmod 3 \\ 8n^{\log_3 2} \end{cases}$$

when $d = 3$.

4. $\sqrt{d} \left(\log_{d-1} \dfrac{(d-2)n + 2}{d} + 1 - d \right) \leq m_v^v \leq n$, *with equality on the*

right if every simple d-polytope admits a Hamiltonian path.

$$d \log_{d-1} \dfrac{(d-2)n + 2 + d}{d} - \lceil d^2/2 \rceil \leq m_v^v \quad \text{when} \quad d \leq 6.$$

$3 \log_2(n + 5) - 9 < m_v^v < 2n^\alpha \ (\text{for some constant } \alpha < 1) \text{ when } d = 3$.

5. $\dbinom{n - \lceil (d+1)/2 \rceil}{n - d} + \dbinom{n - \lceil (d+2)/2 \rceil}{n - d} \leq M_f^v \leq M_f$,

with equality throughout if $d \leq 8$ *or* $n \leq d + 3$ *or* $n \geq (d/2)^2 - 1$ *or the upper bound conjecture is true.*

6. $M_f^f \geq \left\lceil \dfrac{n-2}{d-1} \right\rceil + d \quad \text{if} \quad n \equiv 2 \bmod(d-1)$.

$M_f^f \leq \left\lceil \dfrac{n-2}{d-1} \right\rceil + d \quad \text{if} \quad d \leq 3 \quad \text{or} \quad n \leq d + 3 \quad \text{or} \quad \text{the} \quad \text{lower}$

bound conjecture is true.

7. $m_f \leq m_f^f \leq \begin{cases} 2(d+2)\left(\left\lceil \dfrac{n-2}{d-1} \right\rceil + d \right)^{\log_d 2} \\ a(d)\left(\left\lceil \dfrac{n-2}{d-1} \right\rceil + d \right)^{1/\lceil d/2 \rceil} & \text{if the lower bound conjecture} \end{cases}$

is true.

$$2 \log_2 \frac{n+4}{2} \leq m_f \leq m_f^f \leq \begin{cases} \dfrac{n+15+i}{3} & \textit{for} \quad n/2 \equiv i \bmod 3 \\[2ex] 8\left(\dfrac{n+4}{2}\right)^{\log_3 2} \end{cases}$$

when $d = 3$.

8. $m_f^v \leq (d-1)(n-d) + 2$, *with equality if every simple d-polytope admits a Hamiltonian path and the lower bound conjecture is true.*

$$3 \log_2(2n+1) - 6 < m_f^v < 2(2n-4)^{\alpha}$$

(*for some constant* $\alpha < 1$) *when* $d = 3$.

To justify statements 1 and 5, consider the cyclic polytopes studied by Gale [4] (compare section 4.7). For $d > 3$ each cyclic d-polytope is 2-neighborly and hence admits a Hamiltonian circuit. Gale's characterization of the facets of cyclic polytopes can be used to identify the edges of a cyclic 3-polytope and show it admits a Hamiltonian circuit; statement 1 follows. The same characterization was used by Klee [19] to show all the facets of a cyclic d-polytope can be arranged in a sequence $F_0, F_1, \cdots, F_k$ such that $F_k = F_0$, there is otherwise no repetition, and $F_{i-1} \cap F_i$ is a $(d-2)$-face for $1 \leq i \leq k$. Thus the polars of cyclic polytopes admit Hamiltonian circuits. Statement 5 follows from this fact in conjunction with Gale's [4] count of the facets of a cyclic polytope and the known cases of equality for the upper bound conjecture (Fieldhouse [1], Gale [5], Klee [13]; compare section 10.1).

To establish statement 2 and the first part of 8 we construct, for each $k > d$, a simple d-polytope which has k facets, $(k-d)(d-1) + 2$ vertices, and admits a Hamiltonian circuit. Start with a d-simplex for $k = d + 1$ and then proceed by successive truncation. Observe that if a d-polytope Q admits a Hamiltonian circuit, and if P is a d-polytope formed by truncating Q at a d-valent vertex x, then P admits a Hamiltonian circuit.

If a simplicial d-polytope P has n facets and v vertices, and if the lower bound conjecture applies to P, then $n \geq (d-1)(v-d) + 2$ and consequently $\lambda(P) \leq v \leq [(n-2)/(d-1)] + d$. This establishes part of statement 6. For the other part, start with a Hamiltonian circuit on a d-simplex P_1. Having constructed, for $j \geq 1$, a simplicial d-polytope P_j with $j(d-1) + 2$ facets and $d + j$ vertices in a Hamiltonian circuit, construct P_{j+1} by adding a pyramidal cap over a facet of P_j which is incident to some edge of the circuit. Then P_{j+1} admits a Hamiltonian circuit and the procedure can be continued.

The lower bound in statement 3 is due to Barnette [1] and is based on his theorem asserting that all n vertices of any $(\mathscr{P}^3)$-realizable graph can be covered by a tree T of maximum valence 3. If $(x_1, \cdots, x_k)$ is a longest simple path in T and if $j = [(k + 1)/2]$ then for each i there are at most $3 \cdot 2^{i-1}$ vertices y such that $\delta_T(x_j, y) = i$. Thus $n \le 1 + 3 \sum_{i=1}^{j} 2^{i-1}$, whence $j \ge \log_2((n + 2)/3)$, and since $k \ge 2j - 1$ the desired conclusion follows.

When $d = 3$ the lower bound in statement 8 follows from that in 4. For the lower bounds of statement 4 we combine an earlier result on diameters with a theorem on lengths of simple paths in d-connected graphs of given diameter. In preparation for this let us define an n-*ladder* as a graph formed from two disjoint simple paths $(x_1, x_2, \cdots, x_n)$ and $(y_1, y_2, \cdots, y_n)$ (the *sides* of the ladder) together with n additional edges (the *rungs*) establishing a biunique correspondence between the x_i's and the y_j's. Thus an n-ladder has $2n$ vertices and $3n - 2$ edges, with each vertex of valence 3 except the four *end vertices* x_1, x_n, y_1 and y_n. Let $v(n)$ denote the largest integer k such that every n-ladder admits a simple path using k or more rungs.

9. *For any d-connected graph G of diameter δ,*

$$\lambda(G) > v(d)(\delta + 1 - d);$$

if $v(d) = d$ then

$$\lambda(G) \ge d(\delta + 1) - [d^2/2].$$

PROOF Let x_0 and y_0 be vertices of G such that $\delta_G(x_0, y_0) = \delta$. For $d \le 3$ a simple path of length $d(\delta + 1) - [d^2/2]$ is contained in the subgraph formed by d independent paths from x_0 to y_0. Now suppose $d \ge 4$ with $d = 2u$ or $d = 2u + 1$; we may assume $\delta > u$. Let $(x_0, x_1, \cdots, x_u, \cdots, y_0)$, $(x_0, x_{-1}, \cdots, x_{-u}, \cdots, y_0)$, $(x_0, \cdots, y_u, \cdots, y_1, y_0)$ and $(x_0, \cdots, y_{-u}, \cdots, y_{-1}, y_0)$ be four independent paths from x_0 to y_0, whence $\{x_i \,|\, u - d < i \le u\}$ and $\{y_i \,|\, u - d < i \le u\}$ are disjoint sets of d vertices each. Since G is d-connected a theorem of Whitney (see theorem 11.3.1) guarantees the existence of d disjoint paths $P_1, \cdots, P_d$ joining the vertices of the first set to those of the second set. If P_k joins $x_{i(k)}$ to $y_{j(k)}$ the length of P_k is at least $\delta - |i(k)| - |j(k)|$. Consider the d-ladder whose sides are the two simple paths $(x_{u-d+1}, \cdots, x_{-1}, x_0, x_1, \cdots, x_u)$ and $(y_{u-d+1}, \cdots, y_{-1}, y_0, y_1, \cdots, y_u)$ and whose rungs join $x_{i(k)}$ to $y_{j(k)}$, $1 \le k \le d$. This ladder admits a simple path using $v(d)$ rungs and the corresponding simple path P in G contains $v(d)$ of the paths P_k. When $v(d) = d$ the total number of vertices in P is at least

$$\sum_{k=1}^{d} (\delta + 1 - |i(k)| - |j(k)|) = d(\delta + 1) - [d^2/2]$$

and in any case it is more than

$$\sum_{k=1}^{v(d)} (\delta + 1 - d).$$

The lower bounds in statement 4 follow from 9 in conjunction with statement 16.1.4 and the appropriate parts of the following result:

10. *For $n \le 6$, $v(n) = n$. For all n, $\sqrt{n} \le v(n) \le [2n/3] + 2$.*

PROOF   The assertions are obvious for $n \le 3$. For $n = 4$, verify that every 4-ladder is isomorphic to one of those shown in figure 17.2.1 (ignoring the broken segments) and that every one admits a Hamiltonian path using all four rungs; indeed, such a path may be started at an arbitrary end vertex. Now consider a 5-ladder L_5 with sides $(x_1, \cdots, x_5)$ and $(y_1, \cdots, y_5)$. At least one rung joins x_i to y_j with i and j both odd. If there is such a rung for which x_i or y_j is an end vertex we assume without loss of generality that $i = 1$ and let L_4 denote the 4-ladder formed from L_5 by removing the rung (x_1, y_j) and suppressing x_1 and y_j. Then x_2 is an end vertex of L_4, so L_4 admits a Hamiltonian path $(x_2, \cdots)$ using all four rungs and $(y_j, x_1, x_2, \cdots)$ is the desired path for L_5. In the remaining case (x_3, y_3) is the only candidate for (x_i, y_j) and L_5 is as in one of the last two diagrams of figure 17.2.1 (including the broken segments). These are seen to admit Hamiltonian paths using all five rungs.

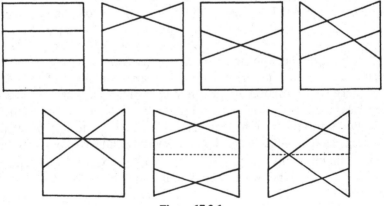

Figure 17.2.1

The proof that $v(6) = 6$, due to J. Folkman, will not be given here. That $\sqrt{n} \leq v(n)$ follows readily from a theorem of Erdös–Szekeres [1] (see also Kruskal [1]) asserting that any sequence of n integers admits a monotone subsequence of at least $\sqrt{n}$ terms. For the inequality, $v(n) \leq [2n/3] + 2$, consider the ladders shown in figure 17.2.2 and others similar to them with the central rung-configuration $(x_{i-1}, y_{i+1}), (x_i, y_i), (x_{i+1}, y_{i-1})$ repeated several times.

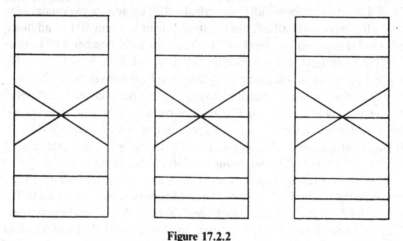

Figure 17.2.2

While establishing the upper bounds in statement 3 we shall prove the following as well.

11. $\left(1 - \dfrac{b(d)}{n^{1-1/[d/2]}}\right)\dfrac{n}{3} \leq M_v^f(\mu, d, n)$ *for some constant* $b(d)$ *depending on d but not on n.*

For $n - 2 \equiv i \pmod 3$,

$$\frac{n - (8 + i)}{3} \leq M_v^f(\mu, 3, n) \leq M_v(\mu, 3, n) \leq \frac{n + 2}{3}.$$

Here $\mu(P)$ *is the minimum number of disjoint simple paths covering the vertices of P.*

Let us first dispose of 11's upper bound on $M_v(\mu, 3, n)$, due to Barnette [1]. Recall Barnette's theorem (see exercise 13.6.3) asserting that every $(\mathscr{P}^3)$-realizable graph can be covered by a tree of maximum valence three. An easy induction on the number of branch points shows that if such a tree has n vertices it can be covered by $(n + 2)/3$ disjoint simple paths, and this implies $M_v(\mu, 3, n) \le (n + 2)/3$. Perhaps a slight improvement can be effected by using the fact that Barnette's theorem applies not only to a 3-polytopal graph G but also to any graph formed by a simple circuit C in G together with all vertices and edges in a component of the complement of C.

Next we shall consider the contribution of Brown [3] to statements 3 and 11. Let S be a simplicial 3-polytope with v vertices and (necessarily) $2v - 4$ facets, and form $S(0)$ [respectively $S(2)$] by adding pyramidal caps over all [respectively all but one] of the facets of S. Form $S(1)$ by adding a pyramidal cap over one facet of $S(0)$. Then the total number n of vertices of $S(i)$ is congruent to $2 + i \pmod 3$, being $3v - 4$, $3v - 5$, and $3v - 3$ for $i = 0, 1, 2$. In any simple path on $S(i)$ there is at least one vertex of S between any two 'new' vertices. Thus if m disjoint simple paths cover w vertices of $S(i)$ we have $w - v \le v + m$. For a single path, $w \le 2v + 1$, and expressing v in terms of n yields the first upper bounds of statement 3 for $d = 3$. If the m paths cover all vertices then $w = n$ and $m \ge n - 2v$; expressing v in terms of n yields the lower bounds of statement 11 for $d = 3$.

For the remaining upper bounds in 3 and lower bounds in 11 we combine the methods of Brown [3] and Moon–Moser [1]. Starting with a simplicial d-polytope P_0, construct a sequence $P_0, P_1, \cdots$ of such polytopes, P_{i+1} being the Kleetope $(P_i)^K$ (see section 11.4) formed by the addition of pyramidal caps over all the facets of P_i. For each i let V_i denote the set of all vertices of P_i, v_i the number of vertices, and f_i the number of facets. Then $v_{i+1} = v_i + f_i$ and $f_{i+1} = df_1$, whence

$$v_j = v_0 + \frac{d^j - 1}{d - 1} f_0.$$

For any integer $n \ge v_0$ let k be determined by the condition that $v_k \le n < v_{k+1}$ and form the polytope $Q(P_0, n)$ by adding pyramidal caps over $n - v_k$ of the facets of P_k. Let V be the set of all vertices of $Q(P_0, n)$. Now consider an arbitrary set of m disjoint simple paths on $Q(P_0, n)$, involving w vertices in all with w_i of them in V_i. In any simple path there is a member of V_k between any two members of $V \sim V_k$, a member of V_{k-1} between any two members of $V_k \sim V_{k-1}$, etc. Hence

$$w - w_k \le w_k + m, \quad w_k - w_{k-1} \le w_{k-1} + m, \cdots, w_1 - w_0 \le w_0 + m,$$

and since $w_0 \le v_0$ it follows that

$$w \le 2^{k+1}(v_0 + m) - m.$$

The choice of k implies

$$v_0 + \frac{d^k - 1}{d - 1} f_0 \leq n,$$

whence

$$k \leq \log_d \left(\frac{(d-1)n - v_0}{f_0} + 1 \right)$$

and

$$w \leq 2(v_0 + m) \left(\frac{(d-1)n - v_0 + f_0}{f_0} \right)^{\log_d 2} - m.$$

When P_0 is a d-simplex this yields

$$w \leq 2(d + 1 + m) \left(\frac{d-1}{d+1} \right)^{\log_d 2} n^{\log_d 2} - m < 2(d + 1 + m) n^{\log_d 2},$$

which for $m = 1$ is essentially the bound given by Moon–Moser [1]. When $d > 4$ an improvement is possible for large n, as we now show.

For $v > d$ let $f(d, v)$ denote the number of facets of a cyclic d-polytope with v vertices, so that

$$f(d, v) = \binom{v - [(d+1)/2]}{v - d} + \binom{v - [(d+2)/2]}{v - d},$$

a polynomial in v of degree $[d/2]$. For any integer $n \geq 2(d + 1)$ let v_0 be defined by the condition that

$$v_0 + f(d, v_0) \leq n < (v_0 + 1) + f(d, v_0 + 1).$$

Let P_0 be a cyclic d-polytope with v_0 vertices and construct $Q(P_0, n)$ as described. Then $f_0 = f(d, v_0)$ and

$$v_1 = v_0 + f(d, v_0) \leq n < (v_0 + 1) + f(d, v_0 + 1) < v_0 + (d + 1)f_0 = v_2,$$

whence $k = 1$ and $w \leq 4v_0 + 3m$. With $n \geq v_0 + f(d, v_0)$, this implies the remaining upper bound in statement 3 and lower bound in 11. The inequalities of statement 7 follow from those of 3.

The upper bound of statement 4 when $d = 3$ is due to Grünbaum–Motzkin [1], whose construction is based on the *merging* of two 3-polytopes P' and P'' by combining a triangular facet of P' with a triangular facet of P'' in the manner of figure 17.2.3. The resulting graph corresponds to a 3-polytope P by Steinitz's theorem and it can be verified that $\lambda(P) \leq \lambda(P') + \lambda(P'')$. Clearly $f_0(P) = f_0(P') + f_0(P'')$. For any simple 3-polytope Q, truncation at a vertex produces a simple 3-polytope Q'

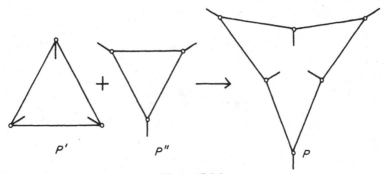

Figure 17.2.3

which has a triangular facet; note that $f_0(Q') = f_0(Q) + 2$ and $\lambda(Q') \leq \lambda(Q) + 2$.

Starting with a simple 3-polytope G which admits no Hamiltonian path, the first aim is to construct a simple 3-polytope P_1 for which $\lambda(P_1) < f_0(P_1) - 5$. This is formed from six copies of G by truncating the first and last copies at one vertex each, the intermediate copies at two vertices each, and then merging appropriately along the new triangular facets. Having constructed P_i, set $v_i = f_0(P_i)$ and $\lambda_i = \lambda(P_i)$. Then $v_1 = 6f_0(G) + 20$ and $\lambda_1 \leq 6\lambda(G) + 20 < v_1 - 5$, so there exists $\beta > 0$ such that $(\lambda_1 + 5)/v_1 < (v_1 + 5)^{-\beta}$.

With P_1 as described, we next produce a sequence of simple 3-polytopes $P_2, P_3, \cdots$ such that $v_{i+1} = v_i(v_i + 5)$ and $\lambda_i < v_i(v_i + 5)^{-\beta}$. To construct P_{i+1} from P_i, first truncate the latter at each of its v_i vertices to obtain P_i^* with $f_0(P_i^*) = 3v_i$; then take v_i copies of P_i, truncate each of them at a single vertex, and merge them with P_i^*. The result is P_{i+1}. Since each simple path on P_i misses at least $v_i - \lambda_i$ vertices of P_i, the same number of truncated copies of P_i (used in constructing P_{i+1}) will be completely missed by any simple path on P_{i-1}. For each of the remaining truncated copies of P_i, the simple path on P_{i+1} determines a simple path on P_i which misses at least $v_i - \lambda_i$ vertices of P_i. Considering also the additional vertices introduced in forming P_i^* from P_i, we see that any simple path on P_{i+1} misses at least

$$(v_i - \lambda_i)(v_i + 5) + \lambda_i(v_i - \lambda_i)$$

vertices, and consequently $\lambda_{i+1} \leq \lambda_i(\lambda_i + 5)$. From $\lambda_i < v_i(v_i + 5)^{-\beta}$ it then follows that $\lambda_{i+1} < v_{i+1}(v_{i+1} + 5)^{-\beta}$. This establishes a strengthened form of the upper bound of statement 4 when n is one of the v_i's, and it remains only to treat the intermediate values of n.

Let $m_i = v_i + 4$, so that $m_{i+1} = m_i(m_i - 3)$ and $\lambda_i + 5 < m_i^{1-\beta}$. Any even integer $n > m_1$ admits a unique expression in the form $n = 2(q_0 - 2) + \sum_{i=1}^{k} q_i m_i$ with $0 \leq q_i < m_i - 3$ for $1 \leq i \leq k, q_k \geq 1$ and $0 \leq 2q_0 < m_1$. A simple 3-polytope $P(n)$ with n vertices is formed by taking q_i copies of P_i (for $1 \leq i \leq k$), truncating and merging at selected vertices, and then truncating the resulting polytope at q_0 of its vertices. It is verified that

$$(m_{j+1} + 3)\lambda_{j+1} \geq m_1 + \sum_{i=1}^{j} m_i(\lambda_i + 2)$$

and

$$\lambda(P(n)) \leq \sum_{i=1}^{k} q_i(\lambda_i + 2) + 2q_0 - 4,$$

whence

$$\frac{\lambda(P(n))}{n} \leq \frac{2q_0 - 4 + \sum_{i=1}^{k} q_i(\lambda_i + 2)}{2q_0 - 4 + \sum_{i=1}^{k} q_i m_i} \leq \frac{q_k(\lambda_k + 2) + (m_{k-1} - 3)\lambda_{k-1}}{q_k m_k}$$

$$\leq \frac{\lambda_{k+2}}{m_k} + \frac{\lambda_{k-1}}{m_{k-1}} \leq 2m_k^{-(1/2)\beta} < 2n^{-(1/4)\beta}.$$

This completes the discussion of statement 4 and takes care of the upper bound in 8 for $d = 3$.

Of the many unsolved problems concerning long paths which are implicit in our discussion, the most important is probably that of determining whether every simple d-polytope admits a Hamiltonian path (for $d \geq 4$) and more generally of finding for each d the minimum $m(d)$ such that each simple d-polytope admits a path which uses all its vertices at least once and does not use any vertex more than $m(d)$ times. Algorithms for finding such paths of minimal multiplicity would be especially interesting. Results of Petersen [1], Tutte [2] and Balinski [1] imply that $m(d) \leq [d/2]$. This was noted by Grünbaum–Motzkin [1], whose paper contains many other interesting results and problems concerning long paths and circuits on polytopes.

17.3 Heights of Polytopes

When φ is a real-valued function defined on the vertices of a polytope P, a path $(x_0, x_1, \cdots, x_k)$ on P is called a φ-path provided that $\varphi(x_0) < \varphi(x_1) < \cdots < \varphi(x_k)$. A φ-path $(x_0, x_1, \cdots, x_k)$ is called a *strict φ-path*, a *steep*

φ-*path*, or a *simplex* φ-*path* provided that for $1 \le i \le k$ the vertex x_i is chosen, among the vertices of P adjacent to x_{i-1}, so as to maximize respectively

> the value of $\varphi(x_i)$,
> the slope $(\varphi(x_i) - \varphi(x_{i-1}))/\|x_i - x_{i-1}\|$,
> the gradient of φ in the space of nonbasic variables (as explained below).

These sorts of paths are all of interest in connection with linear programming, for they all correspond to pivot rules which are used to form a path from an initial vertex to a vertex at which the P-maximum of φ is attained.

The *height* $\eta(P)$ is defined as the largest number realized as the length of a φ-path on P, the maximum being taken over all linear forms φ on the containing space. The *strict height* $\sigma(P)$, the *steep height* $\zeta(P)$, and the *simplex height* $\tau(P)$ are similarly defined. These functions have all been studied in the manner of the function λ in the preceding section (see Klee [15, 16]). However, we shall include here only the results on $M_f^v(\eta, d, n)$, $M_f^v(\sigma, d, n)$, and $M_f^v(\tau, d, n)$. Each of these is the maximum number of iterations which may be required (using the appropriate pivot rule) to solve a nondegenerate linear program whose feasible region is a simple d-polytope with n facets. It seems probable that $M_f^v(\zeta, d, n) = M_f^v(\tau, d, n)$ but this is known only for $d \le 3$ (Klee [15, 16]).

1. $M_f^v(\sigma, 2, n) = n - 2$

$$M_f^v(\sigma, 3, n) = \left[\frac{3n - 1}{2}\right] - 4$$

$$M_f^v(\sigma, d, n) \ge 2(n - d) - 1 \quad \text{for} \quad d \ge 4.$$

2. $M_f^v(\eta, d, n) \ge M_f^v(\tau, d, n) \ge (d - 1)(n - d) + 1$, *with equality throughout for* $d \le 3$.

The first assertion of (1) is obvious. For the second, consider a strict φ-path $(x_0, x_1, \cdots, x_k)$ on a simple d-polytope P with v vertices. Let r denote the number of edges of P which have exactly one endpoint among the x_i's. Then

$$(k + 1)(d - 2) + 2 \le r \le d(v - k - 1),$$

where the first inequality comes from the fact that (by strictness of the path) P has no edges $[x_i, x_j]$ except those used by the path. Thus $k \le [d(v - 2)/2(d - 1)]$. If $d = 3$ and P has n facets, then $v = 2n - 4$ and it follows that $k \leq [(3n - 1)/2] - 4$. This shows

$$M_f^v(\sigma, 3, n) \leq [(3n - 1)/2] - 4.$$

To establish the reverse inequality and to complete the proof of statement 1 we will show that for $n > d \geq 3$ there exists a simple d-polytope P_n with n facets such that $\sigma(P_n) \geq h(n)$, where $h(n) = [(3n - 1)/2] - 4$ if $d = 3$ and $h(n) = 2(n - d) - 1$ if $d \geq 4$.

Let φ be a nontrivial linear form on a d-dimensional vector space E, and let P_{d+1} be a d-simplex in E whose vertices $(x_0, x_1, \cdots, x_d)$ form a φ-path. Then for $n = d + 1$ we have produced a simple d-polytope P_n which has n facets and which satisfies the following two conditions: (i) there is a strict φ-path $(q_0^n, \cdots, q_{h(n)}^n)$ of length $h(n)$ on P_n such that $\varphi(q_{h(n)}^n) = \max \varphi P_n$; (ii) when $h(n + 1) - h(n) = 2$ there is a vertex q_*^n of P_n, adjacent to both $q_{h(n)-1}^n$ and $q_{h(n)}^n$, such that $\varphi(q_{h(n)-1}^n) < \varphi(q_*^n) < \varphi(q_{h(n)}^n)$. To complete the proof of 1 it suffices to show that when such a construction has been made for a given $n > d$ it can be made also for $n + 1$.

Let $z_1, \cdots, z_d$ be the d vertices of P_n adjacent to $q_{h(n)}^n$, with $z_1 = q_{h(n)-1}^n$ and with $z_2 = q_*^n$ when q_*^n exists. Choose the points $y_j \in]z_j, q_{h(n)}^n[$ such that

$$\varphi(q_{h(n)-1}^n) < \varphi(y_1) < \varphi(q_*^n) < \varphi(y_2) < \cdots < \varphi(y_d),$$

where the two inequalities involving $\varphi(q_*^n)$ are replaced by $\varphi(y_1) < \varphi(y_2)$ when $h(n + 1) - h(n) = 1$. Define P_{n+1} as the convex hull of the y_j's together with all vertices of P_n other than $q_{h(n)}^n$. Then P_{n+1} is, as required, a simple d-polytope with $n + 1$ facets. If $h(n + 1) - h(n) = 1$ (that is, if $d = 3$ and n is odd), the sequence $(q_0^n, \cdots, q_{h(n)-1}^n, y_1, y_3)$ is a strict φ-path of length $h(n + 1)$ on P_{n+1} and y_2 can serve as q_*^{n+1}. If $h(n + 1) - h(n) = 2$, $(q_0^n, \cdots, q_{h(n)-1}^n, y_2, y_d)$ is a strict φ-path of length $h(n + 1)$. The point y_3 serves as q_*^{n+1} when $d \geq 4$, and no q_*^{n+1} is required when $d = 3$ for then $h(n + 2) - h(n + 1) = 1$. The proof of statement 1 is complete. (The 3-polytope P_7 is shown in figure 17.3.1, where the arrows indicate a strict φ-path of length 6.)

For $d \leq 3$ a d-polytope with n facets has at most $(d - 1)(n - d) + 2$ vertices and hence does not admit a φ-path of length greater than $(d - 1)(n - d) + 1$. Thus part of statement 2 is obvious. We shall complete the proof of 2 by showing $M_f^v(\tau, d, n) \geq (n - d)(d - 1) + 1$ for $n > d \geq 2$, but of course this requires a clear definition of simplex φ-paths. Let us begin with a coordinate-free description of the simplex algorithm of linear programming (with the most common pivot rule) taken from Klee [15]. It applies only to nondegenerate problems but is adequate for our purpose. It is closely related to the coordinatized descriptions of Dantzig ([1], Chap. 7) and Kuhn–Quandt [1].

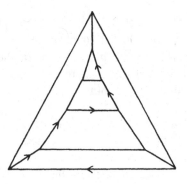

Figure 17.3.1

Suppose P is a simple d-polyhedron* in a d-dimensional vector space E. Any affine form on E which is not constant on P will be called a *variable*. We are concerned with the problem of maximizing a variable φ_0 over P. Let $\mathscr{F}$ denote the set of all facets of P and for each $F \in \mathscr{F}$ let φ_F be a variable such that $F \subset \{x \mid \varphi_F(x) = 0\}$ and $P \subset \{x \mid \varphi_F(x) \geq 0\}$. Let $\Phi = \{\varphi_F \mid F \in \mathscr{F}\}$. Henceforth the discussion is relative to the system (P, Φ, φ_0). A variable $\varphi \in \Phi$ will be called *basic* or *nonbasic* for a vertex x of P according as $\varphi(x) > 0$ or $\varphi(x) = 0$. The set of all nonbasic variables will be denoted by Φ_x, so that $\varphi \in \Phi_x$ if and only if $\varphi = \varphi_F$ for some facet F containing x. Since P is simple, each set Φ_v is of cardinality d and two vertices x and x' are adjacent if and only if there is exactly one variable $\varphi^{x,x'}$ in $\Phi_x \sim \Phi_{x'}$. The *nonbasic (x, x') gradient* of φ_0 is defined as the quotient

$$\frac{\varphi_0(x') - \varphi_0(x)}{|\varphi^{x,x'}(x') - \varphi^{x,x'}(x)|}.$$

Note that the objective function φ_0 admits a unique expression in the form $\sum_{\varphi \in \Phi_x} \gamma(\varphi)\varphi + \gamma_0$ (for real constants $\gamma(\varphi), \gamma_0$) and

$$\frac{\varphi_0(x') - \varphi_0(x)}{|\varphi^{x,x'}(x') - \varphi^{x,x'}(x)|} = \frac{\varphi_0(x') - \varphi_0(x)}{\varphi^{x,x'}(x')} = \gamma(\varphi^{x,x'}).$$

A path $(x_0, x_1, \cdots, x_k)$ on P is called a *simplex φ_0-path* provided that for $1 \leq i \leq k$ it is true that $\gamma(\varphi^{x_{i-1}, x_i}) = \max \gamma(\varphi^{x_{i-1}, w}) > 0$, where the maximum is over the set of all vertices w adjacent to x_{i-1} in P. Thus a φ_0-path is a simplex φ_0-path if and only if it maximizes the nonbasic gradient of

---

* In the present section, *polyhedron* means polyhedral set, i.e. intersection of finitely many halfspaces.

φ_0, or more specifically if and only if (for $1 \leq i \leq k$)

(*) $$\frac{\varphi_0(x_i) - \varphi_0(x_{i-1})}{\varphi^{x_{i-1}, x_i}(x_i)} \geq \frac{\varphi_0(w) - \varphi_0(x_{i-1})}{\varphi^{x_{i-1}, w}(w)}$$

for each vertex w adjacent to x_{i-1} with $\varphi_0(w) > \varphi_0(x_{i-1})$. If inequality always holds in (*) the path will be called an *unambiguous simplex φ_0-path*. These notions are relative to the set Φ of variables corresponding to the facets of P, and they would be affected by nonuniform rescaling of the members of Φ. Thus what we have called $\tau(P)$ is really a function of a particular representation of P; it is not determined by the set P alone.

The following is a more complete statement of the inequality, $M_f^v(\tau, d, n) \geq (d-1)(n-d) + 1$, of theorem 2.

3. *For $n > d > 1$, there exists a simple d-polytope P with exactly n facets, a variable φ_0, and a set Φ of variables corresponding to the facets of P such that the system (P, Φ, φ_0) admits an unambiguous simplex φ_0-path of length $(d-1)(n-d) + 1$ using all the vertices of P.*

The theorem's proof depends on the following lemma, taken from Klee [15]. The lemma's proof is routine and will not be given here.

LEMMA *If $\alpha_1, \gamma_1, \cdots, \gamma_d$ are positive numbers with $1 > \alpha_1$ and $\gamma_2 > \gamma_3 > \cdots > \gamma_d$, then there exist positive numbers $\alpha_2, \alpha_3, \cdots, \alpha_d$ such that*

$$\gamma_1 > \alpha_1 \gamma_1 > \alpha_2 \gamma_2 > \cdots > \alpha_d \gamma_d$$

and

$$C(h, i, j, k): \quad \frac{\alpha_h \alpha_i \gamma_h \gamma_k}{\gamma_j(\alpha_h \gamma_h - \alpha_i \gamma_i + \alpha_i \gamma_k)} < \alpha_j \quad for \quad 1 \leq h < i < j \leq k \leq d.$$

For an arbitrary fixed d, the theorem is proved by induction on n. For the case in which $n = d + 1$, let the numbers γ_i be such that $\gamma_1 > \gamma_2 > \cdots > \gamma_{d+1} > 0$ and let the α_i's be as in the lemma. Let P be the d-simplex in R^{d+1} whose vertices are $\alpha_1 \delta_1, \alpha_2 \delta_2, \cdots, \alpha_{d+1} \delta_{d+1}$, where the points δ_i are the Kronecker deltas, and for each point $x = (x^1, x^2, \cdots, x^{d+1}) \in R^{d+1}$ let $\varphi_0(x) = -\sum_{i=1}^{d+1} \gamma_i x^i$. By the lemma's first condition the sequence $(\alpha_1 \delta_1, \alpha_2 \delta_2, \cdots, \alpha_{d+1} \delta_{d+1})$ is a φ_0-path on P. If F_i is the facet of P determined by all of P's vertices other than $\alpha_i \delta_i$, let the corresponding variable $\varphi_i (= \varphi_{F_i}) \in \Phi$ be the ith coordinate function. Then the nonbasic $(\alpha_h \delta_h, \alpha_j \delta_j)$ gradient of φ_0 is equal to $(\alpha_j \gamma_j - \alpha_h \gamma_h)/\alpha_j$, and the path in question is an unambiguous simplex φ_0-path provided that

$$\frac{\alpha_{h+1} \gamma_{h+1} - \alpha_h \gamma_h}{\alpha_{h+1}} < \frac{\alpha_j \gamma_j - \alpha_h \gamma_h}{\alpha_j} \quad for \quad 1 \leq h < h+1 < j \leq d+1.$$

But this inequality is equivalent to

$$\frac{\alpha_h \alpha_{h+1} \gamma_h}{\alpha_h \gamma_h - \alpha_{h+1} \gamma_{h+1} + \alpha_{h+1} \gamma_j} < \alpha_j,$$

which in turn is equivalent to the inequality $C(h, h+1, j, j)$ of the lemma. (Note that $d+1$ here corresponds to d in the lemma.) This completes the proof for the case in which $n = d + 1$.

Suppose the theorem is known for $n = r \geq d + 1$, and let (P, Φ, φ_0) be the corresponding system. Let $(v_0, v_1, \cdots, v_k)$ be an unambiguous simplex φ_0-path running through all the vertices of P, where $k = (d-1) \times (r-d) + 1$. The P-maximum of φ_0 is attained only at the vertex v_k. We may assume without loss of generality that the d-polytope P lies in R^d, with the vertex v_k at the origin. Since P is simple, there are exactly d variables in Φ which are nonbasic for v_k, and there is a nonsingular linear transformation t of R^d onto R^d such that the variables φt (for $\varphi \in \Phi$) are exactly the d coordinate functions on R^d. The system $(t^{-1}P, \{\varphi t \mid \varphi \in \Phi\}, \varphi_0 t)$ will then have the properties required of the system (P, Φ, φ_0). Thus we may assume without loss of generality that t is the identity transformation, and the d vertices of P which are adjacent to v_k must then lie along the positive coordinate axes. We may assume the P-maximum of φ_0 is 0, whence φ_0 is a linear form on R^d and there are positive numbers γ_i such that $\varphi_0(x) = -\sum_{i=1}^d \gamma_i x^i$ for all $x \in R^d$. The relevant aspects of the situation are clearly unchanged by a sufficiently small perturbation of φ_0, so we may assume the d numbers γ_i are all distinct. By a uniform contraction or dilation together with a suitable permutation of the coordinates we may assume $v_{k-1} = \delta_1$ and $\gamma_2 > \gamma_3 > \cdots > \gamma_d$.

Now let the positive numbers $\alpha_1, \cdots, \alpha_d$ be as in the lemma, and note that for each $\lambda \in [0, 1]$ the conditions of the lemma are also satisfied by the sequence $\lambda\alpha_1, \cdots, \lambda\alpha_d$. Let $v_j(\lambda) = v_j$ for $0 \leq j < k$ and let $v_j(\lambda) = \lambda\alpha_{j-k+1}\delta_{j-k+1}$ for $k \leq j \leq k + d - 1$. Let P_λ denote the convex hull of the set $\{v_j(\lambda) \mid 0 \leq j \leq k + d - 1\}$. Then it can be verified that P_λ is a simple d-polytope having the points $v_j(\lambda)$ as its vertices. Each facet of P_λ is contained in a facet of P, with the sole exception of the facet $F_\lambda = \text{conv}\{\lambda\alpha_i\delta_i \mid 1 \leq i \leq d\}$ of P_λ. Thus P_λ has exactly $r + 1$ facets. Let Φ_λ be obtained from Φ by the addition of a variable φ_{F_λ} corresponding to F_λ. Then the sequence $(v_0(\lambda), v_1(\lambda), \cdots, v_{k+d-1}(\lambda))$ is a path of length $(d-1)(r+1-d) + 1$ which runs through all the vertices of P_λ. Indeed, it is a φ_0-path, for $(v_0(\lambda), v_1(\lambda), \cdots, v_{k-1}(\lambda))$ is identical with the φ_0-path $(v_0, v_1, \cdots, v_{k-1})$ and $(v_{k-1}(\lambda), \cdots, v_{k+d-1}(\lambda))$ is a φ_0-path by the first condition of the lemma (because $\varphi_0(v_{k-1}(\lambda)) = -\gamma_1$ and $\varphi_0(v_j(\lambda)) = -\lambda\alpha_j\gamma_j$

for $k \leq j \leq k + d - 1$). To complete the proof of the theorem, it suffices to show that for λ sufficiently small (in $]0, 1[$) the path $(v_0(\lambda), v_1(\lambda), \cdots, v_{k+d-1}(\lambda))$ is an unambiguous simplex φ_0-path relative to the system $(P_\lambda, \Phi_\lambda, \varphi_0)$.

If $0 \leq h < k \leq j \leq k + d - 1$ and the vertices $v_h(\lambda)$ and $v_j(\lambda)$ are adjacent in P_λ, then v_j and $v_k (= 0)$ are adjacent in P. Further, the variable in Φ_λ which is nonbasic for $v_h(\lambda)$ but basic for $v_j(\lambda)$ is identical with the variable in Φ which is nonbasic for v_h but basic for v_k. Thus for λ very small, the nonbasic $(v_h(\lambda), v_j(\lambda))$ gradient of φ_0 is very close to the nonbasic (v_h, v_k) gradient of φ_0. The desired conclusion then follows from this fact in conjunction with the facts that $(v_0, v_1, \cdots, v_k)$ is an unambiguous simplex φ_0-path in P, that $v_k(\lambda)$ is the only vertex of P_λ which is adjacent to v_{k-1} and gives a greater value to φ_0, and that $(v_k(\lambda), \cdots, v_{k+d-1}(\lambda))$ is (by the first paragraph of the proof of the theorem) an unambiguous simplex φ_0-path relative to the system $(P_\lambda, \Phi_\lambda, \varphi_0)$. The proof of theorem 3 is now complete.

Perhaps we should emphasize, in closing this section, that it has not been our intention to discuss the practical solution of linear programs. Instead, our discussion has related to the important question of determining the maximum number of iterations which may be required to solve a linear program (with feasible region of a given class) by means of various pivot rules. See Klee [15] for interpretation of theorem 3 in terms of linear programs in standard form, Quandt–Kuhn [1, 2] and Kuhn–Quandt [1] for reports of computer experience on the number of iterations, and Dantzig [2] and Klee [18] for other unsolved problems concerning convex polytopes arising in linear programming.

17.4 Circuit Codes

In the most common representation of a d-dimensional cube the 2^d vertices are just the various d-tuples of binary digits. Coupled with the prevalence of two-state components in computers and other electronic devices, this fact has led to some interesting studies and applications of circuits on cubes. Simple circuits on cubes correspond to ways of encoding certain analogue-to-digital conversion systems so as to minimize the errors caused by quantization of continuous data, and the circuit codes of spread s (defined below) have further desirable error-checking properties.

Let $I(d)$ denote the graph of the d-dimensional cube, two vertices being adjacent if and only if they differ in exactly one coordinate. By a d-

dimensional code of spread s we shall mean a subgraph G of $I(d)$ such that for any two vertices x and y of G,

$$\delta_{I(d)}(x, y) \geq \min(\delta_G(x, y), s)$$

or, equivalently,

$$\delta_{I(d)}(x, y) < s \quad \text{implies} \quad \delta_G(x, y) = \delta_{I(d)}(x, y).$$

Every subgraph is of spread 1, and a subgraph G is of spread 2 if and only if it admits no chords (edges of $I(d) \sim G$ joining two vertices of G).

A subgraph of $I(d)$ will be called a *discrete code* provided that it has no edges, and a *circuit code* or *path code* provided that it consists of the vertices and edges of a simple circuit or a simple path. Note that a set of vertices of $I(d)$ forms a discrete code of spread s if and only if the distance between any two vertices is at least s. Such codes have been extensively studied for their error-correcting properties but they will not be discussed here as they do not fit under our general topic of 'long paths and circuits'. The study of path codes (Davies [1], Singleton [1]) has been merely incidental to that of circuit codes and hence our discussion will be confined to circuit codes. Circuit codes of spread 1 have been called *unit distance codes* (Tompkins [1]) or *Gray codes* (Gilbert [1]) and those of spread 2 have been called *snakes* or *SIB* (*snake-in-the-box*) *codes* (Kautz [1]). Circuit codes of spread s have been called SIB_s *codes* (Singleton [1]) and *circuit codes of minimum distance s* (Chien–Freiman–Tang [1]).

The present section is concerned with estimating $\gamma(d, s)$, the maximum length of d-dimensional circuit codes of spread s. The main results can be summarized as follows.

1. $\gamma(d, 1) = 2^d \quad (d \geq 2)$.

2. $\dfrac{7}{4} \dfrac{2^d}{d-1} \leq \gamma(d, 2) \leq 2^{d-1} \quad (d \geq 5)$.

3. $32 \cdot 3^{(d-8)/3} \leq \gamma(d, 3) \leq \dfrac{2^d}{d-2} \quad (d \geq 6)$.

4. $\gamma(d, s) \leq \dfrac{2^d}{\dbinom{d}{r} - 2\dbinom{d-1}{r-1}} \quad \text{for} \quad s = 2r + 1 \quad \text{or} \quad s = 2r + 2 \quad (d \geq s)$.

5. *For odd* $s > 3$,

$$4^{n/s + 1} \precsim \gamma(n, s).$$

6. *For even s > 4,*

$$\delta^n \prec \gamma(n, s) \quad \text{for all positive } \delta < 4^{1/s};$$

$$\delta^n \prec \gamma(n, 4) \quad \text{for all positive } \delta < 3^{1/3}.$$

In results 5 and 6 we write $f(n) \precsim g(n)$ to mean $\liminf_{n \to \infty} g(n)/f(n) > 0$ and $f(n) \prec g(n)$ to mean $\lim_{n \to \infty} g(n)/f(n) = \infty$.

A path $(x_0, x_1, \cdots, x_k)$ in $I(d)$ is uniquely determined by its starting point x_0 together with its *transition sequence* $(t(1), \cdots, t(k))$, where (for $1 \leq i \leq k$) x_{i-1} and x_i differ in the $t(i)^{\text{th}}$ coordinate. If the transition sequence $(t(1), t(2), \cdots, t(2^d - 1), t(2^d))$ corresponds to a Hamiltonian circuit of $I(d)$ then

$$(t(1), t(2), \cdots, t(2^d - 1), d + 1, t(1), t(2), \cdots, t(2^d - 1), d + 1)$$

corresponds to a Hamiltonian circuit of $I(d + 1)$. Thus $I(d)$ admits a Hamiltonian circuit for $d \geq 2$ and this establishes the result 1. For $2 \leq d \leq 4$ the following are transition sequences of Hamiltonian circuits of $I(d)$:

$$(d = 2)\ 1212;\ (d = 3)\ 12131213;\ (d = 4)\ 1213121412131214.$$

A difficult unsolved problem, studied by Gilbert [1] and Abbott [1], is that of determining the number $h(d)$ of Hamiltonian circuits of $I(d)$ and also the number $e(d)$ of equivalence classes of such circuits, where equivalence is with respect to the group of all symmetries of $I(d)$. Considering two circuits to be different only if the corresponding undirected graphs are different, Abbott shows $h(d) > (\sqrt[7]{6})^{2^d}$ and remarks that $e(d) \geq h(d)/(d!2^d)$.

The use of circuits in $I(d)$ to encode certain analogue-to-digital conversion systems has been discussed by Caldwell [1], pp. 391–396, Fifer [1], pp. 308–311, Gray–Levonian–Rubinoff [1], Keister–Ritchie–Washburn [1], chap. 11, Tompkins [1] and many others. We note also that solutions of the 'towers of Hanoi' puzzle correspond to Hamiltonian circuits on cubes (Crowe [1]).

The upper bound in result 2 is due to Singleton [1] and Abbott [1], and has been established also by Danzer–Klee [1]. Suppose C is a d-dimensional circuit code of spread 2 and $I(3)$ is (the graph corresponding to) a 3-dimensional subcube of $I(d)$ such that $I(3)$ includes at least five but not all vertices of C. A simple analysis shows that the intersection $C \cap I(3)$ must be as in figure 17.4.1, and from this it follows that no intersection $C \cap I(4)$ can have more than eight vertices. Hence $\gamma(d, 2) \leq 2^{d-1}$ for $d \geq 4$.

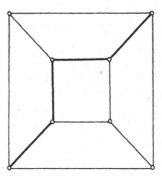

Figure 17.4.1

Upper bounds on the cardinalities of discrete codes (Hamming [1], Gramenapoulos [1], S. M. Johnson [1, 2], Peterson [1], Plotkin [1], Wyner [1]) may be applied to circuit codes by using the fact that a circuit code of spread s and length k must contain s pairwise disjoint discrete codes of spread s and cardinality $[k/s]$. However, the proof of the upper bounds in results 3 and 4 (due to Chien–Freiman–Tang [1], also to Singleton [1] for $d = 3$) does not depend explicitly on discrete codes. Suppose $s = 2r + 1$ or $s = 2r + 2$ and consider a circuit $(x_0, x_1, \cdots, x_k, x_0)$ in $I(d)$ determining a circuit code C of spread s and length $k = \gamma(d, s)$. With $d \geq s$ we have $k \geq 4r + 1$ because the d-dimensional circuit code of length $2d$ and transition sequence $12 \cdots d\, 12 \cdots d$ is of spread s. With $k \geq 4r + 1$ it follows that $\delta_C(x_0, x_i) = i$ for $0 \leq i \leq 2r$ and hence we may assume for this range that the first i coordinates of x_i are 1 and the remainder are 0. For $0 \leq j \leq k$ let U_j denote the set of all vertices y of $I(d)$ such that y is at distance r from x_j but greater than r from each of $x_{j+1}, x_{j+2}, \cdots, x_{j+2r}$, where the indices are reduced modulo k. Clearly U_j misses $U_{j+1} \cup U_{j+2} \cup \cdots \cup U_{j+2r}$, and since C is of spread s we see that U_j intersects $U_{j'}$ only for $j = j'$. Thus the upper bounds for odd s may be established by showing the cardinality of U_j is at least $\binom{d}{r} - 2\binom{d-1}{r-1}$, and for this it suffices to consider U_0. The cube $I(d)$ has $\binom{d}{r}$ vertices whose weight (sum of coordinates) is r, and these include all points of U_0. Let V_0 [respectively V_1] denote the set of all vertices of $I(d)$

which are of weight r, are not in U_0, and have first coordinate 0 [respectively 1]. Then

$$\operatorname{card} U_0 = \binom{d}{r} - \operatorname{card} V_0 - \operatorname{card} V_1$$

and $\operatorname{card} V_1 \leq \binom{d-1}{r-1}$, so the proof can be completed by constructing a biunique mapping of V_0 into V_1.

For each point v of V_0 there exists i between 1 and $2r$ such that $\delta_{I(d)}(x_i, v) \leq r$, and since $\delta_{I(d)}(x_{2r}, v) \geq r$ there must be a smallest i—say $i = i(v)$—such that $\delta_{I(d)}(x_i, v) = r$. If m is the number of 1's among the first $i(v)$ coordinates of v, then since v is of weight r we have

$$(i(v) - m) + (r - m) = \delta_{I(d)}(x_{i(v)}, v) = r$$

and $2m = i(v)$. Thus a biunique mapping φ of V_0 into V_1 can be constructed by defining $\varphi(v)$ as the vertex of $I(d)$ obtained by complementing each of the first $i(v)$ coordinates of v and leaving the remaining coordinates unchanged.

The above reasoning establishes the upper bounds in 3 and 4 when $s = 2r + 1$. For a slight improvement when $s = 2r + 2$, let W_m (for $0 \leq m \leq k$) denote the set of all vertices y of $I(d)$ such that y is at distance $r + 1$ from x_m and at least $r + 1$ from all the x_i's. Then $W_m \cap U_j = \varnothing$ for all m and j, and $W_m \cap W_{m'} = \varnothing$ when m and m' are of different parity. The cardinality of W_m is seen to be $\binom{d-1}{2}$ when $s = 4$, and for $s = 2r + 2$ to have the form

$$\binom{d-r}{r+1} + a_r\binom{d-r}{r} + a_{r-1}\binom{d-r}{r-1} + \cdots + a_0\binom{d-r}{0},$$

where the constants a_i are independent of d. Thus

$$\gamma(d, 4) \leq \frac{2^d - (d^2 - 3d + 2)}{d - 2} \qquad (d \geq 8)$$

and

$$\gamma(d, s) \leq \frac{2^d - P_r(d)}{\binom{d}{r} - 2\binom{d-1}{r-1}} \qquad (d \geq s)$$

when $s = 2r + 2$, $P_r(d)$ being a polynomial of degree $r + 1$ in which the coefficient of d^{r+1} is $2/(r + 1)!$

The lower bounds in results 2 and 6 are based on the following result. It extends a theorem of Danzer–Klee [1], who treated the case $s = 2$.

7. *Suppose that $s \geq 2 \leq c \leq d$, that there exists a c-dimensional circuit code of spread $s - 1$ whose length $m(\geq 2s)$ is divisible by s, and that there exists a d-dimensional circuit code of length $n \geq 2s$. Then*

if $s = 2$ there exists a $(c + d)$-dimensional circuit code of spread s and length $mn/2$;

if s is even there exists a $(c + d + 1)$-dimensional circuit code of spread s and length $m(n + 2)/s$.

PROOF   For each vertex p of $I(c)$, p^* will denote the vertex of $I(d)$ whose first c coordinates are those of p and whose last $d - c$ coordinates are equal to 0. With $k = m/s$, let

$$(a_{11}, a_{12}, \cdots, a_{1s}, a_{21}, a_{22}, \cdots, a_{2s}, \cdots, a_{k1}, a_{k2}, \cdots, a_{ks}, a_{11})$$

and

$$(x_1, x_2, \cdots, x_n, x_1)$$

be circuits corresponding to codes of spread $s - 1$ and s and lengths m and n in $I(c)$ and $I(d)$ respectively. For $1 \leq i \leq k$ the two paths

$$(x_{n-s+1}, x_{n-s+2}, \cdots, x_{n-1}, x_n, x_1)$$

and

$$(a^*_{(i+1)1}, a^*_{is}, \cdots, a^*_{i3}, a^*_{i2}, a^*_{i1})$$

(the paths (x_{n-1}, x_n, x_1) and $(a^*_{(i+1)1}, a^*_{i2}, a^*_{i1})$ when $s = 2$) are both of length s and are such that the $I(d)$-distance between any two vertices of the path is equal to the path-distance. As the symmetry group of $I(d)$ is transitive on paths of this sort, there is a symmetry of $I(d)$ which carries the circuit $(x_1, x_2, \cdots, x_n, x_1)$ onto a circuit $X_i = (x_1^i, x_2^i, \cdots, x_n^i, x_1^i)$ such that

$$x_{n-s+1}^i = a^*_{(i+1)1}, x_{n-s+2}^i = a^*_{is}, \cdots, x_{n-1}^i = a^*_{i3}, x_n^i = a^*_{i2},$$

and

$$x_1^i = a^*_{i1}.$$

(Here $a^*_{(k+1)i} = a^*_{11}$.)

When $s = 2$ we obtain a $(c + d)$-dimensional circuit code C of spread 2 and length $mn/2$ by following the successive rows of the $m/2$-by-n matrix:

$$(a^*_{11}, a^*_{11})(a_{12}, x^1_1)(a_{12}, x^1_2) \cdots (a_{12}, x^1_{n-1})$$

$$(a_{21}, a^*_{21})(a_{22}, x^2_1)(a_{22}, x^2_2) \cdots (a_{22}, x^2_{n-1})$$

$$(a_{k1}, a^*_{k1})\ (a_{k2}, x^k_1)(a_{k2}, x^k_2) \cdots (a_{k2}, x^k_{n-1}).$$

Since $a^*_{i1} = x^i_1$ and $x^i_{n-1} = a^*_{(i+1)1}$, with $x^k_{n-1} = a^*_{11}$, it is clear the above sequence really forms a circuit of length $mn/2$ in $I(c + d)$, where $I(c + d)$ is identified with $I(c) \times I(d)$ in the usual way. Since each sequence $x^i_1, x^i_2, \cdots, x^i_{n-1}$ is part of a circuit of spread 2 in $I(d)$, no chord of C can join two vertices on the same row. Indeed, since it never happens that a_{i1} is adjacent to a_{j1} or a_{i2} to a_{j2}, any chord of C must join (a_{i1}, a^*_{i1}) to (a_{j2}, x^j_k) for some i $\neq$ j and $1 \leq k \leq n$. This requires that $a^*_{i1} = x^i_k$ and a_{i1} is adjacent to a_{j2}, whence of course a^*_{i1} is adjacent to a^*_{j2}. Since the vertices $a^*_{i1} = x^i_k$ and $a^*_{j2} = x^j_n$ lie together in the circuit X_i of spread 2, their adjacency implies $k = 1$ or $k = n - 1$. From $k = n - 1$ it follows that $i = j + 1$, whence (a_{i1}, a^*_{i1}) and (a_{j2}, x^j_k) determine an edge (and not a chord) of C. From $k = 1$ it follows that $i = j$, an impossibility. This settles the case $s = 2$.

For even $s > 2$ the construction and reasoning are similar to those for $s = 2$ but are somewhat more complicated. A $(c + d + 1)$-dimensional circuit code of spread s and length $m(n + 2)/s$ is obtained by following the successive rows of the m/s-by-$(n + 2)$ matrix:

$$(a_{11}, a^*_{11}, 0) \cdots (a_{1r}, a^*_{11}, 0)(a_{1s}, x^1_1, 0)(a_{1s}, x^1_1, 1) \cdots (a_{1s}, x^1_{n-r}, 1)(a_{1s}, x^1_{n-r}, 0)$$

$$(a_{21}, a^*_{21}, 0) \cdots (a_{2r}, a^*_{21}, 0)(a_{2s}, x^2_1, 0)(a_{2s}, x^2_1, 1) \cdots (a_{2s}, x^2_{n-r}, 1)(a_{2s}, x^2_{n-r}, 0)$$

$$(a_{k1}, a^*_{k1}, 0) \cdots (a_{kr}, a^*_{k1}, 0)\ (a_{ks}, x^k_1, 0)(a_{ks}, x^k_1, 1) \cdots (a_{ks}, x^k_{n-r}, 1)(a_{ks}, x^k_{n-r}, 0).$$

See Klee [22] for the proof.

For the lower bound in result 2, due to Danzer–Klee [1], note that if $n \geq 2$ and the inequality

$$(*) \qquad\qquad \gamma(d, 2) \geq \lambda \frac{2^d}{d - 1}$$

is valid for $d = n$ then it is valid also for $d = 2n$ and $d = 2n - 1$. Indeed, since $\gamma(n, 1) = 2^n$ it follows from the first part of theorem 7 that

$$\gamma(n, 2) \geq 2^{n-1}\gamma(n, 2) \geq 2^{n-1}\lambda \frac{2^n}{n - 1} > \lambda \frac{2^{2n}}{2n - 1}$$

and similarly $\gamma(2n - 1, 2) \geq \lambda 2^{2n-1}/(2n - 2)$. An induction then shows that (*) is valid for all $d \geq n$ if it is valid for $n \leq d < 2n$ and hence the lower bound in result 2 is implied by the following table of values of $\gamma(d, 2)$:

d	2	3	4	5	6	7	8	9
$\gamma(d, 2)$	4	6	8	14	26	≥ 48	≥ 64	≥ 112

(To justify these values see Kautz [1], Even [1], and Davies [1] for $d \leq 7$, and note that $\gamma(8, 2) \geq 8\gamma(4, 2)$ and $\gamma(9, 2) \geq 8\gamma(5, 2)$ by theorem 7.)

In addition to papers already mentioned, the reader is referred to Vasiliev [1], Ramanujacharyulu–Menon [1] and Abbott [1] for constructions of circuit codes of spread 2. See Kautz [1] for the error-checking properties of these codes and Žuravlev [1] for their relationship to algorithms for the simplification of disjunctive normal forms.

The lower bounds in results 3 and 5 depend on Singleton's constructions 1, 3a, and 5, which apply to a transition sequence for a circuit code of spread s, dimension d, and length k.

(S1) Divide the sequence into two segments of equal length and within each segment form successive blocks of s transitions, leaving an incomplete block at the end of each segment if s does not divide $k/2$. Insert the new transition number $d + 1$ at the end of each complete block. The new transition sequence corresponds to a circuit code of spread s, dimension $d + 1$, and length $k + 2[k/2s]$.

(S3a) Suppose s is odd. Divide the sequence into two segments of equal length and within each segment form successive blocks of $(s + 1)/2$ transitions, leaving an incomplete block at the end of each segment if $s + 1$ does not divide k. Alternate the new transition numbers $d + 1, \cdots, d + (s + 1)/2$ with the old ones in each complete block, using the new ones in the same order in each case. The new transition sequence corresponds to a circuit code of spread s, dimension $d + (s + 1)/2$ and length $k + (s + 1)[k/(s + 1)]$.

(S5) Suppose $s = 3$, and some number j appears m times in the original transition sequence. Let each occurrence of j be replaced by the pattern

$$x(d + 1)j(d + 2)x(d + 3)j(d + 1)x(d + 2)j(d + 3)x,$$

where the center j is the original one and the x's are other members of the original sequence. (Note that any two of the original j's are separated by at least three x's.) The number j appears $3m$ times in the new transition sequence, which corresponds to a circuit code of spread 3, dimension $d + 3$, and length $k + 8m$.

To facilitate an inductive proof of the lower bound in result 3, consider the strengthened assertion that for each $d \geq 6$ there exists a d-dimensional circuit code of spread 3 and length at least $32 \cdot 3^{(d-8)/3}$ in which a fourth of the transition numbers are the same. To verify this for $d = 6$, apply (S3a) to the code whose transition sequence is 12341234. For $d = 7$ apply (S5) to the same code. For $d = 8$ apply (S1) to the code for $d = 7$. Then the assertion is established for $6 \leq d \leq 8$ and an easy inductive argument based on (S5) shows its validity for all $d \geq 6$.

The lower bound in result 5 follows from an easy induction based on (S3a), and the bounds in result 6 follow from those in 5 by an argument analogous to that used in deriving result 2 from the fact that $\gamma(d, 1) = 2^d$.

The exact values of $\gamma(d, s)$ have been determined in a few cases. We include the following theorem and proof of Singleton [1].

8. *If $d < [3s/2] + 2$ then $\gamma(d, s) = 2d$.*

PROOF The circuit code whose transition sequence is $12 \cdots d12 \cdots d$ is of spread s for all s, whence $\gamma(d, s) \geq 2d$. For the reverse inequality when $d \leq [3s/2] + 1$ we may assume in addition that $s \leq d$, for $s > d$ implies $\gamma(d, s) = \gamma(d, d)$. Suppose there exists a d-dimensional circuit code C of spread s and length $k > 2d$. With $s \leq d$ we have $k > 2s$. If $k = 2s + 2$ then $d < s + 1$ and two vertices of C related by $s + 1$ transitions must have distance at most $s - 1$ in $I(d)$; but that is impossible for their distance in C is $s + 1$ and C is of spread s. Suppose, finally, that $k \geq 2s + 4$ and consider a block of $s + 3$ transitions. If both the end blocks of $s + 2$ transitions (in the given block of $s + 3$) include repetitions, the $I(d)$-distance of the vertices related by the $s + 3$ transitions is $s - 1$, an impossibility. Thus one of the end blocks consists of $s + 2$ distinct transitions. Consider the block of $2s + 3$ transitions formed from that end block and an adjacent block of (necessarily distinct) $s + 1$ transitions. Let i and j denote the number of transitions appearing once and twice respectively in the block of $2s + 3$, whence $i + j \leq d$ and $i + 2j = 2s + 3$. The $I(d)$-distance between the related vertices is

$$i = 2(i + j) - (i + 2j) \leq 2d - (2s + 3) < s,$$

whence the C-distance must be i and the total length of the circuit is $(i + 2j) + i \leq 2d$.

In addition to the equality of theorem 8, Singleton [1] states $\gamma((3s + 3)/2, s) = 4s + 4$ for odd s and conjectures $\gamma((3s + 4)/2, s) = 4s + 6$ for even s.

17.5   Additional notes and comments

The four-color theorem.
Much motivation for investigating the Hamiltonicity of the graphs of simple
3-polytopes has been drawn from the "four-color problem" (see page 364) that
was still unresolved in 1967. The (controversial, computer-based) positive res-
olution in 1977 due to Appel and Haken (see Appel–Haken [a]) has since been
backed by an independent (also computer-based) proof by Robertson, Sanders,
Seymour, and Thomas (see Robertson et al. [a] and Seymour [a]).

Hamiltonicity of polytopal graphs.
Holton–McKay [a] (extending a method of Okamura [a]) proved that every
simple 3-polytope with at most 36 vertices is Hamiltonian. This implies $N_c =
38$ (see pages 359 and 364). Zamfirescu [a] found a simple 3-polytope with 88
vertices not admitting a Hamiltonian path, thus improving the upper bound
on N_p to 88. See Aldred et al. [a] for a recent paper about related problems. A
nice (earlier) survey is in Barnette [h, Chap. 3].

A related problem, posed by Barnette, is still open: Is every cubic 3-connec-
ted *bipartite* planar graph Hamiltonian? This is true for graphs with at most 64
vertices, according to Holton–Manvel–McKay [a].

Furthermore, the "higher-dimensional" problem mentioned on page 375 is
still open (even for $d = 4$): Does every simple d-polytope ($d \geq 4$) have a Hamilton circuit? Paulraja [a] proved that for each cubic 3-connected graph G the
product of G and a single edge (the *prism* over G) is Hamiltonian. A simpler
proof was recently given by Čada et al. [a], who were motivated by the conjecture due to Alspach-Rosenfeld [a] that each such prism can be decomposed into
two Hamiltonian circuits. They proved the conjecture for G being the *bipartite*
graph of a simple 3-polytope.

Checking Hamiltonicity is an NP-complete problem even for graphs that
are planar, 3-connected, and 3-regular (Garey et al. [a]). Grünbaum–Walther
[a] introduced measures for the failure of Hamiltonicity on infinite families of
graphs. See also the recent survey by Owens [a].

Path lengths.
The estimates in 17.2.5–17.2.8 (which were conditioned on the lower or upper
bound conjecture) are valid, due to the proofs of the lower and upper bound
theorems—see the notes in section 10.6.

The Hamiltonian circuits on the duals of cyclic d-polytopes with n vertices,
found by Klee (see page 368), can be described compactly by a "twisted-
lexicographic order" on the d-subsets of $\{1,\dots,n\}$ (Gärtner–Henk–Ziegler [a]).

Heights.

Klee–Minty [a] showed in 1972 that there are simple polytopes—the *Klee–Minty cubes*—of exponential height. In fact, they proved the lower bound $M_f^v(\tau, d, 2d) \geq 2^d$ on the maximal length of a path produced by the simplex algorithm on a simple d-polytope with $2d$ facets that chooses the pivots in order to maximize the gradient (Dantzig's original rule). For fixed d the function $M_f^v(\tau, d, n)$ grows asymptotically like a polynomial (in n) of degree $[d/2]$.

Jeroslow [a] proved $M_f^v(\sigma, d, 3d - 1) \geq 3 \cdot 2^{[d/2]} - 2$ (i.e., for the *greatest increase rule*). For fixed d the function $M_f^v(\sigma, d, n)$ grows asymptotically like a polynomial (in n) of degree $[d/2]$. Goldfarb–Sit [a] rescaled the Klee–Minty cubes in order to derive the analogous results for the function ζ (for the *steepest edge rule*). A unified, comprehensive treatment in the framework of *deformed products* can be found in Amenta–Ziegler [a].

It is still not clear (see Ziegler [a, Problem 3.11*]) whether the upper bound theorem gives a *sharp* upper bound on $M_f^v(\eta, d, n)$. This is the case for $n = d + 2$ by Gärtner et al. [a] and for $d = 4$ by Pfeifle [c]. In particular, the maximal height for duals of cyclic polytopes is unknown.

Randomization.

While still no deterministic pivot rule is known to make the simplex algorithm run in polynomial time, substantial progress has been obtained with *randomized* variants. In 1992, Kalai [h] (see also Kalai [i]) proved that the *random-facet* rule produces (on all instances) an expected number of at most $\exp(\text{const} \cdot \sqrt{n \log d})$ pivot steps. Independently, Matoušek–Sharir–Welzl [a] also found a sub-exponential algorithm. Later the two algorithms turned out to be dual to each other (see Goldwasser [a]).

However, the question about the complexity of the simplex algorithm with the *random-edge* pivot rule is unsolved. For an analysis of randomized pivot rules on the Klee–Minty cubes see Gärtner–Henk–Ziegler [a].

Borgwardt [a] showed that the (deterministic) "shadow vertex" pivot rule has polynomial expected running time on (a certain model of) random linear programs.

Circuit codes.

For a comprehensive treatment of Gray codes in the d-cube (circuit codes of spread 1 of maximum size) see Knuth [a]. The best available bounds (improving 17.4.2) on the maximal length $\gamma(d, 2)$ of an induced cycle in the d-cube graph ("snake in the box") are due to Abbott–Katchalski [a] and Zémor [a]:

$$\frac{77}{128} 2^{d-1} \leq \gamma(d, 2) \leq 2^{d-1}\left(1 - \frac{1}{89\sqrt{d}} + O(\tfrac{1}{d})\right)$$

CHAPTER 18

Arrangements of Hyperplanes

There are many fields which are similar in spirit and related in the methods used and results obtained to the combinatorial theory of polytopes. The present chapter is devoted to one such field: to questions dealing with arrangements of (or partitions by) hyperplanes.

Though arrangements and polytopes are quite analogous in certain aspects, there appears to be sufficient difference between the two topics to justify considering them side by side. The decision to include the chapter on arrangements was made even easier by the fact that there seems to be available in the literature no systematic survey of the topic, despite the multitude of papers written on it and the interest in it shown in recent years by different applied disciplines.

18.1 d-Arrangements

A finite family $\mathscr{A}$ of at least $d + 1$ hyperplanes in the real projective d-space P^d is said to form a *d-arrangement* of hyperplanes provided no point P^d belongs to all the members of $\mathscr{A}$. If $\mathscr{A}$ is a *d-arrangement*, the open set $P^d \sim (\bigcup_{H \in \mathscr{A}} H)$ is the union of a finite number of connected components. The closure of each such component is a polytope in P^d; we shall call it a *d-face* or a *d-cell* of $\mathscr{A}$. The *k-faces* of $\mathscr{A}$, for $-1 \leq k < d$, are defined as the *k*-faces of the *d*-faces of $\mathscr{A}$. We shall denote by $n(\mathscr{A})$ the number of hyperplanes in the *d*-arrangement $\mathscr{A}$, and by $f_k(\mathscr{A})$ the number of different *k*-faces of $\mathscr{A}$. The *f*-vector of $\mathscr{A}$ is defined by $f(\mathscr{A}) = (f_0(\mathscr{A}), \cdots, f_d(\mathscr{A}))$. The faces of any *d*-arrangement may be also considered as forming a *complex* in P^d.

In a quite analogous manner it is possible to define arrangement of hyperplanes in the Euclidean *d*-space. For most purposes it is of small importance whether one considers arrangements in the projective or in the Euclidean space, since the addition of the 'hyperplane at infinity' yields a projective arrangement from each Euclidean one. We prefer the projective setting as being more symmetric. However, the difference between Euclidean and projective arrangements becomes important in some questions, such as the determination of the number of non-equivalent d-arrangements of n hyperplanes.

There is a close analogy between the facial structure of arrangements in projective spaces, and the facial structure of polytopes. This is the reason for the inclusion of the present chapter in the book. However, there are also significant differences between the two topics. It is hoped that the present exposition will induce an exchange of ideas between the two fields.

The subject of d-arrangements goes back at least to Steiner [1], where d-arrangements (for $d = 2, 3$) and the similar concepts in Euclidean space involving planes, or spheres, were considered in detail. The subject was pursued by many authors, some of whom extended the discussion to projective spaces of higher dimensions (see, for example. v. Staudt [1], Eberhard [1, 2], Cahen [1], Roberts [1], Schläfli [1]; see also Pólya [2] (vol. 1, chapter 3) for a stimulating introduction to the field). In the related field of *configurations* a large amount of research has been done; however, this topic is outside the scope of the present book (for detailed accounts see Steinitz [3, 4], Levi [2]. Hilbert–Cohn-Vossen [1]).

A d-arrangement $\mathscr{A}$ will be called *simple* provided no $d + 1$ hyperplanes in $\mathscr{A}$ pass through the same point. The analogy between arrangements and polytopes is strengthened by the similarity of properties of simple arrangements and simple (or simplicial) polytopes.

The f-vectors of d-arrangements satisfy a relation analogous to Euler's, while the f-vectors of simple d-arrangements satisfy equations similar to those of Dehn–Sommerville.

We shall start with Euler's equation (Eberhard [2]):

1. *For every d-arrangement $\mathscr{A}$ the vector $f(\mathscr{A})$ satisfies the equation*

$$\sum_{i=0}^{d} (-1)^i f_i = \tfrac{1}{2}(1 + (-1)^d).$$

For simple d-arrangements $\mathscr{A}$ the inductive proof of theorem 1 is **very simple**: Let $\mathscr{A}^*$ denote the d-arrangement obtained from $\mathscr{A}$ by omitting one hyperplane H_0, and let $\mathscr{A}^{**}$ denote the $(d - 1)$-arrangement determined in H_0 by *its* hyperplanes $H_0 \cap H$, where H is in $\mathscr{A}^*$. Then it is easily seen that $\mathscr{A}$, $\mathscr{A}^*$, and $\mathscr{A}^{**}$ are related by

$$f_0(\mathscr{A}) = f_0(\mathscr{A}^*) + f_0(\mathscr{A}^{**})$$

$$f_i(\mathscr{A}) = f_i(\mathscr{A}^*) + f_i(\mathscr{A}^{**}) + f_{i-1}(\mathscr{A}^{**}) \quad \text{for} \quad 1 \leq i \leq d.$$

Since Euler's relation is easily checked for d-arrangements of only $d + 1$

hyperplanes (in which case $f_i = 2^i \binom{d+1}{i+1}$), the above recursive equations yield at once the inductive proof for the general case.

A different proof of theorem 1, valid for all d-arrangements, follows from the interpretation of the projective d-space as a d-sphere S^d in Euclidean $(d+1)$-space R^{d+1}, with diametral points identified (or from the equivalent interpretation as the set of all lines through the origin of R^{d+1}). Therefore, if diametral points of S^d are not identified, each d-arrangement gives rise to a partition of S^d into (spherically convex) cells, each k-face of the d-arrangement yielding two k-faces of the partition of S^d. Therefore theorem 1 follows from Euler's formula for S^d upon division of the right-hand side by 2.

For simple d-arrangements, a similar approach yields also analogues of the Dehn–Sommerville equations. Indeed, it is sufficient to note that the induced partition of S^d mentioned above has a dual partition which is cubical (see section 9.4). Using theorem 9.4.1, or noting that each i-face of a simple d-arrangement $\mathscr{A}$ is contained in $2^{j-1}\binom{d-i}{j-i}$ j-faces of $\mathscr{A}$ (for $0 \leq i \leq j \leq d$) and applying the method of section 9.1, there results

2. *For each simple d-arrangement $\mathscr{A}$ and for each k with $1 \leq k < d$, the vector $f(\mathscr{A})$ satisfies the equation*

$$\sum_{i=0}^{k} (-1)^i 2^{k-i} \binom{d-i}{d-k} f_i = f_k.$$

Clearly, this system contains $[\frac{1}{2}(d-1)]$ independent equations. In distinction from the cubical polytopes, the f-vectors of simple d-arrangements satisfy additional independent linear relations. The reason for this difference between the two cases is that the derivation of the equations in theorem 2 used only 'local' properties of the d-arrangements which, in fact, would be shared by all partitions of P^d into convex cells induced by identifying diametral points in centrally symmetric partitions of S^d which are dual to cubical partitions of S^d. But in case of simple d-arrangements there is available additional, 'global', information: the fact that different $(d-1)$-faces of the d-arrangement 'fit' together to form hyperplanes.

Using the inductive approach mentioned in the outline of the first proof of theorem 1, it is easy to establish the following facts (see Buck [1]).

3. *The number $f_k(\mathscr{A})$ of k-faces of a simple d-arrangement $\mathscr{A}$ with $n(\mathscr{A}) = n$ depends only on k, d, and n. Denoting this number by $f_k^d(n)$ we have*

$$f_k^d(n) = \sum_{i=0}^{[\frac{1}{2}k]} \binom{d-2i}{d-k}\binom{n}{d-2i} = \binom{n}{d-k}\sum_{i=0}^{k}\binom{n-d-1+k}{i}.$$

4. *The numbers $f_k^d(n)$ satisfy the relations*

$$\sum_{i=0}^{d-k}(-1)^i\binom{d-i}{k}f_i^d(n) = \tfrac{1}{2}\binom{n}{k}(1+(-1)^{d-k})$$

whenever $0 \le k \le d < n$.

Euler's equation is clearly contained in this system for $k = 0$. The equations corresponding to $k \equiv d \pmod 2$ do not depend on n; they form a system equivalent to that of theorem 2 (compare theorems 9.4.1 and 9.4.2).

As an analogue of theorem 8.1.1 we have

5. *The affine hull of the set of all vectors of the form $f(\mathscr{A}) = (f_0(\mathscr{A}), \cdots, f_d(\mathscr{A}))$, where $\mathscr{A}$ varies over all d-arrangements, has dimension d.*

In other words, constant multiples of Euler's equation are the only linear equations in the $f_i(\mathscr{A})$s which hold for all d-arrangements $\mathscr{A}$.

A proof of theorem 5 may be obtained by considering d-arrangements determined by at most $d + 2$ hyperplanes. It is even possible to show that for every $n \ge d + 3$, dim aff$\{f(\mathscr{A}) \mid \mathscr{A}$ a d-arrangement with $n(\mathscr{A}) = n\}$ $= d$.

Similar to the corresponding fact about polytopes is

6. *The functions $f_k(\mathscr{A})$, $0 \le k \le d$, depend in a lower semicontinuous way on the d-arrangement $\mathscr{A}$.*

Therefore, in particular, we have

7. *For every d-arrangement $\mathscr{A}$ with $n(\mathscr{A}) = n$, and for each $k, 0 \le k \le d$,*

$$f_k(\mathscr{A}) \le f_k^d(n).$$

If $\mathscr{A}$ is a d-arrangement with $n(\mathscr{A}) = n$ such that $n - d + 1$ hyperplanes of $\mathscr{A}$ pass through the same $(d - 2)$-dimensional variety, it is easy to show that

$$f_k(\mathscr{A}) = 2^{k-1}\left\{2\binom{d}{k} - \binom{d-1}{k-1}\right\}n - 2^{k-1}(d+1)\left\{\frac{2k}{k+1}\binom{d}{k} - \binom{d-1}{k-1}\right\}.$$

Let $\varphi_k^d(n) = \min\{f_k(\mathscr{A}) \mid \mathscr{A}$ be a d-arrangement with $n(\mathscr{A}) = n\}$. The values of $\varphi_k^d(n)$ are not known for $d \ge 3$. We venture the following:

CONJECTURE

$$\varphi_k^d(n) = 2^{k-1}\left\{2\binom{d}{k} - \binom{d-1}{k-1}\right\}n - 2^{k-1}(d+1)\left\{\frac{2k}{k+1}\binom{d}{k} - \binom{d-1}{k-1}\right\}.$$

Two d-arrangements are called *equivalent* provided there exists a one-to-one incidence-preserving correspondence between their faces. In other words, the d-arrangements are equivalent provided the corresponding d-complexes are combinatorially equivalent. As in section 5.5, it is possible (in principle) to determine all the different classes of nonequivalent d-arrangements of n hyperplanes. However, as with polytopes, the actual determination of the number of different equivalence-classes of d-arrangements with n hyperplanes is quite hopeless unless $n - d$ is very small. The known results deal only with $d < 3$ (see Cummings [1, 2], White [1, 2], R. Klee [1]); they are given in table 18.1.1. The different simple 2-arrangements with $n(\mathscr{A}) \leq 7$ are shown in figure 18.1.1, while figure 18.1.2 shows the nonsimple 2-arrangements with $n(\mathscr{A}) \leq 6$.

Table 18.1.1. Number of nonequivalent 2-arrangements $\mathscr{A}$ with $n(\mathscr{A}) = n$

$n =$	3	4	5	6	7
Simple arrangements in the projective plane	1	1	1	4	11
All arrangements in the projective plane	1	2	4	16	
Simple arrangements in the non-oriented Euclidean plane	1	1	6	43	922*
Simple arrangements in the oriented Euclidean plane	1	1	7	79	1765*

We wish to mention here a rather old conjecture of Cummings [3], to the effect that certain numerical 'indices' of simple 2-arrangements (which depend only on the combinatorial type of the arrangement) have equal values only for equivalent arrangements. Though the truth of this conjecture (which is somewhat analogous to that disproved in exercise 13.6.2) seems rather doubtful (Carver [1]), the conjecture is known to be correct for $n(\mathscr{A}) \leq 7$.

In order to define the *index* $i(\mathscr{A})$ of a simple 2-arrangement $\mathscr{A}$ we proceed as follows. For any vertex V consider the two lines of $\mathscr{A}$ passing through V; they divide the projective plane into two 'halfplanes' H_1 and H_2. Let a_j be the number of vertices of $\mathscr{A}$ contained in the interior of H_j; then clearly $a_1 + a_2 = \binom{n-2}{2}$. The index $i(V)$ of V is defined by $i(V) = \min\{a_1, a_2\}$. The index $i(\mathscr{A})$ is the vector $i(\mathscr{A}) = (i_0, i_1, \cdots, i_m)$,

---

* Due to R. Klee [1], probably not correct (see page 408).

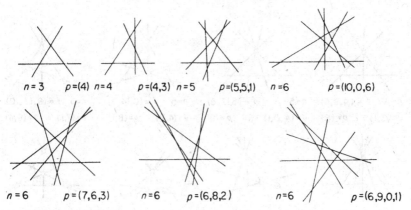

$n=3$ $p=(4)$ $n=4$ $p=(4,3)$ $n=5$ $p=(5,5,1)$ $n=6$ $p=(10,0,6)$

$n=6$ $p=(7,6,3)$ $n=6$ $p=(6,8,2)$ $n=6$ $p=(6,9,0,1)$

Figure 18.1.1 (Part 1). Simple 2-arrangements of at most 6 lines

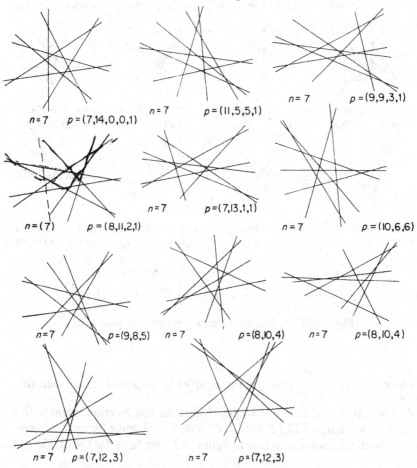

$n=7$ $p=(7,14,0,0,1)$ $n=7$ $p=(11,5,5,1)$ $n=7$ $p=(9,9,3,1)$

$n=(7)$ $p=(8,11,2,1)$ $n=7$ $p=(7,13,1,1)$ $n=7$ $p=(10,6,6)$

$n=7$ $p=(9,8,5)$ $n=7$ $p=(8,10,4)$ $n=7$ $p=(8,10,4)$

$n=7$ $p=(7,12,3)$ $n=7$ $p=(7,12,3)$

Figure 18.1.1 (Part 2). Simple 2-arrangements of 7 lines

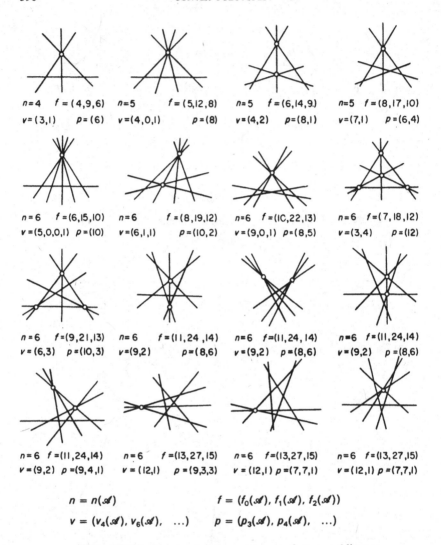

$n = 4 \quad f = (4,9,6)$
$v = (3,1) \quad p = (6)$

$n = 5 \quad\quad f = (5,12,8)$
$v = (4,0,1) \quad\quad p = (8)$

$n = 5 \quad f = (6,14,9)$
$v = (4,2) \quad p = (8,1)$

$n = 5 \quad f = (8,17,10)$
$v = (7,1) \quad p = (6,4)$

$n = 6 \quad f = (6,15,10)$
$v = (5,0,0,1) \quad p = (10)$

$n = 6 \quad\quad f = (8,19,12)$
$v = (6,1,1) \quad\quad p = (10,2)$

$n = 6 \quad f = (10,22,13)$
$v = (9,0,1) \quad p = (8,5)$

$n = 6 \quad f = (7,18,12)$
$v = (3,4) \quad\quad p = (12)$

$n = 6 \quad f = (9,21,13)$
$v = (6,3) \quad p = (10,3)$

$n = 6 \quad f = (11,24,14)$
$v = (9,2) \quad\quad p = (8,6)$

$n = 6 \quad f = (11,24,14)$
$v = (9,2) \quad p = (8,6)$

$n = 6 \quad f = (11,24,14)$
$v = (9,2) \quad p = (8,6)$

$n = 6 \quad f = (11,24,14)$
$v = (9,2) \quad p = (9,4,1)$

$n = 6 \quad\quad f = (13,27,15)$
$v = (12,1) \quad p = (9,3,3)$

$n = 6 \quad f = (13,27,15)$
$v = (12,1) \quad p = (7,7,1)$

$n = 6 \quad f = (13,27,15)$
$v = (12,1) \quad p = (7,7,1)$

$n = n(\mathscr{A}) \qquad\qquad\qquad f = (f_0(\mathscr{A}), f_1(\mathscr{A}), f_2(\mathscr{A}))$

$v = (v_4(\mathscr{A}), v_6(\mathscr{A}), \ldots) \qquad p = (p_3(\mathscr{A}), p_4(\mathscr{A}), \ldots)$

Figure 18.1.2. Nonsimple 2-arrangements of at most 6 lines

where $m = \frac{1}{2}\binom{n-2}{2}$, and i_k is the number of vertices V of $\mathscr{A}$ such that $i(V) = k$, $0 \leq k \leq m$. For example, if $\mathscr{A}$ is the last 2-arrangement with 6 lines shown in figure 18.1.1, then $i(\mathscr{A}) = (6, 0, 3, 6)$, while for the 2-arrangement with six lines shown first in figure 18.1.1 we have $i(\mathscr{A}) = (0, 15, 0, 0)$.

With every d-arrangement $\mathscr{A}$ a graph $\mathscr{G} = \mathscr{G}(\mathscr{A})$ may be associated in the following manner: The vertices of $\mathscr{G}$ correspond to the d-cells of $\mathscr{A}$; two vertices of $\mathscr{G}$ are connected by an edge if and only if the corresponding d-cells in $\mathscr{A}$ have a common $(d-1)$-face. Clearly $\mathscr{G}(\mathscr{A}_1)$ and $\mathscr{G}(\mathscr{A}_2)$ are isomorphic graphs whenever $\mathscr{A}_1$ and $\mathscr{A}_2$ are equivalent d-arrangements. For $d = 2$ it is not hard to show that the isomorphism of $\mathscr{G}(\mathscr{A}_1)$ and $\mathscr{G}(\mathscr{A}_2)$ implies the equivalence of $\mathscr{A}_1$ and $\mathscr{A}_2$. However, already for $d = 3$ it is not known whether this statement is still valid.

18.2 2-Arrangements

In the present section we shall consider certain additional problems and results dealing mostly with 2-arrangements $\mathscr{A}$. Let the number of k-gonal 2-faces of $\mathscr{A}$ be denoted by p_k. Then, as in chapter 13, for simple 3-arrangements $\mathscr{A}$ we have

$$f_2(\mathscr{A}) = \sum_{k \geq 3} p_k,$$

$$4f_0(\mathscr{A}) = 2f_1(\mathscr{A}) = \sum_{k \geq 3} kp_k = 2n(n-1),$$

and

(*) $$p_3 = 4 + \sum_{k \geq 5} (k-4)p_k,$$

where $n = n(\mathscr{A})$ is the number of lines in $\mathscr{A}$.

The following lemma is the source of various other results.

1. *Each line of a 2-arrangement $\mathscr{A}$ is incident with at least three triangular 2-faces of $\mathscr{A}$.*

This result, and the following elegant proof of it, are due to Levi [1]. It was rediscovered in a solution to a problem of Moser [2]. We consider a representation of $\mathscr{A}$ in the Euclidean plane, with L the 'line at infinity'. Let M be the set of vertices of $\mathscr{A}$ which are not in L. If the finite set M consists of a single point, then all lines of $\mathscr{A}$, except L, pass through it and $2n - 2$ triangles have an edge in L. If conv M is a segment (this case was overlooked by Levi [1]) then $\mathscr{A}$ consists of the lines L, $L_0 = \text{aff } M$, and L_i, $1 \leq i \leq k$, where $L_1, \cdots, L_k$ are 'parallel' (i.e. have a common point on L). Thus the 2-arrangement is the same as in the previous case, and L is incident with 4 triangles. If conv M is a polygon it has at least 3 vertices, and for each vertex there are at least two half-lines in A issuing from it and not meeting any other point of M. Each neighboring pair of such halflines determines, together with L, a triangular 2-face of $\mathscr{A}$; thus L is incident with at least 3 such faces, and the proof is complete.

As an immediate corollary we have the following result, mentioned already in Eberhard [1]:

2. *If $\mathscr{A}$ is any 2-arrangement with n lines, then $p_3(\mathscr{A}) \geq n$.*

Indeed, by theorem 1, there are at least $3n$ incidences between triangles and lines of $\mathscr{A}$. On the other hand, there are $3p_3(\mathscr{A})$ such incidences, hence the result.

Together with relation (*), theorem 2 clearly yields a solution to the problem proposed by Moser [1].

It should be noted that the estimate of theorem 2 is best possible of its type. For each $n \geq 4$, the lines determined by the edges of any convex n-gon form such a 2-arrangement. We mention without proof the following result of Roberts [1] which is sometimes stronger than theorem 2:

3. *If $\mathscr{A}$ is a simple 2-arrangement of n lines and if L_0 is any line in $\mathscr{A}$, there are at least $n - 3$ triangles not incident with L_0.*

Theorem 2 naturally leads to the problem of determining the largest and smallest possible values of $p_k(\mathscr{A})$ for 2-arrangements $\mathscr{A}$ with n lines. Each of these questions may be considered for all 2-arrangements, or with $\mathscr{A}$ restricted to simple 2-arrangements. Only very scattered results are known.

Thus, clearly $p_3(\mathscr{A}) \leq \frac{1}{3}n(n-1)$ for all simple 2-arrangements $\mathscr{A}$ with $n \geq 4$ lines. The known maximal values $p_3(n)$ $[p_3^s(n)]$ of $p_3(\mathscr{A})$ for [simple] 2-arrangements $\mathscr{A}$ with n lines are collected in table 18.2.1. The values of $p_3^s(n)$ are less than $[\frac{1}{3}n(n-1)]$ for $n = 5, 7, 8$, but equal this bound for $n = 4, 6, 10$. It is not known whether equality holds for any additional

Table 18.2.1. The known value of $p_3(n)$ and $p_3^s(n)$

n	$p_3(n)$	$p_3^s(n)$	$[\frac{1}{3}n(n-1)]$
3	4	4	2
4	6	4	4
5	8	5	6
6	12	10	10
7		11	14
8		16	18
10		30	30

values of n. To see that $p_3^s(8) \geq 16$ and $p_3^s(10) \geq 30$ it is sufficient to consider the 2-arrangements in figures 18.2.1 and 18.2.2. It is not hard (but somewhat time-consuming) to show that $p_3^s(8) \leq 16$ even without determining all possible simple 2-arrangements $\mathscr{A}$ with $n(\mathscr{A}) = 8$.

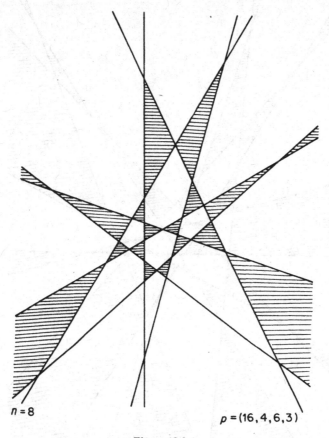

$n = 8$

$p = (16, 4, 6, 3)$

Figure 18.2.1

Another unsolved problem is: For what values of n do there exist simple 2-arrangements $\mathscr{A}$ with $n(\mathscr{A}) = n$ and $p_4(\mathscr{A}) = 0$. The three known cases are $n = 3, 6, 10$ (see figures 18.1.1 and 18.2.2).

In analogy to the determination of the set of f-vectors of all 3-polytopes (section 10.3), we may attempt to characterize the set of f-vectors of all d-arrangements, at least for $d = 2$.

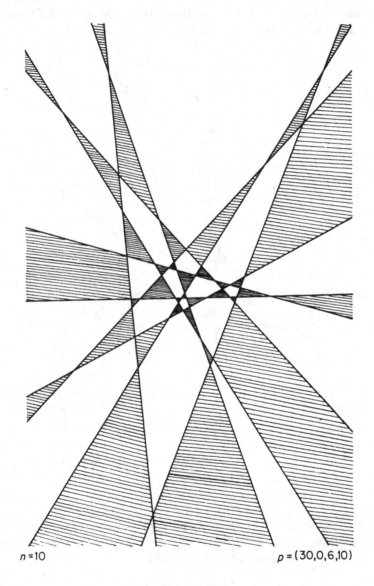

$n = 10$ $p = (30, 0, 6, 10)$

Figure 18.2.2

Let $v_{2k}(\mathscr{A})$ denote the number of $2k$-valent vertices of the 2-arrangement $\mathscr{A}$ (i.e. the number of vertices each of which is on exactly k lines). Then $f_0(\mathscr{A}) = \sum_{k \geq 2} v_{2k}(\mathscr{A})$, $\quad f_1(\mathscr{A}) = \sum_{k \geq 2} k v_{2k}(\mathscr{A})$, and therefore

$$f_2(\mathscr{A}) = f_1 - f_0 + 1 = 1 + \sum_{k \geq 2} (k-1)v_{2k} \geq 1 + \sum_{k \geq 2} v_{2k} = 1 + f_0(\mathscr{A});$$

equality holds if and only if $\mathscr{A}$ is simple. On the other hand, combining $2f_1(\mathscr{A}) = \sum_{i \geq 3} i p_i(\mathscr{A})$ with Euler's equation, there results

$$2f_0(\mathscr{A}) - 2 = 2f_1 - 2f_2 = \sum_{i \geq 3} (i-2)p_i \geq \sum_{i \geq 3} p_i(\mathscr{A}) = f_2(\mathscr{A});$$

equality holds if and only if $f_2(\mathscr{A}) = p_3(\mathscr{A})$.

Hence we have (see figure 18.2.3)

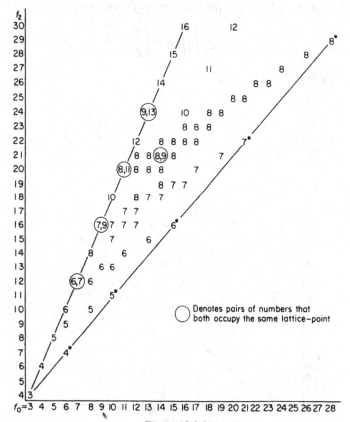

Figure 18.2.3

4. *The convex hull of the set of f-vectors of all 2-arrangements is the set*
$\{(f_0, f_0 + f_2 - 1, f_2) \mid f_0 + 1 \le f_2 \le 2f_0 - 2\}.$

In figure 18.2.3, a number in the position (f_0, f_2) indicates $n(\mathscr{A})$ for 2-arrangements $\mathscr{A}$ with $f_0(\mathscr{A}) = f_0$ and $f_2(\mathscr{A}) = f_2$. As visible from figure 18.2.3, in which all $(f_0(\mathscr{A}), f_2(\mathscr{A}))$ for $n(\mathscr{A}) \le 8$ are represented, the situation in the present case is more complicated than in the analogous case dealt with in theorem 10.3.1. Many lattice-points in the region $f_0 + 1 \le f_2 \le 2f_0 - 2$ do not correspond to any 2-arrangement—but their characterization still eludes us.

As shown by the equation used in the proof of theorem 4, the characterization of f-vectors of 2-arrangements is closely related to the question : Which $(n - 2)$-tuples $(v_4, v_6, \cdots, v_{2n-2})$ may be realized by 2-arrangements $\mathscr{A}$ with $n(\mathscr{A}) = n$?

It is not hard to show that the numbers $v_{2i}(\mathscr{A})$ satisfy the equation

$$\sum_{i=2}^{n-1} \binom{i}{2} v_{2i} = \binom{n}{2}$$

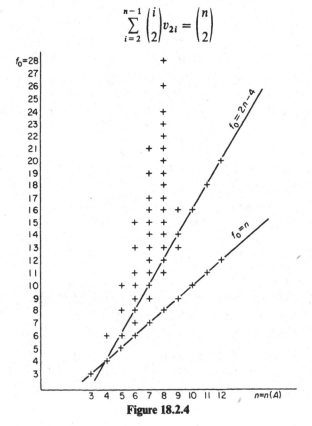

Figure 18.2.4

and the inequalities

$$\sum_{i \geq j} iv_{2i} \leq n + \binom{v^{(j)}}{2} \quad \text{for} \quad j = n - 1, n - 2, \cdots, 2,$$

where

$$v^{(j)} = \sum_{i \geq j} v_{2i}.$$

Additional results in this direction (theorems 5 and 6) are due to Kelly–Moser [1]:

5. *If $n(\mathscr{A}) = n$ and if k is a positive integer such that $v_{2i}(\mathscr{A}) = 0$ for $i > n - k$ and $2n \geq 3(3k - 2)^2 + 3k - 1$, then*

$$f_0(\mathscr{A}) \geq kn - \tfrac{1}{2}(3k + 2)(k - 1).$$

It is obvious that if $n(\mathscr{A}) = n$ and $v_{2n-2}(\mathscr{A}) > 0$, then $v_{2n-2}(\mathscr{A}) = 1$, and $\mathscr{A}$ consists of $n - 1$ lines passing through one point, and one line not through this point. Let this 2-arrangement be denoted $\mathscr{A}_n$. For $k = 2$ theorem 5 implies the following analogue of theorem 10.2.2:

6. *If $n(\mathscr{A}) = n \geq 27$ and if $\mathscr{A}$ is not $\mathscr{A}_n$, then*

$$f_0(\mathscr{A}) \geq 2n - 4.$$

This estimate for $f_0(\mathscr{A})$ is best possible as shown by the 2-arrangement consisting of $n - 3$ 'parallel' lines, two 'parallel' lines (not parallel to the $n - 3$ lines), and the line 'at infinity'. The condition $n \geq 27$ is too restrictive; for example, the estimate is valid for $n = 5, 10, 11$; it is probably valid for all $n \geq 10$. However, for $n = 6, 7, 8, 9$ the best possible estimate is $f_0(\mathscr{A}) \geq 2n - 5$.

A graphic illustration of theorem 6 is shown in figure 18.2.4, in which the possible pairs $(n(\mathscr{A}), f_0(\mathscr{A}))$ are indicated. (The representation is complete for $n \leq 8$, and in the region $f_0 \leq 2n - 4$ for $n \leq 11$.) The 'gaps' at points $(n, \binom{n}{2} - 1)$ and $(n, \binom{n}{2} - 3)$ are not accidental; the reader is invited to prove that no 2-arrangement with $n(\mathscr{A}) = n$ satisfies $f_0(\mathscr{A}) = \binom{n}{2} - 1$ or $f_0(\mathscr{A}) = \binom{n}{2} - 3$. The following conjecture is suggested by figure 18.2.4: For every integer k such that

$$2n - 4 \leq k \leq \binom{n}{2} - 4,$$

there exists a 2-arrangement $\mathscr{A}$ satisfying $n(\mathscr{A}) = n$ and $f_0(\mathscr{A}) = k$.
For a similar phenomenon in a related field see Steinitz [7].

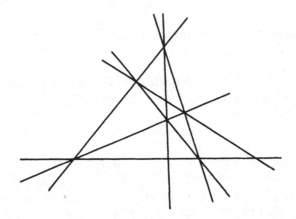

Figure 18.2.5

A famous problem of Sylvester [1] may be formulated as follows:
Is $v_4(\mathscr{A}) > 0$ for every 2-arrangement $\mathscr{A}$?
The affirmative answer to Sylvester's problem was surprisingly late
in coming (see Motzkin [3], Kelly–Moser [1] for the interesting history
of the problem and for references to earlier papers; for higher-dimensional
analogues see Hansen [1] and Bonnice–Edelstein [1]). Kelly–Moser [1]
established the following stronger result:

7. *For every 2-arrangement $\mathscr{A}$,*

$$v_4(\mathscr{A}) \geq \tfrac{3}{7}n(\mathscr{A}).$$

As shown by the 2-arrangement in figure 18.2.5 (with $n(\mathscr{A}) = 7$,
$v_4(\mathscr{A}) = 3$, this is the best possible estimate of the form $v_4(\mathscr{A}) \geq \mathrm{cn}(\mathscr{A})$.
It may be conjectured that $v_4(\mathscr{A}) \geq [\tfrac{1}{2}n(\mathscr{A})]$. This conjecture is true at
least for $n(\mathscr{A}) \leq 9$. On the other hand, the following example of T. S.
Motzkin shows that for even $n(\mathscr{A})$ it is best possible, and incidentally
refutes a conjecture of Erdös [4]: If $\mathscr{A}$ consists of the k lines determined

by the edges of a regular k-gon, and the k axes of symmetry of the k-gon, then $n(\mathscr{A}) = 2k$ and it is easy to see that $v_4(\mathscr{A}) = k$. The conjectured inequality is certainly not best for all odd $n(\mathscr{A})$; thus, for example, it may be shown that if $n(\mathscr{A}) = 9$ then $v_4(\mathscr{A}) \geq 6$.

For an application of theorem 7 to the classification of zonohedra see Coxeter [3]. For some related questions concerning the v_{2i}'s see chapter 4 of Ball [1].

A sequence $(p_k \,|\, k \geq 3)$ of nonnegative integers will be called *projectively realizable* provided there exists an integer n and a simple 2-arrangement $\mathscr{A}$ of n lines such that $p_k(\mathscr{A}) = p_k$ for all k. As in the analogous problem concerning polytopes (see section 13.3), it may be asked whether for every sequence $(p_k \,|\, 3 \leq k \neq 4)$ of nonnegative integers satisfying

(*) $$p_3 = 4 + \sum_{k \geq 5} (k - 4)p_k,$$

there exists a value of p_4 such that the sequence $(p_k \,|\, k \geq 3)$ is projectively realizable.

The following affirmative answer to this problem was mentioned without proof in a footnote in Eberhard [3]:

8. *For every finite sequence $(p_k \,|\, 3 \leq k \neq 4)$ of nonnegative integers satisfying* (*) *there exists values of p_4 such that the sequence $(p_k \,|\, k \geq 3)$ is projectively realizable. Moreover, for every such (p_k) there are only finitely many nonequivalent simple 2-arrangements $\mathscr{A}$ with $p_k(\mathscr{A}) = p_k$ for $k = 3, 5, 6, 7, \cdots$.*

The proof of the last part of the theorem follows at once from theorem 2, which limits the number of lines in $\mathscr{A}$ to at most p_3. In order to prove the first part of theorem 8, we first note that the assertion is obvious if $p_3 = 4$. Then $\mathscr{A}$ may have either 3 or 4 lines, and p_4 may be 0 or 4. In the sequel we shall assume $p_3 \geq 5$ and hence (by (*)) $p_k \geq 1$ for some $k \geq 5$. The idea of the following construction is to have available, at each step, a 'desirable' pair of lines. By this we mean a pair of lines such that one of the 'halves' of the projective plane determined by them consist only of quadrangles and of 2 triangles which have the intersection-point V of the two lines as a common vertex (see the solid lines in figure 18.2.6). If a 2-arrangement containing a 'desirable' pair of lines is given, it is possible to introduce an additional line (dashed in figure 18.2.6) the only effect of which, besides increasing the number of quadrangles, is to

split an 'outer' k-gon containing V into triangle and a $(k + 1)$-gon.
Note that the enlarged 2-arrangement again contains 'desirable' pairs,
neighboring to the $(k + 1)$-gon, and neighboring to quadrangles. Therefore, given a sequence $(p_j \mid 3 \le j \ne 4)$ satisfying (*), we start from the second
2-arrangement in figure 18.1.1 and choose a 'desirable' pair of lines.
By successive addition of $k - 4$ lines we transform one of the quadrangles
containing V into a k-gon (and introduce, besides quadrangles, $k - 4$
new triangles). Repeating this procedure with other 'desirable' pairs of
lines, we easily arrive at a projective realization of the given sequence (p_k).
This completes the proof of theorem 8.

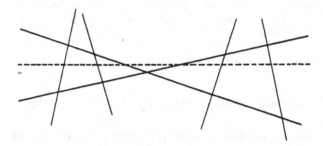

Figure 18.2.6

In contrast to the analogous situation regarding 3-polytopes, the
construction of the above proof yields a definite number of lines in $\mathscr{A}$
(namely p_3) and a definite value of $p_4 = \binom{p_3}{2} + 1 - \sum_{k \ne 4} p_k$.

However, the same sequence $(p_k \mid 3 \le k \ne 4)$ may be projectively
realizable also for other values of p_4 and n. Examples of such 'ambiguous
realizability' are evident from figure 18.1.1. It is not known whether the
set of all values of n for which a given sequence $(p_k \mid 3 \le k \ne 4)$ is realizable
by simple 2-arrangements with n lines, consists in each case of consecutive integers.

Some additional information on the possible values of n is contained
in the next theorem; its first part is due to N. G. Gunderson (see Carver [1];
Gunderson's proof seems not to have been published).

9. *Let $\mathscr{A}$ be a simple 2-arrangement with n lines. If $\mathscr{A}$ contains a p-gon
and a q-gon, then $p + q \le n + 4$; if $\mathscr{A}$ contains a p-gon, a q-gon, and
an r-gon then $p + q + r \le n + 9$.*

The proof of the first part is rather straight forward on observing that if $\mathscr{A}$ contains 5 lines each of which contains an edge of each of two fixed 2-faces of $\mathscr{A}$, then those 5 lines determine a 2-arrangement with at least two pentagons; but the combinatorially unique 2-arrangement of 5 lines (figure 18.1.1) contains only one pentagon. It follows that the p-gon and the q-gon of $\mathscr{A}$ have at most 4 common lines, hence $p + q - 4 \leq n$.

The second part of theorem 9 may be established by similar (but lengthier) arguments, involving configurations of 7 or fewer lines; we shall not give it here.

It would be interesting to generalize theorem 9 to the situation where $\mathscr{A}$ has as faces a q_1-gon, a q_2-gon, $\cdots$, a q_k-gon. Probably there exists an estimate of the form

$$\sum_{i=1}^{k} q_i \leq n + a(k),$$

where $a(k)$ depends only on k. From the above, we have $a(1) = 0, a(2) = 4$, $a(3) = 9$; it seems likely that $a(4) = 14$. The 2-arrangement of figure 18.2.2 shows that $a(5) \geq 20$ and $a(6) \geq 26$.

It would be interesting to find an existence theorem analogous to theorem 8, but dealing with 2-arrangements which are not necessarily simple.

Considering higher-dimensional analogues of theorem 2, using a method similar to that applied in the proof of theorem 1 it is easy to establish the following result (Eberhard [1] for $d = 3$):

10. *For every simple d-arrangement of n hyperplanes at least n of the d-faces are d-simplices.*

Theorem 10 is probably true for all d-arrangements.

18.3 Generalizations

Many properties of d-arrangements, and of simple d-arrangements, are topologically invariant in the sense that they remain valid if hyperplanes are replaced by $(d - 1)$-varieties homotopic to hyperplanes, provided their intersections are homotopic to the corresponding intersections of hyperplanes. For example, all theorems of section 18.1 remain valid for such 'generalized arrangements'. We shall examine more closely only the special case $d = 2$.

Following Levi [1], a system of simple closed curves in the projective plane is called an *arrangement* of *pseudo-lines* provided

(i) Each two curves have precisely one point in common, each of them crossing the other at this point;

(ii) there is no point in common to all the curves.

Arrangements of pseudo-lines have many properties analogous to the properties of 2-arrangements of lines (Levi [1]). For example, it is possible to define 2-*faces* of such arrangements, and they are topological 2-cells; according to the number of pseudo-lines incident to such a 2-face, it is a triangle, quadrangle, etc. Relation (*) of the preceding section holds, and theorem 18.2.2 has a valid analogue: In every arrangement of n pseudo-lines there are at least n triangles.

On the other hand, not every arrangement of pseudo-lines is *stretchable*, i.e. combinatorially equivalent to a 2-arrangement of lines[†]. This fact, mentioned already by Levi [1], was established by examples in Ringel [1]. (See Ringel [2] for some related questions.) We venture the following:

CONJECTURE *There exists an integer n_0 such that an arrangement $\mathscr{A}$ of pseudo-lines is stretchable whenever each sub-arrangement of $\mathscr{A}$ containing at most n_0 pseudo-lines is stretchable.*

Another series of problems arises in connection with the maxima and minima of $p_k(\mathscr{A})$ for arrangements of n pseudo-lines; in particular, are those extrema the same as the corresponding extrema for 2-arrangements of lines? Are all the extremal arrangements stretchable?

Similar questions may be asked regarding the f-vectors and the sequences $(v_4, v_6, \cdots)$.

There is an interesting unsolved problem about the relation of d-arrangements and $(d+1)$-polytopes. Since the problem is open already for $d = 2$, we shall satisfy ourselves with a formulation for this special case.

Any 2-arrangement of lines in the projective plane is naturally associated with a family of planes through the origin in the Euclidean 3-space R^3, the 2-faces of the 2-arrangement being in a bi-unique correspondence with the pairs of antipodal 3-dimensional cones into which R^3 is split

[†]R. Klee [1] gives an invalid argument to show that every arrangement of pseudo-lines is stretchable. The existence of nonstretchable simple arrangements of pseudo-lines implies, among other consequences, that R. Klee [1] counted (in principle) the number of combinatorial types of simple arrangements of pseudo-lines, and not of lines. It is not known whether every simple arrangement of 8 pseudo-lines is stretchable.

by the planes. (For such systems of cones and for related topics and additional references see, for example, Cover [1], Cover–Efron [1], Samelson–Thrall–Wesler [1].) The problem is whether every system of cones obtained in this way may also be obtained by taking a suitable 3-polytope with center 0 and by considering the cones with vertex 0 spanned by the 2-faces of the polytope. If the answer to this should be negative, does it matter if the system of cones arising from the 3-polytope is only required to be combinatorially equivalent to the system of cones obtained from the 2-arrangement? (See Supnick [1] for a related problem.)

It would be interesting to investigate properties of d-arrangements (or 'generalized arrangements') analogous to the properties of polytopes we considered in chapters 11, 12, 13. In particular, what is the analogue for 2-arrangements of Steinitz's theorem 13.1.1; in other words, what graphs imbeddable in the projective plane may be realized by (simple) 2-arrangements?

d-Arrangements may be generalized also in another direction; indeed, this generalization was considered already by Steiner [1].

In Euclidean d-space R^d, a finite family of $(d-1)$- spheres defines in an obvious way a *spherical d-arrangement*. Using the fact that the intersection of k different $(d-1)$-spheres is either a sphere of lower dimension, or a point, it is possible to obtain results analogous to those in section 18.1. (Pólya [2], pp. 224–225). A complication arises from the fact that even for simple spherical d-arrangements $\mathscr{A}$ of n $(d-1)$-spheres, the number $\varphi_k(\mathscr{A})$ of k-cells depends not only on d, n, and k, but varies with $\mathscr{A}$. However, using a simple variant of the inductive approach mentioned in connection with theorems 18.1.1 and 18.1.3, it is possible to show that $\varphi_k^d(n) \leq 2f_k^d(n)$, where $\varphi_k^d(n) = \max\{\varphi_k(\mathscr{A}) \mid \mathscr{A}$ a simple spherical d-arrangement with n spheres$\}$. But even the stronger result $\varphi_k^d(n) = 2f_k^d(n)$ holds.

Indeed, taking any simple d-arrangement $\mathscr{A}$ of n hyperplanes such that $f_k(\mathscr{A}) = f_k^d(n)$, we consider the projective d-space as the d-sphere in R^{d+1}, with antipodal points identified. Disregarding the identification, we obtain an arrangement on S^d determined by n 'great $(d-1)$-spheres', with $2f_k^d(n)$ k-cells. Applying a stereographic projection of S^d onto a Euclidean d-space, the n 'great $(d-1)$-spheres' project onto n $(d-1)$-spheres in R^d; this spherical d-arrangement obviously has $2f_k^d(n)$ k-cells. (For partial results in case $d = k = 3$, see Ruderman [1]; for the general case see Rényi–Rényi–Surányi [1].)

Clearly, most of the problems on arrangements of hyperplanes have valid analogues for spherical d-arrangements. Among problems specific for spherical d-arrangements we mention only the following one, which seems to be open even for $d = 2$.

Is every simple spherical d-arrangement $\mathscr{A}$ of n $(d - 1)$-spheres, with the property $\varphi_k(\mathscr{A}) = \varphi_k^d(n)$, combinatorially equivalent to one which is obtainable by stereographic projection from a d-arrangement of hyperplanes in the projective space?

18.4 Additional notes and comments

Arrangements and zonotopes.
The "interesting unsolved problem" posed on p. 408 has a rather simple positive solution, which establishes that arrangements and zonotopes are equivalent in a very strong sense. Let $\mathscr{A} = \{H_1, \ldots, H_n\}$ be a d-arrangement of $n \geq d+1$ hyperplanes in projective space P^d. Associated with this is an arrangement $\mathscr{A}' = \{H_1', \ldots, H_n'\}$ of n linear hyperplanes in R^{d+1}. (Such an arrangement of hyperplanes through the origin in a Euclidean space is known as a *central* arrangement.) The correspondence is such that each non-empty k-face of $\mathscr{A}$ corresponds to two opposite faces of $\mathscr{A}'$, which are $(k+1)$-dimensional cones.

Let the hyperplanes $H_i' = \{x \in R^{d+1} : \langle z_i, x \rangle = 0\}$ be given in terms of orthogonal unit vectors $z_i \in R^{d+1}$. Then we get a zonotope of dimension $d+1$, associated with the arrangement $\mathscr{A}$, by

$$\mathscr{Z}(\mathscr{A}) := \left\{ \sum_{i=1}^{n} \lambda_i z_i : -1 \leq \lambda_i \leq 1 \text{ for } 1 \leq i \leq n \right\}.$$

The dual $\mathscr{Z}^*$ of this zonotope "spans" the hyperplane arrangement $\mathscr{A}'$ in the way required for the problem on page 408: For this it suffices to note that any two vectors $a, a' \in R^{d+1}$ lie in the same face of $\mathscr{A}'$ if and only if the linear objective functions $\langle a, x \rangle$ and $\langle a', x \rangle$ are maximized on the same face of $\mathscr{Z}$. Thus $\mathscr{A}'$ is the normal fan for $\mathscr{Z}$ (as defined in the notes in section 3.6). Equivalently, the faces of $\mathscr{Z}^*$ span the (closed) faces of $\mathscr{A}'$: The $(k+1)$-faces of $\mathscr{A}'$ are exactly the cones $\text{cone}(F)$ generated by the k-faces F of $\mathscr{Z}^*$ for $k \geq 0$.

With this set-up, there is a bijection between d-arrangements and $(d+1)$-dimensional zonotopes (up to normalization of the lengths of the zones). Here simplicial arrangements correspond to simple zonotopes, while "simple" arrangements (as defined on page 391) correspond to cubical zonotopes.

The conjecture on page 397 was proved by Björner–Edelman–Ziegler [a]. See also the notes in section 12.4.

Line arrangements, pseudo-line arrangements, and oriented matroids.
Grünbaum's [e] own work "Arrangements and Spreads" has been extremely influential for research about line and pseudo-line arrangements. A current survey with many references is Goodman [a].

Substantial theoretical backing for this has arisen since the seventies in the form of oriented matroid theory. Pseudo-line arrangements are in a very precise sense equivalent to oriented matroids of rank 3; line arrangements correspond to the special case of "realizable" oriented matroids. In the same way, general oriented matroids correspond to a well-defined concept of "pseudo-hyperplane arrangements"; the case of realizable oriented matroids corresponds to real hyperplane arrangements. (The basic objects for this are central hyperplane arrangements in which a "positive side" has been fixed for each hyperplane.) This is also the proper setting for results such as Zaslavsky's theorem [a] that the number of (bounded) regions of an affine hyperplane arrangement is determined by the underlying (not oriented) matroid; see Las Vergnas [a]. We refer to the comprehensive discussion in Björner et al. [a], as well as to Bokowski [a], Richter-Gebert–Ziegler [b], and Ziegler [a, Lect. 7].

The Sylvester–Gallai problem.

Sylvester's problem on page 404 (solved by Gallai) has recently been generalized by Chvátal [a] to finite metric spaces (with a proof for metrics induced by connected finite graphs). The related question whether every simple arrangement of n pseudo-lines in the projective plane has at least $[n/2]$ simple intersections (by Grünbaum [c]) is far from being resolved. The best results up to now are due to Czima–Saywer [a]. (See also Fukuda–Finschi [a].)

Simplicial arrangements, simplicial regions.

Simplicial arrangements (i. e., hyperplane arrangements in which all regions are simplicial) have been studied intensively. For $d = 2$, infinite classes may be derived from regular n-gons, and in higher dimensions from finite reflection groups (see the notes in section 19.4), and by taking "direct sums". An unresolved conjecture states that beyond this, there are only finitely many "sporadic" simplicial arrangements—as listed by Grünbaum [c] and Grünbaum–Shephard [c] (with only few later additions, see Wetzel [a]).

The maximal number of triangles in a line resp. pseudo-line arrangement has been the object of intensive studies. In particular, Table 18.2.1 can be extended by $p_3^s(15) = 65$ and $p_3^s(16) = 80$. We refer to Goodman [a, Sect. 5.4].

The claim of theorem 18.2.10 is true for all hyperplane arrangements: There are at least n simplicial regions in every d-arrangement of n hyperplanes. However, Shannon's [a] proof relies on metric constructions that are not available in the more general setting of oriented matroids; see Roudneff–Sturmfels [a]. A still unresolved conjecture of Las Vergnas states that every pseudo-arrangement must contain *at least one* simplicial region. Richter-Gebert [a] produced pseudo-arrangements with "few" simplicial regions.

More on arrangements.

There has been extensive work in computational geometry on the combinatorial complexity of arrangements of curves, surfaces, hyperplanes, etc., by Sharir and others—see Agarwal–Sharir [a].

Complex hyperplane arrangements are discussed in Orlik–Terao [a]. For more general subspace arrangements see Björner [c] and Vassiliev [a].

CHAPTER 19

Concluding Remarks

In the three sections of this, the last, chapter we shall discuss a number of combinatorial-geometric topics which are either directly concerned with polytopes or else, though meaningful for a larger class of sets, reduce in their most interesting aspects to polytopes. Clearly, the number of such topics could be increased almost without bounds, and the selection of those dealt with here reflects only the author's predilections and interests.

Regarding two omitted topics the author feels obliged to explain their absence, and to provide at least some references to the literature.

The first such topic started with Cauchy's [1] famous 'rigidity theorem' and has developed, through interaction with differential geometry, to a large body of knowledge concerning the metric structure of polytopes. Aleksandrov's book [2] gives an excellent and thorough account of the whole field, making superfluous our dwelling on it. (For a recent contribution to the subject see Stoker [1].)

Another topic, even the briefest meaningful discussion of which would by far transcend the place at our disposal, deals with packings, space-fillings, and related subjects. For parts of this material the reader is referred to Fejes Tóth [1,3] and Rogers [1], though even this does not do full justice to the far-flung ramifications of the subject. As one topic not discussed in the above references we mention the following fascinating problem of Keller [1,2] which is, apparently, open for $d \geq 7$ (Perron [1]):

Does there exist, for every space-filling of R^d by congruent d-cubes, a pair of cubes having a common facet?

19.1 Regular Polytopes and Related Notions

Regular polytopes, and different kinds of semiregular polytopes, have been a topic of investigation since antiquity, and during the centuries led to many interesting and important notions and results.

A number of different definitions of regularity are frequently used.

According to the *inductive* definition (see, for example, Coxeter [1], Fejes Tóth [3]) a *d*-polytope is *regular* provided all its facets and all its vertex figures are regular $(d - 1)$-polytopes. (*Vertex figure* is here understood in a more restricted sense than in exercise 3.4.8: the hyperplane determining the vertex figure at a vertex passes through the midpoints of the edges incident to the vertex.)

Another definition (Coxeter [1,5]), equivalent to the former, is: A *d*-polytope $P \subset R^d$ is *regular* provided for every $k, 0 \le k \le d - 1$, and for every $(k + 1)$-face F^{k+1} and $(k - 1)$-face F^{k-1} incident with F^{k+1}, there exists a *symmetry* of P (i.e. an orthogonal transformation T of R^d mapping P onto itself) such that the two *k*-faces of P incident to both F^{k+1} and F^{k-1} are mapped onto each other. (Obviously this implies that for each two *k*-faces of P there is a symmetry of P interchanging them.)

For other definitions of regularity see, for example, Du Val [1], N. W. Johnson [3]. All those definitions are equivalent for (convex) polytopes; most of them are also suitable for nonconvex polytopes, or for tessellations. For detailed accounts of regular polytopes and similar objects, and their properties and history, the reader is referred in particular to Coxeter [1] and Fejes Tóth [3]. (For a characterization of regular nonplanar polygons see Efremovič–Il'jašenko [1].) We mention only the fact that the family of regular polytopes consists, in addition to the infinitely many regular polygons ($d = 2$), of 5 'Platonic solids,' 6 regular 4-polytopes, and just 3 regular *d*-polytopes for each $d \ge 5$ (*d*-simplex, *d*-cube, *d*-octahedron).

There are, nevertheless, many open questions in connection with regular polytopes and allied notions. The aim of the present section is to mention some of the relevant known results, and to formulate a number of still unsolved problems.

Using the second of the above definitions of regularity, Coxeter [5] defined *affinely regular* polytopes by substituting 'affine symmetry' for 'symmetry'. This means that T is allowed to be any affine transformation mapping P onto itself. As pointed out by Coxeter (private communication) it is not hard to show that every affinely regular polytope is affinely equivalent to a regular polytope. The proof uses well known results on finite groups of affine transformations, and is related to theorem 2.4 of Danzer, Laugwitz and Lenz [1].

Similarly, by allowing T to be any projective transformation permissible for P, one may define 'projective symmetries' and *projectively regular* polytopes. Also, using automorphisms of the lattice $\mathscr{F}(P)$ of all faces of P instead of the transformations T, combinatorial symmetries and *combinatorially regular* polytopes may be defined.

It may be conjectured that each projectively (or combinatorially) regular polytope is projectively (or combinatorially) equivalent to a regular polytope. This conjecture may easily be verified for $d \le 3$ (see, for example, Aškinuze [2]), but it seems not to have been solved in higher dimensions.

. The very stringent requirements of the definition of regularity may be (and have been) relaxed in many different ways, yielding a great variety of 'semi-regularity' notions. There is a remarkable lack of completeness in the results, possibly explainable in part by the emphasis many authors place on nonconvex polytopes. We shall mention some of the classes of (convex) polytopes which were investigated; however, the reader should beware of the differing terminologies used by different authors.

A d-polytope P is called *semiregular* (Gosset [1], Coxeter [1, p. 162]) if the facets of P are regular and the vertices of P are *equivalent* (i.e. the group of symmetries of P acts transitively on the vertices of P).

A d-polytope P is called *uniform* (N. W. Johnson [3, p. 31]) if it is regular, or if $d \ge 3$, the facets of P are uniform, and the vertices of P are equivalent.

For $d = 3$ the semiregular (and the uniform) polytopes coincide with the *Archimedean solids*; it is well known that, except for the 5 Platonic solids and the infinite families of 'regular' prisms and antiprisms, there exist 13 Archimedean solids. (Some authors call only those last 13 3-polytopes 'Archimedean solids'.) The Archimedean solids may also be defined by the requirements of equal edges and equivalent vertices. It is remarkable that if the definition is only slightly changed by substituting 'all vertex figures are congruent' for the transitivity of the symmetry group, the Archimedean solids are not the only 3-polytopes allowed (Ball [1], p. 137, Aškinuze [1]). A Schlegel diagram of the only additional polytope is shown in figure 19.1.1.

Though various construction of semiregular and uniform d-polytopes in dimensions $d \ge 4$ have been described (see, for example, Stott [1], and Coxeter [1], where many additional references are given), even for $d = 4$ it is not known whether the enumeration is complete (for any of these classes).

No serious consideration seems to have been given to polytopes in dimensions $d \ge 4$ about which transitivity of the symmetry group is assumed only for faces of suitably low dimensions, and regularity or some variant of it is required only for faces of dimensions $\le d - 2$.

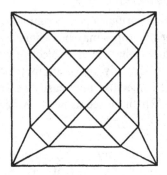

Figure 19.1.1

Similarly uninvestigated are the affine, projective, or combinatorial variants of semi-regularity, uniformity, or the related notions ($d \geq 4$). For $d = 3$ see, for example, Aškinuze [2].

Other generalizations may be derived from the first of the definitions of regularity given above. For example, a d-polytope, $d \geq 3$, is *regular-faced* provided all its facets are regular. The regular-faced 3-polytopes have recently received considerable attention (Freudenthal–van der Waerden [1], N. W. Johnson [1, 2], Zalgaller [1], Zalgaller *et al.* [1], Grünbaum–Johnson [1]). Clearly, all Archimedean solids are regular-faced, and it is not hard to show (N. W. Johnson [1], Zalgaller [1], Grünbaum–Johnson [1]) that there exists only a finite number of regular-faced 3-polytopes which are not Archimedean. N. W. Johnson [2] has found 92 such exceptional regular-faced 3-polytopes. Most of them are parts of Archimedean solids, or unions of smaller regular-faced 3-poly-topes, but 8 of his polytopes cannot be obtained in this way; Schlegel diagrams of three such polytopes are shown in figure 19.1.2. Johnson's list is probably complete, but this has not been established so far. Completely uninvestigated is the problem of higher-dimensional regular-faced poly-topes. An interesting by-product of Johnson's list is his conjecture (true in all known instances) that the group of symmetries of each regular-faced 3-polytope is nontrivial.

A different family of d-polytopes, $d \geq 3$, results by requiring all the facets to be congruent. Clearly the polars of Archimedean solids are such *congruent-faced* 3-polytopes, as are certain rhombic zonohedra (see Coxeter [1], Bilinski [1]). An amusing observation of Steinhaus (see Grünbaum [3]), which could probably be greatly strengthened, is: Each congruent-faced 3-polytope has an even number of facets.

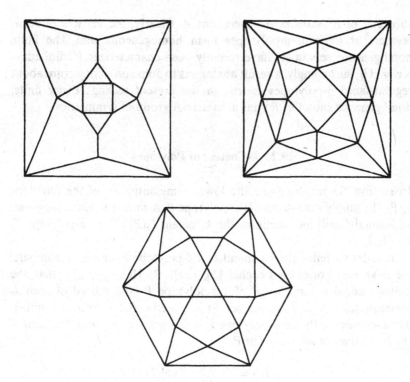

Figure 19.1.2

It would be interesting to investigate whether, in analogy to Johnson's conjecture about regular-faced 3-polytopes, each congruent-faced 3-polytope has a nontrivial symmetry group.

Relaxing the requirement of congruence of facets to that of affine, projective, or combinatorial equivalence, successively larger families of polytopes are obtained. Very interesting results about higher-dimensional polytopes with these properties have been recently obtained by Perles–Shephard [2]; their methods will probably be applicable to many other questions.

For another interesting family of d-polytopes which satisfy certain regularity conditions see Polák [1].

A subset A of a metric space with distance function ρ is said to be *homogeneous* provided for every $a_1, a_2, a_3 \in A$ there exists $a_4 \in A$ such that $\rho(a_1, a_2) = \rho(a_3, a_4)$. It is easily checked that each finite homogeneous subset A of R^d satisfies $A = \text{vert conv } A$. Clearly, the vertices of semi-regular, or uniform d-polytopes form homogeneous sets. The finite homogeneous sets in R^2 have recently been characterized (Grünbaum–Kelly [1]); surprisingly, a result analogous to Johnson's conjecture about regular-faced 3-polytopes is true in the present setting: Every finite, homogeneous subset of R^2 has a nontrivial group of symmetries.

19.2 k-Content of Polytopes

In section 5.3 we discussed the lower semicontinuity of the functions $f_k(P)$, the number of k-faces of the polytope P. A similar result holds for an additional family of functions, the k-content $\zeta_k(P)$ of the d-polytope P, $k = 1, 2, \cdots, d$.

In order to define the k-content of a d-polytope P we recall (compare, for example, Bonnesen–Fenchel [1], chap. 7, Hadwiger [5]) that the d-dimensional volume $\text{vol}_d P$ of a d-polytope P (and indeed of each d-dimensional compact convex set P) is a well-defined positive number. The k-content $\zeta_k(P)$ of a d-polytope P is defined, for each k, as the sum of the k-volumes of all k-faces of P,

$$\zeta_k(P) = \sum_{F \text{ a } k\text{-face of } P} \text{vol}_k F.$$

The function $\zeta_{d-1}(P)$ may easily be defined for all d-dimensional compact convex sets P; it is the $(d-1)$-dimensional 'surface area' of P.

It is well known (see, for example, Hadwiger [5]) that the functions $\zeta_d(P)$ and $\zeta_{d-1}(P)$ depend continuously (in the Hausdorff metric) on P, not only if P ranges over d-polytopes but even if it ranges over all compact convex sets of dimension d. However, it was only recently that the following generalization was established (Eggleston–Grünbaum–Klee [1]):

1. *If a polytope P is the limit of the sequence $P_1, P_2, \cdots$ of polytopes, then, for each k,*

$$\zeta_k(P) \leq \liminf \zeta_k(P_n).$$

In other words, all the functions $\zeta_k(P)$ depend in a lower semicontinuous manner on the polytope P.

In view of theorem 1, it is possible to extend the definition of k-content to all d-dimensional compact convex sets K by putting

$$\zeta_k(K) = \liminf \zeta_k(P_n),$$

for all sequences $P_1, P_2, \cdots$ of d-polytopes converging to K. It is then easily shown that the extended functions $\zeta_k(K)$ depend in a lower semi-continuous manner on the compact convex set K. Clearly, $\zeta_k(K) = \infty$ is possible for $k \leq d - 2$; even in the case $d = 3$ and $k = 1$, no geometric characterization is known for the family $\mathscr{C}_k^d$ of all d-dimensional compact convex sets K such that $\zeta_k(K) < \infty$. (Obviously all d-dimensional compact convex sets belong to $\mathscr{C}_d^d = \mathscr{C}_{d-1}^d$.)

In analogy to the f-vectors of polytopes, one could consider d-tuples of the form $(\zeta_1(P), \zeta_2(P), \cdots, \zeta_d(P))$, where P is a d-polytope or a member of $\mathscr{C}_1^d$. It is more convenient, however, to define the ζ-vector $\zeta(P)$ by

$$\zeta(P) = (\zeta_1(P), \zeta_2(P)^{\frac{1}{2}}, \cdots, \zeta_d(P)^{1/d}).$$

Then it is clear that the sets $\zeta(\mathscr{P}^d) = \{\zeta(P) \,|\, P \in \mathscr{P}^d\}$ and

$$\zeta(\mathscr{C}_1^d) = \{\zeta(K) \,|\, K \in \mathscr{C}_1^d\}$$

are cones with apices at the origin. The sets $\zeta(\mathscr{P}^d)$ and $\zeta(\mathscr{C}_1^d)$ are somewhat analogous to the so-called Blaschke diagrams (see, for example, Hadwiger [3]) of families of convex sets. While the complete description of the Blaschke diagram is not known for any interesting family of sets, in all cases that have been investigated the Blaschke diagram was found not to be a convex set. About the sets $\zeta(\mathscr{P}^d)$ and $\zeta(\mathscr{C}_1^d)$ even this is not known. One of the few affirmative results on those sets is the well known isoperimetric inequality (see, for example, Hadwiger [5])

$$\zeta_d(K)^{1/d} \leq \frac{1}{d\omega_d^{1/d}} \zeta_{d-1}(K)^{1/(d-1)},$$

where ω_d is the d-volume of the unit ball of R^d and $K \in \mathscr{C}_d^d$. The isoperimetric inequality establishes parts of the boundaries of the sets $\zeta(\mathscr{P}^d)$ and $\zeta(\mathscr{C}_1^d)$. Additional information about those cones is contained in the following result of Eggleston–Grünbaum–Klee [1]:

2. *Whenever $1 \leq k \leq s \leq d$ and either $s = d$, or $s = d - 1$, or k divides s, there exists a minimal finite constant $\gamma(d; k, s)$ such that*

$$\zeta_s(K)^{1/s} \leq \gamma(d; k, s)\zeta_k(K)^{1/k}$$

for all $K \in \mathscr{C}_k^d$.

The isoperimetric inequality shows that $\gamma(d; d - 1, d) = 1/d\omega_d^{1/d}$. The only other result known in this direction is $\gamma(3; 1, 2) \leq (6\pi)^{-\frac{1}{2}}$ (Aberth [1]). Melzak [3] conjectured that $\gamma(3; 1, 3) = 2^{-2/3} 3^{-11/6}$, with equality only for the Archimedean 3-sided prism.

It is rather remarkable that even the finiteness of $\gamma(d; k, s)$ is still undecided in all cases not covered by theorem 2.

(A number of similar problems have been frequently considered. Typical of those variants is the following conjecture of Fejes Tóth [1], established by Besicovitch–Eggleston [1]: Of all 3-polytopes which contain a given sphere, the cube has the smallest value of ζ_1. For additional results, and for references to the literature, see Fejes Tóth [1, 3, 4].)

It may be conjectured that the constants $\gamma(d; k, s)$ remain unchanged if $\mathscr{P}^d$ is substituted for $\mathscr{C}_k^d$ in their definition. At least for $s = d$ and $s = d - 1$ it seems probable that equality holds for some d-polytope K in the inequality defining $\gamma(d; 1, s)$. However, even for $d = 3$ this question seems still to be open.

The problem of the constants $\gamma(d; k, s)$ may be refined by considering in their definition—instead of all members of $\mathscr{C}_k^d$, or all d-polytopes— only d-polytopes having a specified number of vertices, or of facets, or which are of a given combinatorial type. Even for the classical isoperimetric problem ($d = 3$, $k = 2$, $s = 3$) many of those questions are still open, despite a variety of contributions and partial solutions ranging from Lhuilier [1] (in 1782) and Steiner [2], through Minkowski [2] and Steinitz [8] to many new results (see, for example, Fejes Tóth [3]).

In particular, no characterization of those combinatorial types of 3-polytopes which have an isoperimetrically 'best' representative is known.

19.3 Antipodality and Related Notions

More than thirty years ago Borsuk [1] formulated the following problem, which rivals the 'four-colors problem' in simplicity of statement and intuitive appeal, as well as in its apparent hopelessness.

Is every bounded set $A \subset R^d$ representable in the form

$$A = \bigcup_{i=0}^{d} A_i,$$

where $\operatorname{diam} A_i < \operatorname{diam} A$ for $i = 0, \cdots, d$?

Borsuk's problem has given rise to, and is connected with, many other problems of a combinatorial-geometric nature. Affirmative solutions of it are known only in some special cases; among them we mention the following two:

(i) $d \leq 3$;

(ii) K is a convex set with sufficiently smooth boundary. (For a summary of known results on Borsuk's problem and related questions, and for references to the quite extensive literature, see Grünbaum [10].)

A somewhat sharper formulation of Borsuk's problem is:

Determine the least β_d such that every bounded set $A \subset R^d$ is representable in the form

$$A = \bigcup_{i=0}^{d} A_i$$

where diam $A_i \leq \beta_d$ diam A for $i = 0, \cdots, d$.

Clearly, $\beta_d < 1$ implies an affirmative solution of Borsuk's problem in R^d. The only information available on β_d is $\beta_2 = \frac{1}{2}\sqrt{3}$ and $\beta_3 < 0.9887$; it has been conjectured that $\beta_3 = \sqrt{(3 + \sqrt{3})/6} = 0.888 \cdots$. It is easy to show that the same numbers β_d would be obtained if in their definition A were restricted to finite sets, and even to sets of the type $A = \text{vert } P$, for all d-polytopes P.

Though this approach to Borsuk's problem has not yielded a solution, it led to a variety of interesting questions about polytopes. Some of them will be discussed presently.

The possibility of decomposing finite subsets of R^d into $d + 1$ parts of smaller diameter (however, without implying $\beta_d < 1$) results for $d = 2, 3$ from the following theorem:

If $A \subset R^d$ and card $A = n$, then the number of pairs $x, y \in A$ for which $\text{diam}\{x, y\} = \text{diam } A$ is at most n if $d = 2$, and $2n - 2$ if $d = 3$. (See Erdös [1] for $d = 2$, Grünbaum [1], Heppes [1], Straszewicz [2] for $d = 3$.)

For $d \geq 4$ the number of such pairs may exceed $[\frac{1}{4}n^2]$ (see Erdös [3]), and the result has no immediate implication for Borsuk's problem (except, possibly, to weaken the seemingly universal belief that the solution to Borsuk's problem is affirmative). The complete determination of the maximal possible number of such pairs in R^d, $d \geq 4$, is still outstanding; Erdös [3] has shown that it is of the form

$$\frac{1}{2}n^2 \left(1 - \frac{1}{[\frac{1}{2}d]}\right) + O(n^{2-\varepsilon})$$

for some $\varepsilon > 0$.

A related question is how many times any fixed distance can occur between the pairs of vertices of a d-polytope with n vertices. It is easy to see that for $d = 2$ there may be as many as $[5(n - 1)/3]$ such pairs, and this is probably the maximum. For $d = 3$ it has been conjectured that the maximal number of such pairs is $3n - 6$; however, in the regular dodecahedron (with $n = 20$ vertices) two different distances occur $60 = 3n$ times each. This naturally implies that for any k, more than $3n + k$ pairs can occur for sufficiently large n.

For additional results and problems of this type see Erdös [3], Altman [1].

Let K be a convex body in R^d and let $c(K)$ denote the least integer c with the property: There exist c translates of int K such that their union contains K. If $c(K) \leq d + 1$ it is clearly possible to cover K by $d + 1$ sets of smaller diameter. Well known procedures allow to show that an affirmative solution to Borsuk's problem would result if it were known that $c(K) \leq d + 1$ for every set of constant width. However, this question seems to be rather hopeless, and therefore some attention has been given to the problem of determining c_d, the maximal value of $c(K)$ for d-dimensional convex bodies K. (Levi [3], Hadwiger [6], Gohberg-Markus [1], Boltyanskii [1], Soltan [2]). The example of the d-cube shows that $c_d \geq 2^d$, and it has been conjectured that $c_d = 2^d$. This problem is also far from a solution (though for $d = 2$ it is easy to see that $c(K) = 3$ unless K is a parallelogram). Some small advances have been made in this direction; to describe them, we use the following notion due to V. Klee:

Let $K \subset R^d$, and let $x, y \in K$. We shall say that x and y are an *antipodal pair* of K provided there exists a pair of parallel (distinct) supporting hyperplanes of K such that x belongs to one of them and y to the other. A d-polytope is said to be *antipodal* provided each two of its vertices form an antipodal pair of the polytope. Clearly c_d is at least as great as $\kappa(d)$, the maximal number of vertices in an antipodal d-polytope.

It has been shown that $\kappa(d) = 2^d$ for each $d \geq 1$ (Danzer–Grünbaum [1]). This result implies also that $\varepsilon(d) = 2^d$ is the answer to the following problem of Erdös [2]: What is the maximal possible number $\varepsilon(d)$ of points in R^d such that all angles determined by triples of them are less than or equal to 90°?

Slightly generalizing the question of $\kappa(d)$ one is led to the problem of determining $e(d, n)$, the maximal number of antipodal pairs among the vertices of a d-polytope with n vertices. It is not hard to show that $e(2, n) = [3n/2]$ (Grünbaum [6]), and that

$$e(3, n) \geq [\tfrac{1}{2}n][\tfrac{1}{2}(n + 1)] + [\tfrac{1}{4}3n] + [\tfrac{1}{4}(3n + 1)].$$

It would be interesting to investigate whether the relation

$$\lim_{n \to \infty} \frac{e(d, n)}{n^2} = \frac{1}{2} - \frac{1}{2^{d-1}}$$

holds for all $d \geq 2$.

Let the notion of an antipodal pair be modified by defining a pair $x, y \in K$ as k-antipodal provided there exist parallel (distinct) supporting hyperplanes of K, each of which intersects K in a set of dimension at most k, such that x belongs to one of the hyperplanes, y to the other. In analogy to the above, we define k-antipodal polytopes and the numbers $\kappa_k(d)$ and $e_k(d, n)$. Clearly, a d-polytope is antipodal if and only if it is $(d - 1)$-antipodal. A number of interesting problems concern 0-antipodality. While it is easy to show that $\kappa_0(2) = 3$, the proof of $\kappa_0(3) = 5$ is rather involved (Grünbaum [6]). For $d \geq 4$, it is known that $\kappa_0(d) \geq 2d - 3$, and it has been conjectured that $\kappa_0(d) = 2d - 3$ (Danzer–Grünbaum [1]). As in the case of antipodal pairs, $\kappa_0(d)$ may be considered as the affine variant of the following Euclidean problem due to Erdös [2]: Determine the maximal possible number $\varepsilon_0(d)$ of points in R^d such that all the angles determined by triples of them are acute. Examples show that $\varepsilon_0(d) \geq 2d - 3$, and clearly $\varepsilon_0(d) \leq \kappa_0(d)$. In contrast to the situation in the case of $\varepsilon(d)$, it is not known whether $\varepsilon_0(d) = \kappa_0(d)$ for $d \geq 4$. (Direct proofs of $\varepsilon_0(3) = 5$ were given by Croft [1] and Schütte [1].)

Still less is known about $\kappa_k(d)$ for $k \geq 1$. Even the conjectured value $\kappa_1(3) = 6$ has not been ascertained so far.

The relationship between $\kappa(d)$ and c_d has an analogue in the relation of $\kappa_0(d)$ and the number $\bar{c}_d$ defined as follows: For a d-dimensional convex body K denote by $\bar{c}(K)$ the minimal number of proper translates of K the union of which covers K; then $\bar{c}_d$ is defined as the maximum of $\bar{c}(K)$ for all d-dimensional convex bodies K. It is easily seen that $\bar{c}_d \geq \kappa_0(d)$, but it is not known whether $\bar{c}_d = \kappa_0(d)$ for $d \geq 3$. As is easily verified, an affirmative solution of Borsuk's problem would be implied by a proof of the conjecture that $\bar{c}(K) = d + 1$ for every d-dimensional K of constant width.

Regarding $e_0(d, n)$ it is easy to prove that $e_0(2, n) = n$ (Grünbaum [6]), but even the 3-dimensional case seems to be very complicated.

The numbers $\kappa_0(d)$ and $e_0(d, n)$ are interesting also from a quite different point of view. If K is a d-polytope with $v = f_0(K)$ vertices, one may consider the possible relations between v and $f_0(K + (-K))$. Clearly $f_0(K + (-K)) \leq v(v - 1)$; however, equality may hold in this relation only if $v \leq \kappa_0(d)$. In general, $f_0(K + (-K)) \leq 2e_0(d, v)$, with equality if and only if K is a d-polytope with v vertices, which form $e_0(d, v)$ 0-antipodal pairs.

This interpretation of $e_0(d, n)$ leads to an additional problem. Let M and N denote d-polytopes with m respectively n vertices, and let $v_d(m, n)$ [$V_d(m, n)$] denote the minimal [maximal] possible value of $f_0(M + N)$ (Grünbaum [6], Danzer–Grünbaum [2]). The determination of $V_d(m, n)$ is quite simple; we have $V_2(m, n) = m + n$ and $V_d(m, n) = mn$ for $d \geq 3$. Regarding $v_d(m, n)$ the following results were obtained by E. T. Poulsen (private communication); for simplicity of formulation we assume that $n \geq m > d \geq 2$.

$$v_2(m, n) = n;$$
$$v_3(m, n) = n \text{ except if } n = 5, m = 4; f_3(4, 5) = 6;$$
$$v_d(m, n) = n \quad \text{if} \quad n \geq 2d \quad \text{or if} \quad m = n;$$
$$v_d(m, n) = m + d - 1 \quad \text{if} \quad n < 2d.$$

The unsolved problem deals, therefore, with the case $4 \leq d < m < n < 2d$.

Problems analogous to $v_d(m, n)$ and $V_d(m, n)$ but dealing with numbers of k-faces, for $0 < k < d$, rather than the numbers of vertices of the polytopes involved, could also be considered. However, no nontrivial results seem to be known, even in the most interesting case $k = d - 1$. Another possibility, also completely uninvestigated, is to substitute the Blaschke addition # (see chapter 15) for the vector addition.

Let K be a d-dimensional convex body in R^d. A set $A \subset \text{bd } K$ [a set U of unit vectors] is called an *inner* [*outer*] *illuminating set* of K provided for every $x \in \text{bd } K$ there exists an $\varepsilon, 0 < \varepsilon < 1$, and a point $a \in A$ [a vector $u \in U$] such that $(1 - \varepsilon)x + \varepsilon a \in \text{int } K$ [respectively $x + \varepsilon u \in \text{int } K$]. An illuminating set is called *primitive* provided none of its proper subsets is an illuminating set. Let $I_i(K)$ [respectively $I_0(K)$] denote the least number of elements in an inner [outer] illuminating set of K, and let $I^i(K)$ [respectively $I^0(K)$] be the supremum of the number of elements in primitive inner [outer] illuminating sets of K. Clearly $I_0(K) \leq c(K)$, and it may be shown that the two numbers coincide for each K.

The determination of the upper and lower bounds for the values of $I_i(K)$ and the other functionals, when K ranges over all d-dimensional convex bodies, leads to a number of interesting problems. Among the known results (except for those on $I_0(K) = c(K)$ which were discussed above) we mention (see Soltan [1], Grünbaum [14]):

$2 \leq I_i(K) \leq d + 1$, with $I_i(K) = d + 1$ if and only if K is the d-simplex;

$I^i(K) \leq 4$ if K is 2-dimensional, with equality if and only if K is a quadrangle;

$I^i(C^d) = 2^d$;

$I^0(K) \leq 6$ if K is 2-dimensional, with equality only for hexagons having opposite edges parallel.

It is easily seen that in the search for the least upper bounds of $I^i(K)$ and $I^0(K)$ there is no loss of generality in assuming the d-dimensional convex body K to be a d-polytope. It is remarkable, however, that even the existence of a finite upper bound for $I^i(K)$ and $I^0(K)$ has not been established, though it may be conjectured that $I^i(K) \leq 2^d$ and $I^0(K) \leq 2(2^d - 1)$ for all d-dimensional K.

Similar open problems exist regarding the *fixing systems* (Fejes Tóth [2], Polák–Poláková [1], Grünbaum [14]). If K is a d-dimensional convex body in R^d, a set $A \subset$ bd K is called a fixing system for K provided for some $\varepsilon > 0$ we have $A \cap (\text{int}(x + K)) \neq \emptyset$ for each nonzero vector x of length at most ε. Let $\varphi(K)$ denote the minimal possible number of points in a fixing system for K, and let $\Phi(K)$ denote the least upper bound for the number of points in *primitive* fixing systems for K. (A fixing system for K is primitive provided none of its proper subsets is a fixing system for K.) It is easy to show that $d + 1 \leq \varphi(K) \leq 2d$ and to relate $\varphi(K)$ to various other problems, such as d-polytopes which are not the $\#$-sum of d-polytopes with a smaller number of facets (section 15.3), or to *primitive polytopes* (see Fejes Tóth [2], Steinitz [2]). The finiteness of $\Phi(K)$, however, is still uncertain; it is known that $\Phi(K) = 2(2^d - 1)$ for some d-polytopes K (Danzer [1]).

19.4   Additional notes and comments

Cube tilings.
Keller's problem (asking whether every tiling of R^d by congruent cubes has two tiles that share a facet, see page 411) was resolved in the negative for $d \geq 10$ by Lagarias–Shor [a], by reduction to a finite graph coloring problem due to Corrádi–Szabó [a]. For $d = 7, 8, 9$ it apparently remains open.

Hajos [a] established an affirmative answer for all d if one considers only *lattice tilings*, i.e., tilings in which the centers of the tiles form a translate of an additive subgroup of R^d.

For an extensive treatment and survey of tilings, including many additional open problems, see Grünbaum–Shephard [b] [d].

Regular polytopes.
The conceptual framework of Coxeter groups has turned out to be very important and fundamental for the study of regular polytopes. We refer to Benson–Grove [a] and Brown [a].

It also allows one to resolve positively the question posed on page 413 whether projectively resp. combinatorially regular convex polytopes are always (projectively resp. combinatorially) equivalent to regular polytopes—see McMullen [a] [b].

Semiregular polytopes (of various kinds and denominations) form a vast field of study. We refer to Schulte's [d] handbook survey and to the references given there.

The enumeration of the uniform polytopes (see page 413) in dimension four was completed by Conway [a] in collaboration with Guy. Johnson [a] will describe these polytopes and many related figures.

The book by McMullen–Schulte [a] on abstract regular polytopes will contain a lot of information (including a large bibliography) about the classical regular convex polytopes and their symmetry groups as well.

Equifacetted polytopes.
A polytope all of whose facets are pairwise combinatorially equivalent (see page 415) is called *equifacetted*. Obviously, simplices and cubes arise as "facet types" of equifacetted polytopes. Going beyond the results of Perles–Shephard [a] (referred to on page 415), Schulte [a] [b] (see also Schulte [c]) found d-polytopes for every $d \geq 3$ that are not even the facet types of any equifacetted manifold and also are not the "tile types" of face-to-face tilings of R^d with combinatorially equivalent tiles. For $d = 3$ the simplest known example is the cuboctahedron.

Kalai [f] shows that the 24-, the 120-, and the 600-cell as well as the d-cross-polytope ($d \geq 4$) are not facet types of equifacetted polytopes; furthermore, there is no face-to-face tiling of R^5 with crosspolytopes.

k-Content.
A different approach to extend the notion of k-content from polytopes to convex bodies is to study directly the k-dimensional Hausdorff measure of the union of the at most k-dimensional extreme faces of a convex body. For a survey on these so called *Hausdorff measures of skeletons* and their relations to $\zeta_k(K)$, see Schneider [b, pp. 68–70].

Borsuk's conjecture in high dimensions.
Borsuk's conjecture is false in high dimensions—it was first disproved by Kahn–Kalai [a] for certain sets of ± 1-vectors. After a number of subsequent simplifications (see the two-page version by Nilli [a]), improvements and new ideas, we now know that Borsuk's conjecture fails for all dimensions $d \geq 298$ (by Hinrichs–Richter [a]). Recent surveys include Boltyanski–Martini–Soltan [a] and Raigorodskii [a].

There is a close connection to the problem of the "chromatic number of R^d" that asks for the smallest number of colors that is sufficient to color R^d such that no two points of distance one get the same color. (For $d = 2$ this is the notorious "Hadwiger–Nelson problem" for which one knows that the answer satisfies $4 \leq \chi(R^2) \leq 7$; see Jensen–Toft [a, Sect. 9.1] and the references given there.)

Borsuk's conjecture in low dimensions.
Borsuk's problem is still unresolved in low dimensions. Let us just mention that $\beta_3 \leq 0.98$, according to Raigorodskii [a, p. 106], while Gale's [a] conjecture from 1953 that the bound $\beta_3 \geq (\frac{1}{6}(3 + \sqrt{3}))^{1/2} \approx 0.888$ should be tight still stands.

In dimension $d = 4$ the best result is that every set may be decomposed into at most 9 sets of smaller diameter, by Lassak [a]. For collections of 0/1-vectors, Borsuk's conjecture is true at least for $d \leq 8$, by Ziegler [c].

Borsuk's conjecture is closely related to the questions about the maximal number of pairs of points of maximal distance resp. the maximal number of pairs of points of any fixed distance in a finite d-dimensional point set V. Of special interest is, of course, the case where V is in convex position (that is, if it is the vertex set of a polytope). We refer to Pach–Agarwal [a, esp. Chap. 13] for an extensive discussion and survey of these questions.

The Borsuk problem for the sphere.

Even the case of the unit ball/sphere poses a substantial problem in view of Borsuk's conjecture: What is the smallest diameter $\delta(d)$ such that the unit sphere S^{d-1} can be decomposed into $d + 1$ parts of diameter at most $\delta(d)$? It has been conjectured that the "obvious" partition induced by the regular simplex is best possible, but that has been verified only for $d \leq 3$; see Croft–Falconer–Guy [a, Problem D14]. Asymptotic results were given by Larman–Tamvakis [a].

Illumination and covering.

An *inner diagonal* of a polytope is a line segment that joins two vertices and lies, except for its ends, in the interior of the polytope. This corresponds to an unordered pair of vertices that do not share any facet, that is, where each vertex is *illuminated* by the other. A d-cube has 2^{d-1} inner diagonals, and von Stengel [a] has conjectured that this is the maximum among all d-polytopes with $2d$ facets. His conjecture arose from a study of Nash equilibrium points of certain bimatrix games; it has been proved only for $d \leq 4$.

Boltyanski–Martini–Soltan [a] found for every $d \geq 4$ a convex body with arbitrarily large primitive inner illuminating sets; in particular, this disproves the conjectured bound $I^i(K) \leq 2^d$ on page 423. For related work on "inner illumination", see also Mani [c] and Bremner–Klee [a].

According to a theorem of Boltyanskii [1], the covering number $c(K)$ agrees for all compact, convex bodies $K \subset R^d$ with the number of directions needed to illuminate the convex body from the outside. The conjecture (by Hadwiger) that $c_d = 2^d$ has not even been proven in full for $d = 3$. See the discussions in Boltyanski–Martini–Soltan [a, p. 270] and Bezdek [a].

For convex bodies K of constant width, Schramm [a] obtained

$$c(K) \leq \left(\sqrt{\tfrac{3}{2}} \right)^d 5d\sqrt{d}(4 + \log d);$$

this result also implies the best known asymptotic bounds for the Borsuk partition problem.

For surveys see Boltyanski–Martini–Soltan [a]; §44 of the same book also provides an extensive discussion of (minimal) fixing systems.

Containment problems.

Given a class $\mathscr{C}$ of convex sets and a convex body K, the quest for a "largest", or "smallest", member of $\mathscr{C}$ contained in K (or containing K) has been considered in lots of variations (see Gritzmann–Klee [c]). For the algorithmic problem of finding the inradius, circumradius, width, and diameter, see Gritzmann–Klee [a] [b] and Brieden et al. [a]. Below, we sketch a few results on largest and smallest j-simplices (with respect to j-dimensional measure).

Simplices containing or contained in a d-polytope.

Among the largest j-simplices contained in a given convex body K, there is one whose vertices are all extreme points of K. When K is a polytope, this leads to an obvious finite algorithm for finding a largest j-simplex in K. For other results concerning largest j-simplices in d-polytopes, see Gritzmann–Klee–Larman [a].

Klee [c] proved that if C is a d-polytope containing a convex body K, and C is a local minimum among polytopes that contain K and have the same number of facets as C, or C is a local minimum among the simple polytopes that contain K and are of the same combinatorial type as C, then the centroid of each facet of C belongs to K. When K is a polytope and $d \in \{2, 3\}$, this fact is useful in designing efficient algorithms that actually find a local or global minimum among the simplices containing K (see O'Rourke et al. [a] and Zhou–Suri [a]). However, Packer [a] shows that the general problem of finding a smallest (largest) d-simplex containing (contained in) a d-polytope, which is specified either in $\mathcal{V}$- or in $\mathcal{H}$-description, is NP-hard.

Largest simplices in cubes.

The problem of finding a largest j-simplex in the d-cube $[0, 1]^d$ amounts to finding optimal weighing designs for spring balances, and there are closely related "largest simplex" problems that amount to finding optimal weighing designs for two-pan balances. See the papers cited below for explanation.

For fixed j, the problem of finding a largest j-simplex in $[0, 1]^d$ has been solved for all $d \geq j$ by Hudelson–Klee–Larman [a] when $j \in \{2, 3\}$ and by Neubauer–Watkins–Zeitlin [b] [c] for $4 \leq j \leq 6$. For each $j \geq 7$, the problem is open for infinitely many d, but when j is even it has been solved for infinitely many d by Neubauer–Watkins–Zeitlin [a]. See Neubauer–Watkins [a] for the case $d = 7$.

The case $j = d$ is of special interest. Define ρ_d as $d!$ times the volume of a largest d-simplex in $[0, 1]^d$. Translated into geometric terms, Hadamard [a] observed $(\rho_d)^2 \leq (d + 1)^{(d+1)}/(4^d)$ and that, for $d > 1$, equality implies d congruent to 3 (mod 4). Hadamard's conjecture that equality holds for all such d has been proved for infinitely many values of d and for all $d < 427$, but it is also still open for infinitely many values of d. For each d, the following three statements are equivalent: (i) a Hadamard matrix of order $d + 1$ exists; (ii) in a d-cube, there is a largest d-simplex that is regular; (iii) the vertex-set of a d-cube contains an equilateral set consisting of $d + 1$ points.

We refer to Hudelson–Klee–Larman [a] for further details and references.

Table 1. Known values of $c_s(v, d)$, the number of different combinatorial types of simplicial d-polytopes with v vertices (or of simple d-polytopes with v facets)

$v - d$	1	2	3	4	5	6	7	8	9
d									
3	1	1	2	5	14	50	233	[a]1249	[b]7616
4	1	2	5	[c]37					
5	1	2	8						
6	1	3	18						
7	1	3	29						
8	1	4	57						
9	1	4	96						
10	1	5	183						
11	1	5	318						
12	1	6	603						
13	1	6	1080						
14	1	7	2047						
15	1	7	3762						
d	1	$[\tfrac{1}{2}d]$	[d]						

[a] Grace [1]; not certain, but probably correct (see section 13.6).
[b] Brückner [4]; probably incorrect (see section 13.6, and Rademacher [1]).
[c] Grünbaum–Sreedharan [1].
[d] Determined by M. A. Perles; see theorem 6.3.2.

Table 2. Known values of $c(v, d)$, the number of different combinatorial types of d-polytopes with v vertices (or with v facets)

$v - d$	1	2	3	4	5
d					
3	1	2	7	34*	257*
4	1	4	31		
5	1	6	116		
6	1	9	379		
d	1	$[\tfrac{1}{4}d^2]$			

The values for $d = 3$ are due to Hermes [1], those for $v = d + 3$ to M. A. Perles (see section 6.3). Asterisks indicate values which have not been checked independently.

Table 3. Relations between numbers of faces of simplicial d-polytopes

$d = 3$
$$f_1 = 3f_0 - 6$$
$$f_2 = 2f_0 - 4$$

$d = 4$
$$f_2 = 2f_1 - 2f_0$$
$$f_3 = f_1 - f_0$$

$d = 5$
$$f_2 = 4f_1 - 10f_0 + 20$$
$$f_3 = 5f_1 - 15f_0 + 30$$
$$f_4 = 2f_1 - 6f_0 + 12$$

$d = 6$
$$f_3 = 3f_2 - 5f_1 + 5f_0$$
$$f_4 = 3f_2 - 6f_1 + 6f_0$$
$$f_5 = f_2 - 2f_1 + 2f_0$$

$d = 7$
$$f_3 = 5f_2 - 15f_1 + 35f_0 - 70$$
$$f_4 = 9f_2 - 34f_1 + 84f_0 - 168$$
$$f_5 = 7f_2 - 28f_1 + 70f_0 - 140$$
$$f_6 = 2f_2 - 8f_1 + 20f_0 - 40$$

$d = 8$
$$f_4 = 4f_3 - 9f_2 + 14f_1 - 14f_0$$
$$f_5 = 6f_3 - 17f_2 + 28f_1 - 28f_0$$
$$f_6 = 4f_3 - 12f_2 + 20f_1 - 20f_0$$
$$f_7 = f_3 - 3f_2 + 5f_1 - 5f_0$$

$d = 9$
$$f_4 = 6f_3 - 21f_2 + 56f_1 - 126f_0 + 252$$
$$f_5 = 14f_3 - 63f_2 + 182f_1 - 420f_0 + 840$$
$$f_6 = 16f_3 - 78f_2 + 232f_1 - 540f_0 + 1080$$
$$f_7 = 9f_3 - 45f_2 + 135f_1 - 315f_0 + 630$$
$$f_8 = 2f_3 - 10f_2 + 30f_1 - 70f_0 + 140$$

$d = 10$
$$f_5 = 5f_4 - 14f_3 + 28f_2 - 42f_1 + 42f_0$$
$$f_6 = 10f_4 - 36f_3 + 78f_2 - 120f_1 + 120f_0$$
$$f_7 = 10f_4 - 39f_3 + 87f_2 - 135f_1 + 135f_0$$
$$f_8 = 5f_4 - 20f_3 + 45f_2 - 70f_1 + 70f_0$$
$$f_9 = f_4 - 4f_3 + 9f_2 - 14f_1 + 14f_0$$

ADDENDUM

The following remarks were added in proof (November 1966).

Page 67. An easy modification of exercise 4.8.25 establishes the following result of Wagner [1]: Every simplicial k-"complex" with at most $\aleph$ vertices has a representation in R^{2k+1} such that all the "simplices" are geometric (rectilinear) simplices.

Page 93. J. H. Conway (private communication) has established the validity of the conjecture mentioned in the second footnote.

Page 126. For $d = 2$, the theorem of Derry [2] given in exercise 7.3.4 was found earlier by Bilinski [1].

Page 183. M. A. Perles (private communication) recently obtained an affirmative solution to Klee's problem mentioned at the end of section 10.1.

Page 204. Regarding the question whether $a(\mathscr{C}) = 3$ implies $b(\mathscr{C}) \leqslant 4$, it should be noted that if one starts from a *topological cell complex* $\mathscr{C}$ with $a(\mathscr{C}) = 3$ it is possible that $\mathscr{C}$ is not a *complex* (in our sense) at all (see exercise 11.1.7). On the other hand, G. Wegner pointed out (in a private communication to the author) that the 2-complex $\mathscr{C}$ discussed in the proof of theorem 11.1.7 indeed satisfies $b(\mathscr{C}) = 4$.

Page 216. Halin's [1] result (theorem 11.3.3) has recently been generalized by H. A. Jung to all complete d-partite graphs. (Halin's result deals with the graph of the d-octahedron, i.e. the d-partite graph in which each class of nodes contains precisely two nodes.)

The existence of the numbers $n(k)$ follows from a recent result of Mader [1]; Mader's result shows that $n(k) \leqslant k.2^{\binom{k}{2}}$.

Page 222. A very elegant, non-computational, construction of the 3-diagram $\mathscr{D}'$ of theorem 11.5.2 was communicated to the author by G. Wegner. His construction is explained in an addendum to Grünbaum–Sreedharan [1]. Concerning the dual $\mathscr{M}^*$ of the 3-complex $\mathscr{M}$ represented by $\mathscr{D}'$ (see page 224), Wegner has shown that it is not representable by a 3-diagram if the basis of the diagram is required to be the 3-face of $\mathscr{M}^*$ which corresponds to the vertex 8 of $\mathscr{M}$.

Page 231, line 4. M. A. Perles has shown (private communication) that the graphs in question are dimensionally ambiguous whenever $d \geqslant n + 3$.

Page 271. In a revised version of Barnette's paper [3], the following improvement of the inequalities of exercise 13.3.13 is established:

If (p_k) is a 3-realizable sequence, then

$$2p_3 + 2p_4 + 2p_5 + 2p_6 + p_7 \geq 16 + \sum_{k \geq 9} (k - 8)p_k.$$

This inequality refutes conjecture 2 (page 268). Indeed, if $p_k = 0$ for $k \neq 3, 6, 6m$, then equation (*) (page 254) implies $p_3 = 4 + (2m - 2)p_{6m}$, hence Barnette's inequality yields $p_6 \geq 4 + (m - 2)p_{6m}$. For $(m - 2)(p_{6m} - 6) \geq 8$ this contradicts conjecture 2.

On the other hand, an affirmative solution of conjecture 1 (page 267) will be established in a forthcoming paper by the author.

Page 315. Recent results have greatly increased the number of known Euler-type relations, and their comparison here seems useful.

We have already discussed Euler's equation (Chapter 8) concerning numbers of faces, Gram's equation (theorem 14.1.1) concerning angle sums, and the equation of theorem 14.3.2 dealing with the Steiner point of a polytope and its faces.

In order to discuss the new results, let $P^d = P_1^d$ denote a d-polytope, let $f_j = f_j(P^d)$, and let the j-faces of P^d be $P_1^j, \ldots, P_{f_j}^j$.

Let $m(K)$ denote the *mean width* of the compact convex set K. Shephard [11] proved that for each d-polytope P^d

$$\sum_{j=0}^{d} (-1)^j \sum_{i=1}^{f_j} m(P_i^j) = -m(P^d).$$

Another function with similar properties is the "angle-deficiency" $\delta(P_k^j, P^d)$ defined below. Let us define $\varphi(P_k^j, P_i^{d-1})$ as the angle (see page 297) spanned by P_k^j in P_i^{d-1} if $P_k^j \subset P_i^{d-1}$, and as 0 if P_k^j is not contained in P_i^{d-1}. Extending the well known case $(d = 3)$, the following results were obtained by Perles–Shephard [1], Perles–Walkup [1], and Shephard [10]:

For each j-face P_k^j of the d-polytope P^d $(0 \leq j \leq d - 1)$,

$$\delta(P_k^j, P^d) = 1 - \sum_{i=1}^{f_{d-1}} \varphi(P_k^j, P_i^{d-1}) \geq 0,$$

with equality for $j \geq d - 2$, and strict inequality for $j \leq d - 3$.

For each d-polytope P^d,

$$\sum_{j=0}^{d-3} (-1)^j \sum_{i=1}^{f_j} \delta(P_i^j, P^d) = 1 + (-1)^{d-1}.$$

An additional relation of the same type (which contains the mean width result as a special case) is due to Shephard [13]. Let P^d be a d-polytope, let $0 < r < d$, and let $K_{r+1}, \ldots, K_d$ be any $d - r$ convex bodies. Let us denote by $v(P_i^j, \ldots, P_i^j, K_{r+1}, \ldots, K_d)$ the *mixed volume* (Bonnesen–Fenchel [1], page 38) of the face P_i^j of P^d (repeated r times) and the sets $K_{r+1}, \ldots, K_d$. Then

$$\sum_{j=0}^{d} (-1)^j \sum_{i=0}^{f_j} v(P_i^j, \ldots, P_i^j, K_{r+1}, \ldots K_d)$$

$$= (-1)^r v(-P^d, \ldots, -P^d, K_{r+1}, \ldots K_d).$$

For proofs of these results, which use a great variety of methods, the reader should consult the papers quoted. They contain also extensions to the case of spherical polytopes, as well as analogues of the Dehn–Sommerville equations (in case of simplicial polytopes, and of other special families of polytopes) for the various quantities considered. One indication of the usefulness of some of these results may be found in Perles–Shephard [2]. This paper contains results of the following type (which were completely unassailable so far); we quote only two very special results which are easy to formulate:

If $d \geqslant 7$, no d-polytope has all facets combinatorially equivalent to the $(d - 1)$-octahedron.

No 5-polytope has all facets combinatorially equivalent to the cyclic polytope $C(8, 4)$.

Page 365. The example (figure 17.1.9) of a 3-valent, 3-connected, cyclically 5-connected planar graph without a Hamiltonian circuit (Walther [1]) was recently improved in some respects. While Walther's example contains 162 nodes, the author has found a similar example with 154 nodes, as well as an example with 464 nodes which does not admit even a Hamiltonian path. The first example was improved still further by Walther, who constructed a 114-node graph of this type which has no Hamiltonian circuit (private communication). A still smaller graph with the same property (with 46 nodes only) has reportedly been found by Kozyrev.

Page 425. The coefficients of the Dehn–Sommerville equations given in table 3 were independently computed by Riordan [1]; however, his table is marred by misprints. In Riordan's notation, the incorrect values are $A_{2,-1}^o(4)$ and $A_{3,2}^e(5)$.

ERRATA FOR THE 1967 EDITION

This is a list of corrections (other than small typos), noted by Marge Bayer, Branko Grünbaum, Michael Joswig, Volker Kaibel, Victor Klee, Carsten Lange, Julian Pfeifle, and Günter M. Ziegler. A minus in front of a line number means "counted from the bottom".

Page	Line	Original	Correction
9	6	Use exercise 2	Use statement 2
9	13	The *relative interior* of a convex set $A \subset R^d$ may be defined as relint $A := \{x \in R^d : (\text{aff} A) \cap (x + \varepsilon B^d) \subset A \text{ for some } \varepsilon > 0\}$. This is empty if and only if $A = \emptyset$.	
13	20	The claim in exercise 7 is not valid. The following counter-example is due to P. McMullen: Let $K := K' + B_3$, where $K' := \{(x,y,0) \in R^3 \mid x > 0, xy \geq 1\}$ and $B_3 := \{(x,y,z) \in R^3 \mid x^2 + y^2 + z^2 \leq 1\}$ is the 3-dimensional unit ball, and let $L := \{(x,0,1) \mid x \in R\}$. (An earlier counterexample, due to T. Botts, appears on p. 459 of Klee [a].)	
35	−15	$P = \text{conv}(F \cup F_1)$	$\text{aff}(P) = \text{aff}(F \cup F_1)$
35	−14	The hypothesis in exercise 3 should include $F_{k-1} \subset F_{k+1}$.	
59	12	section 9.3	section 9.4
61	1	2^{d-1}	2^{d-i}
62	5	$0 \leq i \leq [\frac{1}{2}d]$	$0 \leq i < [\frac{1}{2}d]$
65	−5	K_d^k	$-K_d^k$
68	15	$P(V_j)$	$P_\lambda(V_j)$
68	−3	$\mathscr{F}(K)$	$\mathscr{F}(P)$
77	16	AK	AP
78	−11	In part (ii) of Theorem 1 the "only if" part is not true. Let F be a vertex of a polygon P and let V be the new vertex so that $E^* := \text{conv}(\{V\} \cup F)$ is an edge that contains an original edge E of P. Then E^* is a face of P^*, but (*) is not satisfied, since $V \notin \text{aff} F$, and (**) is not satisfied, because there is no facet (edge) of P containing F for which V is beyond. The (first) error in the proof occurs in lines 4–5 on page 79. See Altshuler–Shemer [a].	
79	−7	$\text{aff} F_0 \subset \text{aff} F$ is false for the same reason as above.	
82	17	The claim in exercise 13(ii) was *not* established by Shephard in [7]; he later found an error in his construction. (See also the notes in section 4.9.)	
100	−8	$[\frac{1}{2}d^2]$	$[\frac{1}{4}d^2]$
113	3	The last product in this formula contains two typos; it should be:	

$$\prod_{j \in \{i \mid \gamma_i < \alpha_i\}} \frac{p_{j-1}}{p_j} \cdot 2^g$$

428a

Page	Line	Original	Correction
115	0	In the figure, the star diagram and the Gale-diagram do not fit together (there should be a ⋆ at position $(2,3)$).	
117	10	e_1, e_2, e_3, e_4	$e_1, e_2, e_3, -e_4$
129	-1	exercise 7.3.4	exercise 7.3.5
129	-3	exercise 7.3.4	exercise 7.3.5
138	3	section 3.3	section 3.2
170	4	Perles and Shephard did *not* prove the existence, for each $d \geq 4$, of infinitely many d-polytopes of type $(2, d-2)$. (See the note above for p. 82.)	
185	17	It is not true that each k-face of P is of (*at least*) one of the listed types. There can be faces of P not contained in F' and containing V_1 but not V_2.	
186	12	(a) P is a d-pyramid over a $(d-1)$-dimensional basis with at least $d+3$ vertices	(a) P has a facet with at least $d+3$ vertices
206	13	$\mathscr{C}(P)$	$\mathscr{B}(P)$
213	23	$F = M \cap P$ is not necessarily a face of P; $M \cap P$ could be a proper subpolytope of a face, intersecting the relative interior of the face.	
217	-14	12.2	12.1
317	-11	The equality in (b) holds only for $\lambda \geq 0$.	
318	10	For the polygon Q to be a summand of P one needs the parallel edges to have the same outer normals. For instance, the reflection of an equilateral triangle is not a summand of the original triangle, even though its edges are parallel to the edges of the original.	
349	-6	$n \geq 2d$	$n < 2d$
392	16	2^{j-1}	2^{j-i}
392	-15	$[(d-1)/2]$	$[d/2]$
394/ 396		Table 18.1.1 and Figure 18.1.2: The number of nonequivalent arrangements of 6 lines in the projective plane should be 17, not 16 (see Grünbaum [e, p. 5]).	
396	-4	$m = \frac{1}{2}\binom{n-2}{2}$	$m = \left[\frac{1}{2}\binom{n-2}{2}\right]$
397	12	simple 3-arrangements	simple 2-arrangements
405	-10	0 or 4	0 or 3
424	5	In Table 1, $c_s(11,3) = 1249$ is correct (as suspected by Grünbaum), while $c_s(12,3) = 7616$ is incorrect (as also suspected by Grünbaum); it should be $c_s(12,3) = 7595$. For the correct values of $c_s(v,d)$ for $4 \leq d \leq 21$ see Royle [a]. Furthermore, in Table 2, the questionable values are correct; the values of $c(v,3)$ up to $v = 13$ are also given by Royle [a].	
424	14	Computation from Perles' formula yields $c_s(15,12) = 604$ rather than 603.	

BIBLIOGRAPHY

For the reader's convenience each item in the bibliography is followed by one or more numbers in square brackets. Those numbers indicate the pages on which this item was mentioned. Only the first mention of an item within a single section of the book is recorded below.

H. L. Abbott
1. *Some problems in combinatorial analysis.* Ph.D. Thesis, University of Alberta, Edmonton 1965. [383]
O. Aberth
1. An isoperimetric inequality. *Proc. London Math. Soc.* (3) **13** (1963), 322–336. [417]
A. D. Aleksandrov (Alexandroff)
1. Zur Theorie der gemischten Volumina von konvexen Körpern. I. *Mat. Sb.* (*N.S.*), **2** (1937), 947–972. [338]
2. *Convex polyhedra.* Moscow 1950. (Russian. German translation *Konvexe Polyeder*, Berlin 1958.) [29, 411]
3. *Intrinsic geometry of convex surfaces.* Moscow 1948. (Russian. German translation *Die innere Geometrie der konvexen Flächen*, Berlin 1955.) [29]
P. Alexandroff and H. Hopf
1. *Topologie.* Berlin 1935. [39, 141, 152, 200]
E. M. Alfsen
1. On the geometry of Choquet simplices. *Math. Scand.*, **15** (1965), 97–110. [52]
E. Altman
1. On a problem of P. Erdös. *Am. Math. Monthly*, **70** (1963), 148–157. [419]
V. G. Aškinuze
1. On the number of semi-regular polyhedra. *Mat. Prosveŝč.*, **1** (1957), 107–118 (Russian). [413]
2. Polygons and polyhedra. In *Encyclop. of elem. math.*, Vol. 4 (Geometry), pp. 382–447. Moscow 1963 (Russian). [413]
E. Asplund
1. A *k*-extreme point is the limit of *k*-exposed points. *Israel J. Math.*, **1** (1963), 161–162. [21]
F. Bagemihl
1. A conjecture concerning neighboring tetrahedra. *Am. Math. Monthly*, **63** (1956), 328–329. [128]
M. Balinski
1. On the graph structure of convex polyhedra in *n*-space. *Pacific J. Math.*, **11** (1961), 431–434. [213, 341, 360, 375]
2. An algorithm for finding all vertices of convex polyhedral sets. *J. Soc. Ind. Appl. Math.*, **9** (1961), 72–88. [356]
W. W. R. Ball
1. *Mathematical recreations and essays.* (11th edition, revised by H. S. M. Coxeter.) London 1940. [356, 364, 405, 413]

D. Barnette
1. Trees in polyhedral graphs. *Can. J. Math.*, **18** (1966), 731–736. [296, 369]
2. W_v paths on 3-polytopes. *J. Combinatorial Theory* (to appear). [352, 354]
3. On k-vectors of simple 3-polytopes. *J. Combinatorial Theory* (to appear). [271, 426]
W. Barthel
1. Zum Busemannschen und Brunn–Minkowskischen Satz. *Math. Z.*, **70** (1959), 407–429. [338]
W. Barthel and G. Franz
1. Eine Verallgemeinerung des Busemannschen Satzes von Brunn–Minkowskischen Typ. *Math. Ann.*, **144** (1961), 183–198. [338]
A. Bastiani
1. Cones convexes et pyramides convexes. *Ann. Inst. Fourier Grenoble*, **9** (1959), 249–292. [52]
V. J. D. Baston
1. *Some properties of polyhedra in Euclidean space.* New York 1965. [128]
C. Berge
1. *Théorie des graphs et ses applications.* Paris 1958. [213]
A. S. Besicovitch
1. On Crum's problem. *J. London Math. Soc.*, **22** (1947), 285–287. [128]
A. S. Besicovitch and H. G. Eggleston
1. The total length of the edges of a polyhedron. *Quart. J. Math.*, *Oxford Ser.*, **8** (1957), 172–190. [418]
C. Bessaga
1. A note on universal Banach spaces of a finite dimension. *Bull. Acad. Polon. Sci.*, **6** (1958), 97–101. [73]
S. Bilinski
1. Über die Rhombenisoeder. *Glasnik Mat. Fiz. Astron. Društvo Mat. Fiz. Hrvatske*, *Ser. 2*, **15** (1960), 251–263. [414, 426]
G. Birkhoff
1. Lattice theory. *Am. Math. Soc. Colloq. Publ.*, Vol. 25. New York 1948. [21]
G. D. Birkhoff
1. The reducibility of maps. *Am. J. Math.*, **35** (1913), 115–128. [365]
W. Blaschke
1. *Kreis und Kugel.* Leipzig 1916. [10, 29, 339]
J. Böhm
1. Simplexinhalt in Räumen konstanter Krümmung beliebiger Dimension. *J. reine angew. Math.*, **202** (1959), 16–51. [313]
G. Bol
1. Über Eikörper mit Vieleckschatten. *Math. Z.*, **48** (1942), 227–246. [74]
V. G. Boltyanskiĭ
1. A problem about the illumination of the boundary of a convex body. *Izv. Moldavsk. Filiala Akad. Nauk SSSR*, **10** (76) (1960), 79–86 (Russian). [420]
T. Bonnesen and W. Fenchel
1. Theorie der konvexen Körper. *Ergeb. Math. Grenzg.*, Vol. 3, No. 1. Berlin 1934. [13, 29, 315, 338, 416]
W. E. Bonnice and M. Edelstein
1. *Flats associated with finite sets in P^d.* (To appear.) [404]
W. Bonnice and V. Klee
1. The generation of convex hulls. *Math. Ann.*, **152** (1963), 1–29. [17]
K. Borsuk
1. Drei Sätze über die n-dimensionale euklidische Sphäre. *Fundam. Math.*, **20** (1933), 177–190. [201, 418]

J. Bosak
   1. *Hamiltonian lines in cubic graphs*. Talk presented to the International Seminar on Graph Theory and its Applications (Rome, July 5–9, 1966).   [359]
C. J. Bouwkamp, A. J. W. Duijvestyn and P. Medema
   1. *Table of c-nets of orders 8 to 19 inclusive*. Phillips Research Laboratories, Eindhoven, Netherlands, 1960.   [48, 289]
T. A. Brown
   1. *Hamiltonian paths on convex polyhedra*. Note P-2069, The RAND Corp., Santa Monica, Calif., 1960.   [360]
   2. *The representation of planar graphs by convex polyhedra*. Note P-2085, The RAND Corp., Santa Monica, Calif., 1960.   [291]
   3. Simple paths on convex polyhedra. *Pacific J. Math.*, **11** (1961), 1211–1214.   [357, 372]
W. G. Brown
   1. Enumeration of triangulations of the disc. *Proc. London Math. Soc.*, (3), **14** (1964), 746–768.   [289]
W. G. Brown and W. T. Tutte
   1. On the enumeration of non-separable planar maps. *Can. J. Math.*, **16** (1964), 572–577.
                                                                                       [289]
M. Brückner
   1. Die Elemente der vierdimensionalen Geometrie mit besonderer Berücksichtigung der Polytope. *Jber. Ver. Naturk. Zwickau* 1893.   [78, 127, 170, 182]
   2. *Vielecke und Vielflache*. Leipzig 1900.   [45, 78, 141, 254, 286, 288]
   3. Ueber die Ableitung der allgemeinen Polytope und die nach Isomorphismus verschiedenen Typen der allgemeinen Achtzelle (Oktatope). *Verhandel. Koninkl. Akad. Wetenschap.* (Eerste Sectie), Vol. 10, No. 1 (1909).
                                                                 [45, 78, 121, 127, 170, 182, 188, 224]
   4. Ueber die Anzahl $\psi(n)$ der allgemeinen Vielflache. *Atti Congr. Intern. Mat.*, *Bologna* 1928, Vol. 4 (Communicazioni), 5–11.   [288, 424]
R. C. Buck
   1. Partition of space. *Am. Math. Monthly*, **50** (1943), 541–544.   [392]
   2. Problem E 923. *Am. Math. Monthly*, **57** (1950), 416 Solution by E. P. Starke, *ibid.*, **58** (1951), 190.   [188]
H. Busemann
   1. A theorem on convex bodies of the Brunn–Minkowski type. *Proc. Nat. Acad. Sci. US*, **35** (1949), 27–31.   [338]
   2. *Convex surfaces*. New York 1958.   [29, 338]
E. Cahen
   1. Théorie des régions. *Nouv. Ann. Math.*, (3) **16** (1897), 533–539.   [391]
S. S. Cairns
   1. Triangulated manifolds which are not Brouwer manifolds. *Ann. Math.*, **41** (1940), 792–795.   [202]
   2. Isotopic deformations of geodesic complexes on the 2-sphere and on the plane. *Ann. Math.*, **45** (1944), 207–217.   [292]
   3. Peculiarities of polyhedra. *Am. Math. Monthly*, **58** (1951), 684–689.   [188]
   4. Skewness, embeddings, and isotopies. (To appear.)   [63]
S. H. Caldwell
   1. *Switching circuits and logical design*. New York 1958.   [383]
C. Carathéodory
   1. Ueber den Variabilitätsbereich der Koeffizienten von Potenzreihen, die gegebene Werte nicht annehmen. *Math. Ann.*, **64** (1907), 95–115.   [14, 29, 63, 127]
   2. Ueber den Variabilitätsbereich der Fourierschen Konstanten von positiven harmonischen Funktionen. *Rend. Circ. Mat. Palermo*, **32** (1911), 193–217.
                                                                               [15, 29, 63, 127]

W. B. Carver
1. The polygonal regions into which a plane is divided by n straight lines. *Am. Math. Monthly*, **48** (1941), 667–675. [394, 406]

A. Cauchy
1. Sur les polygones et les polyèdres. *J. Ecole Polytech.*, **9** (1813), 87. [411]

C. Chabauty
1. Empilement de spheres égales dans R^n et valeur asymptotique de la constante γ_n d'Hermite. *Compt. Rend. Acad. Sci. Paris*, **235** (1952), 529–532. [127]
2. Nouveaux résultats de géométrie des nombres. *Compt. Rend. Acad. Sci. Paris*, **235** (1952), 567–569. [127]
3. Résultats sur l'empilement de calottes égales sur une périsphère de R^n et correction à un travail antérieur. *Compt. Rend. Acad. Sci. Paris*, **236** (1953), 1462–1464. [127]

R. T. Chien, C. V. Freiman, and D. T. Tang
1. *Error correction and circuits on the n-cube.* Proc. 2nd Allerton Conference on Circuit and System Theory, Sept. 28–30, 1964. Univ. of Illinois, Monticello, Ill., pp. 899–912. [382]

G. Choquet
1. Unicité des representations intégrales dans les cônes convexes. *Compt. Rend. Acad. Sci. Paris*, **243** (1956), 699–702. [52]

J. L. Chrislock
1. Imbedding a skeleton of a simplex in Euclidean space. *Am. Math. Monthly*, **73** (1966), 381–382. [201]

J. Chuard
1. Les réseaux cubiques et le problème des quatres couleurs. *Mém. Soc. Vaudoise Sci. Nat.*, No. 25, Vol. 4 (1932), 41–101. [356]

P. Cohen
1. A simple proof of Tarski's theorem on elementary algebra. Lecture notes, Stanford University. [91]

T. M. Cover
1. Geometrical and statistical properties of systems of linear inequalities with applications in pattern recognition. *IEEE Trans. Electron. Computers*, **14** (1965), 326–334. [409]

T. Cover and B. Efron
1. The division of space by hyperplanes with applications to geometrical probability. Mimeographed report, Stanford University 1965. [409]

H. S. M. Coxeter
1. *Regular polytopes.* London 1948. (Second edition, 1963.)
   [49, 51, 56, 65, 142, 169, 258, 357, 412]
2. Problem 16. *Can. Math. Bull.*, **2** (1959), 122. Partial solution by W. Moser, *ibid.*, **6** (1963), 114–117.
3. The classification of zonohedra by means of projective diagrams, *J. Math. pures appl.*, (9) **41** (1962), 137–156. [405]
4. An upper bound for the number of equal nonoverlapping spheres that can touch another of the same size. *Proc. Symp. Pure Math.*, **7** (Convexity), 53–71 (1963). [313]
5. A noninductive definition for a regular polytope. Abstract 603–121. *Notices Am. Math. Soc.*, **10** (1963), 468. [412]
6. *Projective geometry.* New York 1964. [93]

H. S. M. Coxeter and A. Rosenthal
1. Sir William Hamilton's icosian game, Problem E711. *Am. Math. Monthly*, **53** (1946), 593. [357]

H. T. Croft
 1. On 6-point configurations in 3-space. *J. London Math. Soc.*, **36** (1961), 289–306. [421]
 2. Research problems. Dittoed notes, Cambridge University 1965. [73]
D. W. Crowe
 1. The *n*-dimensional cube and the tower of Hanoi. *Am. Math. Monthly*, **63** (1956), 29–30. [383]
A. Császár
 1. A polyhedron without diagonals. *Acta Sci. Math. (Szeged)*, **13** (1949), 140–142. [253]
L. D. Cummings
 1. Hexagonal systems of seven lines in a plane. *Bull. Am. Math. Soc.*, **38** (1933), 105–110. [394]
 2. Heptagonal systems of eight lines in a plane. *Bull. Am. Math. Soc.*, **38** (1933), 700–702. [394]
 3. On a method of comparison for straight-line nets. *Bull. Am. Math. Soc.*, **39** (1933), 411–416. [394]
G. B. Dantzig
 1. *Linear programming and extensions.* Princeton, N.J., 1963. [349, 377]
 2. Eight unsolved problems from mathematical programming. *Bull. Am. Math. Soc.*, **70** (1964), 499–500. [350, 381]
L. Danzer
 1. Review of L. Fejes Tóth [2]. *Math. Rev.*, **26** (1963), 569–570. [423]
L. Danzer and B. Grünbaum
 1. Über zwei Probleme bezüglich konvexer Körper von P. Erdös und von V. L. Klee. *Math. Z.*, **79** (1962), 95–99. [128, 420]
 2. Problem 16. *Proc. Convexity Colloq.*, Copenhagen 1965 (to appear). [422]
L. Danzer, B. Grünbaum and V. Klee
 1. Helly's theorem and its relatives. *Proc. Symp. Pure Math.*, 7 (Convexity), 101–180 (1963). [22, 29, 128]
L. Danzer and V. Klee
 1. Lengths of snakes in boxes. *J. Combinatorial Theory* (to appear). [383]
L. Danzer, D. Laugwitz and H. Lenz
 1. Über das Löwnersche Ellipsoid und sein Analogon unter den einem Eikörper einbeschriebenen Ellipsoiden. *Arch. Math.*, **8** (1957), 214–219. [412]
D. W. Davies
 1. Longest 'separated' paths and loops in an N cube. *IEEE Trans. Electron. Computers*, **14** (1965), 261. [382]
C. Davis
 1. The intersection of a linear subspace with the positive orthant. *Michigan Math. J.*, **1** (1952), 163–168. [73]
 2. Remarks on a previous paper. *Michigan Math. J.*, **2** (1953), 23–25. [73]
 3. Theory of positive linear dependence. *Am. J. Math.*, **76** (1954), 733–746. [348]
 4. The set of non-linearity of a convex piecewise-linear function. *Scripta Math.*, **24** (1959), 219–228. [41]
M. Dehn
 1. Die Eulersche Formel in Zusammenhang mit dem Inhalt in der nicht-Euklidischen Geometrie. *Math. Ann.*, **61** (1905), 561–586. [170, 312]
M. Dehn and P. Heegaard
 1. Analysis Situs. *Enzykl. math. Wiss.*, III AB3 (1907). [253]
D. Derry
 1. Convex hulls of simple space curves. *Can. J. Math.*, **8** (1956), 383–388. [63]
 2. Inflection hyperplanes of polygons. Abstract 633-24. *Notices Am. Math. Soc.*, **13** (1966), 355. [126, 426]

G. A. Dirac
1. Connectedness and structure in graphs. *Rend. Circ. Mat. Palermo, Ser. 2*, **9** (1960), 1–11. [215]
2. In abstracten Graphen vorhandene vollständige 4-Graphen und ihre Unterteilungen. *Math. Nachr.*, **22** (1960), 61–85. [215]
3. Extensions of Menger's theorem. *J. London Math. Soc.*, **38** (1963), 148–163. [213]

H. Durège
1. Ueber Körper von vier Dimensionen. *S.-Ber. Math.-nat. Cl. Kais. Akad. Wiss. Wien, Zweite Abt.*, **83** (1881), 1110–1125. [141]

P. Du Val
1. *Homographies, quaternions and rotations.* Oxford 1964. [412]

V. Eberhard
1. Eine Classification der allgemeinen Ebenensysteme. *J. reine angew. Math.*, **106** (1890), 89–120. [391, 398]
2. Ein Satz aus der Topologie. *Math. Ann.*, **36** (1890), 121–133. [391]
3. *Zur Morphologie der Polyeder.* Leipzig 1891. [254, 290, 405]

V. A. Efremovič and Ju. S. Il'jašenko
1. Regular polygons in E^n. *Vestn. Mosk. Univ. Ser. I, Mat. Mekhan.*, **1962**, No. 5, 18–24. (Russian, English summary.) [412]

H. G. Eggleston
1. On Rado's extension of Crum's problem. *J. London Math. Soc.*, **28** (1953), 467–471. [128]
2. *Problems in Euclidean space: Application of convexity.* New York 1957. [29]
3. *Convexity.* Cambridge 1958. [10, 29, 338]

H. G. Eggleston, B. Grünbaum and V. Klee
1. Some semicontinuity theorems for convex polytopes and cell-complexes. *Comm. Math. Helv.*, **39** (1964), 165–188. [52, 80, 83, 416]

P. Erdös
1. On sets of distances of n points. *Am. Math. Monthly*, **53** (1946), 248–250. [419]
2. Some unsolved problems. *Michigan Math. J.*, **4** (1957), 291–300. [420]
3. On sets of distances of n points in Euclidean space. *Magy. Tud. Akad. Mat. Kut. Int. Közl.*, **5** (1960), 165–169. [419]
4. Some unsolved problems. *Magy. Tud. Akad. Mat. Kut. Int. Közl.*, **6** (1961), 221–254. [404]

P. Erdös and R. Rado
1. A partition calculus in set theory. *Bull. Am. Math. Soc.*, **62** (1956), 427–489. [22]

P. Erdös and G. Szekeres
1. A combinatorial problem in geometry. *Compositio Math.*, **2** (1935), 463–470. [22, 371]
2. On some extremum problems in elementary geometry. *Ann. Univ. Sci. Budapest. Eötvös Sect. Mat.*, **3–4** (1960/61), 53–62. [22]

L. Euler
1. Elementa doctrinae solidorum. *Novi Comm. Acad. Sci. Imp. Petropol.*, **4** (1752/53), 109–140. [141]
2. Demonstratio nonnullarum insignium proprietatum, quibus solida hedris planis inclusa sunt praedita. *Novi Comm. Acad. Sci. Imp. Petropol.*, **4** (1752/53), 140–160. [141]

S. Even
1. Snake-in-the-box codes. *IEEE Trans. Electron. Computers*, **12** (1963), 18. [388]

G. Ewald
1. On Busemann's theorem of the Brunn–Minkowski type. *Proc. Convexity Colloq.*, Copenhagen 1965 (to appear). [338]

G. Ewald and G. C. Shephard
1. Normed vector spaces consisting of classes of convex sets. *Math. Z.*, **91** (1966), 1–19. [317]

F. Fabricius-Bjerre
  1. On polygons of order *n* in projective *n*-space, with an application to strictly convex curves. *Math. Scand.*, **10** (1962), 221–229. [63]
I. Fáry
  1. On straight line representation of planar graphs. *Acta Sci. Math. (Szeged)*, **11** (1948), 229–233. [253]
L. Fejes Tóth
  1. *Lagerungen in der Ebene, auf der Kugel und im Raum*. Berlin 1953. (Russian translation, with remarks and supplements by I. M. Yaglom, Moscow 1958.) [29, 292, 411, 418]
  2. On primitive polyhedra. *Acta Math. Acad. Sci. Hung.*, **13** (1962), 379–382. [423]
  3. *Regular figures*. New York 1964. [29, 411, 418]
  4. On the total area of the faces of a four-dimensional polytope. *Can. J. Math.*, **17** (1965), 93–99. [418]
W. Fenchel
  1. Inegalités quadratiques entre les volumes mixtes des corps convexes. *Compt. Rend. Acad. Sci. Paris*, **203** (1936), 647–650. [338]
  2. Généralizations du théorème de Brunn et Minkowski concernant les corps convexes. *Compt. Rend. Acad. Sci. Paris*, **203** (1936), 746–766. [338]
  3. A remark on convex sets and polarity. *Comm. Semin. Math. Univ. Lund*, Suppl. Vol. 1952, 82–89. [52]
  4. Convex cones, sets, and functions. Lecture notes. Princeton 1953. [29, 52]
W. Fenchel and B. Jessen
  1. Mengenfunktionen und konvexe Körper. *Mat.-Fys. Medd. Danske Vid. Selsk.*, Vol. 16, No. 3 (1938). [339]
M. Fieldhouse
  1. *Linear programming*. Ph.D. Thesis, Cambridge Univ. 1961. (Reviewed in *Operations Res.*, **10** (1962), 740.) [127, 171, 182, 368]
  2. Some properties of simplex polytopes. Dittoed notes. Harvard Univ. 1962. [171, 182, 188]
S. Fifer
  1. *Analogue computation*. Vol. 2., New York 1961. [383]
W. J. Firey
  1. Polar means of convex bodies and a dual to the Brunn–Minkowski theorem. *Can. J. Math.*, **13** (1961), 444–453. [340]
  2. Mean cross-section measures of harmonic means of convex bodies. *Pacific J. Math.*, **11** (1961), 1263–1266. [340]
  3. *p*-Means of convex bodies. *Math. Scand.*, **10** (1962), 53–60. [340]
  4. Some applications of means of convex bodies. *Pacific J. Math.*, **14** (1964), 53–60. [340]
W. J. Firey and B. Grünbaum
  1. Addition and decomposition of convex polytopes. *Israel J. Math.*, **2** (1964), 91–100. [339]
A. Flores
  1. Über *n*-dimensionale Komplexe die im R_{2n+1} absolut selbstverschlungen sind. *Ergebn. math. Kolloq.*, **6** (1933/34), 4–7. [201, 210]
G. Forchhammer
  1. Prøver paa Geometri med fire Dimensioner. *Tidsskr. Math.* (4), **5** (1881), 157–166. [141]
P. Franklin
  1. The four color problem. Galois Lecture, *Scripta Math. Library* No. 5, pp. 49–85. New York 1941. [364]
H. Freudenthal and B. L. van der Waerden
  1. Over een bewering van Euclides. *Simon Stevin*, **25** (1947), 115–121. [414]

J. W. Gaddum
 1. The sums of the dihedral and trihedral angles in a tetrahedron. *Am. Math. Monthly*,
    59 (1952), 370–371.   [312]
 2. Distance sums on a sphere and angle sums in a simplex. *Am. Math. Monthly*, **63**
    (1956), 91–96.   [312]
D. Gale
 1. Irreducible convex sets. *Proc. Intern. Congr. Math.*, Amsterdam 1954. Vol. 2,
    pp. 217–218.   [338]
 2. On convex polyhedra. Abstract 794. *Bull. Am. Math. Soc.*, **61** (1955), 556.   [127]
 3. Neighboring vertices on a convex polyhedron. (In *Linear inequalities and related
    systems*, edited by H. W. Kuhn and A. W. Tucker. Princeton 1956.) pp. 255–263.
                                                                               [85, 127]
 4. Neighborly and cyclic polytopes. *Proc. Symp. Pure Math.*, **7** (Convexity), 225–232
    (1963).   [61, 85, 121, 124, 221, 368]
 5. On the number of faces of a convex polytope. *Can. J. Math.*, **16** (1964), 12–17.
                                                   [80, 85, 121, 182, 368]
D. Gale and V. Klee
 1. Continuous convex sets. *Math. Scand.*, **7** (1959), 379–397.   [29]
H. Gericke
 1. Über stützbare Flächen und ihre Entwicklung nach Kugelfunktionen. *Math. Z.*,
    46 (1940), 55–61.   [314]
E. N. Gilbert
 1. Gray code and paths on the *n*-cube. *Bell System Tech. J.*, **37** (1958), 815–826.   [382]
A. S. Glass
 1. A remark on convex polytopes. *Can. Math. Bull.*, **8** (1965), 829–830.   [51]
I. C. Gohberg and A. S. Markus
 1. A problem on coverings of convex bodies by similar ones. *Izv. Moldavsk. Filiala
    Akad. Nauk SSSR*, **10** (76) (1960), 87–95. (Russian.)   [420]
T. Gosset
 1. On the regular and semi-regular figures in space of *n* dimensions. *Messeng. Math.*,
    29 (1900), 43–48.   [413]
D. W. Grace
 1. Computer search for non-isomorphic convex polyhedra. Report CS 15, Computer
    Science Dept., Stanford Univ. 1965.   [288, 364, 424]
J. P. Gram
 1. Om Rumvinklerne i et Polyeder. *Tidsskr. Math.* (Copenhagen) (3) **4** (1874), 161–163.
                                                                               [297, 312]
N. Gramenapoulos
 1. *An upper bound for error-correcting codes.* M.Sc. thesis, M.I.T. 1963.   [384]
H. J. Gray, Jr.. P. V. Levonian and M. Rubinoff
 1. An analogue-to-digital converter for serial computing machines. *Proc. IRE.*, **41**
    (1953), 1462–1465.   [383]
H. Grötzsch
 1. Zur Theorie der diskreten Gebilde. VII. Ein Dreifarbensatz für dreikreisfreie Netze
    auf der Kugel. *Wiss. Z. Martin-Luther-Univ. Halle-Wittenberg. Math.-Naturw. Reihe*,
    8 (1958/59), 109–120.   [254]
B. Grünbaum
 1. A proof of Vázsonyi's conjecture. *Bull. Res. Council Israel*, **6A** (1956), 77–78.   [419]
 2. On a problem of S. Mazur. *Bull. Res. Council Israel*, **7F** (1958), 133–135.   [73]
 3. On polyhedra in E³ having all faces congruent. *Bull. Res. Council Israel*, **8F** (1960),
    215–218.   [414]
 4. Projection constants. *Trans. Am. Math. Soc.*, **95** (1960), 451–465.   [73]
 5. On a conjecture of H. Hadwiger. *Pacific J. Math.*, **11** (1961), 215–219.   [338]

6. Strictly antipodal sets. *Israel J. Math.*, **1** (1963), 5–10.   [128, 420]
7. On Steinitz's theorem about non-inscribable polyhedra. *Proc. Ned. Akad. Wetenschap. Ser. A*, **66** (1964), 452–455.   [286]
8. Grötzsch's theorem on 3-colorings. *Michigan Math. J.*, **10** (1963), 303–310.   [254]
9. Measures of symmetry for convex sets. *Proc. Symp. Pure Math.*, **7** (Convexity), 233–270 (1963).   [314, 338]
10. Borsuk's problem and related questions. *Proc. Symp. Pure Math.*, **7** (Convexity), 271–284 (1963).   [129, 419]
11. Common secants for families of polyhedra. *Arch. Math.*, **15** (1964), 76–80.   [52]
12. Unambiguous polyhedral graphs. *Israel J. Math.*, **1** (1963), 235–238.   [78, 227, 229]
13. A simple proof of a theorem of Motzkin. *Proc. Ned. Akad. Wetenschap. Ser. A*, **67** (1964), 382–384.   [291]
14. Fixing systems and inner illumination. *Acta Math. Acad. Sci. Hung.*, **15** (1964), 161–163.   [422]
15. On the facial structure of convex polytopes. *Bull. Am. Math. Soc.*, **71** (1965), 559–560.   [200]
16. Diagrams and Schlegel diagrams. Abstract 625–112. *Notices Am. Math. Soc.*, **12** (1965), 578.   [219]

B. Grünbaum and N. W. Johnson
1. The faces of a regular-faced polyhedron. *J. London Math. Soc.*, **40** (1965), 577–586.   [414]

B. Grünbaum and L. M. Kelly
1. *Metrically homogeneous sets* (to appear).   [416]

B. Grünbaum and T. S. Motzkin
1. Longest simple paths in polyhedral graphs. *J. London Math. Soc.*, **37** (1962), 152–160.   [291, 343, 360, 373]
2. On polyhedral graphs. *Proc. Symp. Pure Math.*, **7** (Convexity), 744–751 (1963).   [121, 214, 225, 227, 229]
3. Unsolved problem. *Proc. Symp. Pure Math.*, **7** (Convexity), 498 (1963).   [217, 226]
4. The number of hexagons and the simplicity of geodesics on certain polyhedra. *Can. J. Math.*, **15** (1963), 744–751.   [272]

B. Grünbaum and V. P. Sreedharan
1. An enumeration of simplicial 4-polytopes with 8 vertices. *J. Combinatorial Theory*, **2** (1967) (to appear).   [121, 222, 424, 426]

J. P. de Gua de Malves
1. Propositions neuves, et non moins utiles que curieuses, sur le tétraèdre. *Hist. Acad. R. des Sci.*, *Paris*, 1783.   [312]

A. P. Guinand
1. A note on the angles in an *n*-dimensional simplex. *Proc. Glasgow Math. Assoc.*, **4** (1959), 58–61.   [313]

H. Hadwiger
1. Über eine symbolische Formel. *Elem. Math.*, **2** (1947), 35–41.   [142]
2. Eulers Charakteristik und kombinatorische Geometrie. *J. reine angew. Math.*, **194** (1955), 101–110.   [142]
3. *Altes und Neues über konvexe Körper.* Basel 1955.   [29, 417]
4. Ungelöste Probleme Nr. 17. *Elem. Math.*, **12** (1957), 61–62.   [284]
5. *Vorlesungen über Inhalt, Oberfläche und Isoperimetrie.* Berlin 1957.   [29, 416]
6. Ungelöste Probleme Nr. 38. *Elem. Math.*, **15** (1960), 130–131.   [420]

H. Hadwiger and H. Debrunner
1. *Kombinatorische Geometrie in der Ebene.* Geneva 1960. (English translation *Combinatorial geometry in the plane*, with additions by V. Klee, New York 1964.)   [29]

H. Hadwiger and D. Ohmann
1. Brunn–Minkowskischer Satz und Isoperimetrie. *Math. Z.*, **66** (1958), 1–8.   [338]

R. Halin
 1. Zu einem Problem von B. Grünbaum. *Arch. Math.*, **17** (1966), 566–568.   [215, 426]
P. C. Hammer
 1. Maximal convex sets. *Duke Math. J.*, **22** (1955), 103–106.   [13]
 2. Semispaces and the topology of convexity. *Proc. Symp. Pure Math.*, **7** (Convexity), 305–316 (1963).   [13]
R. W. Hamming
 1. Error-detecting and error-correcting codes. *Bell System Tech. J.*, **29** (1950), 147–160.   [384]

S. Hansen
 1. A generalization of a theorem of Sylvester on the lines determined by a finite point set. *Math. Scand.*, **17** (1966) (to appear).   [404]
H. Hasse
 1. *Proben mathematischer Forschung in allgemeinverständlicher Behandlung.* Frankfurt 1960.   [364]
A. F. Hawkins, A. C. Hill, J. E. Reeve and J. A. Tyrrell
 1. On certain polyhedra. *Math. Gaz.*, **50** (1966), 140–144.   [284]
R. Henstock and A. M. Macbeath
 1. On the measure of sum-sets. I. *Proc. London Math. Soc.*, (3) 3 (1953), 182–194.   [338]
A. Heppes
 1. Beweis einer Vermutung von A. Vázsonyi. *Acta Math. Acad. Sci. Hung.*, **7** (1957), 463–466.   [419]
O. Hermes
 1. Die Formen der Vielflache. *J. reine angew. Math.*, **120** (1899), 27–59, 305–353; **122** (1900), 124–154; **123** (1901), 312–342.   [48, 288, 424]
D. Hilbert and S. Cohn-Vossen
 1. *Anschauliche Geometrie.* Berlin 1932. (English translation *Geometry and the imagination*, New York 1952.)   [391]
W. Höhn
 1. *Winkel und Winkelsumme im n-dimensionalen Euklidischen Simplex.* Ph.D. Thesis, E. T. H. Zurich 1953.   [142, 303, 313]
R. Hoppe
 1. Regelmässige linear begrenzte Figuren von vier Dimensionen. *Arch. Math. Phys.*, **67** (1881), 29–44.   [141]
H. F. Hunter
 1. *On non-Hamiltonian maps and their duals.* Ph.D. Thesis, Rensselaer Polytechnic Institute 1962.   [364]
W. W. Jacobs and E. D. Schell
 1. The number of vertices of a convex polytope. (Abstract) *Am. Math. Monthly*, **66** (1959), 643.   [182]
M. Jerison
 1. A property of extreme points of compact convex sets. *Proc. Am. Math. Soc.*, **5** (1954), 782–783.   [20]
N. W. Johnson
 1. Convex polyhedra with regular faces. Abstract 576–157. *Notices Am. Math. Soc.*, 7 (1960), 952.   [414]
 2. Convex polyhedra with regular faces. *Can. J. Math.*, **18** (1966), 169–200.   [414]
 3. *The theory of uniform polytopes and honeycombs.* Ph.D. Thesis, Univ. of Toronto 1966.   [51, 412]
S. M. Johnson
 1. A new upper bound for error-correcting codes. *IEEE Trans. Information Theory*, **8** (1962), 203–207.   [384]
 2. Improved asymptotic bounds for error-correcting codes. *IEEE Trans. Information Theory*, **9** (1963), 198–205.   [384]

E. Jucovič
  1. Self-conjugate K-polyhedra. (Russian. German summary.) *Mat.-Fyz. Časopis Sloven. Akad. Vied*, **12** (1962), 1–22. [48]
  2. O mnogostenoch bez opisanej gul'ovej plochy. *Mat.-Fyz. Časopis Sloven. Akad. Vied*, **15** (1965), 90–94. [288]
E. Jucovič and J. W. Moon
  1. The maximum diameter of a convex polyhedron. *Math. Mag.*, **38** (1965), 31–32. [347]
J. A. Kalman
  1. Continuity and convexity of projections and barycentric coordinates in convex polyhedra. *Pacific J. Math.*, **11** (1961), 1017–1022. [36]
E. R. van Kampen
  1. Komplexe in Euklidischen Räumen. *Abh. math. Sem. Hamburg*, **9** (1932), 72–78, 152–153. [201, 210]
  2. Remark on the address of S. S. Cairns. 'Lectures in topology', University of Michigan Conference 1940, edited by R. L. Wilder and W. L. Ayres, pp. 311–313. Ann Arbor 1941. [202]
S. Karlin
  1. *Mathematical methods and theory in games, programming, and economics.* Vol. 2. Reading, Mass. 1959. [29]
S. Karlin and L. S. Shapley
  1. Geometry of moment spaces. *Mem. Am. Math. Soc.*, No. 12 (1953). [21, 63]
W. H. Kautz
  1. Unit-distance error-checking codes. *IRE Trans. Electron. Computers*, **7** (1958), 179–180. [382]
W. Keister, A. E. Ritchie and S. H. Washburn
  1. *The design of switching circuits.* New York 1951. [383]
O. H. Keller
  1. Über die lückenlose Erfüllung des Raumes mit Würfeln. *J. reine angew. Math.*, **163** (1930), 231–248. [411]
  2. Ein Satz über die lückenlose Erfüllung des 5- und 6-dimensionalen Raumes mit Würfeln. *J. reine angew. Math.*, **177** (1937), 61–64. [411]
L. M. Kelly and W. O. J. Moser
  1. On the number of ordinary lines determined by *n* points. *Can. J. Math.*, **10** (1958), 210–219. [403]
P. Kelly
  1. On some mappings related to graphs. *Pacific J. Math.*, **14** (1964), 191–194. [234]
R. Klee
  1. *Über die einfachen Konfigurationen der euklidischen und der projektiven Ebene.* Dresden 1938. [394, 408]
V. Klee
  1. The structure of semispaces. *Math. Scand.*, **4** (1956), 54–64. [13]
  2. Some characterizations of convex polyhedra. *Acta Math.*, **102** (1959), 79–107. [20, 26, 29, 37, 74]
  3. Polyhedral sections of convex bodies. *Acta Math.*, **105** (1960), 243–267. [28, 73]
  4. Asymptotes and projections of convex sets. *Math. Scand.*, **8** (1960), 356–362. [29]
  5. The Euler characteristic in combinatorial geometry. *Am. Math. Monthly*, **70** (1963), 119–127. [142]
  6. On a conjecture of Lindenstrauss. *Israel J. Math.*, **1** (1963), 1–4. [73]
  7. Infinite-dimensional intersection theorems. *Proc. Symp. Pure Math.*, **7** (Convexity) 37–51, (1963). [9]
  8. Convexity. *Am. Math. Soc. Proc. Symp. Pure Math.*, **7** (1963), edited by V. Klee. (XV + 516 pp.) [29]
  9. Cyclic polytopes. Lecture notes, Univ. of Washington, Seattle 1963. [61, 127]
  10. Extreme points of convex sets without completeness of the scalar field. *Mathematika*, **10** (1964), 59–63. [29, 76]

11. A combinatorial analogue of Poincaré's duality theorem. *Can. J. Math.*, **16** (1964), 517–531. [141, 145, 150, 171, 182]
12. Diameters of polyhedral graphs. *Can. J. Math.*, **16** (1964), 602–614. [345]
13. The number of vertices of a convex polytope. *Can. J. Math.*, **16** (1964), 701–720. [80, 171, 182, 188, 368]
14. A property of polyhedral graphs. *J. Math. Mech.*, **13** (1964), 1039–1042. [217, 226, 227, 292]
15. A class of linear programming problems requiring a large number of iterations. *Numer. Math.*, **7** (1965), 313–321. [376]
16. Heights of convex polytopes. *J. Math. Anal. Appl.*, **11** (1965), 176–190. [376]
17. Paths on polyhedra. I. *J. Soc. Ind. Appl. Math.*, **13** (1965), 946–956. [354]
18. Convex polytopes and linear programming. *Proc. IBM Scientific Computing Symposium on Combinatorial Problems*, March 16–18, 1964. pp. 123–158 (1966). [51, 141, 291, 381]
19. Paths on polyhedra. II. *Pacific J. Math.*, **16** (1966), 249–262. [354, 368]
20. Problem size in linear programming. *Proc. Convexity Colloq.*, Copenhagen 1965. pp. 177–184 (1966). [347]
21. A comparison of primal and dual methods of linear programming. *Numer. Math.*, **8** (1966) (to appear). [188, 349]
22. A method for constructing of circuit codes. Memorandum RM-5112-PR, The RAND Corporation, Santa Monica, Calif., 1966. [387]

V. Klee and D. Walkup
1. The *d*-step conjecture for polyhedra of dimension *d* < 6. *Acta Math.*, **117** (1967) (to appear). [347, 354]

H. Kneser
1. Eine Erweiterung des Begriffes 'konvexer Körper'. *Math. Ann.*, **82** (1921), 287–296. [30]

L. Kosmák
1. A remark on Helly's theorem. (Czech. Russian and English summaries.) *Spisy Přirodovědecké. Fak. Univ. Brně*, **1963**, 223–225. (See the review in *Math. Rev.*, **29**, #6390.) [17]

A. Kotzig
1. The construction of Hamiltonian graphs of degree 3. *Čas. Pěst. Math.*, **87** (1962), 148–168. (Russian. Czech and German summaries.) [364]
2. Coloring of trivalent polyhedra. *Can. J. Math.*, **17** (1965), 659–664. [291]

J. B. Kruskal
1. Monotonic subsequences. *Proc. Am. Math. Soc.*, **4** (1953), 264–274. [371]
2. The number of simplices in a complex. Symposium on 'Mathematical optimization techniques', Berkeley 1960. pp. 251–278 (1963). [178]

T. Kubota
1. Über die Schwerpunkte der konvexen geschlossenen Kurven und Flächen. *Tohoku Math. J.*, **14** (1918), 20–27. [314]

H. Kuhn and R. E. Quandt
1. An experimental study of the simplex method. *Proc. Symp. Appl. Math.*, **15** (Experimental Arithmetic), 107–124 (1963). [377]

H. Kuhn and A. W. Tucker (editors)
1. *Linear inequalities and related systems.* Princeton 1956. [29]

H. Lebesgue
1. Quelques conséquences simples de la formule d'Euler. *J. Math. pures appl.*, (9) **19** (1940), 27–43. [254]

J. Lederberg
1. Topological mapping of organic molecules. *Proc. Nat. Acad. Sci. US.*, **53** (1965), 134–139. [362]

2. Systematics of organic molecules, graph topology and Hamilton circuits. Instrumentation Res. Lab. Rept. No. 1040, Stanford Univ., 1966. [359]

S. Lefschetz

1. Topology. Am. Math. Soc. Colloq. Publ. No. 12 (1930). (Second edition, Chelsea, New York 1956.) [39, 141, 199]

F. Levi

1. Die Teilung der projektiven Ebene durch Gerade oder Pseudogerade. Ber. math.-phys. Kl. sächs. Akad. Wiss. Leipzig, 78 (1926), 256–267. [397, 408]

2. Geometrische Konfigurationen. Leipzig 1929. [391]

3. Überdeckung eines Eibereiches durch Parallelverschiebung seines offenen Kerns. Arch. Math., 6 (1955), 369–370. [420]

S. A. J. Lhuilier

1. De relatione mutua capacitatis et terminorum figuram. Varsaviae 1782. [418]

L. A. Lyusternik

1. Convex figures and polyhedra. Moscow 1956 (Russian; English translation, New York 1963). [29, 291]

W. Mader

1. Homomorphieeigenschaften und mittlere Kantendichte von Graphen (to appear).[426]

A. Marchaud

1. Sur les ensembles lineairement connexes. Ann. Mat. pura appl., (4) 56 (1961), 131–158. [30]

P. H. Maserik

1. Convex polytopes in linear spaces. Illinois J. Math., 9 (1965), 623–635. [52]

Z. A. Melzak

1. Limit sections and universal points of convex surfaces. Proc. Am. Math. Soc., 9 (1958), 729–734. [73]

2. A property of convex pseudopolyhedra. Can. Math. Bull., 2 (1959), 31–32. [73]

3. Problems connected with convexity. Can. Math. Bull., 8 (1965), 565–573. [417]

H. Minkowski

1. Allgemeine Lehrsätze über die konvexe Polyeder. Nachr. Ges. Wiss. Göttingen, 1897, 198–219. [332, 339]

2. Gesammelte Abhandlungen. (2 vols.) Berlin 1911. [29, 51, 418]

G. J. Minty

1. Problem E 1667. Am. Math. Monthly, 71 (1964), 205. Solution by J. W. Moon, ibid., 72 (1965), 81–82. [291]

H. Mirkil

1. New characterizations of polyhedral cones. Can. J. Math., 9 (1957), 1–4. [28]

J. W. Moon and L. Moser

1. Simple paths on polyhedra. Pacific J. Math., 13 (1963), 629–631. [357, 372]

L. Moser

1. Problem 65. Can. Math. Bull., 6 (1963), 113. [398]

2. Problem 77. Can. Math. Bull., 7 (1964), 137. Solution by B. Grünbaum, ibid., 477–478. [397]

T. S. Motzkin

1. Beiträge zur Theorie der linearen Ungleichungen. Ph.D. Thesis, Basel 1933 (Jerusalem 1936). [29, 52, 73]

2. Linear inequalities. Mimeographed lecture notes, University of California, Los Angeles, 1951. [13, 29, 52, 76, 85]

3. The lines and planes connecting the points of a finite set. Trans. Am. Math. Soc., 70 (1951), 451–464. [404]

4. Comonotone curves and polyhedra. Abstract 111. Bull. Am. Math. Soc., 63 (1957), 35. [63, 85, 124, 127, 166, 182]

5. The evenness of the number of edges of a convex polyhedron. *Proc. Nat. Acad. Sci. US.*, **52** (1964), 44–45. [283, 291]
6. A combinatorial result on maximally convex sets. Abstract 65T-303. *Notices Am. Math. Soc.*, **12** (1965), 603. [85, 126, 129]
7. Extension of the Minkowski–Carathéodory theorem on convex hulls. Abstract 65T-385. *Notices Am. Math. Soc.*, **12** (1965), 705 [38]
8. Generation and combinatorics of convex polyhedra. Talk at the Colloq. on Convexity, Copenhagen 1965. [283, 291]

R. C. Mullin
1. On counting rooted triangular maps. *Can. J. Math.*, **17** (1965), 373–382. [289]

H. Naumann
1. Beliebige konvexe Polytope als Schnitte und Projektionen höherdimensionaler Würfel, Simplices und Masspolytope. *Math. Z.*, **65** (1956), 91–103. [73]
2. Über Vektorsterne und Parallelprojektionen regulärer Polytope. *Math. Z.*, **67** (1957), 75–82. [73]

O. Ore
1. *Theory of graphs.* Am. Math. Soc. Colloq. Publ. No. 38. (1962). [213]

M. A. Perles
1. *Critical exponents of convex bodies.* Ph.D. Thesis. Hebrew University, Jerusalem 1964. (Hebrew. English Summary.) See also *Proc. Convexity Colloq.*, Copenhagen 1965 (to appear). [37]
2. *f*-Vectors of polytopes with a given group of symmetries (in preparation). [171]

M. A. Perles and G. C. Shephard
1. Angle sums of convex polytopes. *Math. Scand.* (to appear). [313, 427]
2. Facets and nonfacets of convex polytopes. *Acta Math.* (to appear). [415, 427]

M. A. Perles and D. W. Walkup
1. Angle inequalities for convex polytopes (in preparation). [427]

O. Perron
1. Über lückenlose Ausfüllung des *n*-dimensionalen Raumes durch kongruente Würfel. I, II. *Math. Z.*, **46** (1940), 1–26, 161–180. [411]

E. Peschl
1. Winkelrelationen am Simplex und die Eulersche Charakteristik. *Bayer. Akad. Wiss., Math.-Nat. Kl.S.-B.*, **1955** (1956), 319–345. [313]

J. Petersen
1. Die Theorie der regulären Graphs. *Acta Math.*, **15** (1891), 193–220. [375]

W. W. Peterson
1. *Error correcting codes.* M.I.T. Technology Press, Cambridge, Mass. 1961. [384]

M. Plotkin
1. Binary codes with specified minimum distance. *IRE Trans. Information Theory*, **6** (1960), 445–450. [384]

H. Poincaré
1. Sur la generalisation d'un theoreme d'Euler relatif aux polyedres. *Compt. Rend. Acad. Sci. Paris*, **117** (1893), 144–145. [142]
2. Complément a l'Analysis Situs. *Rend. Circ. Mat. Palermo*, **13** (1899), 285–343. [142]
3. Sur la generalization d'un theoreme élémentaire de Geometrie. *Compt. Rend. Acad. Sci. Paris*, **140** (1905), 113–117. [304, 312]

V. Polák
1. On a problem concerning convex polytopes. *Čas. Pěst. Math.*, **87** (1962), 169–179. (Russian, Czech and English summaries.) [415]

V. Polák and N. Poláková
1. Some remarks on indispensable elements in the systems of half-spaces in E_n. *Spisy Přírodovědecké Fak. Univ. Brně*, **1963**, 229–262. [423]

G. Pólya
1. Kombinatorische Anzahlbestimmungen für Gruppen, Graphen und chemische Verbindungen. *Acta Math.*, **68** (1937), 145–254. [289]
2. *Mathematics and plausible reasoning*. (2 vols.). Princeton 1954. [391, 409]
I. V. Proskuryakov
1. A property of *n*-dimensional affine space connected with Helly's theorem (Russian). *Usp. Mat. Nauk*, **14** (1959), No. 1 (85), 215–222. [17]
R. E. Quandt and H. Kuhn
1. On some computer experiments in linear programming. *Bull. Inst. Intern. Statist.*, **1962**, 363–372. [381]
2. On upper bounds for the number of iterations in solving linear programs. *Operations Res.*, **12** (1964), 161–165. [381]
H. Rademacher
1. On the number of certain types of polyhedra. *Illinois J. Math.*, **9** (1965), 361–380. [289, 424]
H. Rademacher and I. J. Schoenberg
1. Helly's theorem on convex domains and Tchebycheff's approximation problem. *Can. J. Math.*, **2** (1950), 245–256. [17]
R. Rado
1. A sequence of polyhedra having intersections of specified dimensions. *J. London Math. Soc.*, **22** (1947), 287–289. [128]
J. Radon
1. Mengen konvexer Körper, die einen gemeinsamen Punkt enthalten. *Math. Ann.*, **83** (1921), 113–115. [16]
C. Ramanujacharyulu and V. V. Menon
1. A note on the snake-in-the-box problem. *Publ. Inst. Statist. Univ. Paris*, **13** (1964), 131–135. [388]
F. P. Ramsey
1. On a problem of formal logic. *Proc. London Math. Soc.*, (2) **30** (1930), 264–286. [22]
J. R. Reay
1. Generalizations of a theorem of Carathéodory. *Mem. Am. Math. Soc.*, No. 54 (1965). [17]
2. A new proof of the Bonnice–Klee theorem. *Proc. Am. Math. Soc.*, **16** (1965), 585–587. [17]
3. An extension of Radon's theorem (to appear). [16]
A. Rényi, C. Rényi and J. Surányi
1. Sur l'independence des domains simples dans l'espace euclidean à *n* dimensions. *Colloq. Math.*, **2** (1951), 130–135. [409]
G. Ringel
1. Teilungen der Ebene durch Geraden oder topologische Geraden. *Math. Z.*, **64** (1956), 79–102. [408]
2. Über Geraden in allgemeiner Lage. *Elem. Math.*, **12** (1957), 75–82. [408]
3. Färbungsprobleme auf Flächen und Graphen. Berlin 1959. [254, 284, 364]
J. Riordan
1. The number of faces of simplicial polytopes. *J. Combinatorial Theory*, **1** (1966), 82–95. [171, 428]
S. Roberts
1. On the figures formed by the intercepts of a system of straight lines in a plane, and on analogous relations in space of three dimensions. *Proc. London Math. Soc.*, **19** (1889), 405–422. [391, 398]
C. A. Rogers
1. *Packing and covering*. Cambridge 1964. [29]

K. Rudel
 1. *Vom Körper höherer Dimension.* Kaiserlautern 1882. [141]
H. D. Ruderman
 1. Problem E 1663. *Am. Math. Monthly*, **71** (1964), 204. Partial solutions by M. Goldberg and J. D. E. Konhauser, *ibid.*, **72** (1965), 78–79. [409]
M. E. Rudin
 1. An unshellable triangulation of a tetrahedron. *Bull. Am. Math. Soc.*, **64** (1958), 90–91. [142]
H. J. Ryser
 1. *Combinatorial Mathematics.* New York 1963. [22]
T. L. Saaty
 1. The number of vertices of a polyhedron. *Am. Math. Monthly*, **62** (1955), 326–331. [182]
G. T. Sallee
 1. A valuation property of Steiner points. *Mathematika*, **13** (1966), 76–82. [315]
 2. Incidence graphs of polytopes (to appear). [216]
H. Samelson, R. M. Thrall and O. Wesler
 1. A partition theorem for Euclidean *n*-space. *Proc. Am. Math. Soc.*, **9** (1958), 805–807. [409]
J. A. Šaškin
 1. A remark on adjacent vertices of a convex polyhedron. *Usp. Mat. Nauk*, **18** (1963), No. 5 (113), 209–211 (Russian). [63]
L. Schläfli
 1. *Theorie der vielfachen Kontinuität.* *Denkschr. Schweiz. naturf. Ges.*, **38** (1901), 1–237. [51, 141, 391]
V. Schlegel
 1. *Theorie der homogen zusammengesetzten Raumgebilde.* *Nova Acta Leop. Carol.*, **44** (1883), 343–459. [43]
P. H. Schoute
 1. *Mehrdimensionale Geometrie. Zweiter Teil: Die Polytope.* Leipzig 1905. [121, 127, 142]
H. P. Schultz
 1. Topological organic chemistry. Polyhedranes and prismanes. *J. Org. Chem.*, **30** (1965), 1361–1364. [364]
K. Schütte
 1. Minimale Durchmesser endlicher Punktmengen mit vorgegebenem Mindestabstand. *Math. Ann.*, **150** (1963), 91–98. [421]
R. F. Scott
 1. Note on a theorem of Prof. Cayley. *Messeng. Math.*, **8** (1879), 155–157. [67]
A. Seidenberg
 1. A new decision method for elementary algebra. *Ann. Math.*, **60** (1954), 365–374. [91]
H. Seifert and W. Threlfall
 1. *Lehrbuch der Topologie.* Leipzig 1934. [253]
G. C. Shephard
 1. Inequalities between mixed volumes of convex sets. *Mathematika*, **7** (1960), 125–138. [338]
 2. Decomposable convex polyhedra. *Mathematika*, **10** (1963), 89–95. [338]
 3. On a conjecture of Melzak. *Can. Math. Bull.*, **7** (1964), 561–563. [73]
 4. Approximation problems for convex polyhedra. *Mathematika*, **11** (1964), 9–18. [338]
 5. Reducible convex sets. *Mathematika*, **13** (1966), 49–50. [338]
 6. Steiner points of convex polytopes. *Can. J. Math.* (to appear). [314]
 7. A proof of Walkup's conjecture (in preparation). [82]
 8. Steiner points of convex polytopes and cell complexes. *Proc. London Math. Soc.* (to appear). [314]

9. Approximations by polytopes with projectively regular facets. *Mathematika* (to appear). [82]
10. Angle deficiencies for convex polytopes (to appear). [427]
11. The mean width of a convex polytope. *J. London Math. Soc.* (to appear). [315, 427]
12. Projective polytopes (to appear). [68]
13. On Euler-type relations and mixed volumes of convex polytopes (in preparation). [427]

G. C. Shephard and R. J. Webster
1. Metrics for sets of convex bodies. *Mathematika*, **12** (1965), 73–88. [314]

F. W. Sinden
1. Duality in convex programming and in projective space. *J. Soc. Ind. Appl. Math.*, **11** (1963), 535–552. [30]

R. C. Singleton
1. Generalized snake-in-the-box codes. *IEEE Trans. Electron. Computers*, **15** (1966). [382]

T. Skolem
1. Ein kombinatorischer Satz mit Anwendung auf ein logisches Entscheidungsproblem. *Fundam. Math.*, **20** (1933), 254–261. [22]

P. S. Soltan
1. Illumination from within of the boundary of a convex body. *Mat. Sb. (N.S.)*, **57** (99) (1962), 443–448. (Russian.) [422]
2. On the problems of covering and illuminating convex bodies. *Izv. Akad. Nauk Moldavsk. SSR*, **1963**, No. 1, 49–57. (Russian.) [420]

D. M. Y. Sommerville
1. The relations connecting the angle-sums and volume of a polytope in space of *n* dimensions. *Proc. Roy. Soc. London*, Ser. A, **115** (1927), 103–119. [140, 145, 148, 170, 312]
2. *An introduction to the geometry of n dimensions*. London 1929. [43, 51, 63, 121, 127, 142, 171, 309, 312]

D. A. Sprott
1. A combinatorial identity. *Math. Gaz.*, **40** (1956), 207–209. [313]

G. K. C. von Staudt
1. *Geometrie der Lage.* Nürnberg 1847. [391]

S. K. Stein
1. Convex maps. *Proc. Am. Math. Soc.*, **2** (1951), 464–466. [291]

J. Steiner
1. Einige Gesetze über die Theilung der Ebene und des Raumes. *J. reine angew. Math.*, **1** (1826), 349–364. [391, 409]
2. Sur le maximum et le minimum des figure dans le plan, sur la sphère et dans l'espace en général. *J. reine angew. Math.*, **24** (1842), 93–152, 189–250. [418]
3. *Gesammelte Werke.* (2 vols.) Berlin 1881, 1882. [284, 313]

E. Steinitz
1. Über die Eulersche Polyderrelationen. *Arch. Math. Phys.*, (3) **11** (1906), 86–88. [142, 190]
2. Über diejenigen konvexen Polyeder mit *n* Grenzflächen, welche nicht durch *n* − 4 ebene Schnitte aus einen Tetraeder abgeleitet werden können. *Arch. Math. Phys.*, (3) **14** (1909), 1–48. [423]
3. Konfigurationen der projektiven Geometrie. *Enzykl. math. Wiss.*, Vol. 3 (Geometrie), Part I.1, pp. 481–516 (1910). [391]
4. Über Konfigurationen. *Arch. Math. Phys.*, (3) **16** (1910), 289–313. [391]
5. Bedingt konvergente Reihen und konvexe Systeme. *J. reine angew. Math.*, **143** (1913), 128–175; **144** (1914), 1–40; **146** (1916), 1–52. [17, 29]
6. Polyeder und Raumeinteilungen. *Enzykl. math. Wiss.*, Vol. 3 (Geometrie), Part 3AB12, pp. 1–139 (1922). [127, 141, 188, 235, 288]

7. Über die Maximalzahl der Doppelpunkte bei ebenen Polygonen von gerader Seitenzahl. *Math. Z.*, **17** (1923), 116–129. [404]

8. Über isoperimetrische Probleme bei konvexen Polyedern. *J. reine angew. Math.*, **158** (1927), 129–153; **159** (1928), 133–143. [285, 292, 418]

E. Steinitz and H. Rademacher
1. *Vorlesungen über die Theorie der Polyeder.* Berlin 1934. [78, 141, 188, 235, 290]

J. J. Stoker
1. Uniqueness theorems for polyhedra. *Proc. Nat. Acad. Sci. US.*, **55** (1966), 1398–1404. [411]

A. B. Stott
1. Geometrical deduction of semiregular from regular polytopes and space fillings. *Verhandel. Koninkl. Akad. Wetenschap.* (Eerste sectie), **11**, No. 1 (1910). [413]

S. Straszewicz
1. Über exponierte Punkte abgeschlossener Punktmengen. *Fundam. Math.*, **24** (1935), 139–143. [19]
2. Sur un problème geométrique de P. Erdös. *Bull. Acad. Polon. Sci. Sér. Sci. Math. Astron. Phys.*, **5** (1957), 39–40. [419]

W. I. Stringham
1. Regular figures in *n*-dimensional space. *Am. J. Math.*, **3** (1880), 1–14. [141]

F. Supnick
1. On the perspective deformation of polyhedra. *Ann. Math.*, **49** (1948), 714–730; **53** (1951), 551–555. [292, 409]

W. Süss and H. Kneser
1. Aufgabe 298. *Jber. dtsch. MatVer.*, **51** (1941), *298.* Solutions by G. Bol, *ibid.*, **53** (1943), *35–37*, and by O. Baier, *ibid.*, **53** (1943), *38–40.* [74]

J. J. Sylvester
1. Mathematical question 11851. *Educational Times*, **59** (1893), 98. [404]

G. Szász
1. *Introduction to lattice theory.* New York 1963. [21]

B. Sz.-Nagy
1. Sur un problème pour les polyèdres convexes dans l'espace *n*-dimensionnel. *Bull. Soc. Math. France*, **69** (1941), Commun. et Conférer. 1938–1939, pp. 3–4. [127]

P. G. Tait
1. Remarks on the colouring of maps. *Proc. Roy. Soc. Edinburgh*, **10** (1880), 729 [356]
2. Note on a theorem in geometry of position. *Trans. Roy. Soc. Edinburgh*, **29** (1880), 657–660. [356]
3. On Listing's 'Topologie'. *Phil. Mag.*, **17** (1884), 30–46. [356]

A. Tarski
1. *A decision method for elementary algebra and geometry.* Berkeley 1951. [91]

H. Tietze
1. Über das Problem der Nachbargebiete im Raum. *Monatsh. Math.*, **16** (1905), 211–216. [127]
2. *Gelöste und ungelöste mathematische Probleme aus alter und neuer Zeit.* Munich 1949. (English translation '*Famous problems of mathematics*', New York 1965.) [128]

H. E. Tompkins
1. Unit distance binary codes. Report No. 58-09, Moore School of Elec. Eng., Univ. of Pennsylvania, Philadelphia 1957. [382]

W. T. Tutte
1. On Hamiltonian circuits. *J. London Math. Soc.*, **21** (1946), 98–101. [356, 358]
2. The factorization of linear graphs. *J. London Math. Soc.*, **22** (1947), 107–111. [375]
3. A theorem on planar graphs. *Trans. Am. Math. Soc.*, **82** (1956), 99–116. [364]
4. A non-Hamiltonian planar graph. *Acta Math. Acad. Sci. Hung.*, **11** (1960), 371–375. [365]

5. Convex representations of graphs. *Proc. London Math. Soc.*, **10** (1960), 304–320. [291]
6. A theory of 3-connected graphs. *Proc. Ned. Akad. Wetenschap. Ser. A*, **64** (1961), 441–455. [290]
7. A census of planar triangulations. *Can. J. Math.*, **14** (1962), 21–38. [289]
8. A census of Hamiltonian polygons. *Can. J. Math.*, **14** (1962), 402–417. [289]
9. A new branch of enumerative graph theory. *Bull. Am. Math. Soc.*, **68** (1962). 500–504. [289]
10. A census of planar maps. *Can. J. Math.*, **15** (1963), 249–271. [289]
11. How to draw a graph. *Proc. London Math. Soc.*, **13** (1963), 743–767. [291]

H. Tverberg
1. A generalization of Radon's theorem. *J. London Math. Soc.*, **41** (1966), 123–128. [16]

S. Ulam
1. The Scottish Book. Mimeographed notes, Los Alamos 1957. [73]

P. Ungar
1. On diagrams representing maps. *J. London Math. Soc.*, **28** (1953), 336–342. [292]

F. A. Valentine
1. *Convex sets*. New York 1964. [29, 74]

J. L. Vasilev
1. On the length of a cycle in an *n*-dimensional unit cube. *Dokl. Akad. Nauk SSSR*, **148** (1963), 753–756. (Russian; English translation in *Soviet Math.*, **4** (1963), 160–163.) [388]

O. Veblen and J. W. Young
1. *Projective geometry*. Volume 2. Boston 1918. [30]

K. Wagner
1. Einbettung von überabzählbaren, simplicialen Komplexen in euklidische Räume. *Arch. Math.*, **17** (1966), 169–171. [426]

C. T. C. Wall
1. Arithmetic invariants of subdivisions of complexes. *Can. J. Math.*, **18** (1966), 92–96. [426]

H. Walther
1. Ein kubischer, planarer, zyklisch fünffach zusammenhängender Graph, der keinen Hamiltonkreis besitzt. *Wiss. Z. Hochschule Elektrotech. Ilmenau*, **11** (1965), 163–166. [365, 428]

H. Weyl
1. Elementare Theorie der konvexen Polyeder. *Comm. Math. Helv.*, **7** (1935/36), 290–306. (English translation in 'Contributions to the theory of games', *Ann. Math. Studies* No. 24, Princeton 1950, pp. 3–18.) [29, 51, 76]

H. S. White
1. The plane figures of seven real lines. *Bull. Am. Math. Soc.*, **38** (1932), 59–65. [394]
2. The convex cells formed by seven planes. *Proc. Nat. Acad. Sci. US.*, **25** (1939), 147–153. [394]

H. Whitney
1. A theorem on graphs. *Ann. Math.*, **32** (1931), 378–390. [364]
2. Congruent graphs and the connectivity of graphs. *Am. J. Math.*, **54** (1932), 150–168. [213, 234]

H. Wiesler
1. Despre convexitatea unor coordonate baricentrice generalizate. *Stud. Cerc. Mat.*, **15** (1964), 369–373. [37]

A. D. Wyner
1. Capabilities of bounded discrepancy decoding. *Bell System Tech. J.*, **44** (1965), 1061–1122. [384]

I. M. Yaglom and V. G. Boltyanskii
1. *Convex figures*. Moscow 1951. (Russian; German translation Berlin 1956; English translation New York 1961.) [29]

M. Zacharias
  1. Elementargeometrie und elementare nicht-euklidische Geometrie in synthetischer
     Behandlung. *Enzykl. math. Wiss.*, Vol. 3 (Geometrie), Part 1, pp. 862–1172 (1913). [141]
V. A. Zalgaller
  1. Regular-faced polyhedra. *Vestn. Leningr. Univ. Ser. Mat. Mekhan. Astron.*, **18** (1963),
     No. 7, 5–8 (Russian).   [414]
V. A. Zalgaller *et al.*
  1. On regular-faced polyhedra. *Vestn. Leningr. Univ. Ser. Mat. Mekhan. Astron.*,
     20 (1965), No. 1, 150–152 (Russian).   [414]
J. I. Žuravlev
  1. Set-theoretic methods in a Boolean algebra. *Probl. Kibernetiki.*, **8** (1962), 5–44
     (Russian).   [388]

ADDITIONAL BIBLIOGRAPHY

H. L. Abbott and M. Katchalski
  a. On the construction of snake in the box codes, *Util. Math.*, **40** (1991), 97–116.   [389b]

H. Achatz and P. Kleinschmidt
  a. Reconstructing a simple polytope from its graph, in *Polytopes – Combinatorics and Computation*, G. Kalai and G. M. Ziegler, eds., vol. 29 of DMV Semin., Birkhäuser, Basel, 2000, 155–165.   [234b]

R. M. Adin
  a. A new cubical *h*-vector, *Discrete Math.*, **157** (1996), 3–14.   [171b]

P. K. Agarwal and M. Sharir
  a. Arrangements and their applications, in *Handbook of Computational Geometry*, J.-R. Sack and J. Urrutia, eds., North-Holland, Amsterdam, 2000, 49–119.   [410b]

O. Aichholzer, H. Alt, and G. Rote
  a. Matching shapes with a reference point, *Int. J. Comput. Geom. Appl.*, **7** (1997), 349–363.   [315b]

M. Aigner and G. M. Ziegler
  a. *Proofs from THE BOOK*, Springer, Heidelberg Berlin, second ed., 2000.   [129b]

R. E. L. Aldred, S. Bau, D. A. Holton, and B. D. McKay
  a. Nonhamiltonian 3-connected cubic planar graphs, *SIAM J. Discrete Math.*, **13** (2001), 25–32.   [389a]

N. Alon
  a. The number of polytopes, configurations and real matroids, *Mathematika*, **33** (1986), 62–71.   [121a]

B. Alspach and M. Rosenfeld
  a. On Hamilton decompositions of prisms over simple 3-polytopes, *Graphs Comb.*, **2** (1986), 1–8.   [389a]

A. Altshuler
  a. Neighborly 4-polytopes and neighborly combinatorial 3-manifolds with 10 vertices, *Can. J. Math.*, **29** (1977), 400–420.   [96b, 129a]

  b. The Mani-Walkup spherical counterexamples to the W_v-path conjecture are not polytopal, *Math. Oper. Res.*, **10** (1985), 158–159.   [355b]

A. Altshuler, J. Bokowski, and L. Steinberg
  a. The classification of simplicial 3-spheres with nine vertices into polytopes and nonpolytopes, *Discrete Math.*, **31** (1980), 115–124.   [96b]

A. Altshuler and P. McMullen
  a. The number of simplicial neighborly *d*-polytopes with $d + 3$ vertices, *Mathematika*, **20** (1973), 263–266.   [121a]

A. Altshuler and I. Shemer
  a. Construction theorems for polytopes, *Isr. J. Math.*, **47** (1984), 99–110.   [428a]

A. Altshuler and L. Steinberg

a. Neighborly 4-polytopes with 9 vertices, *J. Comb. Theory, Ser. A*, **15** (1973), 270–287. [129a]

b. Enumeration of the quasisimplicial 3-spheres and 4-polytopes with eight vertices, *Pac. J. Math.*, **113** (1984), 269–288. [96b]

c. The complete enumeration of the 4-polytopes and 3-spheres with eight vertices, *Pac. J. Math.*, **117** (1985), 1–16. [96b]

N. Amenta and G. M. Ziegler

a. Deformed products and maximal shadows, in *Advances in Discrete and Computational Geometry*, B. Chazelle, J. E. Goodman, and R. Pollack, eds., vol. 223 of Contemp. Math., Amer. Math. Soc., Providence RI, 1998, 57–90. [389b]

K. Appel and W. Haken

a. *Every planar map is four colorable*, vol. 98 of Contemp. Math., Amer. Math. Soc., Providence, RI, 1989. [389a]

J. Ashley, B. Grünbaum, G. C. Shephard, and W. Stromquist

a. Self-duality groups and ranks of self-dualities, in *Applied Geometry and Discrete Mathematics. The Victor Klee Festschrift*, P. Gritzmann and B. Sturmfels, eds., vol. 4 of DIMACS Series in Discrete Mathematics and Theoretical Computer Science, Amer. Math. Soc., Providence RI, 1991, 11–50. [52d]

D. Avis

a. A C implementation of the reverse search vertex enumeration algorithm, http://cgm.cs.mcgill.ca/~avis/C/lrs.html, 1992+. [52b]

D. Avis, D. Bremner, and R. Seidel

a. How good are convex hull algorithms?, *Comput. Geom.*, **7** (1997), 265–301. [52b]

D. Avis and K. Fukuda

a. A pivoting algorithm for convex hulls and vertex enumeration of arrangements and polyhedra, *Discrete Comput. Geom.*, **8** (1992), 295–313. [52a]

E. K. Babson, L. Finschi, and K. Fukuda

a. Cocircuit graphs and efficient orientation reconstruction in oriented matroids, *Eur. J. Comb.*, **22** (2001), 587–600. [234b]

K. Ball

a. An elementary introduction to modern convex geometry, in *Flavors of Geometry*, S. Levy, ed., vol. 31 of Publ. MSRI, Cambridge University Press, 1997, 1–58. [30b, 96a]

I. Bárány

a. A short proof of Kneser's conjecture, *J. Comb. Theory, Ser. A*, **25** (1978), 325–326. [129b]

b. Intrinsic volumes and f-vectors of random polytopes, *Math. Ann.*, **285** (1989), 671–699. [129b]

I. Bárány and A. Pór

a. 0-1 polytopes with many facets, *Adv. Math.*, **161** (2001), 209–228. [69a]

D. W. Barnette

a. Projections of 3-polytopes, *Isr. J. Math.*, **8** (1970), 304–308. [296b]

b. The minimum number of vertices of a simple polytope, *Isr. J. Math.*, **10** (1971), 121–125. [198a]

c. Graph theorems for manifolds, *Isr. J. Math.*, **16** (1973), 62–72. [198a]

d. A proof of the lower bound conjecture for convex polytopes, *Pac. J. Math.*, **46** (1973), 349–354. [198a]

e. The projection of the f-vectors of 4-polytopes onto the (E,S)-plane, *Discrete Math.*, **10** (1974), 201–216. [198c]

f. An upper bound for the diameter of a polytope, *Discrete Math.*, **10** (1974), 9–13. [355a]

g. A family of neighborly polytopes, *Isr. J. Math.*, **39** (1981), 127–140. [129a]

h. *Map Coloring, Polyhedra, and the Four Color Theorem*, vol. 8 of Dolciani Mathematical Expositions, Mathematical Association of America, Washington DC, 1983. [389a]

i. A 2-manifold of genus 8 without the W_v-property, *Geom. Dedicata*, **46** (1993), 211–214. [355b]

D. W. Barnette and B. Grünbaum
a. On Steinitz's theorem concerning convex 3-polytopes and on some properties of 3-connected graphs, in *The Many Facets of Graph Theory*, vol. 110 of Lecture Notes in Mathematics, Springer, 1969, 27–40. [296a]

b. Preassigning the shape of a face, *Pac. J. Math.*, **32** (1970), 299–302. [296b]

D. W. Barnette and J. R. Reay
a. Projections of f-vectors of four-polytopes, *J. Comb. Theory, Ser. A*, **15** (1973), 200–209. [198c]

A. I. Barvinok
a. On equivariant generalization of Dehn-Sommerville equations, *Eur. J. Comb.*, **13** (1992), 419–428. [171b]

b. Lattice points and lattice polytopes, in *Handbook of Discrete and Computational Geometry*, J. E. Goodman and J. O'Rourke, eds., CRC Press, Boca Raton, 1997, ch. 7, 133–152. [52a]

M. M. Bayer
a. The extended f-vectors of 4-polytopes, *J. Comb. Theory, Ser. A*, **44** (1987), 141–151. [198c]

M. M. Bayer and L. J. Billera
a. Generalized Dehn-Sommerville relations for polytopes, spheres and Eulerian partially ordered sets, *Invent. Math.*, **79** (1985), 143–157. [198c]

M. M. Bayer and A. Klapper
a. A new index for polytopes, *Discrete Comput. Geom.*, **6** (1991), 33–47. [198c]

M. M. Bayer and C. W. Lee
a. Combinatorial aspects of convex polytopes, in *Handbook of Convex Geometry*, P. Gruber and J. Wills, eds., North-Holland, Amsterdam, 1993, 485–534. [xii]

E. A. Bender and N. C. Wormald
a. The number of rooted convex polyhedra, *Can. Math. Bull.*, **31** (1988), 99–102. [296c]

C. T. Benson and L. C. Grove
a. *Finite Reflection Groups*, vol. 99 of Graduate Texts in Mathematics, Springer, New York, second ed., 1985. [423a]

J. D. Berman and K. Hanes
a. Volumes of polyhedra inscribed in the unit sphere E^3, *Math. Ann.*, **188** (1970), 78–84. [296d]

b. Optimizing the arrangement of points on the unit sphere, *Math. Comput.*, **31** (1977), 1006–1008. [296d]

K. Bezdek

a. Hadwiger-Levi's covering problem revisited, in *New Trends in Discrete and Combinatorial Geometry*, J. Pach, ed., vol. 10 of Algorithms and Combinatorics, Springer, Berlin Heidelberg, 1993, ch. VIII, 199–233. [423c]

L. J. Billera and A. Björner

a. Face numbers of polytopes and complexes, in *Handbook of Discrete and Computational Geometry*, J. E. Goodman and J. O'Rourke, eds., CRC Press, Boca Raton, 1997, ch. 15, 291–310. [198a]

L. J. Billera and C. W. Lee

a. A proof of the sufficiency of McMullen's conditions for f-vectors of simplicial polytopes, *J. Comb. Theory, Ser. A*, **31** (1981), 237–255. [198a]

L. J. Billera and B. Spellman Munson

a. Polarity and inner products in oriented matroids, *Eur. J. Comb.*, **5** (1984), 293–308. [96a]

A. Björner

a. Face numbers of complexes and polytopes, in *Proc. of the International Congress of Mathematicians (Berkeley CA, 1986)*, 1986, vol. 2, 1408–1418. [198a]

b. Partial unimodality for f-vectors of simplicial polytopes and spheres, in *Jerusalem Combinatorics '93*, H. Barcelo and G. Kalai, eds., vol. 178 of Contemp. Math., Providence RI, 1994, Amer. Math. Soc., 45–54. [198b]

c. Subspace arrangements, in *Proc. of the First European Congress of Mathematics (Paris 1992)*, vol. I, Birkhäuser, 1994, 321–370. [410b]

A. Björner, P. H. Edelman, and G. M. Ziegler

a. Hyperplane arrangements with a lattice of regions, *Discrete Comput. Geom.*, **5** (1990), 263–288. [234b, 410a]

A. Björner and G. Kalai

a. An extended Euler-Poincaré theorem, *Acta Math.*, **161** (1988), 279–303. [198c]

A. Björner, M. Las Vergnas, B. Sturmfels, N. White, and G. M. Ziegler

a. *Oriented Matroids*, vol. 46 of Encyclopedia of Mathematics, Cambridge University Press, Cambridge, second (paperback) ed., 1999. [7b, 30b, 96a, 340c, 410b]

A. Björner and S. Linusson

a. The number of k-faces of a simple d-polytope, *Discrete Comput. Geom.*, **21** (1999), 1–16. [198b]

G. Blind and R. Blind

a. Convex polytopes without triangular faces, *Isr. J. Math.*, **71** (1990), 129–134. [198d]

b. The almost simple cubical polytopes, *Discrete Math.*, **184** (1998), 25–48. [69b]

c. Shellings and the lower bound theorem, *Discrete Comput. Geom.*, **21** (1999), 519–549. [198a]

R. Blind and P. Mani-Levitska

a. On puzzles and polytope isomorphisms, *Aequationes Math.*, **34** (1987), 287–297. [234a]

A. I. Bobenko and B. A. Springborn

a. Variational principles for circle patterns, and Koebe's theorem, Tech. Rep. No. 545 (Sfb 288 preprint series), TU Berlin, 2002; arXiv: math.GT/0203250, 38 pages. [296a]

J. Bochnak, M. Coste, and M.-F. Roy

a. *Real Algebraic Geometry*, Springer, New York, 1998. [52b]

J. Bokowski
  a. Oriented matroids, in *Handbook of Convex Geometry*, P. Gruber and J. Wills, eds., North-Holland, Amsterdam, 1993, 555–602. [410b]

J. Bokowski and K. Garms
  a. Altshuler's sphere M_{425} is not polytopal, *Eur. J. Comb.*, **8** (1987), 227–229. [96b, 129a]

J. Bokowski and A. Guedes de Oliveira
  a. Simplicial convex 4-polytopes do not have the isotopy property, *Port. Math.*, **47** (1990), 309–318. [224a]

J. Bokowski and J. Richter[-Gebert]
  a. On the finding of final polynomials, *Eur. J. Comb.*, **11** (1990), 21–34. [96b]

J. Bokowski, J. Richter-Gebert, and W. Schindler
  a. On the distribution of order types, *Comput. Geom.*, **1** (1992), 127–142. [129b]

J. Bokowski, J. Richter[-Gebert], and B. Sturmfels
  a. Nonrealizability proofs in computational geometry, *Discrete Comput. Geom.*, **5** (1990), 333–350. [96b]

J. Bokowski and I. Shemer
  a. Neighborly 6-polytopes with 10 vertices, *Isr. J. Math.*, **58** (1987), 103–124. [96b, 129a]

J. Bokowski and B. Sturmfels
  a. Polytopal and nonpolytopal spheres – an algorithmic approach, *Isr. J. Math.*, **57** (1987), 257–271. [96b, 129a]
  b. *Computational Synthetic Geometry*, vol. 1355 of Lecture Notes in Mathematics, Springer, 1989. [96b]

E. Bolker
  a. A class of convex bodies, *Trans. Am. Math. Soc.*, **145** (1969), 323–345. [340b]

V. G. Boltianskii
  a. *Hilbert's Third Problem*, V. H. Winston & Sons (Halsted Press, John Wiley & Sons), Washington DC, 1978. [315b]

V. Boltyanski, H. Martini, and P. S. Soltan
  a. *Excursions into Combinatorial Geometry*, Universitext, Springer, 1997. [30b, 423b, 423c]

V. Boltyanski, H. Martini, and V. Soltan
  a. On Grünbaum's conjecture about inner illumination of convex bodies, *Discrete Comput. Geom.*, **22** (1999), 403–410. [423c]

K. H. Borgwardt
  a. *The Simplex Method. A Probabilistic Analysis*, vol. 1 of Algorithms and Combinatorics, Springer, Berlin Heidelberg, 1987. [389b]

J. Bourgain, J. Lindenstrauss, and V. Milman
  a. Approximation of zonoids by zonotopes, *Acta Math.*, **162** (1989), 73–141. [340b]

U. Brehm
  a. A nonpolyhedral triangulated Möbius strip, *Proc. Am. Math. Soc.*, **89** (1983), 519–522. [224a]

U. Brehm and G. Schild
  a. Realizability of the torus and the projective plane in R^4, *Isr. J. Math.*, **91** (1995), 249–251. [224a]

D. Bremner
  a. Incremental convex hull algorithms are not output sensitive., *Discrete Comput. Geom.*, **21** (1999), 57–68. [52b]

D. Bremner and V. Klee
   a. Inner diagonals of convex polytopes, *J. Comb. Theory, Ser. A*, **87** (1999), 175–197. [423c]

A. Brieden, P. Gritzmann, R. Kannan, L. Lovász, and M. Simonovits
   a. Deterministic and randomized polynomial-time approximation of radii, *Mathematika*, to appear. [423c]

G. Brightwell and W. T. Trotter
   a. The order dimension of convex polytopes, *SIAM J. Discrete Math.*, **6** (1993), 230–245. [296d]

G. R. Brightwell and E. R. Scheinerman
   a. Representations of planar graphs, *SIAM J. Discrete Math.*, **6** (1993), 214–229. [296a]

G. Brinkmann and A. W. Dress
   a. A constructive enumeration of fullerenes, *J. Algorithms*, **23** (1997), 345–358. [296b]

G. Brinkmann and B. McKay
   a. `plantri`. A program for generating planar triangulations and planar cubic graphs, `http://cs.anu.edu.au/people/bdm/plantri/`. [296c]

M. Brion
   a. The structure of the polytope algebra, *Tôhoku Math. J.*, **49** (1997), 1–32. [340a]

A. Brøndsted
   a. *An Introduction to Convex Polytopes*, vol. 90 of Graduate Texts in Mathematics, Springer, New York Berlin, 1983. [xii]

H. Brönnimann
   a. Degenerate convex hulls on-line in any fixed dimension, *Discrete Comput. Geom.*, **22** (1999), 527–545. [96b]

K. S. Brown
   a. *Buildings*, Springer, New York, 1989. [423a]

H. Bruggesser and P. Mani
   a. Shellable decompositions of cells and spheres, *Math. Scand.*, **29** (1971), 197–205. [142a, 198a]

C. Buchta, J. Müller, and R. Tichy
   a. Stochastical approximation of convex bodies, *Math. Ann.*, **271** (1985), 225–235. [129b]

G. R. Burton
   a. The non-neighbourliness of centrally symmetric convex polytopes having many vertices, *J. Comb. Theory, Ser. A*, **58** (1991), 321–322. [121b]

R. Čada, T. Kaiser, M. Rosenfeld, and Z. Ryjáček
   a. Hamiltonian decompositions of prisms over cubic graphs, Tech. Rep., University of West Bohemia, Plzeň, Czech Republic, February 2002. [389a]

C. Chan
   a. Plane trees and *h*-vectors of shellable cubical complexes, *SIAM J. Discrete Math.*, **4** (1991), 568–574. [171b]

B. Chazelle
   a. An optimal convex hull algorithm in any fixed dimension, *Discrete Comput. Geom.*, **10** (1993), 377–409. [52a]

V. Chvátal
   a. Sylvester-Gallai theorem and metric betweenness, DIMACS Tech. Rep. 2002-19, Rutgers University, April 2002. [410b]

C. Collins
  a. Quantifier elimination for real closed fields by cylindrical algebraic decomposition, in *Automata Theory and Formal Languages*, H. Brakhage, ed., vol. 33 of Lecture Notes in Computer Science, Springer, Heidelberg, 1975, 134–163. [96b]

J. H. Conway
  a. Four-dimensional Archimedean polytopes, in *Proc. Colloquium on Convexity (Copenhagen 1965)*, W. Fenchel, ed., Copenhagen, 1967, Københavns Universitets Matematiske Institut, 38–39. [423a]

W. J. Cook, W. H. Cunningham, W. R. Pulleyblank, and A. Schrijver
  a. *Combinatorial Optimization*, Wiley-Interscience, 1998. [69a]

W. A. Coppel
  a. A theory of polytopes, *Bull. Aust. Math. Soc.*, **52** (1985), 1–24. [30b]
  b. *Foundations of Convex Geometry*, Cambridge University Press, Cambridge, 1998. [30b]

R. Cordovil and P. Duchet
  a. Cyclic polytopes and oriented matroids, *Eur. J. Comb.*, **21** (2000), 49–64. [7b]

K. Corrádi and S. Szabó
  a. A combinatorial approach for Keller's conjecture, *Period. Math. Hung.*, **21** (1990), 95–100. [423a]

H. H. Corson
  a. A compact convex set in E^3 whose exposed points are of the first category, *Proc. Am. Math. Soc.*, **16** (1965), 1015–1021. [30a]

H. S. M. Coxeter
  a. The evolution of Coxeter–Dynkin diagrams, in *Polytopes: Abstract, Convex and Computational (Proc. NATO Advanced Study Institute, Toronto 1993)*, T. Bisztriczky, P. McMullen, and A. Weiss, eds., Kluwer Academic Publishers, Dordrecht, 1994, 21–42. [171b]

H. T. Croft, K. J. Falconer, and R. K. Guy
  a. *Unsolved Problems in Geometry*, vol. II of Unsolved Problems in Intuitive Mathematics, Springer, New York, corrected reprint ed., 1994. [xii, 423c]

J. Csima and E. T. Sawyer
  a. There exist $6n/13$ ordinary points, *Discrete Comput. Geom.*, **9** (1993), 187–202. [410b]

J. Dancis
  a. Triangulated n-manifolds are determined by their $[n/2] + 1$-skeletons, *Topology Appl.*, **18** (1984), 17–26. [234a]

L. Danzer
  a. Finite point-sets on S^2 with minimum distance as large as possible, *Discrete Math.*, **60** (1986), 3–66. [296d]

M. de Berg, M. van Kreveld, M. Overmars, and O. Schwarzkopf
  a. *Computational Geometry. Algorithms and Applications*, Springer, Berlin Heidelberg, second ed., 2000. [52a, 96b, 142a]

J. De Loera, B. Sturmfels, and R. R. Thomas
  a. Gröbner bases and triangulations of the second hypersimplex, *Combinatorica*, **15** (1995), 409–424. [69a]

M. M. Deza and M. Laurent
  a. *Geometry of Cuts and Metrics*, vol. 15 of Algorithms and Combinatorics, Springer, Berlin Heidelberg, 1997. [69a]

M. M. Deza and S. Onn
   a. Lattice-free polytopes and their diameter, *Discrete Comput. Geom.*, **13** (1995), 59–75.
                                                                                        [355a]

R. Diestel
   a. *Graph Theory*, Springer, New York, second ed., 2000.   [224b]

M. B. Dillencourt
   a. Polyhedra of small orders and their Hamiltonian properties, *J. Comb. Theory, Ser. B*, **66**
      (1996), 87–122.   [52d, 296c]

D. P. Dobkin, H. Edelsbrunner, and C. K. Yap
   a. Probing convex polytopes, in *Autonomous Robot Vehicles*, I. J. Cox and G. T. Wilfong, eds.,
      Springer, 1990, 328–341.   [52c]

D. H. Doehlert and V. L. Klee
   a. Experimental designs through level reduction of the d-dimensional cuboctahedron, *Discrete
      Math.*, **2** (1972), 309–334.   [340b]

M. E. Dyer and A. M. Frieze
   a. On the complexity of computing the volume of a polyhedron, *SIAM J. Comput.*, **17** (1988),
      967–974.   [340b]

M. E. Dyer, A. M. Frieze, and R. Kannan
   a. A random polynomial-time algorithm for approximating the volume of convex bodies, *J.
      ACM*, **38** (1991), 1–17.   [340b]

M. E. Dyer, P. Gritzmann, and A. Hufnagel
   a. On the complexity of computing mixed volumes, *SIAM J. Comput.*, **27** (1998), 356–400.
                                                                                         [340c]

J. Eckhoff
   a. Helly, Radon, and Carathéodory type theorems, in *Handbook of Convex Geometry*, P. Gru-
      ber and J. Wills, eds., North-Holland, Amsterdam, 1993, 389–448.   [30b]

P. H. Edelman and R. E. Jamison
   a. The theory of convex geometries, *Geom. Dedicata*, **19** (1985), 247–270.   [30b]

M. R. Emamy-K. and C. Caiseda
   a. An efficient algorithm for characterization of cut-complexes, in Proc. of the Twenty-second
      Southeastern Conference on Combinatorics, Graph Theory, and Computing (Baton Rouge,
      LA, 1991), *Congr. Numerantium*, **85** (1991), 89–95.   [121b]

M. R. Emamy-K. and L. Lazarte
   a. On the cut-complexes of the 5-cube, *Discrete Math.*, **78** (1989), 239–256.   [121b]

P. Engel
   a. The enumeration of four-dimensional polytopes, *Discrete Math.*, **91** (1991), 9–31.   [96b]

D. Eppstein, G. Kuperberg, and G. M. Ziegler
   a. Fat 4-polytopes and fatter 3-spheres, in: W. Kuperberg Festschrift, A. Bezdek, ed., Marcel-
      Dekker, to appear; arXiv: math.CO/0204007, 12 pages.   [69a, 142b, 198d]

P. Erdős and Z. Füredi
   a. The greatest angle among n points in the d-dimensional Euclidean space, *Ann. Discrete
      Math.*, **17** (1983), 275–283.   [129b]

J. Erickson
   a. Arbitrarily large neighborly families of congruent symmetric convex 3-polytopes, in: W.
      Kuperberg Festschrift, A. Bezdek, ed., Marcel-Dekker, to appear; arXiv: math.CO/
      0106095, 9 pages.   [129b]

G. Ewald
  a. *Combinatorial Convexity and Algebraic Geometry*, vol. 168 of Graduate Texts in Mathematics, Springer, New York, 1996.   [xii, 52a, 52c, 171b, 198b, 224a]

S. Felsner
  a. Convex drawings of planar graphs and the order dimension of 3-polytopes, *Order*, **18** (2001), 19–37.   [296d]

L. Finschi and K. Fukuda
  a. Complete combinatorial generation of small point configurations and hyperplane arrangements, in *Proc. 13th Canadian Conference on Computational Geometry (CCCG 2001)*, 2001, 97–100.   [410b]

J. C. Fisher
  a. An existence theorem for simple convex polyhedra, *Discrete Math.*, **7** (1974), 75–97. [296b]

K. Fritzsche and F. B. Holt
  a. More polytopes meeting the conjectured Hirsch bound, *Discrete Math.*, **205** (1999), 77–84.   [355a]

K. Fukuda
  a. CDD – a C-implementation of the double description method,
  `http://www.cs.mcgill.ca/~fukuda/soft/cdd_home/cdd.html`.   [52b]

  b. Frequently asked questions in polyhedral computation, Preprint, October 2000, 30 pages;
  `http://www.cs.mcgill.ca/~fukuda/soft/soft.html`.   [52b]

W. F. Fulton
  a. *Introduction to Toric Varieties*, vol. 131 of Annals of Math. Studies, Princeton University Press, Princeton, 1993.   [52c, 198b]

D. Gale
  a. On inscribing n-dimensional sets in a regular n-simplex, *Proc. Am. Math. Soc.*, **4** (1953), 222–225.   [423b]

R. J. Gardner
  a. The Brunn-Minkowski inequality, *Bull. Am. Math. Soc., New Ser.*, **39** (2002), 355–405.   [340c]

M. R. Garey, D. S. Johnson, and R. E. Tarjan
  a. The planar Hamiltonian circuit problem is NP-complete, *SIAM J. Comput.*, **5** (1976), 704–714.   [389a]

B. Gärtner, M. Henk, and G. M. Ziegler
  a. Randomized simplex algorithms on Klee-Minty cubes, *Combinatorica*, **18** (1998), 349–372.   [389a, 389b]

B. Gärtner, J. Solymosi, F. Tschirschnitz, P. Valtr, and E. Welzl
  a. One line and n points, *Proc. 23rd ACM Symposium on the Theory of Computing (STOC)*, ACM Press, 2001, 306–315.   [389b]

E. Gawrilow and M. Joswig
  a. Polymake: A software package for analyzing convex polytopes,
  `http://www.math.tu-berlin.de/diskregeom/polymake/`.   [xii, 52b, 52c]

  b. Polymake: A framework for analyzing convex polytopes, in *Polytopes – Combinatorics and Computation*, G. Kalai and G. M. Ziegler, eds., vol. 29 of DMV Semin., Birkhäuser, Basel, 2000, 43–73.   [xii, 52b, 52c]

I. M. Gel'fand, M. Goresky, R. D. MacPherson, and V. Serganova
    a. Combinatorial geometries, convex polyhedra and Schubert cells, *Adv. Math.*, **63** (1987), 301–316.   [69a]

I. M. Gelfand, M. M. Kapranov, and A. V. Zelevinsky
    a. *Discriminants, Resultants, and Multidimensional Determinants*, Birkhäuser, Boston, 1994.   [69a]

A. A. Giannopoulos and V. D. Milman
    a. Euclidean structure in finite dimensional normed spaces, in *Handbook of the geometry of Banach spaces, Vol. I*, North-Holland, Amsterdam, 2001, 707–779.   [30b, 96a]

D. Girard and P. Valentin
    a. Zonotopes and mixtures management, in *New methods in optimization and their industrial uses (Pau/Paris, 1987)*, vol. 87 of Internat. Schriftenreihe Numer. Math., Birkhäuser, Basel, 1989, 57–71.   [340c]

D. Goldfarb and W. Y. Sit
    a. Worst case behavior of the steepest edge simplex method, *Discrete Appl. Math.*, **1** (1979), 277–285.   [389b]

M. Goldwasser
    a. A survey of Linear Programming in randomized subexponential time, *SIGACT News*, **26** (1995), 96–104.   [389b]

P. R. Goodey
    a. Some upper bounds for the diameters of convex polytopes, *Isr. J. Math.*, **11** (1972), 380–385.   [355a]

J. E. Goodman
    a. Arrangements of pseudolines, in *Handbook of Discrete and Computational Geometry*, J. E. Goodman and J. O'Rourke, eds., CRC Press, Boca Raton, 1997, ch. 5, 83–109. [410a, 410b]

J. E. Goodman and R. Pollack
    a. Upper bounds for configurations and polytopes in R^d, *Discrete Comput. Geom.*, **1** (1986), 219–227.   [121a]

D. W. Grace
    a. Search for largest polyhedra, *Math. Comput.*, **17** (1963), 197–199.   [296d]

R. L. Graham, D. E. Knuth, and O. Patashnik
    a. *Concrete Mathematics. A Foundation for Computer Science*, Addison-Wesley, Reading MA, second ed., 1994.   [171b]

R. L. Graham, B. L. Rothschild, and J. Spencer
    a. *Ramsey Theory*, J. Wiley & Sons, New York, 1990.   [7b]

P. Gritzmann and A. Hufnagel
    a. On the algorithmic complexity of Minkowski's reconstruction theorem, *J. Lond. Math. Soc., II. Ser.*, **59** (1999), 1081–1100.   [340c]

P. Gritzmann and V. Klee
    a. Inner and outer *j*-radii of convex bodies in finite-dimensional normed spaces, *Discrete Comput. Geom.*, **7** (1992), 255–280.   [423c]

    b. Computational complexity of inner and outer *j*-radii of polytopes in finite-dimensional normed spaces, *Math. Program.*, **59** (1993), 163–213.   [423c]

    c. On the complexity of some basic problems in computational convexity. I. Containment problems, *Discrete Math.*, **136** (1994), 129–174.   [423c]

d. On the complexity of some basic problems in computational convexity: II. Volume and mixed volumes, in *Polytopes: Abstract, Convex and Computational (Proc. NATO Advanced Study Institute, Toronto 1993)*, T. Bisztriczky, P. McMullen, and A. Weiss, eds., Kluwer Academic Publishers, Dordrecht, 1994, 373–466. [340b]

e. Computational convexity, in *Handbook of Discrete and Computational Geometry*, J. E. Goodman and J. O'Rourke, eds., CRC Press, Boca Raton, 1997, ch. 27, 491–515. [52a]

P. Gritzmann, V. Klee, and D. Larman
a. Largest *j*-simplices in *n*-polytopes, *Discrete Comput. Geom.*, 13 (1995), 477–515. [423d]

P. Gritzmann, V. Klee, and J. Westwater
a. Polytope containment and determination by linear probes, *Proc. Lond. Math. Soc., III. Ser.*, 70 (1995), 691–720. [52b]

M. Grötschel and M. Henk
a. On the representation of polyhedra by polynomial inequalities, Tech. Rep., TU Wien, March 2002; arXiv: math.MG/0203268, 19 pages. [52b]

M. Grötschel, L. Lovász, and A. Schrijver
a. *Geometric Algorithms and Combinatorial Optimization*, vol. 2 of Algorithms and Combinatorics, Springer, Berlin Heidelberg, 1988. [30b, 52b, 340b]

M. Grötschel and M. Padberg
a. Polyhedral theory, in *The Traveling Salesman Problem*, E. Lawler, J. Lenstra, A.H.G. Rinnooy Kan, and D. Shmoys, eds., Wiley-Interscience Series in Discrete Mathematics and Optimization, John Wiley and Sons, Chichester NY, 1985, ch. 8, 251–360. [69a]

P. M. Gruber and C. G. Lekkerkerker
a. *Geometry of Numbers*, North-Holland, Amsterdam, 1987. [30b]

P. M. Gruber and J. Wills, eds.
a. *Handbook of Convex Geometry*, North-Holland, Amsterdam, 1993, two volumes. [30a]

B. Grünbaum
a. Some analogues of Eberhard's theorem on convex polytopes, *Isr. J. Math.*, 6 (1968), 398–411. [296b]

b. Planar maps with prescribed types of vertices and faces, *Mathematika*, 16 (1969), 28–36. [296c]

c. The importance of being straight (?), in *Time Series and Stochastic Processes; Convexity and Combinatorics. Proc. of the Twelfth Biannual Intern. Seminar of the Canadian Math. Congress (Vancouver 1969)*, R. Pyke, ed., Montreal, 1970, Canadian Math. Congress, 243–254. [121b, 410b]

d. Polytopes, graphs, and complexes, *Bull. Am. Math. Soc., New Ser.*, 76 (1970), 1131–1201. [xii]

e. *Arrangements and Spreads*, vol. 10 of Regional Conference Series in Mathematics, Amer. Math. Soc., Providence RI, 1972. [410a, 428b]

B. Grünbaum and G. C. Shephard
a. Convex polytopes, *Bull. London Math. Soc.*, 1 (1969), 257–300. [xii]

b. Tilings with congruent tiles, *Bull. Am. Math. Soc., New Ser.*, 3 (1980), 951–973. [423a]

c. Simplicial arrangements in projective 3-space, *Mitt. Math. Sem. Gießen*, 166 (1984), 49–101. [410b]

d. *Tilings and Patterns*, W. H. Freeman, New York, 1987. [423a]

B. Grünbaum and V. P. Sreedharan
  a. An enumeration of simplicial 4-polytopes with 8 vertices, *J. Combinatorial Theory*, **2** (1967), 437–465.   [129a]

B. Grünbaum and H. Walther
  a. Shortness exponents of families of graphs, *J. Comb. Theory, Ser. A*, **14** (1973), 364–385.   [389a]

H. Günzel
  a. On the universal partition theorem for 4-polytopes, *Discrete Comput. Geom.*, **19** (1998), 521–552.   [96a]

C. Haase and G. M. Ziegler
  a. Examples and counterexamples for the Perles conjecture, *Discrete Comput. Geom.*, **28** (2002), 29–44.   [234a]

J. Hadamard
  a. Résolution d'une question relativ aux déterminants, *Bull. Sci. Math.*, **28** (1893), 240–246.   [423d]

G. Hajos
  a. Über einfache und mehrfache Bedeckung des n-dimensionalen Raumes mit einem Würfelgitter, *Math. Z.*, **47** (1941), 427–467.   [423a]

J. Håstad
  a. On the size of weights for threshold gates, *SIAM J. Discrete Math.*, **7** (1994), 484–492.   [121b]

M. Henk, J. Richter-Gebert, and G. M. Ziegler
  a. Basic properties of convex polytopes, in *Handbook of Discrete and Computational Geometry*, J. E. Goodman and J. O'Rourke, eds., CRC Press, Boca Raton FL, 1997, ch. 13, 243–270.   [69b]

P. Hersh and I. Novik
  a. A short simplicial h-vector and the upper bound theorem, *Discrete Comput. Geom.* **28** (2002), 283–289.   [198a]

A. Hinrichs and C. Richter
  a. New sets with large Borsuk numbers. Preprint, Jena University, February 2002, 10 pages.   [423b]

W. V. D. Hodge and D. Pedoe
  a. *Methods of Algebraic Geometry*, Cambridge University Press, Cambridge, 1947/1952; paperback reprint 1968.   [7b]

C. D. Hodgson, I. Rivin, and W. D. Smith
  a. A characterization of convex hyperbolic polyhedra and of convex polyhedra inscribed in a sphere, *Bull. Am. Math. Soc., New Ser.*, **27** (1992), 246–251.   [296c]

F. Holt and V. Klee
  a. Counterexamples to the strong d-step conjecture for $d \geq 5$, *Discrete Comput. Geom.*, **19** (1998), 33–46.   [69b]

  b. Many polytopes meeting the conjectured Hirsch bound, *Discrete Comput. Geom.*, **20** (1998), 1–17.   [355a]

  c. A proof of the strict monotone 4-step conjecture, in *Advances in Discrete and Computational Geometry (Mount Holyoke 1996)*, B. Chazelle, J. E. Goodman, and R. Pollack, eds., vol. 223 of Contemporary Mathematics, Providence RI, 1998, Amer. Math. Soc., 201–216.   [224a]

D. A. Holton, B. Manvel, and B. D. McKay
   a. Hamiltonian cycles in cubic 3-connected bipartite planar graphs, *J. Comb. Theory, Ser. B*, **38** (1985), 179–197. [389a]

D. A. Holton and B. D. McKay
   a. The smallest non-hamiltonian 3-connected cubic planar graphs have 38 vertices, *J. Comb. Theory, Ser. B*, **45** (1988), 305–319; erratum **47** (1989), 248. [389a]

A. Höppner and G. M. Ziegler
   a. A census of flag-vectors of 4-polytopes, in *Polytopes – Combinatorics and Computation*, G. Kalai and G. Ziegler, eds., vol. 29 of DMV Semin., Birkhäuser, Basel, 2000, 105–110. [198c]

M. Hudelson, V. Klee, and D. Larman
   a. Largest j-simplices in d-cubes: Some relatives of the Hadamard maximum determinant problem, *Linear Algebra Appl.*, **241-243** (1996), 519–598. [423d]

S. Jendrol'
   a. On symmetry groups of selfdual convex polyhedra, in *Proc. of Fourth Czechoslovakian Symposium on Combinatorics, Graphs, and Complexity*, J. Nešetřil and M. Fiedler, eds., Amsterdam, 1992, Elsevier, 129–135. [52d]
   b. On face vectors and vertex vectors of convex polyhedra, *Discrete Math.*, **118** (1993), 119–144. [296c]

S. Jendrol' and E. Jucovič
   a. On a conjecture by B. Grünbaum, *Discrete Math.*, **2** (1972), 35–49. [296c]

T. R. Jensen and B. Toft
   a. *Graph Coloring Problems*, John Wiley & Sons, New York, 1995. [423b]

R. G. Jeroslow
   a. The simplex algorithm with the pivot rule of maximizing criterion improvement, *Discrete Math.*, **4** (1973), 367–377. [389b]

M. R. Jerrum and A. J. Sinclair
   a. The Markov chain Monte Carlo method: An approach to approximate counting and integration, in *Approximation Algorithms for NP-Hard Problems*, D. Hochbaum, ed., PWS, 1996, ch. 12, 482–520. [340b]

W. Jockusch
   a. An infinite family of nearly neighborly centrally symmetric 3-spheres, *J. Comb. Theory, Ser. A*, **72** (1995), 318–321. [121b]

N. W. Johnson
   a. *Uniform Polytopes*, Cambridge University Press, Cambridge, to appear. [423a]

M. Joswig
   a. Reconstructing a non-simple polytope from its graph, in *Polytopes – Combinatorics and Computation*, G. Kalai and G. M. Ziegler, eds., vol. 29 of DMV Semin., Birkhäuser, Basel, 2000, 167–176. [234b]

M. Joswig, V. Kaibel, and F. Körner
   a. On the k-systems of a simple polytope, *Isr. J. Math.*, **129** (2002), 109–118. [234a]

M. Joswig and G. M. Ziegler
   a. Neighborly cubical polytopes, *Discrete Comput. Geom.*, **24** (2000), 325–344. [69b, 234b]

E. Jucovič
   a. On face-vectors and vertex-vectors of cell-decompositions of orientable 2-manifolds, *Math. Nachr.*, **73** (1976), 285–295. [296c]

J. Kahn and G. Kalai
  a. A counterexample to Borsuk's conjecture, *Bull. Am. Math. Soc., New Ser.*, **29** (1993), 60–62.   [423b]

V. Kaibel and M. Pfetsch
  a. Some algorithmic problems in polytope theory, in: *Algebra, Geometry and Software Systems*, M. Joswig and N. Takayama, eds., Springer, Heidelberg, 2003, to appear; arXiv: math.CO/0202204, 25 pages.   [52b]

V. Kaibel and A. Schwartz
  a. On the complexity of polytope isomorphism problems. *Graphs Comb.*, to appear; arXiv: math.CO/01060093, 16 pages.   [52b]

V. Kaibel and M. Wolff
  a. Simple 0/1-polytopes, *Eur. J. Comb.*, **21** (2000), 139–144.   [340a]

M. J. Kaiser, T. L. Morin, and T. B. Trafalis
  a. Centers and invariant points of convex bodies, in *Applied Geometry and Discrete Mathematics. The Victor Klee Festschrift*, P. Gritzmann and B. Sturmfels, eds., vol. 4 of DIMACS Series in Discrete Mathematics and Theoretical Computer Science, Amer. Math. Soc., Providence RI, 1991, 367–385.   [315b]

G. Kalai
  a. Rigidity and the lower bound theorem, I, *Invent. Math.*, **88** (1987), 125–151. [198a, 198b]

  b. Many triangulated spheres, *Discrete Comput. Geom.*, **3** (1988), 1–14.   [121a]

  c. A new basis of polytopes, *J. Comb. Theory, Ser. A*, **49** (1988), 191–209.   [198c]

  d. A simple way to tell a simple polytope from its graph, *J. Comb. Theory, Ser. A*, **49** (1988), 381–383.   [234a, 234b]

  e. The number of faces of centrally-symmetric polytopes (Research Problem), *Graphs Comb.*, **5** (1989), 389–391.   [69b, 224b]

  f. On low-dimensional faces that high-dimensional polytopes must have, *Combinatorica*, **10** (1990), 271–280.   [198d, 224b, 423b]

  g. The diameter of graphs of convex polytopes and f-vector theory, in *Applied Geometry and Discrete Mathematics. The Victor Klee Festschrift*, P. Gritzmann and B. Sturmfels, eds., vol. 4 of DIMACS Series in Discrete Mathematics and Theoretical Computer Science, Amer. Math. Soc., Providence RI, 1991, 387–411.   [198b, 355b]

  h. A subexponential randomized simplex algorithm, in *Proc. 24th ACM Symposium on the Theory of Computing (STOC)*, ACM Press, 1992, 475–482.   [355a, 355b, 389b]

  i. Linear programming, the simplex algorithm and simple polytopes, *Math. Program.*, **79** (1997), 217–233.   [389b]

  j. Polytope skeletons and paths, in *Handbook of Discrete and Computational Geometry*, J. E. Goodman and J. O'Rourke, eds., CRC Press, Boca Raton, 1997, ch. 17, 331–344.   [234a]

  k. Algebraic shifting, in *Computational Commutative Algebra and Combinatorics (Osaka, 1999)*, Math. Soc. Japan, Tokyo, 2002, 121–163.   [198c]

G. Kalai, P. Kleinschmidt, and G. Meisinger
  a. Flag numbers and FLAGTOOL, in *Polytopes – Combinatorics and Computation*, G. Kalai and G. Ziegler, eds., vol. 29 of DMV Semin., Birkhäuser, Basel, 2000, 75–103.   [224b]

G. Kalai and D. J. Kleitman
  a. A quasi-polynomial bound for the diameter of graphs of polyhedra, *Bull. Am. Math. Soc., New Ser.*, **26** (1992), 315–316.   [355a]

M. Kallay
    a. Decomposability of polytopes is a projective invariant, in *Convexity and Graph Theory (Jerusalem, 1981)*, North-Holland, Amsterdam, 1984, 191–196.   [340a]

D. A. Klain and G.-C. Rota
    a. *Introduction to Geometric Probability*, Lezioni Lincee, Cambridge University Press, 1997.
                    [315b]

V. Klee
    a. Convex sets in linear spaces, *Duke Math. J.*, **18** (1951), 443–466.   [428a]

    b. Paths on polyhedra II, *Pac. J. Math.*, **17** (1966), 249–262.   [355b]

    c. Facet-centroids and volume minimization, *Stud. Sci. Math. Hung.*, **21** (1986), 143–147.
                    [423d]

V. Klee and P. Kleinschmidt
    a. The d-step conjecture and its relatives, *Math. Oper. Res.*, **12** (1987), 718–755. [355a, 355b]

    b. Convex polytopes and related complexes, in *Handbook of Combinatorics*, R. Graham, M. Grötschel, and L. Lovász, eds., North-Holland/Elsevier, Amsterdam, 1995, 875–917.
                    [xii, 296c, 355a]

V. Klee and M. Martin
    a. Semicontinuity of the face-function of a convex set, *Comment. Math. Helv.*, **46** (1971), 1–12.   [30a]

V. Klee and G. J. Minty
    a. How good is the simplex algorithm?, in *Inequalitites, III*, O. Shisha, ed., Academic Press, New York, 1972, 159–175.   [389b]

V. Klee and D. W. Walkup
    a. The d-step conjecture for polyhedra of dimension $d < 6$, *Acta Math.*, **117** (1967), 53–78.
                    [69b, 355a]

P. Kleinschmidt and S. Onn
    a. On the diameter of convex polytopes (Communication), *Discrete Math.*, **102** (1992), 75–77.
                    [355a]

D. E. Knuth
    a. *Generating all n-tuples*, ch. 7.2.1.1, prefascicle 2A of "The Art of Computer Programming" (vol. 4), released September 2001,
    `http://www-cs-faculty.stanford.edu/~knuth/fasc2a.ps.gz`.   [389b]

J. Komlós and E. Szemerédi
    a. Topological cliques in graphs II, *Comb. Probab. Comput.*, **5** (1996), 79–90.   [224b]

U. H. Kortenkamp
    a. Every simplicial polytope with at most $d+4$ vertices is a quotient of a neighborly polytope, *Discrete Comput. Geom.*, **18** (1997), 455–462.   [129b]

W. Kühnel
    a. Higherdimensional analogues of Császár's torus, *Result. Math.*, **9** (1986), 95–106. [142b]

J. C. Lagarias and P. W. Shor
    a. Keller's cube-tiling conjecture is false in high dimensions, *Bull. Am. Math. Soc., New Ser.*, **27** (1992), 279–283.   [423a]

D. G. Larman
    a. Paths on polytopes, *Proc. Lond. Math. Soc., III. Ser.*, **20** (1970), 161–178.   [355a]

b. On a conjecture of Klee and Martin for convex bodies, *Proc. Lond. Math. Soc., III. Ser.*, **23** (1971), 668–682. [30a]

D. G. Larman and N. K. Tamvakis
a. The decomposition of the *n*-sphere and the boundaries of plane convex domains, *Ann. Discrete Math.*, **20** (1984), 209–214. [423c]

M. Las Vergnas
a. Acyclic and totally cyclic orientations of combinatorial geometries, *Discrete Math.*, **20** (1977), 51–61. [410b]

M. Lassak
a. An estimate concerning Borsuk partition problem, *Bull. Acad. Polon. Sci. Sér. Sci. Math.*, **30** (1982), 449–451. [423b]

J. Lawrence
a. Polytope volume computation, *Math. Comput.*, **57** (1991), 259–271. [340b]

C. W. Lee
a. Winding numbers and the generalized lower-bound conjecture, in *Discrete and Computational Geometry (New Brunswick, NJ, 1989/1990)*, Amer. Math. Soc., Providence, RI, 1991, 209–219. [198b]

b. Subdivisions and triangulations of polytopes, in *Handbook of Discrete and Computational Geometry*, J. E. Goodman and J. O'Rourke, eds., CRC Press, Bocà Raton FL, 1997, 271–290. [96b]

K. Leichtweiß
a. *Affine geometry of convex bodies*, Johann Ambrosius Barth Verlag, Heidelberg, 1998. [30b]

J. Lindenstrauss and R. R. Phelps
a. Extreme point properties of convex bodies in reflexive spaces, *Isr. J. Math.*, **6** (1968), 39–48. [30a]

B. Lindström
a. On the realization of convex polytopes, Euler's formula, and Möbius functions, *Aequationes Math.*, **6** (1971), 235–240. [96b]

E. K. Lloyd
a. The number of *d*-polytopes with *d* + 3 vertices, *Mathematika*, **17** (1970), 120–132. [121a]

L. Lovász
a. *An Algorithmic Theory of Numbers, Graphs and Convexity*, vol. 50 of CMBS-NSF Regional Conference Series in Applied Mathematics, Society for Industrial and Applied Mathematics (SIAM), Philadelphia, 1986. [30b]

b. Steinitz representations of polyhedra and the Colin de Verdiére number, *J. Comb. Theory, Ser. B*, **82** (2001), 223–236. [296a]

F. H. Lutz
a. *Triangulated manifolds with few vertices and vertex-transitive group actions*, PhD thesis, TU Berlin, 1999, 134 pages. [121b]

W. Mader
a. 3*n* − 5 edges do force a subdivision of K_5, *Combinatorica*, **18** (1998), 569–595. [224b]

P. Mani
a. Automorphismen von polyedrischen Graphen, *Math. Ann.*, **192** (1971), 279–303. [296a]

b. Spheres with few vertices, *J. Comb. Theory, Ser. A*, **13** (1972), 346–352. [198b]

c. Inner illumination of convex polytopes, *Comment. Math. Helv.*, **49** (1974), 65–73. [423c]

P. Mani and D. W. Walkup
 a. A 3-sphere counterexample to the W_v-path conjecture, *Math. Oper. Res.*, **5** (1980), 595–598.
[355b]

H. Martini
 a. Shadow-boundaries of convex bodies, *Discrete Math.*, **155** (1996), 161–172. [296b]

J. Matoušek
 a. *Using the Borsuk–Ulam Theorem. Lectures on Topological Methods in Combinatorics and Geometry*, book to appear; preliminary version, 151 pages, August 2002,
 http://kam.mff.cuni.cz/~matousek/lectnotes.html. [224a]

 b. *Lectures on Discrete Geometry*, vol. 212 of Graduate Texts in Mathematics, Springer, New York, 2002. [30b, 96a]

J. Matoušek, M. Sharir, and E. Welzl
 a. A subexponential bound for linear programming, *Algorithmica*, **16** (1996), 498–516. [389b]

P. McMullen
 a. Combinatorially regular polytopes, *Mathematika*, **14** (1967), 142–150. [423a]

 b. Affinely and projectively regular polytopes, *J. Lond. Math. Soc., II. Ser.*, **43** (1968), 755–757. [423a]

 c. The maximum numbers of faces of a convex polytope, *Mathematika*, **17** (1970), 179–184. [171a, 198a]

 d. The numbers of faces of simplicial polytopes, *Isr. J. Math.*, **9** (1971), 559–570. [198a, 198b]

 e. The number of neighbourly d-polytopes with $d+3$ vertices, *Mathematika*, **21** (1974), 26–31. [121a]

 f. Transforms, diagrams and representations, in *Contributions to Geometry (Proc. Geometry Symposium, Siegen 1978)*, J. Tölke and J. Wills, eds., Birkhäuser, Basel, 1979, 92–130. [96a]

 g. The polytope algebra, *Adv. Math.*, **78** (1989), 76–130. [340a]

 h. On simple polytopes, *Invent. Math.*, **113** (1993), 419–444. [198b, 340a]

 i. Separation in the polytope algebra, *Beitr. Algebra Geom.*, **34** (1993), 15–30. [198b, 340a]

 j. Valuations and dissections, in *Handbook of Convex Geometry*, P. Gruber and J. Wills, eds., North-Holland, Amsterdam, 1993, 933–988. [315b]

 k. Weights on polytopes, *Discrete Comput. Geom.*, **15** (1996), 363–388. [340a]

 l. Triangulations of simplicial polytopes, preprint, UC London, 2002, 10 pages. [198b]

P. McMullen and E. Schulte
 a. *Abstract Regular Polytopes*, Cambridge University Press, Cambridge, 2003, to appear. [423a]

P. McMullen and G. C. Shephard
 a. Diagrams for centrally symmetric polytopes, *Mathematika*, **15** (1968), 123–138. [121b]

 b. *Convex Polytopes and the Upper Bound Conjecture*, vol. 3 of London Math. Soc. Lecture Notes Series, Cambridge University Press, Cambridge, 1971. [xii]

P. McMullen and D. W. Walkup
 a. A generalized lower-bound conjecture for simplicial polytopes, *Mathematika*, **18** (1971), 264–273. [198b]

G. Meisinger, P. Kleinschmidt, and G. Kalai
   a. Three theorems, with computer-aided proofs, on three-dimensional faces and quotients of
      polytopes, *Discrete Comput. Geom.*, **24** (2000), 413–420.   [224b]

T. W. Melnyk, O. Knop, and W. R. Smith
   a. Extremal arrangements of points and unit charges on a sphere: Equilibrium configurations
      revisited, *Can. J. Chem.*, **55** (1971), 1745–1761.   [296d]

K. Menger
   a. Zur allgemeinen Kurventheorie, *Fundam. Math.*, **10** (1927), 96–115.   [224b]

J. Mihalisin
   a. *Polytopal Digraphs and Non-Polytopal Facet Graphs*, PhD thesis, University of Washing-
      ton, Seattle, 2001.   [296b]

J. Mihalisin and V. Klee
   a. Convex and linear orientations of polytopal graphs, *Discrete Comput. Geom.*, **24** (2000),
      421–436.   [296b]

J. Mihalisin and G. Williams
   a. Nonconvex embeddings of the exceptional simplicial 3-spheres, *J. Comb. Theory, Ser. A*, **98**
      (2002), 74–86.   [224a]

E. Miller
   a. Planar graphs as minimal resolutions of trivariate monomial ideals, *Doc. Math.*, **7** (2002),
      43–90.   [296d]

N. E. Mněv
   a. The universality theorems on the classification problem of configuration varieties and con-
      vex polytopes varieties, in *Topology and Geometry – Rohlin Seminar*, O. Y. Viro, ed.,
      vol. 1346 of Lecture Notes in Mathematics, Springer, 1988, 527–544.   [96a]

   b. The universality theorem on the oriented matroid stratification of the space of real matri-
      ces, in *Discrete and Computational Geometry: Papers from the DIMACS Special Year*, J. E.
      Goodman, R. Pollack, and W. Steiger, eds., vol. 6 of DIMACS Series in Discrete Mathemat-
      ics and Theoretical Computer Science, Providence RI, 1991, Amer. Math. Soc., 237–243.
      [96a]

B. Mohar and C. Thomassen
   a. *Graphs on Surfaces*, Johns Hopkins Studies in the Mathematical Sciences, Johns Hopkins
      University Press, Baltimore, MD, 2001.   [234a]

R. Morelli
   a. A theory of polyhedra, *Adv. Math.*, **97** (1993), 1–73.   [340a]

S. Muroga
   a. *Threshold Logic and its Applications*, Wiley-Interscience, New York, 1971.   [121b]

D. Naddef
   a. The Hirsch conjecture is true for $(0,1)$-polytopes, *Math. Program.*, **45** (1989), 109–110.
      [355a]

M. G. Neubauer and W. Watkins
   a. D-optimal designs for seven objects and a large number of weighings, *Linear Multilinear
      Algebra*, **50** (2002), 61–74.   [423d]

M. G. Neubauer, W. Watkins, and J. Zeitlin
   a. Maximal j-simplices in the real d-dimensional unit cube, *J. Comb. Theory, Ser. A*, **80**
      (1997), 1–12.   [423d]

b. *D*-optimal weighing designs for four and five objects, *Electron. J. Linear Algebra*, **4** (1998), 48–73. [423d]

c. *D*-optimal weighing designs for six objects, *Metrika*, **52** (2000), 185–211 (electronic). [423d]

A. Nilli
a. On Borsuk's problem, in *Jerusalem Combinatorics '93*, H. Barcelo and G. Kalai, eds., vol. 178 of Contemporary Math., Providence RI, 1994, Amer. Math. Soc., 209–210. [423b]

I. Novik
a. Upper bound theorems for homology manifolds, *Isr. J. Math.*, **108** (1998), 45–82. [198a]

b. Lower bounds for the *cd*-index of odd-dimensional simplicial manifolds, *Eur. J. Comb.*, **21** (2000), 533–541. [198c]

c. A note on geometric embeddings of simplicial complexes in a Euclidean space, *Discrete Comput. Geom.*, **23** (2000), 293–302. [224a]

H. Okamura
a. Every simple 3-polytope of order 32 or less is Hamiltonian, *J. Graph Theory*, **6** (1982), 185–196. [389a]

P. Orlik and H. Terao
a. *Arrangements of Hyperplanes*, vol. 300 of Grundlehren der math. Wissenschaften, Springer, Berlin Heidelberg, 1992. [410b]

J. O'Rourke, A. Aggarwal, S. Maddila, and M. Baldwin
a. An optimal algorithm for finding minimal enclosing triangles, *J. Algorithms*, **7** (1986), 258–269. [423d]

P. J. Owens
a. Shortness parameters for polyhedral graphs, *Discrete Math.*, **206** (1999), 159–169. [389a]

J. Pach and P. K. Agarwal
a. *Combinatorial Geometry*, J. Wiley and Sons, New York, 1995. [423b]

A. Packer
a. NP-hardness of largest contained and smallest containing simplices for V- and H-polytopes, *Discrete Math.*, to appear. [423d]

P. Paulraja
a. A characterization of Hamiltonian prisms, *J. Graph Theory*, **17** (1993), 161–171. [389a]

M. A. Perles
a. At most 2^{d+1} neighborly simplices in E^d, *Ann. Discrete Math.*, **20** (1984), 253–254. [129b]

M. A. Perles and G. C. Shephard
a. Facets and nonfacets of convex polytopes, *Acta Math.*, **119** (1967), 113–145. [423a]

M. Petkovšek, H. S. Wilf, and D. Zeilberger
a. $A = B$, A. K. Peters, Wellesley, MA, 1996. [171b]

J. Pfeifle
a. Centrally-symmetric neighborly fans with $2d + 4$ rays do not exist, preprint, TU Berlin 2000, 12 pages. [121b]

b. Kalai's squeezed 3-spheres are polytopal, *Discrete Comput. Geom.*, **27** (2002), 395–407. [121b]

c. Work in progress, TU Berlin, 2002. [389b]

J. Pfeifle and G. M. Ziegler
   a. Many triangulated 3-spheres. TU Berlin 2002, in preparation.   [121b]

K. Polthier, S. Khadem-Al-Charieh, E. Preuß, and U. Reitebuch
   a. JavaView visualization software, 1999-2002, http://www.javaview.de.   [xii]

W. Prenowitz and J. Jantosciak
   a. *Join Geometries: A Theory of Convex Sets and Linear Geometry*, Undergraduate Texts in Math., Springer, 1979.   [30b]

A. M. Raigorodskii
   a. Borsuk's problem and the chromatic number of some metric spaces, *Russ. Math. Surv.*, **56** (2001), 103–139.   [423b]

J. Richter-Gebert
   a. Oriented matroids with few mutations, *Discrete Comput. Geom.*, **10** (1993), 251–269.   [410b]

   b. *Realization Spaces of Polytopes*, vol. 1643 of Lecture Notes in Mathematics, Springer, 1996.   [52c, 96a, 296a, 296b, 296d]

   c. The universality theorems for oriented matroids and polytopes, in *Advances in Discrete and Computational Geometry (Mount Holyoke 1996)*, B. Chazelle, J. E. Goodman, and R. Pollack, eds., vol. 223 of Contemporary Mathematics, Providence RI, 1998, Amer. Math. Soc., 269–292.   [96a]

J. Richter-Gebert and G. M. Ziegler
   a. Realization spaces of 4-polytopes are universal, *Bull. Am. Math. Soc., New Ser.*, **32** (1995), 403–412.   [96a]

   b. Oriented matroids, in *Handbook of Discrete and Computational Geometry*, J. E. Goodman and J. O'Rourke, eds., CRC Press, Boca Raton FL, 1997, ch. 6, 111–132.   [410b]

F. J. Rispoli
   a. On the diameter of polyhedra associated with network optimization problems, Tech. Rep. 94–48, DIMACS, September 1994.   [355b]

   b. The monotonic diameter of traveling salesman polytopes, *Oper. Res. Lett.*, **22** (1998), 69–73.   [355b]

F. J. Rispoli and S. Cosares
   a. A bound of 4 for the diameter of the symmetric traveling salesman polytope, *SIAM J. Discrete Math.*, **11** (1998), 373–380.   [355b]

N. Robertson, D. Sanders, P. Seymour, and R. Thomas
   a. The four-colour theorem, *J. Comb. Theory, Ser. B*, **70** (1997), 2–44.   [389a]

R. M. Robinson
   a. Arrangements of 24 points on a sphere, *Math. Ann.*, **144** (1961), 17–48.   [296d]

   b. Finite sets of points on a sphere with each nearest to five others., *Math. Ann.*, **179** (1969), 296–318.   [296d]

R. T. Rockafellar
   a. *Convex Analysis*, Princeton University Press, Princeton, 1970; reprint 1997.   [30b]

C. A. Rogers and G. C. Shephard
   a. The difference body of a convex body, *Arch. Math.*, **8** (1957), 220–233.   [340b]

J.-P. Roudneff
   a. An inequality for 3-polytopes, *J. Comb. Theory, Ser. B*, **42** (1987), 156–166.   [296c]

J.-P. Roudneff and B. Sturmfels
   a. Simplicial cells in arrangements and mutations of oriented matroids, *Geom. Dedicata*, **27** (1988), 153–170.   [410b]

G. Royle
   a. Planar graphs, tables and databases, March 2001, http://www.cs.uwa.edu.au/~gordon/remote/planar/.   [296c, 428b]

K. Rybnikov
   a. Stresses and liftings of cell-complexes, *Discrete Comput. Geom.*, **21** (1999), 481–517. [52c]

E. B. Saff and A. B. J. Kuijlaars
   a. Distributing many points on a sphere, *Math. Intell.*, **19** (1997), 5–11.   [296d]

G. T. Sallee
   a. A non-continuous "Steiner point", *Isr. J. Math.*, **10** (1971), 1–5.   [315b]

   b. Minkowski decomposition of convex sets, *Isr. J. Math.*, **12** (1972), 266–276.   [340a]

K. Sarkaria
   a. Tverberg's theorem via number fields, *Isr. J. Math.*, **79** (1992), 317–320.   [30b]

R. G. Schneider
   a. Über eine Integralgleichung in der Theorie der konvexen Körper, *Math. Nachr.*, **44** (1970), 55–75.   [121b]

   b. *Convex Bodies: The Brunn-Minkowski Theory*, vol. 44 of Encyclopedia of Mathematics, Cambridge University Press, Cambridge, 1993.   [30a, 52c, 340a, 340c, 423b]

   c. Polytopes and Brunn-Minkowski theory, in *Polytopes: Abstract, Convex and Computational (Proc. NATO Advanced Study Institute, Toronto 1993)*, T. Bisztriczky, P. McMullen, and A. Weiss, eds., Kluwer Academic Publishers, Dordrecht, 1994, 273–299.   [340c]

R. G. Schneider and W. Weil
   a. Zonoids and related topics, in *Convexity and Its Applications*, P. Gruber and J. Wills, eds., Birkhäuser, Basel, 1983, 296–317.   [340b]

O. Schramm
   a. Illuminating sets of constant width, *Mathematika*, **35** (1988), 180–189.   [423c]

   b. How to cage an egg, *Invent. Math.*, **107** (1992), 543–560.   [296a]

A. Schrijver
   a. *Theory of Linear and Integer Programming*, Wiley-Interscience Series in Discrete Mathematics and Optimization, John Wiley and Sons, Chichester NY, 1986; reprint 1998.   [52a]

   b. Polyhedral combinatorics, in *Handbook of Combinatorics*, R. Graham, M. Grötschel, and L. Lovász, eds., vol. II, North-Holland, Amsterdam, 1995, ch. 30, 1649–1704.   [69a]

E. Schulte
   a. Nontiles and nonfacets for the Euclidean space, spherical complexes and convex polytopes, *J. Reine Angew. Math.*, **352** (1984), 161–183.   [423a]

   b. The existence of nontiles and nonfacets in three dimensions, *J. Comb. Theory, Ser. A*, **38** (1985), 75–81.   [423a]

   c. Tilings, in *Handbook of Convex Geometry*, P. Gruber and J. Wills, eds., North-Holland, Amsterdam, 1993, 899–932.   [423a]

   d. Symmetry of polytopes and polyhedra, in *Handbook of Discrete and Computational Geometry*, J. E. Goodman and J. O'Rourke, eds., CRC Press, Boca Raton, 1997, ch. 16, 311–330.   [423a]

C. Schulz
   a. An invertible 3-diagram with 8 vertices, *Discrete Math.*, **28** (1979), 201–205.   [52c]
   b. Dual pairs of non-polytopal diagrams and spheres, *Discrete Math.*, **55** (1985), 65–72. [52c]

R. Seidel
   a. The upper bound theorem for polytopes: an easy proof of its asymptotic version, *Comput. Geom.*, **5** (1995), 115–116.   [142a]

P. Seymour
   a. Progress on the four color theorem, in *Proc. International Congress of Mathematicians (ICM'94, Zürich)*, Birkhäuser, Basel, 1995, vol. I, 183–195.   [389a]

R. W. Shannon
   a. Simplicial cells in arrangements of hyperplanes, *Geom. Dedicata*, **8** (1979), 179–187.   [410b]

A. Shapiro
   a. Obstructions to the imbedding of a complex in a Euclidean space. I. The first obstruction, *Annals of Math.*, (1957), 256–269.   [224a]

I. Shemer
   a. Neighborly polytopes, *Isr. J. Math.*, **43** (1982), 291–314.   [129a, 129b]

G. C. Shephard
   a. An elementary proof of Gram's theorem for convex polytopes, *Can. J. Math.*, **19** (1967), 1214–1217.   [315a]

D. D. Sleator, R. E. Tarjan, and W. P. Thurston
   a. Rotation distance, triangulations, and hyperbolic geometry, *J. Am. Math. Soc.*, **1** (1988), 647–681.   [30b]

Z. Smilansky
   a. An indecomposable polytope all of whose facets are decomposable, *Mathematika*, **33** (1986), 192–196.   [340a]
   b. Decomposability of polytopes and polyhedra, *Geom. Dedicata*, **24** (1987), 29–49.   [340a]

W. Smith
   a. A lower bound for the simplexity of the n-cube via hyperbolic volumes, *Eur. J. Comb.*, **21** (2000), 131–137.   [30b]

R. P. Stanley
   a. The upper bound conjecture and Cohen-Macaulay rings, *Stud. Appl. Math.*, **54** (1975), 135–142.   [171a, 198a]
   b. The number of faces of simplicial convex polytopes, *Adv. Math.*, **35** (1980), 236–238.   [171b, 198a]
   c. The number of faces of simplicial polytopes and spheres, in *Discrete Geometry and Convexity (New York 1982)*, J. E. Goodman, E. Lutwak, J. Malkevitch, and R. Pollack, eds., vol. 440 of Annals of the New York Academy of Sciences, 1985, 212–223.   [171b, 198a]
   d. Generalized h-vectors, intersection cohomology of toric varieties, and related results, in *Commutative Algebra and Combinatorics*, M. Nagata and H. Matsumura, eds., vol. 11 of Advanced Studies in Pure Mathematics, Kinokuniya, Tokyo, 1987, 187–213.   [171b, 198c]
   e. A monotonicity property of h-vectors and h^*-vectors, *Eur. J. Comb.*, **14** (1993), 251–319.   [198b]
   f. A survey of Eulerian posets, in *Polytopes: Abstract, Convex and Computational (Proc. NATO Advanced Study Institute, Toronto 1993)*, T. Bisztriczky, P. McMullen, and A. Weiss, eds., Kluwer Academic Publishers, Dordrecht, 1994, 301–333.   [142b]

g. *Combinatorics and Commutative Algebra*, vol. 41 of Progress in Mathematics, Birkhäuser, Boston, second ed., 1996. [171a, 198c]

h. *Enumerative Combinatorics, Volume I*, vol. 49 of Cambridge Studies in Advanced Mathematics, Cambridge University Press, Cambridge, second ed., 1997. [142b, 198c]

i. Positivity problems and conjectures in algebraic combinatorics, in *Frontiers and Perspectives*, V. I. Arnold et al., ed., Amer. Math. Soc., Providence RI, 2000, 295–319. [198c]

J. Steiner
a. *Systematische Entwicklung der Abhängigkeit geometrischer Gestalten von einander*, Fincke, Berlin, 1832; also in: Gesammelte Werke, Vol. 1, Reimer, Berlin 1881, 229–458. [296b]

J. Stoer and C. Witzgall
a. *Convexity and Optimization in Finite Dimensions I*, vol. 163 of Grundlehren der mathematischen Wissenschaften, Springer, Berlin Heidelberg, 1970. [30b]

B. Sturmfels
a. Cyclic polytopes and d-order curves, *Geom. Dedicata*, **24** (1987), 103–107. [7b, 129b]

b. On the decidability of Diophantine problems in combinatorial geometry, *Bull. Am. Math. Soc., New Ser.*, **17** (1987), 121–124. [96b]

c. Neighborly polytopes and oriented matroids, *Eur. J. Comb.*, **9** (1988), 537–546. [129b]

d. Some applications of affine Gale diagrams to polytopes with few vertices, *SIAM J. Discrete Math.*, **1** (1988), 121–133. [96a, 129b]

e. Totally positive matrices and cyclic polytopes, *Linear Algebra Appl.*, **107** (1988), 275–281. [129b]

T. Tarnai
a. Spherical circle-packing in nature, practice and theory, *Structural Topology*, (1984), 39–58. [296d]

T. Tarnai and Z. Gáspár
a. Arrangement of 23 points on a sphere (on a conjecture of R. M. Robinson), *Proc. R. Soc. Lond., Ser. A*, **433** (1991), 257–267. [296d]

T.-S. Tay
a. Lower-bound theorems for pseudomanifolds, *Discrete Comput. Geom.*, **13** (1995), 203–216. [198a, 198b]

A. C. Thompson
a. *Minkowski Geometry*, vol. 63 of Encyclopedia of Mathematics and Its Applications, Cambridge University Press, Cambridge, 1996. [30a]

M. J. Todd
a. The monotonic bounded Hirsch conjecture is false for dimension at least 4, *Math. Oper. Res.*, **5** (1980), 599–601. [355b]

K. Truemper
a. On the delta-wye reduction for planar graphs, *J. Graph Theory*, **13** (1989), 141–148. [296a]

V. A. Vassiliev
a. Topology of plane arrangements and their complements, *Russ. Math. Surv.*, **56** (2001), 365–401. [410b]

O. Veblen and J. W. Young
a. *Projective Geometry*, Ginn and Co., Boston, Vol. I 1910; Vol. II 1917. [7b]

B. von Stengel
a. New maximal numbers of equilibria in bimatrix games, *Discrete Comput. Geom.*, **21** (1999), 557–568. [423c]

D. W. Walkup
a. The lower bound conjecture for 3- and 4-manifolds, *Acta Math.*, **125** (1970), 75–107. [198b]

R. Weismantel
a. On the 0/1 knapsack polytope, *Math. Program.*, **77** (1997), 49–68. [121b]

E. Welzl
a. Gram's equation – a probabilistic proof, in *Results and Trends in Theoretical Computer Science*, J. Karhumäki, H. Maurer, and G. Rozenberg, eds., vol. 812 of Lecture Notes in Computer Science, Springer, 1994, 422–424. [315a]
b. Entering and leaving *j*-facets, *Discrete Comput. Geom.*, **25** (2001), 351–364. [198b]

R. Wenger
a. Helly-type theorems and geometric transversals, in *Handbook of Discrete and Computational Geometry*, J. E. Goodman and J. O'Rourke, eds., CRC Press, Boca Raton, 1997, ch. 4, 63–82. [30b]

J. E. Wetzel
a. Which Grünbaum arrangements are simplicial?, *Expo. Math.*, **11** (1993), 109–122. [410b]

H. S. Wilf
a. *Generatingfunctionology*, Academic Press, Boston, second ed., 1994. [171b]

W.-T. Wu
a. *A Theory of Imbedding, Immersion, and Isotopy of Polytopes in a Euclidean Space*, Science Press, Beijing, 1965. [224a]

V. A. Yemelichev, M. M. Kovalev, and M. K. Kravtsov
a. *Polytopes, Graphs and Optimization*, Cambridge University Press, Cambridge, 1984. [xii]

J. Zaks
a. Arbitrarily large neighborly families of symmetric convex polytopes, *Geom. Dedicata*, **20** (1986), 175–179. [129b]
b. No nine neighborly tetrahedra exist, *Mem. Am. Math. Soc.*, **91** (1991). [129b]

T. Zamfirescu
a. Three small cubic graphs with interesting Hamiltonian properties, *J. Graph Theory*, **4** (1980), 287–292. [389a]

T. Zaslavsky
a. Facing up to arrangements: Face count formulas for partitions of space by hyperplanes, *Mem. Am. Math. Soc.*, No. 154, **1** (1975), 102 pages. [410b]

G. Zémor
a. An upper bound on the size of the snake-in-the-box, *Combinatorica*, **17** (1997), 287–298. [389b]

Y. Zhou and S. Suri
a. Algorithms for minimum volume enclosing simplex in $\mathbf{R}^3$, in *Proc. of the 11th annual ACM-SIAM Symposium on Discrete Algorithms (SODA)*, 2000, 500–509. [423d]

G. M. Ziegler
a. *Lectures on Polytopes*, vol. 152 of Graduate Texts in Mathematics, Springer, New York, 1995; revised edition, 1998; "Updates, corrections, and more" at http://www.math.tu-berlin.de/~ziegler.
[xii, 7b, 30b, 52b, 52c, 69b, 96a, 96b, 142a, 296a, 355a, 355b, 389b, 410b]

b. Lectures on 0/1-polytopes, in *Polytopes – Combinatorics and Computation*, G. Kalai and G. Ziegler, eds., vol. 29 of DMV Semin., Birkhäuser, Basel, 2000, 1–41.   [69a]

c. Coloring Hamming graphs, optimal binary codes, and the 0/1-Borsuk problem in low dimensions, in *Computational Discrete Mathematics Lectures*, H. Alt, ed., vol. 2122 of Lecture Notes in Computer Science, Springer, 2001, 159–171.   [423b]

d. Face numbers of 4-polytopes and 3-spheres, in *Proc. International Congress of Mathematicians (ICM 2002, Beijing)*, Li Tatsien, ed., Higher Education Press, Beijing, 2002, vol. III, 625–634.   [198d]

R. T. Živaljević

a. Topological methods, in *Handbook of Discrete and Computational Geometry*, J. E. Goodman and J. O'Rourke, eds., CRC Press, Boca Raton, 1997, ch. 11, 209–224.   [30b]

INDEX OF TERMS

INDEX OF SYMBOLS

Graduate Texts in Mathematics

(continued from page ii)